The Lafayette Flying Corps

The American Volunteers in the French Air Service in World War One

The Lafayette Flying Corps

The American Volunteers in the French Air Service in World War One

DENNIS GORDON

Schiffer Military History
Atglen, PA

Acknowledgements

A special thanks to Fred Andrews, George Banister, Bill Foley, Tim Gordon, and Kitty Herrin of Arrow Graphics, and Steve St. Martin and Dr. Jim Parks for their assistance in compiling this history.

Dedication

For those survivors and families murdered by the War.

Library of Congress Catalog Number: 99-68886.

Printed in China.
ISBN: 0-7643-1108-5

We are always looking for people to write books on new and related subjects. If you have an idea for a book, please contact us at the address below.

Published by Schiffer Publishing Ltd.
4880 Lower Valley Road
Atglen, PA 19310
Phone: (610) 593-1777
FAX: (610) 593-2002
E-mail: Schifferbk@aol.com.
Visit our web site at: www.schifferbooks.com
Please write for a free catalog.
This book may be purchased from the publisher.
Please include $3.95 postage.
Try your bookstore first.

In Europe, Schiffer books are distributed by:
Bushwood Books
6 Marksbury Ave.
Kew Gardens
Surrey TW9 4JF
England
Phone: 44 (0)208 392-8585
FAX: 44 (0)208 392-9876
E-mail: Bushwd@aol.com.
Free postage in the UK. Europe: air mail at cost.
Try your bookstore first.

Contents

The Layfayette Flying Corps

In August of 1914 when war broke out between France and Germany, young American citizens who were residing in France rushed to the aid of the Republic. The first volunteers quickly joined the French Foreign Legion. Soon after, other Americans living in France and in the United States, heeded the call for drivers for the American volunteer motor-ambulance organizations. As early as 31 August 1914, still others joined France's *Service Aeronautique*, the vanguard of volunteer aviators in an organization to be founded later, the Lafayette Flying Corps.

On 21 August America's first volunteers offered themselves up for France at the Foreign Legion's great court at the *Hotel des Invalides* in Paris. Following a cursory medical exam, they swore their allegiance to the Legion and marched from the *Palais Royal* to the *Gare Saint Lazare* behind unfurled American flags. From there they were sent to Rouen, then to Toulouse, to be trained to fight in the *2nd Regiment de Marche* of the *2nd Etranger*. Two other groups of Americans were later incorporated into the *2nd Regiment de Marche* of the *First Etranger*, and into the *3rd Regiment de Marche* of the *First Etranger*.

Following a brief period of instruction, the Americans were transported to the Front to fight at Champagne, to help to hold the lines before Craonne, and to defend a sector on the Rheims Heights. Thus began the exploits of the hundred American volunteers in the Foreign Legion. Eventually, most would be wounded, and more than half would sacrifice their lives for France. Listed among the twenty-one Legion volunteers who became Lafayette Flying Corpsmen were Bach, Bouligny, Chapman, Chatkoff, Dowd, Dugan, Genet, Hall, Lufbery, Rockwell, Rocle, Scanlon, Soubiran, Thaw, Trinkard, and Zinn.

As Americans in the Foreign Legion were entering the trenches to fight for France, 115 other young Americans sought to serve France as non-combatants in the Volunteer Ambulance Services. The first of America's two largest Ambulance services to reach the Front was Richard Norton's American Volunteer Motor-Ambulance Corps. Richard Norton, its founder, was an educator and archaeologist who had been working as a correspondent in Paris when the French wounded from the Battle of the Marne poured into Paris. Deeply touched by this suffering human tide, Norton vowed to establish a system to transport the wounded, a service which was non-existent at that time.

Norton immediately left Paris for London where he had financial contacts whom he had hoped would provide the funds needed to form an Ambulance Corps. With $17,000 Norton had solicited from them, he bought his first ambulance cars and returned to Paris. On 15 October 1914 he left for the Front as head of his American Volunteer Motor-Ambulance Corps (SSU#7).

In 1917 Norton sought to expand his Corps's capacity to transport wounded. To finance this effort, he formed an alliance with banker Henry Harjes of the Paris firm of Morgan, Harjes & Co. Harjes had previously financed SSU#5 of his Morgan-Harjes Ambulance Corps. Both men joined forces to form the Norton-Harjes Ambulance Corps under the Red Cross. By 1917 there were 5 sections in the Norton-Harjes Corps in which a number of future Lafayette Flying Corps pilots would first see service as volunteer drivers.

The American Ambulance Field Service, later the American Field Service, was the largest of the volunteer American ambulance services to provide humanitarian aid to the French cause. This unit's light Ford ambulances were used to transport wounded French soldiers from the trenches to the safety of the field hospitals in the rear. The majority of the vehicles'

Members of the "Legion Etrangere" on leave in Paris, 7 July 1915. Future Escadrille Lafayette *pilot Victor Chapman is standing in the center. Edmond Genet and William Dugan, two other future* Escadrille Lafayette *pilots are seated in the center and on the right. Standing on either side of Chapman are future Lafayette Flying Corps pilots Marius Rocle and Eugene Bullard, America's first black combat aviator.*

conducteurs (drivers) were former students or graduates of America's colleges and universities, young men who had hoped to aid the cause of France as non-combatants.

The American Field Service traced it origin to the American Hospital of Paris in Neuilly. In 1908 Dr. Edmund Gros, an American-born physician, had helped to found the hospital which was staffed and run by Americans who tended to the medical needs of Paris's wealthy and influential American colony. In August of 1914, at the outbreak of World War One, Dr. Gros had been serving as the hospital's governor and chief of medical staff.

France's first great military conflict, the Battle of the Marne, was fought between September 6-11, 1914. Germany's swift advance against retreating British and French forces was finally halted along a 150-mile curving front between Paris and Verdun. As 150,000 wounded Poilus flooded Paris from the battlefield, Dr. Gros immediately stepped forward to assist his beleaguered French comrades by founding the American Ambulance Hospital of Paris in a converted high school at Lycee Pastuer. He also devised a rudimentary transportation system to carry the wounded to the hospital.

By March of 1915 A. Piatt Andrew, former assistant Secretary of the U.S. Treasury, instituted at the American Ambulance Hospital a transportation department called the American Ambulance Hospital Field Service, later renamed the American Field Service. Andrew was given the title Inspector-General of Ambulances with the task of gathering the equipment, staffing the volunteers into units, and assigning them duties.

During his tenure, Andrew's Field Service ambulances would be driven by 2,437 American volunteers who carried to safety 400,000 wounded "poilus." One-hundred and twenty-seven men would give their lives for France. Twenty-one unit citations were bestowed upon Andrew's ambulance sections, and 252 individual decorations of valor were awarded its men in the form of the Legion of Honor, the Medaille Militaire, and the Croix de Guerre.

Dr. Edmund Gros, as head of the American Ambulance Hospital, held an ideal position to help to found, and later to direct the Lafayette Flying Corps. However, the concept of young Americans flying as volunteers in France's *Service Aeronautique* cannot be credited to him.

In the latter part of 1914, shortly after the outbreak of war between France and Germany, Norman Prince and William Thaw, two young Americans, each had a vision of forming an all-American squadron of volunteer pilots to fight for the French Republic. Initially, neither man had known of the other's efforts, but they eventually combined forces to see their dream realized. As they worked toward its fruition, both men had become *breveted* aviators in France's *Service Aeronautique* and had been assigned as combat pilots with French squadrons at the Front.

Prince and Thaw had sought out powerful allies both within the French government and inside the rich and influential American colony centered in Paris at that time. In the Spring of 1915 Norman Prince had contacted Dr. Gros, and the two had met with Jarousse de Sillac, under secretary of the French Ministry of Foreign Affairs, an enthusiastic Prince ally. Together, the three men had created a French/American committee and had drawn up plans for a future American air squadron. At the time William Thaw had been serving at the Front and was unable to attend the meeting. Aware of the efforts of Prince and Gros, he had given them his full support.

Once de Sillac and the two Americans had mapped out their plans they approached France's Minister of War, Etienne Millerand. However, the Minister was reluctant to lend his approval to the effort, fearing a violation of international law and the usurpation of America's guarded neutrality.

A U.S. leave in December of 1915 by William Thaw, Norman Prince, and a third American, Elliot Cowdin, by then all pilots in the French Air Service, convinced

Future Escadrille Lafayette *pilot Henry Jones at the wheel of a Ford ambulance while serving as a driver in the American Ambulance Field Service with S.S.U.I., summer of 1916.*

the Republic of the support it might gain from the United States by the formation of an all-American squadron of volunteers. The sympathetic reaction of the American press toward the three young pilots serving for France produced the desired results. And upon their return to France, the Americans suddenly found a more favorable atmosphere existing in Paris toward their concept.

Sgt. Norman Prince (r.l.), Lt. William Thaw, and Sgt. Elliot Cowdin, three founders of the Escadrille Americaine, *standing before a Nieuport fighter on the Behonne airfield, May 1916.*

In order to resolve the misgivings of France's Minister of War, de Sillac had arranged a meeting between former U.S. Ambassador Robert Bacon and Leon Bourgeois, France's Minister of State, and General Hirschauer, head of French aviation, at the Paris residence of French Senator Gaston Menier. At this meeting Bacon and the French officials reached the conclusion that there was no international law preventing Americans from enlisting in a foreign army as long as their recruitment had not taken place in the United States.

The issue was soon resolved with General Hirschauer giving the order to gather into one *Escadrille* under the command of a French captain the various American aviators then in service in the French army. Seven young Americans were ultimately selected. Along with Prince and Thaw, there were Elliot Cowdin, Victor Chapman, James McConnell, Kiffin Rockwell, and Bert Hall. *Capitaine* Georges Thenault, William Thaw's old commander in *C.42* was made commanding officer of the squadron. Lieutenant Alfred deLaage deMeux was appointed his adjutant.

The new all-American pursuit squadron was to be designated *Nieuport 124* and was appropriately named the *Escadrille Americaine* on 18 April 1916. But on 16 November 1916, because of German diplomatic protests, the name was changed to *Escadrille de Volontaires* at the request of the U.S. Secretary of State. At the issued instructions of France's Minister of War it was again changed on 6 December 1916 to *L'Escadrille Lafayette*, a name created by Dr. Gros himself. However, this name was not used officially at the Front until January of 1917.

Once the squadron had been founded and manned, a committee of prominent American businessmen was appointed, most of them wealthy patients of Dr. Gros, whose duty it was to obtain the necessary funds to pay for the pilots' monthly allowances and for the accompanying cash prizes which went with the receipt of awards for valor and the downing of enemy aircraft.

The American colony was much quicker than the French to realize how an American squadron could galvanize their country's support for France. As the thunder of German guns drew ever closer to Paris they became keenly aware of what they might lose to Prussia should the French Republic collapse.

Norman Prince's father, who had granted the French government extensive war loans, had considerable property in France. William Thaw's father also held private property in France. And while neither father had initially approved of his son's involvement in French aviation, each knew what he might lose if France should fall.

Railroad tycoon William K. Vanderbilt was the first American contacted by Dr. Gros for a donation for the *Lafayette Escadrille* fund. He wrote out a check for $20,000, an initial contribution which eventually totalled more than $50,000 for the fund's coffers. Two incentives for Vanderbilt's contributions had been the large sums he had invested in French war relief and his magnificent chateau and horse-breeding farm which stabled 125 thoroughbreds. Vanderbilt, for his generous donations, was made honorary president of the Franco-American Flying Corps (later renamed the Lafayette Flying Corps.)

J. Pierpont Morgan, Jr., the son of an old Vanderbilt ally, became another *Lafayette* patron when he placed one of his Morgan-Drexel banking firms at the Fund's disposal. Morgan, like Vanderbilt, had much at stake in France. In April of 1915 he had loaned the French government $50,000,000 on a 5% note, and in September of that same year had formed a syndicate which had loaned France and Britain another $500,000,000 on five-year bonds. Clearly, it was in Morgan's interest to see the *Lafayette Escadrille* succeed. While Morgan was providing the banking service for the Lafayette fund, his sister, Anne, periodically motored down from their home in Versailles to the Buc aviation school to entertain American trainees there.

The war legacy of the American pilots of the *Lafayette Escadrille* was not found merely in the enemy aircraft they destroyed, but in the sympathy they created through their example which helped to awaken America to France's war effort. They also helped to bond the two countries together and to make America realize that the War was more than "a local conflict between European nations." Their example inspired others of their countrymen to enlist in France's *Service Aeronautique*.

During its 23 months of combat flying, the squadron served over every sector of the Western Front. During that time its 38 American pilots and five French officers flew more than 3,000 combat patrols, scoring approximately 39 confirmed victories and as many as 100 unconfirmed victories over enemy aircraft.

When the exploits of the *Escadrille Americane* became publicized by the French and American press, it created a large body of young American volunteers who wished to join the Squadron. The demand far exceeded the Squadron's few vacancies that appeared on its 12-19-man roster. So in mid-June 1916 the Franco-American executive committee, which had formed the *Escadrille American*, decided to form the Franco-American Flying Corps (later renamed the Lafayette Flying Corps) in hopes of utilizing the new volunteers by placing them in other French squadrons. By then, dozens of Americans had been serving at the Front in the Lafayette Escadrille and with other French bombardment and chasse escadrilles. Their previous service also qualified those men to become members of the Flying Corps.

The three-man executive committee was expanded to twenty-four individuals both in France and in the

The Escadrille Americaine *at Luxeuil, May 1916. (l to r) Victor, Chapman, Elliot Cowdin, Bert Hall, William Thaw,* Capitaine *Thenault, Lt. deLaage deMeux, Norman Prince, Kiffin Rockwell, James McConnell.*

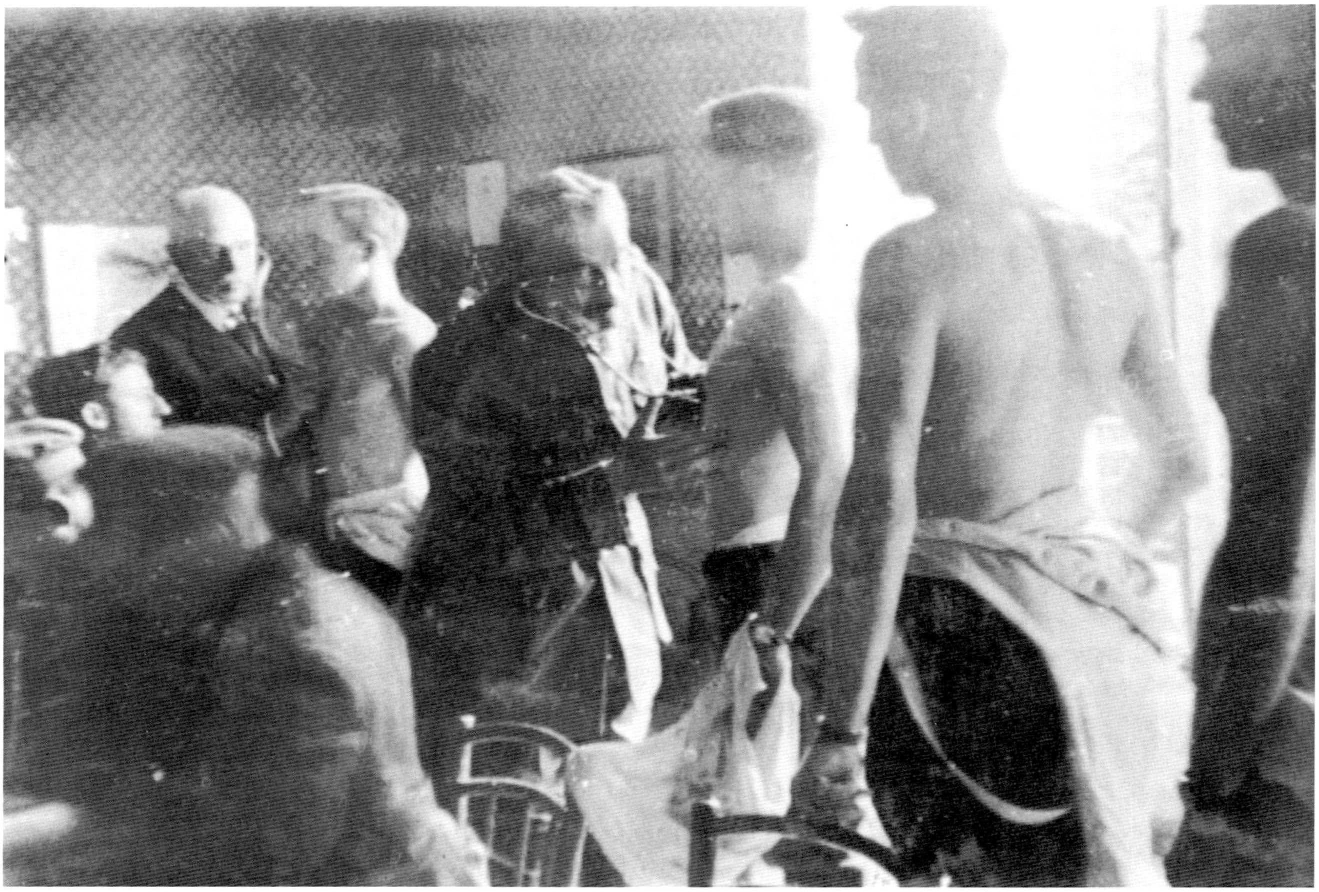

Dr. Gros (center) and Dr. Du Bouchet examining American volunteers. BILL FOLEY COLLECTION

U.S. who were to handle the finances and business of the Corps. The expanded committee appointed Dr. Gros, Director for France, and Examining Physician.

American millionaire William K. Vanderbilt posted sizeable funds at the disposal of the Corps at the Morgan/Drexel bank of Bonbright and Co. Corps headquarters was established at 15 avenue des Champ Elyses. There, one of the committee's first tasks was the composition of a recruiting pamphlet, listing the requirements for enlistment in French aviation, for distribution in the United States through its American representatives.

Money from the Vanderbilt contribution (500,000 francs) was used to help volunteers to pay their passage from the United States. Once the candidate had been enrolled in France's *Service Aeronautique*, he was paid a monthly allowance of one-hundred francs (later increased to two-hundred francs).

The Corps' first recruits came from the Foreign Legion and the volunteer ambulance services. Once its American representatives had set up a U.S. organization, recruits were screened and processed in New York City before being sent on to France.

Upon his arrival in Paris, the candidate presented himself to Dr. Gros, who oversaw all aspects of the Corps's overseas operation. He first examined the man's credentials then gave him a physical before sending him on to the *Bureau de Recrutement* at the Invalides. There, the candidate signed the Foreign Legion enlistment document in order to protect his U.S. citizenship. This was a right awarded to France at the Hague Conference of 1907, whereby any man could enter the French Army via the Legion and still retain his citizenship. When this formality was met, the candidate was then detached to Aviation.

The American candidate signed a special enlistment form which would allow him to be released outright should he prove inapt at flying so he would not be forced to serve with the Legion in the trenches. Dr. Gros's approval of a candidate assured his acceptance in aviation. As soon as he had signed his enlistment document, the prospective pilot journeyed to the supply depot near Dijon to sign more papers, to be outfitted in a uniform, and to pick up his flying equipment. Finally, he was given a railroad ticket to the French flying school where he had been assigned.

Upon completion of their flight training, 179 men served at the Front, assigned singly, and in twos and threes to a total of 92 French squadrons, spread from the Channel coast to the Swiss frontier. Later on 128 men would transfer to the U.S. aviation forces, becoming the group and squadron commanders, patrol leaders, and pilots of America's fledgling air service.

It is confusing when trying to establish the exact number of volunteers who enrolled in the Lafayette Flying Corps. In the October 1917 issue of *Flying* magazine, Dr. Edmund Gros, who listed himself as "Founder Lafayette Flying Corps" published a roster of "The Members Of Lafayette Flying Corps" which contained the name of 317 men. At that time it appears that the minimum requisite was that the volunteer had applied for and had been "accepted" by the Corps' representatives either in the United States or in France.

In late November 1918 the French Ministry of War presented a blue souvenir ribbon and an accompanying letter to those men that department addressed in an accompanying document as "the 214 pilots of the Lafayette Flying Corps."

But with the publication in 1920 of Hall and Nordhoff's official history, *The Lafayette Flying Corps*, the number of volunteers had been expanded to 268 men. Those men were separated into two categories: the 212 volunteers whose biographies appear in Volume I of the history, the men who had served at the Front; and the 56 men appearing in Volume II who were classified "Members of the Lafayette Flying Corps Released By The French Government Before Serving At The Front."

By May of 1928 the membership was drastically reduced to "209 pilots," when the Association's members published a booklet for distribution at the dedication of the Lafayette Escadrille Memorial near Paris. At that date the roster contained only the names of the volunteers from Volume I of the history, but with the elimination of three men who had appeared in that volume: Bert Hall, an original member of the *Escadrille Lafayette*, for desertion; Negro Corpsman Eugene Bullard, whom Dr. Gros had a particular dislike; and Harold E. Wright, whose service had been judged unacceptable by his colleagues. Also eliminated from the Corps were all but one of the 56 pilots from Volume II of Hall & Nordhoff's history, the sole exception being Samuel Wiggins Skinner, a suicide at the *G.D.E.*, who was listed among the Honor Roll Of The Dead.

It is puzzling why the Association chose to overlook the service of the rest of the men of Volume II, despite the fact that several of them had been awarded the Lafayette Flying Corps document and service ribbon. In fact, one of the 56 did serve at the Front with the *Service Aeronautique*, and five others later served at the Front with the U.S. Air Service. Two, Stafford Brown and Edwin Schreiber, died in service in plane crashes and should have been listed among the honored dead. It is especially puzzling when the Corps records reveal that at least 32 of the 212 men in Volume I saw no service at the Front with the *Service Aeronautique*.

During World War One Dr. Gros compiled a list with information on 269 volunteers at the Corps headquarters. The names included all the men in Volumes I and II of Hall and Nordhoff's history, and one other individual, William S. Gilmore, who did not make the list in Volume II.

The requisites for making the list were that the volunteer had to have been processed by Dr. Gros through

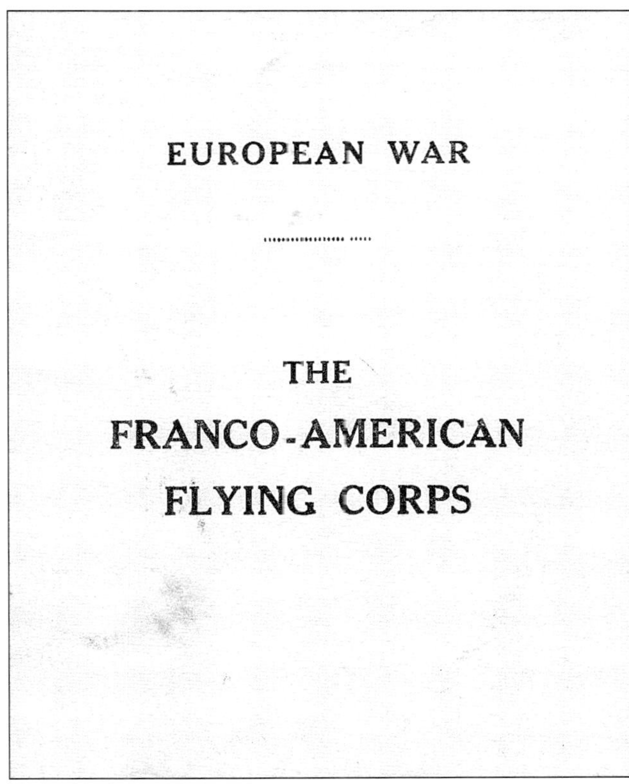

EUROPEAN WAR

THE

FRANCO-AMERICAN

FLYING CORPS

Recruiting pamphlet distributed in the United States and in France in an effort to attract young Americans to the Franco-American FlyingCorps. The pamphlet contained a brief description of the Corps' history, how and where to enlist, and an application form.

Future Lafayette Escadrille *pilots (l to r) Paul Pavelka, Laurence Rumsey, Dudley Hill, and James McConnell at the Military Aviation School at Pau, January 1916.*

his Paris office, that he had enlisted in the Corps and in France's *Service Aeronautique*, and that he had entered at least the primary level of French aviation training. Two-hundred and sixty-nine volunteers met the minimum requirement prior to the Corps's conclusion date in early August 1917, and only those men should have been considered members of the Lafayette Flying Corps. This must explain why a handful of Americans who served as volunteers in French squadrons were not a part of the Corps, most notably Ewart S. Miller of *N.49/SPA49* who is listed as a five-victory ace on some World War One scoring tallies.

Fewer than half the Corps' 269 volunteers enlisted directly into France's *Service Aeronautique*. Twenty-one had come out of the Foreign Legion, and the horror of the trenches, to join aviation, men who sought a cleaner, less strenuous way to serve; 115 had been members of the various volunteer ambulance services who were seeking the more active role of combatant; one had served in the French Cavalry; and two had served in the British Army.

All Corpsmen appear to have been citizens of the United States with the exception of Canadian Alfred Dighby Pelton. By birth, they represented 32 States and four foreign countries. At least fourteen were born outside the U.S., with ten having come from France.

By far the majority of the volunteers (188 men) were born in the Eastern United States, with just three states, Massachusetts, New York, and Pennsylvania, furnishing 128 of that number. The only other state with such sizeable numbers was Illinois with 26.

Although the Corpsmen represented all socio-economic backgrounds, many were from privileged families, which would explain why 167 of the 269 volunteers either had attended or had graduated from America's colleges and universities. Again, the Eastern United States was heavily represented with 149 of that number. Harvard University supplied 35 volunteers; Yale 20; Princeton 12; and Dartmouth 10.

It has been said with some justification that the Lafayette Flying Corps was a millionaire's club. True, the names of the scions of America's wealthiest industrialists and bankers can be found there: Cowdin, Drexel, Prince, Thaw, Turnure. Frederick Henry Prince, father of pilots Norman and Frederick Prince, was one of the wealthiest men of his age, having palatial estates both in the United States and in France. But there were other volunteers who served who came from the bottom of the socio-economic scale: miners, soldiers of fortune, taxi-cap drivers, etc. And from both ranks came airmen of exceptional character and valour, along with a handful of scoundrels and cowards.

American student-aviator Stuart Walcott wrote the following to his parents in a 4 July 1917 letter from the Avord school. "There are some 150 Americans learning to fly now in France, besides the ones the Government may have sent over-more than a hundred at this one school, and the oddest combination I've ever been thrown with: chauffeurs, second-story men, ex-college athletes, racing drivers, salesmen, young bums of leisure, a colored prize fighter, ex-Foreign Legionnaires, ball players, millionaires and tramps. Not too good a crowd according to most standards, but the worst bums may make the best aviators."

Pilot-trainee George Dock painted a different profile of the men at the Avord school in a 27 October 1917 letter to his aunt in the States. He was responding to a newspaper clipping she had sent him from a Pennsylvania paper containing the letter of another Corpsman which criticized the caliber of the Lafayette trainees.

"You may imagine from his revelations that the Americans here are a gang of devil-may-care thugs. This is not true at all. There are in the crew of 200 Americans, a lot of boors-perhaps 15 or 20, of course. They naturally are more conspicuous than the rest, so that a new man here would notice them first. But to say that the "majority of men here are a pretty hard lot although college men" is a silly remark.

"Most of the men here are from Harvard, Princeton, Pennsylvania, Yale, Dartmouth, or Amherst, and of them, the greater number have just left college last year. Several of them are given to drinking in occasional celebrations. They are not hardened tipplers, strange to say, but college boys who are so young that they think it is 'cosume il faut' to imbibe plenty of champagne whenever they drink at all. There are about 20 such, in addition to perhaps 20 adventurous birds who are unscrupulous, if not dishonest. The 160 other men are of as fine a caliber as I ever want to know, famous athletes, young business men, and men of money and distinguished antecedents, whom no one would take for snobs, sissies, or tramps. It makes me rather ill to think of the majority of these fellows being termed a pretty hard lot. I am sorry to waste so much time over an affair like this, but as you forwarded the clipping I presume that you gathered from it ideas that are as erroneous as they are ungrammatically couched."

Why did these young Americans, whose average age was 24, trade a life of leisure and security for the uncertainty of war? There were several reasons. For many, there was the lure of adventure. Others were prompted by higher motives, a desire to serve France, America's longtime friend and ally. Still others had physical, cultural, and educational ties to the Republic. But what is certain about most of the volunteers is that few had a true sense of the slaughter and carnage they would soon face.

Of the Corps' 269 volunteers, nearly 25% paid the supreme sacrifice. Forty-two were killed in combat or died later as a result of wounds received in combat; twenty-one were killed in airplane accidents; four died in service from illness; and two committed suicide.

Meanwhile the survivors, unlike the dead, paid an ongoing price for their service to France. At least eight other volunteers died following the war as a direct result of their military service, as the war's physical and psychological legacy scarred and disrupted many Corpsmen's lives to the end. And yet, most looked back at those terrible times as a "Great Adventure" and a noteworthy effort to a worthy cause.

American eleve *pilots receiving flying instructions from famous French ace, Jean Navarre. Three future* Lafayette Escadrille *pilots are in the class: Edward Hinkle (2, l), Harold Willis who stands directly behind Navarre, and Walter Lovell (6, l). Avord School of Aviation, October 1916.*

The fifty-horsepower Bleriot trainer, which the American eleve *pilot had to master in order to receive his* brevet. *While the aircraft was fragile in appearance, it was remarkably dependable, and few Americans actually lost their lives at its controls.*

1 • Avion de Chasse Training

The aviation career of the prospective *Lafayette Flying Corps* pilot began with his arrival at 23 avenue du Bois de Boulogne, the Paris office of Dr. Edmund Gros. There Dr. Gros personally examined each candidate "medically and morally." If he passed Dr. Gros's scrutiny, the candidate was sent to the *Bureau de Recrutement* at the *Hotel des Invalides*, the headquarters for the French Foreign Legion, where he was given an optical test which consisted of reading an eye chart "the size of a billboard." He then signed numerous papers, among them the *Legion Etrangere* document of engagement which pledged fidelity to the Legion. In this way, the candidate was protecting his U.S. citizenship, since he was technically not placing himself in the service of France.

The candidate was then made a *soldat, deuxieme classe* (second-class) before making a five-hour train ride to the Dijon Aviation depot to draw his flying gear which consisted of leather jacket, pants, gloves, goggles, and crash helmet. He was also issued an ill-fitting second class French poilu uniform and shoes. These he had to purchase at a cost of $50 in francs, the sum to be taken from his allotment expenses from the France-American Flying Corps fund.

After signing more papers, the candidate then boarded the train for the *Ecole d'Aviation*. The early candidates went to Pau, near the Spanish border, or to Buc, near Paris. But most went to Avord, in central France, where many future *Flying Corps* pilots received their *brevet militaires*. It was there that the candidate began his course of study to become a pilot of an *avion de chasse*.

This was a risky world the young American was about to enter into with its omnipresent threat of injury or death. His day would be composed of long hours of waiting to accomplish short periods of flight, a routine which too often was only interrupted by lightening-quick crashes of breaking wood and rising dust.

With the exception of the earliest *Lafayette* pilots whose training period was rudimentary, the usual progression through the French flying schools followed a route from Avord or Pau, on to Cazeaux, then back to Pau, culminating at the G.D.E.(*Groupe des Divisions d'Entrainement*) pilot pool. This route could, and often would be altered, depending upon the influx of students at any particular school.

The normal time to receive a *brevet militaire* was approximately three months. This, of course depended upon the weather, the size of the student's class, and his ability. Candidate Julian Biddle, who held a civilian pilot's license, earned his *brevet* in 18 days. As many as 800 students were training at a school at any given time, and the air overhead was constantly filled with the drone of engines.

The French system for training their *chasse*, or pursuit pilots was most unique. It was the only system among Allied aviation which never used dual-control aircraft. In other words the *eleve pilote* was alone in his plane from the first moment he climbed into the cockpit through all stages of his *brevet* training.

His *moniteurs*, or instructors, were usually older pilots who either had been wounded at the Front and were convalescing, or who had been physically and mentally used up from the strain of combat flying. The instructor's job was to calmly and clearly explain to the student pilot his instructions, to point out any mistakes, and order the student to correct those he might have made. This French system, while most innovative, was nevertheless effective in training the prospective combat pilot. And in spite of the worn out conditions of the training aircraft, fatalities were infrequent.

Upon his arrival at Avord, the student entered a world of far-reaching rows of low wooden barracks, repair shops, and barren landscape. The inside of the barracks had been whitewashed and had accommodation for about forty beds. In reality these "beds" were two wooden saw horses joined together with slats. Upon these slats was placed a *paillasse*, a large bag of straw which served as a mattress. The fortunate American student with the financial where-with-all had the option of living in civilian quarters at a nearby village.

In the mess hall, the students sat on long benches partaking in a menu ladled from great steaming vats of lentil soup, boiled tripe, and potatoes. Following the meal, each student cleaned his pewter utensils with a hugh piece of bread which was then discarded. Most Americans shortly transferred their affections to restaurants outside of camp, such as "The Old Lady's", the "Turco" and the "Cafe des Aviateurs" where quality food was found at a reasonable price. In the evening following instruction he was free to gather over coffee with his fellow Americans at the village cafes at Farges and Savigny.

The student's day began before dawn when he was awakened by a bugle blast and was given a cup of lukewarm chicory by one of the Indo-Chinese Annamites who were quartered at the camp for guard and fatigue duty. His chicory was to be his sole sustenance until his first meal at 11 a.m. While gulping it down, the student rushed to the wash house where he showered and shaved in cold water before rushing back to his unheated barracks. There, he quickly donned his flying outfit which consisted of leather jacket and pants, gloves, goggles, and a leather-covered cork crash helmet.

By daylight he would stand with a class of 20 to 40 other students on one of the various landing fields shivering against the cold as he awaited his turn to begin his flying career. If the class was cancelled because of poor weather, or during mid-day when the air was full of heat-waves, the student usually returned to his barracks to doze on his bed, since his schedule left him only five or six hours at night for sleep. Sundays were *repos* days (days off) and leave was usually granted.

The first class the student *chasse* pilot entered into was the *Penguin* class. Here, he was to train on a Bleriot monoplane with a wing area so small in proportion to its weight that it was impossible to fly. It merely rolled along the ground at a speed of fifteen to thirty miles an hour. All flying school instructions were given in French. An interpreter had been provided for by the Franco-American Committee, but he could not be everywhere at once. Still, the Americans had little difficulty in understanding their French *monituers* who were skilled in making their instructions understood through elaborate gestures.

From inside the "penguin" the student drove the craft through use of hand and foot controls and the regulation of his three-cylinder Anzani engine. He accomplished this by blipping the *coupe-contact* button on and off which rested on the side of the triangular-shaped grip located on top of the stick. His *moniteur* ordered the pupil to steer the machine by its rudder as nearly as possible in a straight line across a ten-acre field. This was to accustom the student to the handling of the stick and rudder while also introducing him to an airplane's noise and wind blast. He was also taught how to lift the tail of his bus.

Running the "penguin" on a straight path was no easy task. The ship seemed to possess a will of its own and without apparent warning would swerve suddenly, or fall into a *chevaux de bois* (merry-go-round loop) much like a dog chasing its tail, often colliding with another "penguin" coming from the opposite direction. As a result, the casualty rate among the "penguins" was high. When not flying, the students often helped the Annamites laborers put the Bleriots back into service.

The student gradually learned to master the machine and to steer it along a straight path from one end of the field to the other to an Annamite waiting to turn his bus around. Once he had conquered the "penguin" "grass cutter" he was ordered inside a forty-five horsepower, six-cylinder Bleriot *Rouleur* with a twenty-foot wing span. This plane could fly after a fashion, but the student's job was to keep it on the ground and to take it in a straight line across the field. The usual result was that the student would forget his instructions and would soon be crow-hopping across the field in fifteen to twenty foot-high jumps. The terror of this experience often wiped the emergency instructions from his memory, that he was to "blip" his motor off, and invariably he would crash his plane. However, this inadvertent "hopping" was exactly what his instructor had wanted his student to do, as it was his first step in learning how to take off and land his aircraft.

In time the student mastered this course and was then assigned to the fifty-horsepower, six-cylinder Bleriot. For the first time he would experience the sensation of sustained flight. Now that the student would become airborne for the first time, flying was restricted to only those periods of the day when the air was calm. This was before 10 a.m. and after 5 p.m. when it was less likely that there would be air breezes.

The *eleve pilote's* first duty was to learn to control the Bleriot in the air. In his *Decollet* class he accomplished this by skimming the ground at an altitude no greater than eight feet and by following a path marked off with flags.

His instructor next ordered him into the *Piquet* class, there to increase his altitude to twenty feet. He was also taught left and right spirals, figure eights, and dead-stick spot landings. He was warned that if he experienced motor trouble either while taking off or at a low altitude, he was to immediately nose his plane down and land straight ahead. Under no circumstances was he to attempt a turn back to the field, because his lack of airspeed would invariably cause a fatal wing slip. Upon completion of this phase of his instruction the student was ready to make a *tour de piste*, or tour around the airfield, using flat turns.

With his growing confidence and ability to master the plane's controls and crude instruments, the student's altitude was increased until he had attained a height of 1,000 feet. He then practiced flying at differing altitudes under varying weather conditions, and practiced landing his plane with his motor on and off. He also began to fully utilize his compass, altimeter, tachometer, oil pulsator, clock, and map.

Eventually he was graduated to the sixty-horsepower Bleriot with a top speed of sixty miles an hour. Following a minimum of thirty hours in the air, he was ready to begin his attempt for his military *brevet*, or pilot's license.

In order for the *eleve-pilote* to earn his *brevet* he had to perform three tests to the satisfaction of his *moniteur*. First he must make two *petits voyages* (short voyages) of approximately forty miles each, then three triangle voyages, and finally he must pass an altitude test. His instructor imposed conditions and time limits upon these tests. With the exception of the earliest members of the *Lafayette Flying Corps*, the majority of the Squadron's future pilots took their *brevet* tests either on the Bleriot or on the Caudron G-3. The Caudron was preferred over the Bleriot because it was considered sturdier and more dependable. It was powered by either a sixty or an eighty horsepower Gnome or LeRhone rotary engine, or by a 110 horsepower Anzani engine.

Eventually the day arrived when the student pilot was given a map of the country adjoining the Avord school. His *monituer* marked a point on the map and ordered him to leave the aerodrome and fly to the spot so indicated. There, he was to land and have his *Carnet* (logbook) officially validated. He must then return to the aerodrome before a specified time. If the preconditions were not met, the test would not count. When the student had satisfactorily made two such *petits voyages*, he could then attempt his two triangular voyages

The restrictions placed on the two triangular voyages were these: each must consist of a three-lap flight; the combined distance of these laps must total 150 miles; a minimum altitude of 3,000 feet must be maintained; the student pilot must land at two previously specified intermediate points, so marked on his map, before returning to his starting point. Finally, he must complete the voyage within forty-eight hours.

The normal triangle route was an air journey from Avord to the two intermediate points, and back to Avord. When the student reached these locations, usually landing fields near towns, he was required to land and have his papers examined and signed by a local official. He must also present his tamper-proof sealed barograph and have its time and altitude read and certified. When he had accomplished both landings, he was to return to the aerodrome.

The triangular voyage was particularly difficult to accomplish since the student-pilot was forced to find his way with map and compass, and by ground landmarks, providing these were not obscured by clouds or fog from his three-thousand-foot minimum altitude. Many student pilots lost their way in an attempt to find their bearings or suffered engine trouble and were often forced into emergency landings with disastrous results.

Once the triangular voyages had been successfully completed, the student-pilot was ready to attempt his altitude test. To pass this test, he must reach an altitude of 6000 feet within a given time and maintain that altitude for more than one hour before returning to the aerodrome. The student's barograph recorded whether he had accomplished the test. When the student had successfully passed his altitude test, he received his Aero Club of France pilot's *brevet* and France's International Aeronautical Federation license.

At this point in his instruction the *eleve pilote* was classified as a *pilote aviateur* and was promoted from a *soldat* second-class to the rank of *caporal* unless he had already attained that rank through previous service. Also his monthly allowance from the Lafayette Flying Corps was doubled in amount.

The novice pilot was granted a four-day Paris leave to properly celebrate this event. His first action usually would be a trip to the nearest tailor's shop where he would have his new wing-and-star insignias sewn to his uniform collar and his corporal *galons* (stripes) stitched upon his sleeve.

If the novice *aviateur* chose to purchase a new uniform at the tailor's at the equivalent of $50 in Francs, as a member of France's *Service Aeronautique* he had the option of choosing whatever style and cut of uniform he desired. The color also varied, depending upon branch of service, and ranged from an horizon blue to deep blue tunic and breeches, to a black tunic with vivid red breeches. Only the insignias sewn to the uniform were standard.

Invariably a number of student pilots who entered *chasse* school failed to complete the course. The most common reason was ineptitude. If the student pilot had seriously damaged his machine three times, he was "radiated" out of *chasse* school and was assigned to "double-command" or dual control school and was taught to fly by aid of an instructor who flew with him in a two-seater Caudron or Farman dual-control aircraft. This student, upon completion of his instruction, would then be assigned to pilot a two-seater *avions de bombardment* (bombing) or an *avions de reglage* (artillery regulating, photographing, or observing) machine.

Once the *chasse* pilot had completed Bleriot school and had earned his *brevet militaire* he turned in his flight gear then took his theoretical exams on map reading, astronomy, the study of winds, and on the proper use of the compass. His course of instruction would now continue either at Avord or at the Aviation school at Pau, a six-week course of instruction on the twin-motor Caudron, on the double-command Nieuport, and on the 23-meter Nieuport biplane and Morane monoplane. Initially he flew with his *monituer* but was soon switched to single-seater aircraft and continued on alone, flying the swifter, more powerful 80 hp. *Bebe* (Baby) Nieuport. During this phase of his instruction he was again in danger of being radiated to two-seaters if he crashed more than twice. Following fifteen hours of flight time, he was given a half-hour altitude test with a minimum level of 9000 feet.

Once the *chasse* pilot had completed this phase of his instruction, he was sent to a two to three-week course in aerial gunnery at Cazeaux. There, he was taught the nomenclature of machine guns and their workings and learned how to disassemble and reassemble them. He practiced shooting at stationary and moving targets with pistol, rifle, and shotgun, and took machine gun practice at various ranges at disappearing and surprise targets.

The novice pilot then began shooting practice in the air from the back of a two-seater, first at stationary marks, then at objects floating on the water, then at balloons floating in the air, and finally at targets towed by another plane. Once the pilot had become proficient at hitting a moving target with a swivel-mounted machine gun, he began his gunnery training in a single-seater machine. At this point the pilot had to become proficient at transforming his plane into a giant rifle. He aimed the plane, rather than the swivel machine gun, at the same targets he had been firing at from the back of the two-seater. He then fired his machine gun, which was mounted on the upper wing plane, shooting over the propeller, or on the motor cowl, shooting through the propeller.

The final school the novice pilot entered into was the four-week course at the School of Acrobatics and Combat at Pau, near the Basses Pyrenees. It was here that most accidents and fatalities occurred as the pilot's judgement was severely tested in a 110 h.p Baby Nieuport fighter while performing a series of bewildering maneuvers. For this reason he was taught to obey his instructor's commands without question, or he could place his life in jeopardy. The stunting maneuvers-high speed twists, turns, and spins-required to pass the test had to be performed exactly or the pilot would lose control of his fighter and crash to his death.

The stunting courses conducted in both 13 and 15-meter Nieuports were followed by courses in formation and group flying, a necessity if the pilot was to properly perform patrol duty at the Front. During the twelve-hour course the novice pilot also learned the French system of signals, specific maneuvers with definite meaning which pilots used to communicate with each other while in the air.

With the rush of flyers being put through the schools, it was not unlikely that a student would invariably miss a phase of instruction which would later put him at jeopardy in combat situations. For example *Lafayette Escadrille* pilot Carl Dolan missed the aerial gunnery course at Cazeaux and for some unexplained reason Christopher Ford was sent to the Front without ever having taken Combat and Acrobacy at Pau. However, both pilots managed to survive, and to overcome these shortcomings through diligence at the Front.

An upended Nieuport trainer, a familiar scene. Harold Willis stands nearest the plane.

Approximately six months after entering into the *Service Aeronautique* the pilot had one final stop before journeying to the Front. This was a the G.D.E. (*Groupe des Divisions d'Entrainement*), which was located on the bleak plains near *Plessis-Belleville*, some thirty miles from Paris. The G.D.E. served as a pilot pool or clearing station for aviators destined for the Front, and for pilots from forward squadrons who had been reassigned.

The G.D.E. swelled with activity as pilots arrived and departed daily by train. The usual stay for the novice pilot was a period of from ten days to three weeks. During this time he was to fly officer observers on long trips over friendly territory while they trained themselves at taking photographs, spotting dummy batteries, and perfecting the wireless code. The novice pilot also added time to his log book by practicing group and formation flying while familiarizing himself with the variety and peculiarities of French aircraft, including the newer SPAD fighter which had been appearing in greater numbers on the Front.

At the G.D.E. his only duty was to fly, to refine the skills he had been taught in his previous schools. If he had learned them well enough during his six to nine months of training, he might survive long enough at the Front to become a productive member of his *Escadrille*.

(Left) *The Fokker E.III Eindecker fighter with its forward-firing Parabellum or Spandau machine gun quickly earned it title "Fokker scourge" after its introduction into combat in the winter of 1915–1916.* (Below) *The Albatros D series fighter was introduced into service by Germany in the autumn of 1916 and soon proved a formidable foe against the French Nieuport with its Mercedes power plant and twin 7.92 Spandau machine guns.*

(Above) *The German L.V.G. reconnaissance aircraft was a frequent target for the American* Lafayette *pilots. But it was no easy prey with its observer-manned, free-firing machine gun in the rear cockpit, causing one American pilot to lament, "If you want to go to heaven . . . dive on a two-seater."* (Left) *The Fokker D.VII, Germany's ultimate fighting aircraft of World War One, arrived at the Front in Arpil of 1918. It performed exceptionally well at high altitudes and could out climb and out maneuver anything the Allies could throw against it.*

2 • Enemy

Whenever the Lafayette Flying Corps pilot faced his German adversary in the skies above the Western Front, invariably the odds stood in the enemy pilot's favor. It was not that the German foe was a braver or more skillful fighter pilot than his American counterpart. However, he was often better trained, with a minimum of six months fighter training over and above his training in two-seater machines. This was followed by three weeks of close scrutiny under the critical eye of top German aces. What gave him the decisive edge was his superiority of aircraft armament; the fact that he did battle over his own airspace; and the strategy employed by his squadron.

The French chasse patrols usually employed the v-shaped wedge formation, whereas, the Germans invariably flew in single-file, with the leader occupying the lowest position. At the time of attack, the German *equipe* would spread out on a wide front, as the leader moved slightly in advance of the rest of the patrol. The German single-file formation, when engaged in combat, was more flexible than the French wedge, although it was more vulnerable to attack from above.

The varying camouflage patterns that were employed by both sides seemed to balance out. Lafayette Flying Corps pilot George Dock recalled that the "Boche planes were cleverly camouflaged" and only a keen-eyed pilot could pick them out when flying above them. Conversely, the bright tri-color cockades on the French machines made them conspicuous, despite the fact that the plane's camo coloring of dark green, black, and yellow splotches made the rest of the plane blend well with the ground over which it flew.

However the French *avions*, with their sky-gray undersurfaces, were less visible from below. In contrast, the Germans utilized the same dark, crazy-quilt camouflage throughout the entire undersurface of their fighters which made them easy to see from below. Thus, both sides preferred to "hide in the sun" before attacking, hoping to use surprise and invisibility over their opponents.

The German Flying Corps remained one step ahead of the Allies in the development of its aircraft and armament. It had first gained air supremacy in the summer of 1915 with the introduction of its Fokker E.I and E.III fighters with their belt-fed machine guns firing through the propeller disc. Because of the Fokker's high ceiling, the German pilot could pick his target then make a diving attack, while fully utilizing the additional advantage of extra ammunition and longer bursts of fire. The enemy's machine gun belts held 1100 rounds while the French *Aeronautique*'s Lewis gun magazines held 47 or 97 rounds.

By the spring of 1916, at about the time the *Escadrille Americaine* had arrived at the Front, the French *Service Aeronautique* had regained the edge in fighter supremacy with the introduction of the Nieuport 11 fighter. But by fall that edge had dissipated when the Germans introduced in great numbers their Albatros D.I, and the improved D.II fighter with its twin 7.92 Spandau guns and 160-h.p. Mercedes engine. Again, Germany stood unmatched in speed, altitude, and fire power until the autumn of 1917 when France began employing in large numbers its SPAD S.XIII fighter armed with twin belt-fed Vickers machine guns.

Captain Oswald Boelcke and his six-machine Jagdstaffel may have done battle with the Escadrille Americaine *in June of 1916 when Boelcke's unit was operating from the airfield at Sirvy-sur-Meuse opposite the American squadron at Behonne. Boelcke may personally have been responsible for Victor Chapman's deep scalp wound on 17 June 1916 in an aerial engagement above the right bank of the Meuse River. Boelcke was born, one of six children of a professor at Giebichenstein near Halle, on 9 May 1891. He was one of the first German pilots to be chosen to fly the Fokker 'E' series monoplanes. His E.IV scout appears in the photo below. During his war service he was credited with the destruction of 40 allied aircraft. He was killed in an air collision with his wingman on 28 October 1916 while in pursuit of a British machine.*

But even with this improved weapon, the German Spandau and Parabellum machine guns still proved more dependable and smoother to operate than the Lewis and Vickers guns which France employed on its own fighters. And by 1917 the enemy held additional advantages when it began to utilize as standard equipment electrically heated flying suits, parachutes, and liquid oxygen. In April of 1918 the German air force introduced its most formidable fighter to the Front. It was the Fokker D.VII biplane, a fighter which could outclimb the best Allied planes, and was easily maneuvered with deadly precision at high altitudes.

Another advantage the German Air Force held was in the method of its tactics. Unlike their French counterparts, the German fighter pilots flew no regular patrols, but would be quickly moved about in field tent hangars to any part of the Front. This allowed them to confront the small French patrols with large numbers of aircraft.

The pilots of the "Flying Circuses"(*) would remain on the ground scanning the sky with binoculars for approaching enemy fighters. Or they would sit in their deck chairs awaiting orders to take to their fighters the instant forward observation posts phoned in information of approaching enemy.

Throughout most of their service on the Western Front, the duty of the German fighter *Jagdstaffeln* (squadrons) was to protect their air space from enemy incursion. Few of their fighters sought combat over Allied territory. Since they were defensively deployed and accomplished the majority of their combats over their own lines, they held an added advantage in fuel capacity and in the direction of the prevailing westerly wind which often carried their aerial melees with Allied pilots deep into their own lines.

Finally, whenever the German fighters attacked their adversaries, it was only when they held the advantage in numbers and in altitude, and through the use of ruse and decoy, with a lumbering two-seater often serving as bait. Canadian ace William Bishop wrote how his foe had developed pack hunting into a refined art.

"Almost every evening we would find well-laid traps set for us. It required careful manoeuvering to avoid falling into them. Several times we did and it took a lot of trouble to get out safely. Four or five Huns would come along and we would engage them. Then suddenly as many as fifteen or twenty would appear from all angles and join in the fight. This happened every day and the Huns were evidently out to get us."

But even the straight-flying German two-seaters which performed reconnaissance patrols within allied lines were no easy prey for the American pilots. They often flew too high for the French Nieuports and SPADs to reach them, or they were so well-armed they posed a serious threat to the Lafayette pilot who chose to attack them. As one American pilot lamented: "If you want to go to heaven, the easiest way I know is to dive on a two-seater."

*The term "Flying Circus" was created by the Allied pilots to identify the enemy's "habitual tactics" of flying large numbers of fighters round and round in a great circle like horses in the ring at a fair. The Germans would bait the French pilots into attacking them then would dive toward the interior of the circle which allowed the enemy fighter next in order of the circle to attack the French plane from behind.

The Nieupot 11 C. 1 fighters of the Escadrille Americaine *at Luxeuil-les-Bains, May 1916. Nicknamed the "Bebe" because of its diminutive size, it had an 80 h.p. rotary engine with a top speed of 100 miles per hour.*

3 • The Front

The Lafayette Flying Corps pilot's first act at the *G.D.E.* upon rising in the morning was to anxiously check the bulletin board to see if he had been posted to the Front. When his name finally appeared for assignment to a *Group de Combat*, he felt relief and a hightened sense of anticipation. In following the invariable custom for French pilots, he would tarry a day or two in Paris on his way to the Front, usually attending the frivolities of the Follies Bergeres or some other diversion.

The American pilot would arrive at his squadron's aerodrome either by flying in a new fighter plane from the factory, or by taking a French military train to the village nearest the squadron's aerodrome. Often this village from which the aerodrome had taken its name was merely a row of makeshift structures with rough pine walls and corrugated iron roofs facing the length of a dusty street. Off in the distance stood the battered remains of the original French village, its stone edifices long pulverized into heaps of rubble.

A squadron chauffeur with a *camion* greeted the replacement pilot as he stepped off the train. In his valise he probably carried a new fur-lined flying helmet, Meyrowitz flying goggles, and a set of gauntlet gloves, all recently purchased in a Paris outfitter's shop. With his gear stored in the camion, the American pilot was transported down the dusty road toward the aerodrome through rows of stooped and sweating *permissionnaires* returning to the trenches.

The squadron's aerodrome was located fifteen miles from the Front between the war zone and the infantry rest zone. It usually consisted of crude barracks and a cluster of portable canvas hangars squatting on the edges of a 300-yard-square airfield. The chances were that the Lafayette Flying Corps pilot would be the only American assigned to his squadron. Upon his arrival there, he would be warmly greeted by his French comrades who were lounging in the *Popote*, or mess hall, at one end of the long barracks. The pilots' rooms took up the remainder of the building, and once he had been assigned his quarters, he would deposit his gear near his cot.

The squadron's Nieuport and SPAD fighters occupied hangars on the same airfield with three other pursuit squadrons. Together the four squadrons formed the *Groupe de Combat* with a specific sector of Front to cover. In flying weather, through a series of interlapping patrols, they protected their sector from daylight to dusk. Circumstances dictated the number of patrols flown and the amount of aircraft in each patrol.

The replacement pilot soon learned his duties. In a *chasse* squadron his daily mission was often pre-designated, such as protecting the aircraft of the *Corps-d'Arm'ee*: the allied photographic machines, the artillery reglage machines, or the bombing formations. But with no specific assignment, he patrolled the lines unrestrained. And if he had a malingering nature, he had access to a myriad of excuses- a boiling radiator, a stuttering motor, a jammed gun block-which could keep him from harm's way. However, the majority suppressed their fear and served dutifully, attacking enemy aircraft at every opportunity, or ground strafing enemy trenches, transports, and troop concentrations.

Under normal conditions a pilot was required to fly two two-hour daily patrols, the approximate limit of his petrol supply. During an allied offensive he might be required to fly six hours daily, but rarely more than that. The constant strain in the air was too taxing for the normal pilot to function beyond this limit.

Invariably a new pilot, filled with enthusiasm and unfamiliar with the strain of combat flying, would ask to fly voluntary patrols. The veteran pilots, upon hearing this oft-repeated question, merely shook their heads and told their newly-arrived comrade that within six weeks time he would fly no more than the required limit, and would fly voluntary patrols only when it was necessary, when the situation demanded it.

But first the novice pilot had to survive his most dangerous time, his first month at the Front. He had only a fifty-fifty chance of living through that period and was of little value to the squadron because of his inability to spot the enemy in the air. But if he survived those first thirty days his odds of living improved considerably, since eighty percent of the casualties were suffered by pilots with fewer than twenty missions.

When not flying patrols, the pilots often slept. One German pilot called sleep "the food of the nerves" and it was not uncommon for pilots to sleep up to twelve hours a day as their bodies demanded relief from the intense fatigue and nerve strain brought on by combat flying. Unlike the British Royal Flying Corps which allowed a three-month rest period following six months of combat flying, the pilots in France's *Service Aeronautique* served until they were killed, seriously wounded, or were physically and emotionally used up. When the pilots were not sleeping, they read, played bridge or billiards, or took walks. If the squadron was not flying around the clock during an offensive, nights were devoted to drink and music from the piano and phonograph located in the mess hall.

When dining, all officer and non-commissioned pilots sat at a long table. Dinner was well-prepared and might consist of thick soup, juicy steak, fried potatoes, salad, and cream-cheese, all served on clean china by a solicitous orderly.

Every American replacement arrived at the Front as a *Caporal* pilot unless he had earned a higher rank from previous service. After having flown thirty hours over the Front lines he was promoted to *Sergent*. When he had accumulated 100 air hours and had participated in a number of combats and victories, he was ususally promoted to *Adjudant*.

The pilot's pay consisted of nine *sous* a day and an additional *franc* a day for flight pay. This sum totalled twenty cents in U.S. money. In addition a pilot was paid one-hundred *francs* (later increased to two-hundred *francs*) a month from the Lafayette Flying Corps Fund as his mess fund and to see to other necessities. The Fund also distributed cash prizes and Paris leave whenever a pilot received a French decoration or was credited with the destruction of an enemy aircraft. The prizes were distributed as follows:

Legion of Honor	1500 *Francs* ($300)
Medaille Militaire	1000 *Francs* ($200)
Croix de Guerre	500 *Francs* ($100)
Citation(palm)	250 *Francs* ($50)
Downing an enemy aircraft	1000 *Francs* ($200)

The French standards for confirming a victory over enemy aircraft were rigid when compared to its allies. This stemmed from bogus victory claims by its pilots early in the war. To stem this abuse the *Service Aeronautique* established a rule whereby the pilot alone could not claim a victory. His claim had to be verified by a second source, either from a ground or balloon observer. If it could not be so verified, the pilot's claim was disallowed. Also, the foe had to fall inside French lines to be counted. Due to these strict standards, the Lafayette Flying Corps pilots destroyed many more enemy aircraft than they were ever credited for since a majority of their combats and subsequent victories took place over enemy lines, and beyond the view of allied observers.

For example, *Lafayette Escadrille* pilot Raoul Lufbery, the Lafayette Flying Corps' top ace, was credited with the official destruction of sixteen enemy aircraft during his tenure with the French Air Service. In truth, he may have had as many as fifty victories over enemy aircraft behind German lines, with seven coming in one single day.

A French squadron at any one time ususally carried no more than 20 pilots on its rolls. Early in the war, when the rosters were small, a patrol consisted of one-to-three pilots. But later on as rosters grew, the pilots were divided into three flights of six or seven pilots each including the flight commander. However, it was rare that the patrol ever consisted of more than four pilots due to the unreliability of their aircraft.

The total personnel of the French Squadron consisted of approximately eighty men. This included the pilots, the armament officer and his assistants, the medical officer, two mechanics for each plane, twelve chauffeurs to drive the squadron trucks and *camions*, squadron and officers' orderlies, the mess cooks, and secretaries.

Every night at approximately 10 p.m. orders reached the squadron from Headquarters of the Army containing the following day's assignments. These assignments were coordinated with the General in charge of the Army in the *Groupe*'s sector.

Under normal conditions the squadron's daily routine of three flights consisted as follows: When A Flight patrolled the Front, B Flight was on alert. This meant that a pilot of B Flight was dressed in his flying clothes, his plane out of the hangar and its engine kept warm by his two mechanics who continually turned it over. He passed the time while on alert sitting in the operations tent, reading, writing letters, or playing cards. Or he might journey out to his fighter to hand gauge and load the cartridges into his machine gun belts; or to discuss his aircraft's performance with his mechanics. (It was customary for a pilot to pay his mechanics a monthly stipend as an added guarantee that his engine would run smoothly.)

Lt. Thomas Cassady in his 180 hp SPAD scout at La Ferme de la Noblette while assigned to SPA 163, *May 1918.*

Meanwhile the pilots of C flight were on *repos* (sleeping or resting). This dull routine was followed day after day, including Sundays and holidays except for special occasions when Paris leave might be granted.

The only time the routine was altered was during an allied offensive, and then the pilots flew from daylight to dusk, pushed physically and mentally to their limits, until the conclusion of the assault. Gasoline flares were put on the airfield for the pilots to land by who came in after dusk.

Weather dictated whether the pilots were to fly patrols. Often the squadron was kept grounded from inclement weather, rain, sleet, or an impassable airstrip of alternately frozen and oozing clay.

When weather permitted, the pilots flew two types of daily patrols: *offensive* and *defensive* patrols. Offensive patrols were patrols flown over enemy territory to protect allied observation or bombing aircraft, to seek out and destroy enemy aircraft, to spot enemy troop movements, and to conduct ground strafing. *Defensive* patrols were flown to protect allied balloons, aircraft, and ground installations, and to knock down enemy aircraft which penetrated allied air space.

A *defensive* patrol was normally flown at three different altitudes. The selected squadron, when flying a defensive patrol, was accompanied by two other patrols flown by other squadrons of the *Groupe de Combat*.

The area over which the squadrons patrolled was divided into three zones. The *rear* zone consisted of a five-mile area extending from the allied observation-balloon lines to the allied front-line trenches. The zone of *advance* consisted of the area between the German front-line trenches and five miles inside enemy territory to their observation-balloon lines. The third zone was that area over German territory which was behind the enemy's observation-balloon lines.

A normal patrol over these zones usually lasted two hours, the fuel limit of the allied fighter. The *low patrol* was the first patrol off the airfield. It flew at 3,000 feet just in front of the allied balloon lines, and its duty was to protect the balloons and to intercept enemy bombing, observation or pursuit aircraft. The second patrol off the field was the *intermediate patrol* which covered an altitude between ten and twelve-thousand feet. The *intermediate patrol* was to protect the *low patrol*.

The last patrol off the airfield flew *high patrol*, which was at the highest altitude the fighter could reach, depending upon the air pressure and temperature of any given day. For the Nieuport 11C.1 this was at approximately 15,000 feet. For the Nieuport 17C.1 and the SPAD S.VII fighter this was between 17,000 and 18,000 feet, although some pilots coaxed their SPADs beyond 20,000 feet under favorable atmospheric conditions.

The *high patrol* left the airfield last since it normally took a half hour to reach its altitude before it could level off and begin patrolling back and forth across its sector. The duty of the *high patrol* was to protect the *intermediate patrol*. A fighter plane grew mushy and unstable at high altitude. The air was bitterly cold, "a pure agony of mind and body," and because the pilots carried no oxygen supply, the slightest exertion, such as the cocking of a machine gun, caused one to pant. Fingers became like "frozen bananas", and the pilot's bladder, which reached a bursting point soon after takeoff, swelled with unrelieved agony.

Patrolling was done in a V-wedge formation, with the flight leader taking the point and with the deputy leader nearest him, both in clearly marked fighters with flight streamers. The newest pilots were usually placed toward the front of the patrol with the experienced pilots covering the rear, the likely point of enemy attack. A patrol never flew in a straight line, but continually altered its path to prevent surprise attack by enemy fighters and bracketing by enemy antiaircraft fire.

No pilot was placed on *high patrol* for two successive days because of the physical stress and the resulting severe headaches. When a pilot came off high patrol, his next assignment following *repos* and *alert* status was to *low patrol*, then to *intermediate patrol*, before being assigned again to *high patrol*.

When allied ground gunners spotted German formations from their positions they signaled the enemy's location to the patrol overhead through antiaircraft fire. Using a high-explosive shell which emitted white smoke, the allied battery fired in front of the patrol, setting the round to explode either above or below the formation to let their comrades know that the enemy patrol was either above or below them. The battery used armor-piercing shells which emitted black smoke to indicate which sector the enemy patrol was flying over. A single black explosion indicated sector one; two explosions indicated sector two, etc.

Whenever a patrol spotted enemy aircraft the usual response was to climb above the German formation with the sun at the pilots' backs. This gave the advantage of striking an adversary while coming out of the blinding rays of the sun. But if both patrols spotted each other simultaneously, both would climb for advantage. As *Lafayette Escadrille* pilot Carl Dolan remembered:

"We were interested in getting up behind them and getting a crack at them. But they had the same idea, you see. So if we spotted each other at the same time, we'd both climb. I have never been high enough to be above the Germans. No matter how high we'd go, the Germans always seemed to be a couple of hundred meters above us. They'd come diving at us. Of course, if we could get the sun at our backs we'd turn and face them. Otherwise we'd run and attempt to outmaneuver them. We had been trained in school to be aviators, not fighters."

The greatest threat to the Lafayette Flying Corps pilot and his French comrades was what their German adversaries described as "airweariness." This manifested itself in a number of ways. Mind and body would go on strike and refuse to function. The pilot could not sleep; he would become quarrelsome as his nerves were laid raw; or he might seek escape in excessive drinking.

The pilot's mind and body would rebel against the deadly monotony of his work, flying a seemingly endless series of patrols above a tortured and murdered earth. A sense of fatalism and apathy would overtake him, and his oft-repeated actions would come to appear senseless. When this happened, he was most vulnerable to enemy attack. One Lafayette Flying Corps pilot best summed up the conditon of airweariness and how to overcome it:

"I've had five hundred hours over the lines. You don't know what that means, not yet. I'm no good any more. It's strain. Let me give you some advise. Save your nervous energy. You will need all you have and more. Above everything else, don't think at the Front. The best pilot is the best machine."

4 • Paris

"In a way it seems right for us to get a little real taste of war. We are here in Paris, in comfortable houses with theaters and restaurants and every luxury while for a year and a half poor devils have been living in trenches."

—Henrietta Thaw Slade
30 January 1916

In August of 1914 Paris appeared through the windows of the luxurious apartments of America's wealthy and privileged as a beautiful carefree city at the zenith of its *belle epoque*. This new aristocracy, who had made their fortunes in railroads, shipping, and mining, frequently commuted to Paris in search of beauty and pleasure. But behind their satin blindfolds they held no idea that their beautiful era of self indulgence which had begun in 1890 was about to come to a brutal end.

The American colony viewed the good life they experienced in Paris as their consummate reward purchased through diligence and acumen in America's industrial and financial heartland. And it was to the world's upper-class playground that these Americans had come, to Europe's center of art and fashion, in pursuit of fine cuisine, exquisite wine, and the sun at Biarritz.

For the next two years, until the raging battle of Verdun had cast a pall over their festive spirit, most Paris-based Americans remained untouched by the nearby war which mutilated growing numbers of the city's former inhabitants. The shortages of fuel and foodstuffs which brought privation to many households would not be felt in those American domiciles with ample supplies of francs until commodities could not be purchased at any price. Meanwhile many Americans continued a lavish lifestyle of sumptuous dining and entertaining long into the war.

In the summer of 1914 madames and mademoiselles in bright dresses and lacy parasols strolled the *Champs Elysees* and the avenues and riverbanks of Paris. The city was staunchly defended both on the ground and in the air. And despite increasingly frequent German air raids and bombardments, it would survive basically intact throughout more than four years of war.

But as the war dragged on, an ominous change overtook the city's inhabitants. By the spring of 1915 most had adopted an attitude of soldierly stoicism, their dark, plain clothes indicative of lost loved ones as well as an expression of their common grief. For a woman to dress brightly in the latest fashion was to invite the scornful glances of passersby who well knew the price in blood that France had paid each day to keep the city free.

The streams of men and munitions moving out of the city and the shattered bodies of soldiers returning served as a constant reminder to the American in Paris that the trenches were not far off. As one American observer wrote to her son in the United States:

"Yesterday I saw some more rolling kitchens. . . . The horses look used up and I pity them. You have no idea of the amount of food that goes to the front every day. Wagons pass with soldiers for hours and you get tired of looking at them."

The darkness engulfing the city at night served up another reminder of the nearby war. Paris was blacked out with the exception of searchlights probing the sky for German Taubes and zeppelins which rarely entered the city's airspace. In the early months of the war an entrepreneur had built seats on the Montmartre to watch the "fireworks" in the impending attack on Paris by enemy Taubes. But when the attack had failed to materialize, the irate Parisians had demanded their money back.

In the early fall of 1914 women sobbed openly as their uniformed husbands boarded troop trains for the Front. But within a year of the anniversary of the war's outbreak tears were stifled at the choked railroad stations as lovers concealed their emotions under a veneer of stoicism.

At first there had been few inconveniences, and most commodities were still available if one could pay the

Lafayette Escadrille *pilots in Paris, spring of 1917.* (l.r.) *Robert Rockwell, Charles Johnson, Dudley Hill, Edgar Bouligny (L.F.C.), Didier Masson, William Thaw, Raoul Lufbery, Robert Soubiran, Paul Rockwell*

price as Mrs. Alice Weeks noted in a 17 August 1915 letter to her brother in America.

"Meat is very expensive and a small chicken costs $1.25, coal $20. a ton, but there seems to be plenty of vegetables still. You say you thought all the automobiles were requisitioned. They were in the beginning of the war, but not now. We really have very little inconvenience from living in the war zone, so far."

But in the quiet of night the boom and vibration of cannon and the muffled stutter of machine guns coming through the open window continually reminded the American in Paris what price was being expended at the Front to keep the city free.

For those city dwellers who could not afford the escalating cost of food and fuel, the arrival of the bitter winter brought on suffering. The spring of 1916 saw the awful beginning of the Battle of Verdun and soon Parisian boulevards filled with the wreckage of men.

"There will be no end of artificial legs needed," Mrs. Weeks wrote in a 20 March 1916 letter to her son in the United States. "You see sometimes five or six men walking together, all with one leg. Paul (Rockwell) saw fourteen the other day. Somehow you don't see as many armless. This is not so bad, it is only when one sees part of a face gone, that is terrible. As I write a regiment is passing going to the front with the band playing the Marseillaise. Poor fellows, but I am glad to see they are not boys. There were also horses with rapid firing guns strapped on their backs, ending up with a string of ambulances."

Paris had become a city of contrasts. And even the appearance of mutilated soldiers could not mute its beauty in the springtime.

"A beautiful day, we have had four of them," Mrs. Weeks wrote her brother in a 2 April 1916 letter. "Spring is in the air. Windows are open and people hanging over their balconies. Someone is playing the piano very well, and life smiles for the moment...I find myself at times very happy, coming home in the river steamer with the sun setting behind the Tour Eiffel and the bridges arching across the gold of the sunset."

Mrs. Weeks continued to describe the beauty of a city coming to life. "The trees are fast leaving out," she wrote, "and the cherry trees are in bloom in the Bois. Yesterday being Sunday, the avenues and the Bois were crowded with people and automobiles and there was a jam at the entrance." And yet sixty miles away men were fighting and dying in great numbers to keep the city open.

War seemed incapable of intruding on the beauty of Paris throughout the summer of 1916. But by early fall a bitter chill had settled over Paris which heaped further misery on the city's occupants. Many apartments were without heat. Soon shops were ordered to close early and the theaters limited to playing one night a week in order to conserve the coal and electricity needed to run the munitions factories.

But not everyone had become a victim of the rationing system. Profiteers who had made enormous fortunes off the battle raging near the city either via the black market or by providing the arms and material to keep the cauldron boiling soon discovered a new symbol with which to flaunt their recent wealth. Mrs. Weeks made note of this in a 26 November 1916 letter to her son.

"Every place is full of automobiles. Paris this winter has so many new cars that I am shocked. I stand waiting for the tram near my house and the new automobiles that come by as I wait, is a wonder. When I came over we saw none at all, as they had all been requisitioned, but as in America, some have made money out of the war and like all nouveau riche, they come at once to Paris to spend it. I am glad to say that the Government is to do away with all private automobiles, but now you will see that many will evade the law by becoming a Red Cross, or take some military position."

Private motor cars, taxis, and military transports choked the streets of Paris. And at night as they were driven about with muted headlights they became a great risk to the city's populous as a number of people were run down and killed.

The winter of 1916–1917 provided the worst weather Paris and the country had suffered since 1870. With the requisition of wood and coal, the capital endured further agony. In the spring of 1917 there was new hope as America entered the war, but shortages became more acute and rationing compounded. The city's occupants had hot water but twice a week, and meat, sugar, and butter was strictly rationed. For the first time a coarse bread made of grain appeared in Paris in large quantity. But with every part of the grain ground to flour, many Parisians could not eat it.

Since the beginning of the war France had lost ten percent of her territory and fourteen per cent of her industrial potential. And agricultural productivity had been on a continual decline as the existing soil had grown depleted from overuse, and as the workers needed to cultivate it had been slaughtered at the Front. Food and coal had to be purchased with ration cards. Milk and sugar had all but disappeared, and rabbit had become the only available meat. Eventually the allies had set up a system to supply the capital, but by the time food began coming in prices had soared beyond the reach of many households.

Fall again brought an early chill and renewed agony from the cold. Some Americans left their frigid apartments and took daily walks on empty boulevards in order to warm themselves. Every Parisian had war work to do, either in a munitions factory, in a canteen, or in a hospital, and streets that had once bustled with humanity now stood deserted. Nights were filled with the rumble of trucks and trains as convoys carried their deadly hardware to the Front while the railroads returned with war's awful byproduct, torn and shattered men.

Early in 1918 the Germans stepped up their air raids against Paris. Empty coal bins at the bottom of buildings were converted to makeshift bomb shelters. Soon antiaircraft emplacements bristled throughout the city which wreaked havoc on the enemy invaders.

But on 23 March the German high command employed a new and awesome weapon. Seventy-five miles from Paris in the forest of Saint Gobain German artillerymen began lobbing 275-pound projectiles through an 118 foot-long cannon into the suburbs of Paris. Named "Fraulein Bertha" by French school children, three of these cannons launched a total of 370 shells into the city before the guns were finally removed by the Germans at the end of July And while no French military installation was ever struck by them, the guns further eroded the city's morale.

As more and more troops poured into Paris, the capital became a polyglot of nationalities. "Paris is most cosmopolitan in these days," Mrs. Weeks wrote on 10 March 1918, "and yesterday I noticed in the crowd that there were English, Americans, East Indians in uniforms with big turbans, Spanish and Swiss. English is spoken everywhere. One hears it on the streets as much as French. I do not wonder that the ignorant people think that France has been absorbed by the English and Americans and fear they will not go home after the war."

By the end of May food shortages had become acute. On 20 May Mrs. Weeks wrote: "Chickens are almost prohibitive, milk is no longer inspected when we can get it, bread full of chaff, and we have three meatless days a week, even if you can get it at all."

But even with all its shortages Paris remained ever popular for the *Lafayette Flying Corps* pilot on *repos*. The aviation *Groupe* bar and the cafes in the villages near the aerodrome provided pale relief from war's drudgery. And as pilot Edwin Parsons noted, they were no substitute for the good food, fine wine, and bad women of Paris. Parsons later wrote of Paris:

"Taking it all in all, our squadron life was very quiet, and we existed only for the times we could wangle a leave to go to Paris. Then we tried to crowd all the excitement and general hell-raising we could into a few short hours. I'm far from being the only man who has mournfully remarked after a hectic forty-eight hours among the houris and bright lights of the Big Sinful City, "Gosh, I gotta go back to the front for a rest." Despite its hazards, it was a haven of rest for the weary."

Many a young Lafayette Flying Corpsman experienced his first initiation into sex at the nude revues and in the bordellos of Paris. The Lafayette Escadrille pilots exclusively frequented "The Lafayette whore house" at 22 Rue Rrey, near the Arch de Triumph, having chosen it because it was licensed, clean, and had "nice young girls." As Henry Jones later recalled, one young pilot lost his virginity there. "I encouraged him," Jones said, "and he took to it very well."

But not all young Flying Corps pilots were able to escape the consequences of their sexual escapades. Several contracted venereal disease which resulted in their expulsion from the French Air Service, and one young candidate committed suicide, unable to face the consequences of his condition.

As Carl Dolan remembered of Paris "You couldn't walk a block without fifty girls coming up to you and saying 'I sleep with you tonight, yes?' Everybody, anybody. The woods were full of them. We were at the Countess of Betune's one weekend. She was having a party for the *Escadrille*. I asked her, why all this lack of morality. She said, 'You don't realize. See that girl over there, and this applies to most of the girls in this room,' there were about ten of them she had invited to entertain us. 'She's one of fifty in her family, and there's only four men left. Their husbands, their fathers, their brothers are all dead.'"

But not all aviators sought escape through female companionship or through drink and conversation at the old rendezvous spots, Harry's or the Chatham bar. For some, Paris leave provided the opportunity to visit relatives who were a part of the American colony, or who were involved in some capacity with Paris-based relief agencies such as the Red Cross or the American Ambulance Service. For Edmond Genet it presented the opportunity to attend religious services. While Victor Chapman would go to 148 Rue De Longchamp, to the apartment of Henrietta Thaw Slade, sister of pilot William Thaw and there spend many hours of his leave reading books from her library.

LAFAYETTE FLYING CORPS ACES

Aviation historians credit the American pilots of the Lafayette Flying Corps with the destruction of between 199–206 enemy aircraft. These are the numbers given for official confirmations by French and American authorities. However, there were many victories that went unconfirmed due to France's stringent confirmation requirements which were also adopted by the General Headquarters of the AEF for use by the U.S. Air Service.

The British Royal Flying Corps adopted a victory credit-system, which accepted "probable" confirmations for "out of control" aircraft that were not actually seen to crash. This was due to the fact that most aerial combats were fought deep within enemy lines where confirmation proved difficult. However, the French system demanded confirmation of the actual destruction of aircraft by independent observers before assigning credit for aerial victories.

Listed here are the thirteen Americans of the Lafayette Flying Corps who were classified as aces for having officially downed five or more enemy aircraft while at the Front. It is reasonable to assume that the pilots' victory credits in most cases would have been higher than indicated. Raoul Lufbery's victories were probably more than fifty if all his unconfirmed had been counted, and David Putnam may have had twenty-one unconfirmed victories, which would have pushed his total victory credits to thirty-four.

	SQUADRONS	VICTORIES
Major Lufbery, Raoul	(VB 106, N124,US 94th)	16
Lt. Putnam, David	(SPA94, MS156, SPA 38, US 139TH)	13
Lt. Baylies, Frank	(SPA 73, SPA 3)	12
Lt. Baer, Paul	(SPA 80, US 103d)	9
Capt. Cassady, Thomas	(SPA 157, 163, US 28th)	9
Lt. Parsons, Edwin	(N124, SPA 3)	8
Major Biddle, Charles	(SPA 73, N124, US 103d, US 13th, 4th PG)	7
Adj. Connelly, James	(SPA 157, SPA 163)	7
Capt. Larner, Gorman	(SPA 86, US 103d)	7
Major Peterson, David	(N124, US103d, US94th, US95th)	6
Lt. Ponder, William	(SPA 67, SPA 163, US 103d)	6
Capt. Grey, Charles	(SPA 93, US 213th)	5
LTC Thaw, William	(D6, C42, N65, N124, US 103d, 3rdPG)	5

5 • Lafayette Flying Corps Roster

269 Men

ABBOTT, WAINWRIGHT
ADAMS, JOHN RUSSELL
ALLEN, SIDNEY THAYER
APPLETON, JR, WALTER K.
ASH, ALAN NEWTON (DIED IN SERVICE)
ATEN, ARTHUR MILLS
BACH, JULES JAMES
BAER, PAUL FRANK
BAIRD, BENJAMIN HESTER
BALSLEY, HORACE CLYDE (LAFAYETTE ESCADRILLE)
BARCLAY, LEIF NORMAN (DIED IN SERVICE)
BASSETT, JR., CHARLES CHESTER
BATCHELOR, HENRY AUGUSTUS
BAUGHAM, JAMES HENRY (DIED IN SERVICE)
BAYLIES, FRANK LEAMAN (DIED IN SERVICE)
BAYNE, JAMES ALEXANDER (DIED IN SERVICE)
BENNY, PHILIP PHILLIPS (DIED IN SERVICE)
BENOIT, LEO E. (BENWAY, ERNEST L.)
BIDDLE, CHARLES JOHN
BIDDLE, JULIAN CORNELL (DIED IN SERVICE)
BIGELOW, STEPHEN SOHIER (LAFAYETTE ESCADRILLE)
BLAKE, CHARLES RAYMOND
BLUETHENTHAL, ARTHUR (DIED IN SERVICE)
BOAL, PIERRE
BOGGS, ELLISON CONVERSE
BOOTH, JR., WILLIAM VERNON (DIED IN SERVICE)
BOSWORTH, CLARENCE MARSH
BOULIGNY, EDGAR JEAN
BOYESEN, ALGERNON
BRADY, LESTER STRAYER
BRIDGMAN, RAY CLAFLIN (LAFAYETTE ESCADRILLE)
BROWN, JASPER CORNISH
BROWN, STAFFORD LEIGHTON (DIED IN SERVICE)
BUCKLEY, EVERETT TIMOTHY
BUFFUM, THOMAS BRADLEY
BULLARD, EUGENE JACQUES
BULLEN, RICHARD NIXON
BULLEN, WILLIAM GRAHAM
BUSH, PHILIP NELSON
BYERS, LOUIS LESLIE
CAMPBELL, ANDREW COURTNEY, JR. (LAFAYETTE ESCADRILLE-DIS)
CAMPBELL, HUGH GORDON
CARRERE, JR, JOSEPH MAXWELL
CASSADY, THOMAS GANTZ
CHADWICK, OLIVER MOULTON (DIED IN SERVICE)
CHAMBERLAIN, CYRUS FOSS (DIED IN SERVICE)
CHAPMAN, CHARLES WESLEY (DIED IN SERVICE)
CHAPMAN, VICTOR EMMANUEL (LAFAYETTE ESCADRILLE-DIS)
CHARTON, LOUIS
CHATKOFF, HERMAN LINCOLN
CLAPP, ROGER HARVEY (DIED IN SERVICE)
COATSWORTH, CALEB JAMES
COLLIER, EDWARD M.
COLLINS, PHELPS (DIED IN SERVICE)
CONNELLY, JAMES ALEXANDER
COOK, ALAN AUGUSTUS
COOKSON, LINN PALMER (DIED IN SERVICE)
COREY, RUSSELL BRACKEN
CORSI, EDWARD CHARLES
COTTON, JOHN ROWELL
COURT, ISADORE
COWDIN, ELLIOT CHRISTOPHER (LAFAYETTE ESCADRILLE)
CREHORE, AUSTEN BALLARD
CUNNINGHAM, ARTHUR LAWRENCE
CURTIS, FRAZIER
CUSHMAN, ALVAN ALEXANDER
DAVIS, PHILIP WASHBURN (DIED IN SERVICE)
DOCK, JR, GEORGE

DOLAN, CHARLES HEAVE (LAFAYETTE ESCADRILLE)
DONZE, ROBERT LOUIS
DOOLITTLE, JAMES RALPH (LAFAYETTE ESCADRILLE-DIS)
DOWD, DENNIS (DIED IN SERVICE)
DOWD, MEREDITH LOVELAND (DIED IN SERVICE)
DREW, JR, SIDNEY RANKIN (DIED IN SERVICE)
DREXEL, JOHN ARMSTRONG (LAFAYETTE ESCADRILLE)
DUFFY, NATHANIEL EDMUND
DUGAN, JR, WILLIAM EDWARD (LAFAYETTE ESCADRILLE)
DULON, LOWELL RICHARDS
EATON, SHERBURNE
EDGAR, STUART EMMET (DIED IN SERVICE)
ELDREDGE, DONALD HERBERT
ELLIOTT, CHESTER ARTHUR
ELY, DINSMORE (DIED IN SERVICE)
ENDICOTT, JOHN
EOFF, ROBERT GRIMSHAW
FAIRCHILD, EDWIN BRADLEY
FAITH, CLARENCE HENRY
FAUNTLEROY, CEDRIC ERROLL
FERGUSON, IAN FEARCHAR
FLYNN, JOSEPH
FORD, CHRISTOPHER WILLIAM (LAFAYETTE ESCADRILLE)
FORD, TOD
FORSTER, HENRY
FOWLER, ERIC ANDERSON (DIED IN SERVICE)
FREY, WILLIAM
GENET, EDMOND CHARLES CLINTON (LAFAYETTE ESCADRILLE-DIS)
GIBSON, WILLIAM WALLACE
GILL, JOSEPH FRANCIS
GILMORE, WILLIAM SMITH
GLOVER, CLARENCE MERRITT
GOURARD, REGINALD G.
GREY, CHARLES GOSSAGE
GRIEB, HENRY NORMAN (DIED IN SERVICE)
GRIER, JAMES MURRAY
GUEST, DAVID PORTER
GUNDELACH, ANDRE (DIED IN SERVICE)
GUY, DAVID WADE
HALL, JAMES NORMAN (LAFAYETTE ESCADRILLE)
HALL, WESTON BIRCH (BERT) (LAFAYETTE ESCADRILLE)
HAMILTON, EDGAR GUERARD
HANFORD, ROBERT MARSHALL (DIED IN SERVICE)
HARRISON, JR, JOHN B.
HAVILAND, WILLIS BRADLEY (LAFAYETTE ESCADRILLE)
HEILBUTH, JOHN RAYNOLDS
HEWITT, JR, THOMAS MOSES (LAFAYETTE ESCADRILLE)
HICKSON, LESLIE MATHESON
HILL, DUDLEY LAWRENCE (LAFAYETTE ESCADRILLE)
HINKLE, EDWARD FOOTE (LAFAYETTE ESCADRILLE)
HITCHCOCK, JR., THOMAS
HOBBS, WARREN TUCKER (DIED IN SERVICE)
HOEBER, ROBERT BENTLEY
HOLDEN, MILTON WHITELY
HORTON, CHARLES DABNEY
HOSKIER, RONALD WOOD (LAFAYETTE ESCADRILLE-DIS)
HOUGH, EDWIN A.
HUFFER, JOHN WILLIAM
HUGER, DANIEL ELLIOTT
HUGHES, EARL WAYLAND
HULL, MARK LESLIE
JACOB, SERENO THORP
JOHNSON, CHARLES CHOUTEAU (LAFAYETTE ESCADRILLE)
JOHNSON, HARRY FIRMSTONE (DIED IN SERVICE)
JOHNSTON, ARCHIBALD BURTT
JONES, CHARLES MAURY
JONES, HENRY SWEET (LAFAYETTE ESCADRILLE)
JUDD, EDWARD DAVID
KENYON, HUGO ALDEN
KERWOOD, CHARLES WAYNE
KINSOLVING, CHARLES MCILVAINE
KIRKWOOD, WILLIAM FRANCIS
KOWALL, JOHN ROBERT
KRUIJFF, THEODORE DE
KYLE, GEORGE MARION
LARNER, GORMAN DE FREEST
LEE, HENRY S.
LEE, SCHUYLER (DIED IN SERVICE)
LEHR, MANDERSON (DIED IN SERVICE)
LEWIS, DAVID WILBUR
LITTAUER, KENNETH PROCTER
LOOMIS, RALPH LANE
LOOMIS, WILLIAM FITCH
LOUGHRAN, EDWARD (DIED IN SERVICE)
LOVELL, WALTER (LAFAYETTE ESCADRILLE)
LUDLAM, W. LESLIE
LUFBERY, RAOUL GERVAIS (LAFAYETTE ESCADRILLE-DIS)
McCALL, GEORGE ARCHIBALD
McCONNELL, JAMES ROGERS (LAFAYETTE ESCADRILLE-DIS)
McCREARY, JR. JAMES B.
McGINN, WILLIAM
McKEE, HERSHEL
McKERNESS, WILLIAM JOHN (DIED IN SERVICE)
McMILLEN, JAMES HAITT
MACKE, GORDON B.
MACMONAGLE, DOUGLES (LAFAYETTE ESCADRILLE-DIS)
MAGLEY, GUY BERTRAM
MALONE, CHARLES THOMAS
MANIERRE, HAROLD L.
MARR, KENNETH ARCHIBALD (LAFAYETTE ESCADRILLE)
MASSON, PIERRE DIDIER (LAFAYETTE ESCADRILLE)
MEEKER, WILLIAM HENRY (DIED IN SERVICE)
MILLER, ALVIN FORD
MILLER, GEORGE
MILLER, WALTER BERNARD (DIED IN SERVICE)
MILLS, GORDON R.
MOLTER, BENNET ARTHUR
MOORE, ROBERT LOUIS
MOSELEY, GEORGE CLARK
MOUVET, OSCAR
MUNSON, CURTIS B.

NICHOLS, ALAN HAMMOND (DIED IN SERVICE)
NORDHOFF, CHARLES BERNARD
OAKES, JR, NATHAN PRINCE
OVINGTON, CARTER LANDRAM (DIED IN SERVICE)
PADEN, DAVID
PALMER, HENRY BREWSTER (DIED IN SERVICE)
PARKER, AUSTEN GILETTE
PARSONS, EDWIN CHARLES (LAFAYETTE ESCADRILLE)
PAVELKA, PAUL (LAFAYETTE ESCADRILLE-DIS)
PELTON, ALFRED DIGBY (DIED IN SERVICE)
PETERSON, DAVID MCKELVY (LAFAYETTE ESCADRILLE-DIS 16 MARCH 1919)
POLLOCK, GRANVILLE ALEXANDER
PONDER, WILLIAM THOMAS
POTTER, THOMAS WINDEATT
PRINCE, JR, FREDERICK HENRY (LAFAYETTE ESCADRILLE)
PRINCE, NORMAN (LAFAYETTE ESCADRILLE-DIS)
PUTNAM, DAVID ENDICOTT (DIED IN SERVICE)
RAND, RUFUS RANDALL
RANDALL, JOHN FRANCIS
READ, ROBERT EMERY
RENO, LEONARD MINOR
RHENO, WALTER DAVIS (DIED IN SERVICE)
RIDLON, HUGH OWEN
ROCKWELL, JR, GEORGE J.
ROCKWELL, KIFFIN YATES (LAFAYETTE ESCADRILLE-DIS)
ROCKWELL, ROBERT LOCKERBIE (LAFAYETTE ESCADRILLE)
ROCLE, MARIUS ROMAIN
RODGERS, JR., WILLIAM BLACKSTOCK
ROLF, JR, JOHN F.
ROODE, CLIFFORD DE
ROSS, RAYMOND THOMAS
ROTHARMEL, KENNETH
ROUNDS, LELAND LASELL
RUMSEY, JR, LAURENCE DANA (LAFAYETTE ESCADRILLE)
SAUL, JOSEPH ROE
SAXON, HAROLD YOUNG
SCANLAN, LAWRENCE
SCHREIBER, EDWIN BOOTH (DIED IN SERVICE)
SEAVER, HORACE
SHAFFER, WALTER JOHN
SHIPLEY, WALTER B.
SHONINGER, CLARENCE BERNARD
SINCLAIRE, REGINALD
SITTERLY, GLENN NELSON
SKINNER, SAMUEL WIGGINS (DIED IN SERVICE)
SOUBIRAN, ROBERT (LAFAYETTE ESCADRILLE)
SPEERS, WALLACE
SPENCER, DUMARESQ (DIED IN SERVICE)
STANLEY, ALFRED HOLT
STARRET, JR, FRANK ELMER (DIED IN SERVICE)
STEARNS, RUSSELL FALCONER
STEHLIN, JOSEPH CHARLES
STICKNEY, HENRY ELMER
STONE, DONALD EDWARD (DIED IN SERVICE)
STONE, GERALD STARR
SULLIVAN, UPTON SUPPLE
TABER, LESLIE RAY
TAILER, WILLIAM HALLET (DIED IN SERVICE)
TAYLOR, ELMER BOWDEN (DIED IN SERVICE)
TERRES, HUGH (DIED IN SERVICE)
THAW, WILLLIAM (LAFAYETTE ESCADRILLE)
THOMPSON, CLIFTON BADLAM
TRINKARD, CHARLES (DIED IN SERVICE)
TUCKER, DUDLEY GILMAN (DIED IN SERVICE)
TURNURE, JR, GEORGE EVANS
TYSON, STEPHEN MITCHELL (DIED IN SERVICE)
VAN FLEET, JR, WILLIAM CAREY
VEIL, CHARLES HERBERT
WAINWRIGHT, NEAL
WALCOTT, BENJAMIN STUART (DIED IN SERVICE)
WASS, WILLIAM ETHELBERT VAN
WELLMAN, WILLIAM AUGUSTUS
WELLS, FRANK WILLARD
WHITMORE, HERMAN KOTZCHMER
WHITMORE, JOHN JOYCE
WILCOX, CHARLES HERBERT
WILD, MARCELLUS EDWARD
WILLARD, GEORGE GALE
WILLIS, HAROLD BUCKLEY (LAFAYETTE ESCADRILLE)
WILLOUGHBY, WESTEL ROBINSON
WILSON, JOSEPH VOLNEY (DIED IN SERVICE)
WILSON, PIERRE MARIE
WINSLOW, ALAN FRANCIS
WINSLOW, CARROLL DANA
WINTER,JR, CHARLES WALLACE (DIED IN SERVICE)
WOODWARD, HENRY HOUSTON (DIED IN SERVICE)
WORTHINGTON, WARWICK
WRIGHT, HAROLD EVERETT
YORK, WALTER RAYMOND (DIED IN SERVICE)
ZINN, FREDERICK WILHELM

Wainwright Abbott

Wainwright Abbott was born in Pittsburgh, Pennsylvania on 27 April 1891, the son of William Lathan Ballard Abbott, an executive with Carnegie Steel Corporation, and Annie (Wainwright) Abbott. Wainwright was one of six offspring. He spent much of his youth in Europe with his parents where the family had gone to seek a cure for one of his sisters who was stricken with polio. Wainwright was still in Europe with the family when war broke out in August of 1914.

Abbott attended St. Paul's School, Concord, New Hampshire, and Phillips Academy, Andover, Massachusetts before entering Yale University (Sheffield Scientific), Class of 1914. Desirous of entering the European War, in February of 1916 he joined the Norton-Harjes Ambulance section 62 and served at Verdun with that organization until April of 1917.

At the urging of his friend William Thaw, on 2 April 1917 Abbott enlisted in France's *Service Aeronautique* in Paris. From 3 April to 15 September he attended the aviation schools at Avord and Pau. He received his *brevet* on the Nieuport on 7 August 1917.

From 8 September 1917 to 14 August 1918 *Caporal* Abbott flew at the Front with *Escadrille SPA 154*. Soon promoted to *Sergent*, he flew numerous combat patrols over the Aisne and Marne sectors and in April and August 1918 he was officially credited with the destruction of two enemy aircraft. Abbott always tried to fly aircraft with the number 13 on them, considering that number lucky.

On 15 August 1918 Abbott was commissioned a First Lieutenant in the Aviation Section of the Signal Corps, with duty as test pilot at Air Service HQ, Paris. He continued in the same capacity at Colombey-les-Belles from 3 September 1918 to the 11 November Armistice. Abbott was awarded France's *Croix de Guerre*, with Palm and Star for his victories over enemy aircraft.

Following his World War Service, Abbott joined the U.S. Foreign Service in 1920, serving in Athens, Dublin, London, Stockholm, Belgrade, Hamburg, Caracas, the Fiji Island, and Washington, D.C. In 1949 he was appointed Consul General in Belfast, Ireland. He retired in 1951. Abbott married Mary Alice Unanda-Scareen of Sweden. She passed away around 1975. Wainwright Abbott died in his home in Dublin, Ireland, on 15 January 1985 at age 93.

JOHN RUSSELL ADAMS

John Russell Adams, Jr. was born on 21 October 1897 in New York City, New York, the son of John Adams and Annette F.(Ridgeway) Adams. He attended the State Agricultural School of Connecticut before leaving that institution to join the Norton-Harjes Ambulance service in 1916.

Adams served with SSU 5 before enlisting in France's *Service Aeronautique* in Paris on 20 February 1917. From 27 February to 10 August 1917 he attended the aviation schools at Avord, Pau, and the *G.D.E.* He received his *brevet* on the Nieuport on 14 June 1917.

From 12 August to 12 October 1917 *Caporal* Adams served at the Front with *Escadrille SPA 95*. He was involved in several aerial combats and participated in his flight's unconfirmed destruction of two enemy aircraft. In mid-October Adams was granted leave to the United States. He returned to France in late November and from 31 December 1917 to 13 March 1918 he served with *SPA 81*.

On 29 March 1918 *Caporal* Adams was commissioned a Second Lieutenant in the Aviation Section of the Signal Corps. He served as ferry pilot at the American Acceptance Park, Orly from 4 April 1918 to the Armistice, at which time he was in charge of the school for ferry-pilots. Lieutenant Adams was honorably discharged on 6 August 1919 at Mitchell Field, Long Island, New York.

Following World War One Adams relocated to California where he married Romona Shorb. Adams died at the Livermore Sanitarium, California on 17 July 1947 at age 50. His cremated remains were inurned at the Chapel of Memories, Oakland, California. His ashes were later transferred to Golden Gate National Cemetery on 22 October 1963.

SIDNEY ALLEN

Sidney Thayer Allen was born in St. Louis, Missouri on 11 September 1893, the son of George L. Allen and Lillie H. Allen. He attended Columbia University but left that institution in early 1917 to join the American Ambulance Field Service.

On 17 January 1917 Allen sailed for France aboard the *Chicago*. He served in France with SSU 4 for four months before enlisting in France's *Service Aeronautique* on 19 July 1917.

Allen was released from French Aviation in September 1917 because of defective eyesight. He then enlisted in the U.S. Army and was commissioned a 2nd Lieutenant in the Field Artillery. From 5 February to 26 April 1918 Lieutenant Allen was assigned to the 12th Observation Squadron, USAS. From 26 April to 8 June 1918 he served with French (SPA) or *SOP 228*.

WALTER APPLETON

Walter K. Appleton, Jr. was born in New York City, New York in 1896, the son of Walter K. Appleton, a publisher. He attended Geneva College, Beaver Falls, Pennsylvania. At the outbreak of World War One he left school and sailed for France. Appleton enlisted in the French Foreign Legion (Infantry) in Paris in early August 1914. On his enlistment papers he gave his occupation as that of jockey. He later transferred to the 170th Infantry Regiment. In March of 1916 Appleton was evacuated from the Battle of Verdun, ill from exposure. Suffering from typhoid fever, he was sent to hospital.

Upon his recovery, on 15 July 1916 Appleton enlisted in France's *Service Aeronautique*. He was released on 16 December 1916 because of inaptitude. He returned to the 170th French Infantry Regiment. There he remained until February 19, 1918 when he joined the United States Army in Paris.

Appleton served with Company H of the 29th Engineers until 14 November 1918. He was promoted to sergeant in May of that year, but was reduced to private the following month. Following the Armistice, Appleton was promoted to corporal and served with the Headquarters detachment of the division of Light Railway and Roads until his honorable discharge on 6 March 1919.

ALAN ASH

Alan Newton Ash was born in Chicago, Illinois, on 6 December 1890, the son of Isaac Newton Ash, a prominent Chicago businessman and member of the Board of Trade, and Eleanor (Tenney) Ash. When Alan was five years old, his father died at age 63 of a heart attack, the same day he lost $500,000 on the Chicago Board of Trade. However, he still left a substantial estate which adequately saw to the needs of his widow and four children.

Alan Ash, nicknamed "Skinny," attended public schools in Chicago. He pursued a four-year curriculum at the Hackley College, Terrytown, New York then took a course in agriculture at the University of Illinois, Urbana, where he was a high jump and pole vaulting champion. Upon leaving school, he made his home with his brothers on their farm at Henry, Illinois.

On a visit to Chicago, Alan had been stirred by a propaganda display dealing with German atrocities toward Belgian civilians. When President Wilson declared war on Germany, Alan enlisted in the American Ambulance Field Service at Chicago, listing his occupation as farmer. After taking a course in the construction and operation of the Ford motor, he sailed for France on 26 May 1917.

Ash left the Field Service shortly after his arrival in Paris. On 15 June 1917 he enlisted in France's *Service Aeronautique*. From 19 June 1917 to 20 February 1918 he attended the aviation schools at Avord, Crotoy, and the *G.D.E.* He received his *brevet* on the Caudron on 3 November 1917.

From 23 February to 31 May 1918 *Caporal* Ash served at the Front with *Escadrille Br. 134*. On that date he was killed in aerial combat north of Oulchyle-Chateau (Aisne.) He was 28 years of age at the time of his death. His body was never found. Ash was posthumously awarded France's *Medaille Militaire* and *Croix de Guerre*, with Palm.

ARTHUR ATEN

Arthur Mills Aten was born in Brooklyn, New York. He attended the Westchester School and was involved in mining prior to World War One. On 7 September 1916 Aten enlisted in France's *Service Aeronautique*. He received his *brevet* on 30 April 1917. On 10 May 1917 he received permission to return briefly to the United States. When his permission expired he did not return to France and was considered a deserter by the French Army.

On 18 May 1918 officers of the U.S. Army intelligence Service arrested Arthur Aten in New York City at a club on East Sixty-Seventh Street. He was charged with impersonating a French officer and using the French uniform to cash checks at a bank where he had no funds. Aten was carrying with him a list of socially prominent individuals, and he had just left the house of a society woman. The intelligence officers told reporters that Aten had cashed checks in hotels in several eastern cities, including Washington., D.C. and Baltimore, Maryland, and that he had failed to pay a hotel bill in Baltimore.

Aten claimed the right to wear the French uniform because of his short service in France's aviation section. He also falsely claimed to the officers that he had gone abroad as a member of the Ambulance Service. Prior to going to Europe Aten said he had taken part in an effort to show that Dr. Cook had not scaled Mount McKinley.

James Bach

Jules James Bach was born in New Orleans, Louisiana on 27 March 1885, the son of Mr. & Mrs. Jules Bach. He was educated in England and in France, receiving a mechanical engineering degree in 1904 from the Sorbonne University, Paris. In 1908 he received a degree in Civil Engineering from the *Ecole Centrale*, Paris. He then worked as a civil engineer in England for three years before being employed in the U.S. by the Utah Copper Company.

At the outbreak of World War One, on 24 August 1914, James Bach became one of the first Americans to enlist in the French *Foreign Legion*. He left the *Legion* on 10 December to enlist in the *Service Aeronautique*. From 10 March to 29 August 1915 Bach attended the aviation schools at Pau and the *R.G.A.* He received his *brevet* on the Caudron on 4 July 1915.

Bach served at the Front with *Escadrille M.S. 38* from 29 August to 23 September 1915, attaining the rank of *Caporal*. On 23 September, while on a spy mission flying two French soldiers behind enemy lines, Bach crashed his aircraft and was captured by the enemy. Thus, he became the first American in the war to be taken prisoner. He was held in various prison camps for more than three years. For his war service, Bach was awarded France's *Legion of Honor*, *Medaille Militaire*, and *Croix de Guerre*.

Following the War Bach changed his name to the less-Germanic J. James Buck. He operated the Packard Motors Car Agency in Paris until 1940. It was in that city on 2 December 1927 that he married Adrienne DeLeon, a Parisian fashion designer, whom he had known prior to World War One. Mrs. Buck had graduated from the French State Nurses school and had served as a nurse captain in World War One. She had previously graduated from the *Academic Julian*, a French art school, and had a master's degree from the Sorbonne University.

The Buck's had an elaborate home in Paris. With the invasion of France by Germany in 1940, they fled Paris with only four suitcases, and on the road to Bordeaux they were frequently strafed and bombed by German aircraft. They left Bordeaux by ship just three hours before the Germans occupied that city.

The Bucks eventually settled in Salt Lake City where he was hired on 13 September 1943 to teach Italian and French at the University of Utah. He remained in that capacity for fifteen years. Mrs. Bach served for 25 years as the French Consulate agent for Utah and also taught French at the University. She was awarded the Legion of Honor by France for her service as a nurse in World Wars One and Two.

James Bach died at the Latter-Day Saints Hospital at Salt Lake City, Utah on 23 February 1960 at age 74. He is buried at Salt Lake's Mt. Calvary Cemetery.

PAUL BAER

Paul Frank Baer was born in Fort Wayne, Indiana on 29 January 1894, the son of Alvin E. and Emma (Pasent) Baer. He attended Fort Wayne public schools. In 1913 he became a machinist at Detroit's Cadillac Motor Car Co. following graduation from its applied mechanics school. In 1916 Baer served with the Indiana National Guard as a truck driver in the Mexican Border campaign. In early 1917 Baer sailed for France aboard the liner, *Rochambeau*. On 20 February 1917 he enlisted in France's *Service Aeronautique*. From 27 February to 12 August 1917 he attended the schools at Avord and the *G.D.E.* He received his *brevet* on the Caudron on 15 June 1917.

From 14 August 1917 to 10 January 1918 *Caporal* Baer served at the Front with *Escadrille SPA 80*. From 10 January to 18 February 1918 he flew with *SPA 124*, the date the squadron evolved into the U.S. 103d Pursuit Squadron. Previously, on 5 November 1917 Baer had been commissioned a First Lieutenant in the Aviation Section Signal Corps. From 18 February to 22 May 1918 he flew with the 103d, the day he was shot down and wounded near Armentieres. He remained a prisoner until the Armistice. The first of his nine credited victories came on 11 March 1918 and his was the first U.S. victory by a U.S. officer. Baer was awarded the U.S. Distinguished Service Cross, with bronze oak leaf, France's *Legion of Honor*, *Croix de Guerre*, with seven Palms, and the Aero Club of America medal.

Following the War Baer settled in Detroit and became a civilian aviator, engaging in experimental work at the aeronautical laboratories there. He later became an inspector in the Aeronautic Branch, Department of Commerce, at Washington, DC. He then involved himself in air shows and worked in Hollywood as a stunt pilot. On 25 June 1925 Fort Wayne dedicated the city's airport in his honor. Baer also worked as a pilot in New York, Michigan, California, and Texas. In 1929 he joined Pan American Airways, establishing airmail routes in South America.

On 2 October 1930 Baer, then 36, set sail for Shanghai, China aboard the *Empress of Canada* to work as a transport pilot. On 28 October he made his first flight for China Airways in a Hornet-powered Leoning amphibian from Shanghai's Lung Hwa amphibian port. He soon began making the five hundred-sixteen-mile Shanghai-Hankow run, occasionally coming under fire by Chinese bandits. On 9 December 1930, shortly after Baer had taken off from Shanghai in his amphibian aircraft, he struck the mast of a Chinese junk, mortally injuring himself and killing his Chinese co-pilot and a passenger, a Mrs. Andrieva, a Russian.

Elaborate military services were held for Baer in Shanghai. His body was returned to the United States, and on 3 January 1931 one of the largest military funerals in Fort Wayne's history was held in his honor. Baer was buried at Fort Wayne's Lindenwood cemetery. On 14 June 1941 the War Department designated the new air field to be constructed south of Fort Wayne as "Baer Field" in recognition of Paul Baer's World War service.

BENJAMIN BAIRD

Benjamin Hester Baird, Jr. was born in San Francisco, California on 12 June 1887, the son of Benjamin H. Baird, and Veronica (Ireland) Baird. Benjamin attended California public schools before studying mechanical engineering for two years at the University of Santa Clara. Baird was working in New York City as a mechanical engineer when he decided to join Richard Norton's American Volunteer Motor Ambulance Corps in France.

On 2 June 1917 Baird sailed for France aboard the *Louraine*. On 25 June 1917 Baird enlisted in France's *Service Aeronautique*. From 29 June to 1 March 1918 he attended the aviation schools at Avord and the *G.D.E.* He received his *brevet* on the Caudron on 26 September 1917.

On 2 March 1918, while Baird was at the *G.D.E.* awaiting assignment to the Front, he received his commission from Paris as an Ensign in U.S. Naval Aviation. On 3 April Baird was designated Naval Aviator (HTA) No. 554 at Moutchic, France. Baird served at Ile Tudy, Pauillac, Bolsena, Italy, and Brest, France.

Following the Armistice, Baird returned to the United States and was honorably discharged on 1 March 1922. He returned to California where he worked as a machinist until his retirement. He died in San Francisco on 19 August 1962 at age 75.

CLYDE BALSLEY

ESCADRILLE LAFAYETTE

Horace Clyde Balsley, the 9th volunteer of the *Escadrille Americaine*, was born in Carbondale, Pennsylvania on 27 July 1893, the son of the Reverend Enos Josiah Balsley and Elizabeth (Phelps) Balsley. Rev. Balsley had been rector of the Episcopal Church at Carbondale before health reasons forced him to move his family to San Antonio, Texas around 1907.

In 1910 Clyde Balsley graduated with honors from San Antonio's West Texas Military Academy. He then enrolled in the University of Texas at Austin but was forced to withdraw after one year when his father became invalided with arteriosclerosis, a disease which brought about his death in 1914.

Mrs. Balsley, upon her husband's illness, had started the *Novel Bakery* in her home to support her family. Clyde had helped out by making deliveries to customers in a Ford wagon.

The Rev. Balsley had hoped Clyde would follow him into the ministry. But Clyde's interests were drawn toward France. In a letter written to his mother while overseas in the fall of 1915, he explained why he had wished to enter the European conflict. Foremost was a desire "to see the war, to see it well."

"I can learn aviation, the newest game in the world, which will mean that I have learned a paying profession while on this trip. I will be made a sous-officer, and very probably will get a commission as a lieutenant in the French Army, which means I can transfer after the war to the United States Army and of which I have always dreamed. And, finally, I can take a man's part in this war for humanity..."

Clyde discovered through an overseas friend that the quickest way into the European war was as an ambulance driver. On 23 January 1915 he departed San Antonio for New Orleans, carrying a want-ad he had clipped from the *Express* which sought men to care for horses and mules bound by ship for Europe. On the 26th Balsley signed on as a muleteer for his passage and $65.00 and by the 27th he had boarded the S.S. *Dundin* and was headed for England.

Balsley arrived at Bairy, Wales on 18 February 1915. He did not receive his promised wage and was forced to sell his boots to survive. He was finally paid off and he arrived in London on the 21st. He worked as a timekeeper for Harrod's LTD while he waited for his application to be accepted by the British Ambulance Committee. His application was ultimately rejected about the same time he quit his job in a dispute with a supervisor. On 17 April he secured a passport at the French Consulate, booked passage for Paris, and arrived there on the 20th.

On 1 May Clyde was accepted for work at the American Ambulance Hospital at Paris where he worked as a mechanic and ambulance driver until his resignation on 25 September, the date of his acceptance into French aviation. On the 30th he began his pilot training at the aviation field at Pau.

On 2 January 1916 Balsley received his *brevet* on the Bleriot. Following further training on the Voisin, on 26 February he was sent to LeBourget, twelve kilo-

Military Aviation School at Pau, January 1916. Left to right: Jean Huffer, Clyde Balsley, Charles Johnson, Dudley Hill, Paul Pavelka, and Laurence Rumsey.

meters from Paris, to *Escadrille V.97* as a member of the Air Guard of Paris. On 3 April he arrived at the G.D.E. (*Groupe des Division d'Entrainement*) at Plesses-Belleville for training on the Nieuport 23. On 27 May *Caporal* Balsley reached the Front with the *Escadrille Americaine* at Bar-le-Duc. Balsley officially joined the squadron as its ninth American pilot on the 29th.

And it was there that he was destined to log the briefest combat time yet endure the greatest physical pain of any of Nieuport 124's thirty-eight American pilots. His actual combat time was measured in minutes while his physical suffering would be logged in years.

In the second week of June Balsley's Nieuport fighter arrived at Bar-le-Duc, and he chose a white star as his personal aircraft insignia to represent the "Lone Star" of Texas. On 18 June 1916, in the pre-dawn hours of a Sunday, Balsley began his first and only combat mission, flying fighter protection for "artillery reglage machines." Accompanying Balsley had been squadron commander Thenault, Norman Prince, and Kiffin Rockwell. At 3500 meters above the forts of Verdun they had met and immediately engaged a large enemy patrol. Balsley's target became an Aviatik bi-place. He dove down on the machine and closed to within 50 meters before firing his Lewis gun which instantly jammed after expelling one round. Balsley jockeyed his 15-meter Nieuport out of the fracas to clear the gun breech. He was immediately attacked by three enemy fighters.

"I was then about twelve-thousand feet up. It was while I was standing completely on my head, the belly of my machine skyward, that something struck me. It felt like a kick of a mule. With the sensation of losing a leg, I put my hand down to learn if it was still there. I was strapped in too tightly to be knocked overboard, and I had the presence of mind to cut the motor. But as my right foot went back with the shock of the bullet, my left foot sprang forward. So, with my commands reversed, my leg knocked out, still standing on my head, I fell into a spinning nose dive."

Balsley fought desperately to right his plane to prevent crashing it into enemy territory. Working his damaged leg with his hands, he managed to level out. His landing site became the space between the Allied frontline and reserve trenches. As he dropped down, barbed wire snagged his wheels and his bus flipped over, catapulting him to earth. Shells exploded around him as German cannons bracketed in on his plane. Unable to crawl, Balsley grabbed the grass before him and pulled himself forward. Four French *poilu* crawled through the wire and dragged the young American toward their trench.

Clyde's flight suit was removed at the dressing station. He was injected with morphine and an antitoxin. Once his trousers had been cut away he saw the "grotesque" hole in his right hip. He was placed on a stretcher then transferred via limousine to a French evacuation hospital ten miles from the Verdun front, "a long narrow shed" filled with mutilated suffering men. The surgeon soon learned that the German bullet which had struck his right thigh had splintered into fragments, riddling his intestines, kidneys, and lungs.

The hospital, so near the Front, was frequently bombed by German aircraft. With the Verdun seige raging, Balsley was afforded little attention from the five nurses ministering to the swell of wounded filling the fifty portable sheds which made up the hospital. Balsley was denied water because of the nature of his wounds. He later wrote of his suffering:

"My cry for water was so intense that it absorbed every other sense. I wanted to hear water, to see water, to feel it trickling through my fingers. So violent was this one longing that I was actually blinded. I did not at first see a man standing beside my bed.

"Then came to me one by one-the heavy, black hair, the great arms, and the sincerest eyes in the world. When I put them all together, I gave a groan of joy. It was Victor Chapman, flown over from Bar-le-Duc."

Clyde Balsley beside his Nieuport showing his personal insignia, the "Lone Star" of Texas, 8 June 1916.

Chapman, from New York City, was one of the first Americans to enlist in the French Foreign Legion, and an original member of the *Escadrille Americaine*. He had brought Balsley his toothbrush.

"Anything I can get for you, old man?" said he, meeting my thirsty eyes.

"You bet," said I. "They won't let me have any water." The way I kept moistening my lips finished the appeal..

"How about oranges?" said he, and turned to my doctor, just at that moment come in."

The doctor said he would allow it, and Chapman promised to return with oranges, even if it meant flying to Paris to get them. Once Chapman had gone, Balsley fell into a fitful sleep.

Clyde awoke next morning in the presence of three French officers. One, a colonel, decorated him with France's *Medaille Militaire* and *Croix de Guerre*, with Palm, stating: "You are the first American aviator to be severely wounded-for France." Balsley viewed the decorations less as an honor and more as a sign that the officers believed he would soon die since the two soldiers who had been decorated the previous day had passed away shortly after they had received their awards.

"But I'm not going to die," Balsley protested. Yet in his suffering he could feel his will to live leave him. As he studied the two medals pinned to his pillow, he thought how willingly he would have traded them for oranges to slake his thirst.

The following day oranges did arrive, not brought by Chapman but by another New Yorker, *Sergent* Elliot Cowdin, one of the founding members of *N.124*. When Clyde asked him about Chapman, Cowdin told him he hadn't come because his plane had broken down. The next day Clyde learned the horrible truth from a French officer in his ward who read from a Paris paper how Chapman had been killed on 23 June 1916 while flying a patrol near Verdun. In his fighter was a bag of oranges he had planned to deliver to Balsley after his mission.

On the fourth day of his confinement, Clyde again petitioned the doctor for his return to his Squadron. Instead, he was promised a transfer to the American hospital at Paris. The nights in the ward became unbearable for him. In spite of the summer heat, the windows and doors were kept tightly shut and the stench of unwashed wounded produced the smell of "a dog-kennel."

Balsley endured a "haunted imprisonment" at the Verdun hospital. The ward became a charnel house, a butcher shop, as mutilated men were lugged in from the battlefield, expired, then were lugged out. The cots, when vacated, were never empty longer than five minutes. As Clyde watched this hellish procession, smelled the stench of the wounded who "never had a drop of water on their bodies," he grew too weary to brush the flies from his face. He realized the fault lay not with the overworked doctors and calloused orderlies. They simply could not keep up with the casualties pouring in from Verdun. Clyde's only relief from his suffering came with the frequent visits of his fellow Americans from the Squadron who always tried to drop in and cheer him up in spite of their rigorous flying schedules.

Clyde Balsley at the American Hospital in Paris.

Clyde Balsley and wife Mariam taken shortly after their marriage in Los Angeles. Note that Balsley still uses a cane. Circa 1920.

The July heat evoked Clyde's worst suffering. Festering bedsores broke out on his hips, the result of lying day after day on a sweat-soaked cot.

"For days I would lie without moving an inch in a reeking mass of blood, pus, and perspiration. Only when my temperature went so many degrees above normal would my dressing be changed. This was usually about every four days. Then as I was lifted up for the trip to the dressing table, the hole in the bed made by my body would be black with blood and matter . . . I was also getting such cramps that I asked the doctor if he couldn't straighten my leg.

Balsley underwent two operations and nearly died. Only after his attending doctor was transferred to Paris did he manage to escape the field hospital at Verdun. By then Clyde had recognized the surgeon's incompetence. For the first time the injured aviator was given the care he required under the head surgeon at the American Hospital at Paris.

Clyde spent a year and a half mending at that hospital. He underwent six more operations to remove bullet fragments and to straighten his leg. On 18 May 1917 Balsley was commissioned a Captain in the U.S. Air Service. On Christmas Day 1917 he was told he would walk again, but with a heavy limp and with the support of a cane. Still, he would walk, and he felt elated.

When his wounds had adequately healed Captain Balsley returned to the United States where he served as an instructor in the School of Aviation at Lake Charles, Louisiana. He was then assigned to the Pursuit Division at Washington, D.C. where he took part in recruiting and war bond drives with such celebrities as the actress Mary Pickford. He travelled across the United States in this capacity and was finally stationed at Rockwell Field, California where he served from 1 January to 13 October 1919. On 10 November 1919 he married silent screen actress Miriam MacDonald. He was discharged from the military on 14 June 1920 while a patient at the Letterman General Hospital, the Presidio of San Francisco. He was later awarded France's *Legion d'Honneur*.

For a time Balsley ran a transcontinental airmail service. He then formed a movie production company with his sister-in-law, Katherine MacDonald. Around 1923 he settled in Los Angeles and eventually entered the real estate business. But he gave that up when he met Merle Norman to become a business manager for her cosmetic firm. He remained in that capacity until he died of a heart attack in Los Angeles on 23 July 1942 at age 48.

LEIF BARCLAY

Leif Norman Barclay was born on 21 May 1895 in New York City, the son of Dr. H.V. and Hansine O. Barclay. As a youth he lived in East Elmhurst, New York and was educated in Flushing High School. He briefly attended Yale University before spending two years in business.

In August of 1914 Barclay was in Norway visiting his ancestral country when war broke out in Europe. He shortly returned to the United States, and on 24 November 1914 he joined the American Ambulance Service.

Barclay was attached to Section 2 until August of 1915, at which time he took three months leave in the United States. He re-enlisted in the Ambulance Service in November of 1915 and was again attached to Section 2 until June of 1916.

On 26 June 1916 Barclay enlisted in France's *Service Aeronautique*. He trained at Buc, Avord, Cazeaux, Pau, and the *G.D.E.* He experienced some delay in aviation school from an accidental pistol wound. He received his *brevet* on the Bleriot on 6 October 1916.

On 12 April 1917 *Caporal* Barclay was assigned at the Front to *Escadrille N.82.* He was promoted to *Sergent* on 31 May 1917. On 1 June 1917, while doing acrobacy over the Chaux field, north of Belfort, the muzzle cup from his machine gun vibrated loose. striking and shattering his propeller. The fragments struck and tore loose the wing bracings, collapsing the wings and causing Barclay's fighter to crash. Fatally injured, Barclay was 21 years of age at the time of his death. He was posthumously awarded France's *Croix de Guerre*, with Palm for his war service and the Aero Club of America medal. He was buried at Caux,

Territorie de Belfort. His body was later transferred to the American Cemetery, Belleau Wood. In 1928 his remains were transferred to the Lafayette Flying Corps Monument near Paris.

CHARLES BASSETT

Charles Chester Bassett, Jr. was born in Washington, DC on 22 December 1896, the son of Major Charles Bassett, a Civil Engineer, and Fanny (Rice) Bassett. He attended Harvard University, Class of 1915–1917, spending two years in a languages curriculum. While at Harvard he was a member of the varsity track team.

Shortly after America entered World War One, Bassett left Harvard University in May of 1917 and volunteered as a sergeant in the aviation section of the U.S. Army Signal Corps to prepare for a flying commission. Finding the process too slow, he obtained an honorable discharge in order to join the Lafayette Flying Corps.

Bassett sailed for France on 2 June 1917 and enlisted in the *Service Aeronautique* on 17 June. He attended the aviation schools at Avord, Pau, Cazeaux, and the *G.D.E.*, training on the Nieuport.

Caporal Bassett was released from French aviation to accept an appointment as an Ensign in the Naval Reserve Flying Corps at Paris on 4 January 1918. On 1 February 1918 at Moutchick he was made Naval Aviator (HTA) No. 1316. From 14 February to 3 September 1918 he flew hydro aircraft, serving at Dunkirk, Clermont, and with the Northern Bombing Group at St. Inglevert. While on detached service with RAF Squadron 218, he flew day bombing missions against Ostend, Zeebrugge, and Bruges. Bassett was awarded the U.S. Navy Cross. His citation is as follows:

"For distinguished and heroic service as an aviator of land planes attached to the Northern Bombing Group in active operations co-operating with the Allied Armies on the Belgian front during September, October and November, 1918, bombing enemy bases, aerodromes, submarine bases, ammunition dumps, railroad junctions."

On 6 November 1918 Bassett returned to the U.S. as flight instructor at Miami. He was discharged as a Lieutenant(JG) on 4 March 1919. He re-entered in Class 6 as Ensign and was honorably discharged on 13 May 1928 as an Ensign in the U.S. Naval Reserve Force.

Following his war service Bassett worked for Burlington Mills then became a vice-president for Hartford Fiber Co., a division of the Bigelow-Sanford Carpet Co. He married Rinetta Heath, then Julia Hebard, and on 24 August 1932 he married Marian Shaw. Bassett was the father of a son, Charles, and daughter, Brooke. During World War Two Basset served as a Major in the U.S. 8th Air Force in England.

Bassett retired from Bigelow-Sanford, Inc. in 1962. He died in a Greensboro, North Carolina, nursing home on 22 March 1978 at age 81.

HENRY BATCHELOR

Henry Augustus Batchelor, III was born in Saginaw, Michigan on 27 January 1895, the son of Jacob F. Batchelor, owner of the Batchelor Timber Co., and Gertrude (Barr) Batchelor. Henry's boyhood home was Detroit, Michigan. He attended the Elstow School, Bedford, England, and the Worchester Academy, Worchester, Massachusetts, graduating in 1913. He attended Brown University, Providence, Rhode Island, Class of 1917, studying an English curriculum.

Batchelor left Brown University at the end of his junior year to join the American Ambulance Service in September 1916. He served with Section 2 at Verdun until December of that year then transferred to Section 10 and worked in Macedonia and Albania for six months.

On 1 August 1917 Batchelor enlisted at Paris in France's *Service Aeronautique*. From 2 August to 24 December 1917 he attended the aviation schools at Avord, Tours, Pau, and the *G.D.E.* He received his *brevet* on the Caudron on 14 October 1917.

From 26 December 1917 to 1 March 1918 *Caporal* Batchelor flew at the Front with *Escadrille SPAD 103*. On 3 March 1918 he enrolled at Paris in U.S. Naval Aviation as an Ensign. He was made Naval Aviator (HTA) No. 755 and served as chief pilot at Moutchic, France until December 1918. Batchelor served on Admiral Wilson's staff at Brest until the end of January 1919 then transferred to the American Red Cross's Balkan Commission as Aide to the Commissioner, working in Roumania and Serbia until September 1919. He was honorably discharged from the Navy as a Lieutenant(JG) on 30 September 1921.

That same year Batchelor began work as production Manager for the Strable Lumber & Salt Co of Saginaw, Michigan, serving as the New York and Foreign Sales Representative for that Company and for the Tennessee Oak Flooring Co. of Nashville, Tennessee. In 1925 he was associated with various firms in the development of Florida real estate, including the All-Florida Realty Co. and the H.A. Batchelor Co. In 1928 he married Alice Stearns. In 1936 he worked as a real estate broker in Chicago.

During World War Two Batchelor re-entered the Navy as a Lieutenant. He served on the Naval Transport Ship *Memphis* and was Commanding officer of CASU 3. He served at Tarawa and at Okinawa. He was honorably discharged as a LCDR on 8 October 1945.

Following the war he returned to his civilian occupation of real estate developer. He later married and moved to Los Angeles, California where he died on 29 October 1971 at age 76. His ashes are buried in Saginaw, Michigan.

JAMES BAUGHAM

James Henry Baugham was born in Washington, North Carolina on 12 February 1900, the son of William Patrick Baugham, a banker and gentleman farmer, and Mary Anna (Hannaford) Baugham, a musician and patriotic-song writer. James attended Washington public schools and the Bingham Military Academy in Asheville. As a youth he was given to daredevil acts involving great risk. Once he jumped into the street before a team of galloping horses and brought them to an abrupt halt. Another time, he and a companion were trapped for two days and nights in a small motorboat on Pamlico sound under steady rain and turbulent waves. When they finally reached safety, they were so weakened from their ordeal that they had to be carried from their craft.

In 1915 Baugham attended the Agricultural and Mechanical College of North Carolina. In the fall of 1916 he attended North Carolina State College at Raleigh, Class of 1920, but left that institution after two months, desiring to enter the European War. On 13 April 1917 he enrolled in the Curtiss School of Aviation at Newport News, Virginia. Having mastered flying boats and land planes, on 20 June 1917 he was awarded a civilian pilot's license.

That same month, after reading an advertizement in the Raleigh *News and Observer* asking for volunteers to fly for France, Baugham sailed for France and on 10 July 1917 he enlisted in the *Service Aeronautique* . At age 17, he was the youngest candidate in the Lafayette Flying Corps, one day younger than fellow American Tommy Hitchcock. From 19 July to 24 December 1917 Baugham attended the aviation schools at Avord, Juvisy, Pau, and the *G.D.E.* He received his *brevet* No. 9360 on the Caudron on 18 October 1917. Just before he left for the Front, he wrote the following to his mother: "I go to the front in a few days, and although I am going to do my best, I don't expect to last very long. I am not afraid of death or anything else that the Germans might do to me. I have had so many of my comrades killed that I look forward to it as a matter of course."

From 26 December 1917 to 27 June 1918 *Caporal* Baugham flew at the Front with *Escadrille N.157.* While with that squadron he won the France's *Medaille Militaire* for landing his crippled aircraft in no-man's-land before safely escaping under fire to French lines. From 27 June to 2 July 1918 *Sergent* Baugham flew with *SPA 98*. In June he applied for commission as Ensign in U.S. Naval Aviation

Baugham was seriously wounded in the leg on 1 July 1918 over Villers-Cotterets while doing battle with three enemy aircraft. He managed to reach French lines but crashed to earth, severely injuring himself. He died in hospital on 2 July 1918. His decorations included France's *Croix de Guerre*, with Palm, and the American Harmon Trophy. His remains are interred at the Lafayette Monument at St. Cloud, near Paris.

FRANK BAYLIES

Frank Leaman Baylies was born in New Bedford, Massachusetts on 23 September 1895, the son of Charles S. Baylies, a grain merchant, and Lydia Terry (Paige) Baylies. He was educated at the Friends Academy, the Josiah Swift School, New Bedford, and the Moses Brown School, Providence, Rhode Island. Following graduation he worked for his father as a salesman.

On 26 February 1916 "Jules" Baylies joined the American Ambulance Service. From October of 1916 to 11 May 1917 he served in France with Section One at Verdun, and with Section Three in Solonika. On 25 March 1917 Baylies received France's *Croix de Guerre*, with Palm for rescuing wounded while under fire.

On 21 May 1917 Baylies enlisted in France's *Service Aeronautique*. From 26 May to 15 November 1917 he attended the aviation schools at Avord, Pau, and the *G.D.E.* He received his *brevet* on the Caudron 20 September 1917.

From 17 November to 18 December 1917 *Caporal* Baylies flew at the Front with *Escadrille SPA 73*. From 18 December to 17 June 1918 *Sergent* Baylies flew with *SPA 3*. While with that Squadron he was credited with the destruction of 12 enemy aircraft. He was awarded France's *Legion of Honor*, the *Croix de Guerre*, with six Palms and one Star, the *Medaille Militaire*, and the Aero Club of America Medal. On 13 May 1918 he was transferred to the U.S. Air Service and was commissioned a Second Lieutenant, although he had earlier declined a captaincy. Baylies remained on detached service with his old French squadron.

On 17 June 1918 Baylies was killed in combat with two German triplanes of *Jasta 19* over enemy lines near Rollet, Oise. It is believed he was shot down by Leutnant Rudolph Rienau. Baylies was 22 years of age at the time of his death. He was buried at Courcelles-Epayelles, Oise. He remains were later transferred to the Lafayette Flying Corps Memorial near St. Cloud.

JAMES BAYNE

James Alexander Bayne was born in Grand Rapids, Michigan 18 November 1892, the son of James Bayne, a furniture manufacturer. He graduated from Grand Rapids Central High School in 1908 then attended the Chicago Art Institute for four years. When he left that institution he pursued an artist's career in Chicago for the next five years. While there, he created and painted the first Kellogg's Corn Flakes baby for that company's cereal box.

Following the example of his close friend, Andrew Courtney Campbell, Bayne sailed for France on 2 June 1917. On 10 July he enlisted in France's *Service Aeronautique*. From 19 July 1917 to 26 February 1918 he attended the aviation schools at Avord, Tours, Pau, Cazeaux, and the *G.D.E.* Bayne received his *brevet* on the Caudron on 19 October 1917.

From 1 March to 3 March 1918 *Caporal* Bayne flew at the Front with *Escadrille SPA 85*. From 3 March to 29 March 1918 he was with *SPA 81*. While a member of *SPA 81*, Bayne re-designed that unit's running greyhound insignia and painted it on the aircraft.

On 29 March 1918 *Caporal* Bayne was commissioned a First Lieutenant in the U.S. Air Service and was re-assigned to his old squadron, *SPA 81*, until the day of his death on 8 May 1918. On that date Bayne had been testing a 220 H.P. SPAD at 2000 meters above the flying field when his fighter was seen to go into a steep dive which lasted for 1000 meters. The four wings of his fighter tore away at that altitude, and the fuselage plunged into the ground, instantly killing Bayne. He was 25 years of age at the time of his death. His remains are interred at the Lafayette Flying Corps monument outside Paris.

PHILLIP BENNY

Philip Phillips Benney was born in Pittsburgh, Pennsylvania on 28 June 1895, the son of George Andrew Benny, president of the Benalite company, and Eugenia (Hill) Benny. Benny was educated at the Shady Side Academy, Pittsburgh. In 1916 he served as a volunteer on the U.S. Navy's civilian cruise. Benny was involved in the automobile business as a salesman in Pittsburgh when he decided to enter the European War. On 8 January 1917 he joined the American Ambulance Field Service in France. He served with Section 12 until the end of May.

On 31 May 1917 Benny enlisted in France's *Service Aeronautique*. From 18 June to 10 December 1917 he attended the flying schools at Avord, Tours, Pau, and the *G.D.E.* He received his *brevet* on the Caudron on 16 October 1917.

From 12 December 1917 to 26 January 1918 *Caporal* Benney was assigned at the Front with *Escadrille SPA 67*. Of his service there, his commanding officer said:

"So ardent was he that I had long delayed the moment of sending him against the enemy, fearing a little too much audacity and too little experience." On 25 January 1918 *Caporal* Benny set out on his first combat patrol. Flying with four comrades deep inside enemy lines near Verdun, his bus was attacked from beneath by a German patrol. While over Montfaucon, Benney was grievously wounded in the calf and thigh but still managed to pilot his fighter to a safe landing within allied lines despite a severe loss of blood. He was rushed by ambulance to the hospital at Glorieux where heroic efforts by Dr. Henriot and his comrades failed to save his life.

Benney succumbed to his wounds the following day, 26 January 1918. He was 22 years of age at the time of his death. He was posthumously awarded France's *Croix de Guerre*, with Palm for his 25 January action. He was buried at the Glorieux cemetery. His remains were later transferred to the Lafayette Flying Corps Memorial near Paris.

LEO BENOIT

Ernest Leon Benway (Leo Benoit) was born in Foxboro, Massachusetts on 1 August 1898, the son of Louis A. Benway, a jeweler, and Delia (Roy) Benway. He attended schools in Attleboro, Massachusetts. In the spring of 1917 Benway decided to enter the European War as an ambulance driver. Lacking his parents' approval, he applied for a passport under the name Leo Ernest Benoit, the French spelling of his surname. Benway was accepted for service with the American Ambulance Service in France, and on 5 May 1917 he sailed for Bordeaux aboard the SS *Espagne*.

On 16 May Benway arrived in Paris and decided to enter French aviation instead of the Ambulance Service. On 10 June 1917 he joined France's *Service Aeronautique* under the name of Leo Benoit. From 22 June to 15 November 1917 he attended the aviation schools at Avord, Juvisy, and the G.D.E. He received his *brevet* (#8941) on the Caudron on 22 September 1917.

On 18 November 1917 Benway was assigned to *Escadrille SPA 84* and flew at the Front with that squadron until 13 December. On that date, while on patrol, he became lost and was shot down in aerial combat near Laon when he was jumped by seven German Fokkers. Benway's plane crashed just inside allied lines and burst into flames. French troops pulled the unconscious pilot from the wreckage, and he was transported to the hospital at Villers-Cotterets where he remained comatose for 37 hours. Doctors treated Benway for a broken nose, a broken left leg, and a broken left arm. He also suffered a dislocated hip.

Following a period of recuperation, Benway was returned to the G D.E. and trained on the SPAD bi-place. On 31 January 1918 he was sent back to the Front to *SPA 228*, an observation squadron, and remained with that unit until 25 August 1918. On 29 March *Sergent* Benway was commissioned a Second Lieutenant in the U.S. Aviation Section Signal Corps. On 6 April 1918 he was awarded France's *Croix de Guerre*, with Star for continuing on a photographic patrol over enemy lines after having been wounded.

Benway had to fight his way home from this mission. His Spad bi-place again came under heavy attack by swarming enemy fighters. His observer was killed, yet Benway managed to shoot one enemy chasse down despite a painful arm wound. Weakened from loss of blood, he crash-landed within allied lines, severely wrenching his shoulder.

While recovering from these injuries, the now-Lieutenant Benway, USAS, served as a ferry pilot at the 5th Air Depot before being assigned on 25 August to the 213th Pursuit Squadron, USAS where he served until the 11 November 1918 Armistice. Along with his other war decorations, he was also awarded the French fourragere for having served with a unit that had won two citations in French Army Orders.

When Lieutenant Benway returned to the United States in early 1919, he was assigned to the Air Service Flying School, North Island, San Diego, California. In May he was transferred to the 2nd Aero Squadron, and on 27 June he married Sarah Frances Bell of Los Angeles. From that union four children were born.

On 30 October 1919 Lieutenant Benway was ordered with the 2nd Aero Squadron to Fort Mills, Corregidor Island, in Manila Bay, the Philippines. Their orders were to aerially map Luzon and the other islands of the Philippine archipelago. Once there, Lieutenant Benway began the mapping of the islands in anticipation of the construction of military airfields, flying one of the Army's first pontoon-equipped aircraft. Upon completion of the project, Benway decided to leave military service and enter commercial aviation.

In 1921, following his release from the officer's reserve corps, Benway relocated to Los Angeles and began what was to be a distinguished career in commercial aviation and flying instruction. He also served as a stunt pilot for the Hollywood studios. The list of Benway's numerous aviation accomplishments is wide and varied. He served as the private pilot for millionaire banker and broker, August Belmont, and set a commercial aviation record while in that capacity. He also served as aerial chauffeur for Charles and Anne Lindberg on their honeymoon. He later flew mail and passengers in Central and South America along challenging air routes. In 1934 he was placed in charge of Lockheed's "aerobot" and ejection-seat projects.

A pilot of vast experience, Benway flew nearly every type of aircraft in service on the American continent while amassing a total commercial flying time of more than 11,000 hours.

In 1940 Benway became a recruiter and Pacific Coast technical head of the Clayton Knight Committee, an RAF alumni association which was enrolling qualified American pilots in the Royal Canadian Air Force. He also served as the western regional director for the Canadian Aviation Bureau which was helping to induct Americans into the RAF Eagle Squadron. He subsequently was made an honorary Captain in the Canadian Air Force.

Following the Japanese attack on Pearl Harbor, Benway helped to transfer the Eagle Squadron volunteers back into the US Army and Naval Air Forces. When this effort was completed, he left the RCAF to take a commission as a Lieutenant in the U.S. Naval Reserve. He was placed in charge of the PBY base at the San Diego U.S. Naval Air Station. His duty assignment was submarine patrol along the west coast of the United States.

In the fall of 1943 Lieutenant-Commander Benway was forced to take medical retirement from the military. In December he began flying Catalina PBYs under contract in Brazil. In the spring of 1944 he returned to the United States and flew as a private contractor with the military. In 1945 Benway learned that his son, William, an Army Air Corps lieutenant who was stationed on Corrigedor when it had fallen, had been beaten to death by Japanese troops while on the Bataan death march.

Following the war, Benway worked briefly at the Hollywood studios as a flyer, flying consultant, and as an aviation technical writer. Bothered by lingering health problems, he concluded an aviation career which had spanned three decades. In 1952 Benway retired with his family to Oahu, Hawaii and became involved in the import/export business. While out sailing his boat alone off Oahu one calm spring day, on 28 May 1956, "Benny" Benway suffered a fatal heart attack. He was 57 years of age at the time of his death.

Charles Biddle

Charles John Biddle was born in Andalusia, Bucks County, Pennsylvania on 13 March 1890, the son of Charles and Letitia (Glenn) Biddle. Biddle was educated at the Blight School, Philadelphia, Princeton University, A.B. 1911, and Harvard University, LL.B., 1914. That same year he was admitted to the Bar and joined the Philadelphia firm of Driniker and Reoth.

Biddle left his law practice in the spring of 1917 to enter the European War. On 8 April 1917 Biddle enlisted in France's *Service Aeronautique* in Paris. From 13 April to 26 July 1917 he attended the aviation schools at Avord, Pau, and the *G.D.E.* He received his *brevet* on the Caudron on 2 June 1917.

From 28 July 1917 to 10 January 1918 *Caporal* Biddle flew at the Front with *Escadrille SPA 73.* On 7 November 1917 he was commissioned a Captain in the U.S. Air Service. From 10 January to 18 February 1918 Biddle served with *SPA 124.* On 18 February 1918 that squadron evolved into the 103rd Pursuit Squadron, USAS. Captain Biddle served at the Front with the 103rd until 22 June 1918. On 15 May 1918, while on patrol over St. Jean, he was shot above the left knee in an aerial combat with a German two-seater. He left the squadron on June 22nd to become Commanding Officer of the 13th Pursuit Squadron, USAS. He served in that capacity until 24 October 1918. The following day he was made Commanding Officer of the 4th Pursuit Group. On 1 November 1918 he was promoted to Major. He remained in charge of the 4th Pursuit Group until the 11 November Armistice.

Major Biddle officially destroyed seven enemy aircraft and unofficially destroyed three others. He was awarded the U.S. Distinguished Service Cross for extraordinary heroism. That action took place near Damvillers, France on 26 September 1918. His citation is as follows: "During an engagement between 11 Spads and 12 enemy Fokkers, Captain Biddle, perceiving a comrade in distress from the attack of two planes, dived upon them and by his fire forced them to withdraw. His prompt action saved the life of his comrade, who was in imminent danger of being shot to the ground." Biddle was also awarded France's *Croix de Guerre*, with four palms, the *Legion of Honor*, Belgium's Order of Leopold, and, in 1933, the U.S. Purple Heart.

In 1919 Biddle published his memoirs under the title *Way of the Eagle*, which was later reprinted as *Fighting Airman*. Following his World War One service, Biddle rejoined his Philadelphia law firm. On 10 February 1923 he married Katherine J. Legendre of New Orleans. From this union two sons, Charles and James, were born. Besides being a practicing attorney and senior partner of his law firm, Biddle was also director of the Provident Title Company of Philadelphia and was a member of the Pennsylvania State Aeronautiques Commission. He died in Andalusia, Bucks County, Pa. on 22 March 1972 at age 82.

JULIAN BIDDLE

Julian Cornell Biddle was born in Philadelphia, Pennsylvania on 19 April 1890, the son of Arthur Biddle, a lawyer at the firm of Biddle & Ward, and Julia Cox (Biddle) Biddle, Arthur Biddle's second cousin. Julian prepared for university at Mochmann's School in Dresden, Germany, at the DeLancey School, Philadelphia, and in 1903 at St. Mark's School at Southboro, Massachusetts, where he established an enviable academic and athletic record.

Biddle graduated from Yale University with a B.A., Class of 1912. He then served for six months as secretary at the American Embassy at Tokyo, Japan under Col. O'Brian, the U.S. Minister. He returned to the States to pursue a banking career with the Philadelphia banking house of Montgomery, Clothier & Tyler. He was a member of St. Peter's Episcopal Church, Philadelphia.

When war broke out in Europe in 1914, Biddle realized that America would one day enter the conflict. With this in mind, he spent his 1915 summer vacation at the Curtiss Aeroplane Works near Buffalo, New York. In October, 1916 Biddle received his Aero Club of America Hydroaeroplane certificate #54 at Essington, Pennsylvania. On 28 April 1917 he made arrangements with the American representative of the Lafayette Flying Corps to join French Aviation and sailed for Bordeaux two days later.

Biddle joined the *Service Aeronautique* on 25 May 1917. From 2 June to 8 August he attended the aviation schools at Avord, Pau, and the *G.D.E.* Drawing on his civilian flying experience, Biddle received his *brevet* on the Bleriot on 20 June 1917, after only five hours of flying, and only 18 days after entering the service. On 31 July he was sent to the Plessis-Belleville to await assignment to the Front.

On 11 August 27-year-old *Caporal* Biddle was ordered to the Front to serve with *Escadrille SPA 73, Groupe de Combat 12*. On 18 August, at 10:45 a.m. Biddle left the aerodrome on a practice flight and evidently crashed into the sea just west of Dunkirk. Fragments of his SPAD were found floating in the water, and his superiors believed that he had flown into a flight of German bombers returning from a raid against England's south coast and was shot down. On 26 August his body washed ashore at Egmond-aan-Zee, Holland. Dutch authorities reported that he had been shot in the abdomen. His remains were buried there in the village churchyard but were later transferred to the United States.

Sergent Biddle was posthumously awarded the Aero Club of America medal in January, 1918. On 7 September 1922 he was posthumously conferred the *Medaille Militaire*, and on 26 September 1922 he was awarded the *Croix de Guerre*, with Star.

STEPHEN BIGELOW

ESCADRILLE LAFAYETTE

Stephen Sohier Bigelow, the 22nd volunteer of the *Escadrille Lafayette*, was born on 18 March 1894 at Boston, Massachusetts, the son of investment banker, Joseph S. Bigelow, and Mary C. (Bryant) Bigelow. Stephen Bigelow graduated from Groton School in 1911 and from Harvard College, A.B. '15. The summer following graduation he trained at the reserve officer military camp at Plattsburg, New York.

In early 1916 Bigelow booked ship's passage to France after following the efforts there of former Harvard alumni Frazier Curtis and Norman Prince at assembling an all-American aviation squadron of volunteers. On 13 April 1916 Bigelow enlisted in the Foreign Legion at Paris. In June he began his air service training, progressing through the schools at Dijon, Buc, Avord, Cazeaux, Pau, and the G.D.E. Bigelow received his *brevet militaire* No. 4435 on the Bleriot on 7 September 1916.

From 24 January to 8 February 1917 *Caporal* Bigelow flew at the Front with *Escadrille Nieuport 102*. On 8 February he joined *Nieuport 124* (Escadrille Lafayette) at Ravenel, near Saint-Juste-en-Chausse.

When *Sergent* Bigelow wasn't flying patrols, he was entertaining his squadron mates with his musical talents. As Edwin Parsons remembered of him "Steve Bigelow (was) the pleasure-loving ... piano-playing fool whenever we could beg, borrow or steal a piano."

On 21 June Bigelow barely escaped the guns of five German fighters while flying deep inside enemy territory as he protected a French machine photographing the Boche's positions. He and his escort returned safely to their airfield in spite of a sizeable hole blown through the wing of Bigelow's fighter. On 24 June and on 30 June Bigelow engaged enemy aircraft. On the 25th he had earned mention in the official communique on American flyers.

By the summer of 1917 there existed a growing need for experienced American pilots to train and to man the fledgling A.E.F. squadrons. Bigelow submitted his application to join the U.S. Air Service.

On 20 August *Sergent* Bigelow flew an escort mission as part of a patrol protecting Sopwith bombers on a raid over Dun-sur-Meuse. Eight enemy fighters tore into the formation, and he mixed it up with six of them in an attempt to save a targeted Sopwith bomber. Bigelow dove down on the swarming Germans. As he levelled off, he met a Hun fighter head on. The enemy's machine guns obliterated his windscreen, and he was struck twice in the face with bullet fragments. He wobbled out of the fight, shook off the pursuing enemy pack, then made a run for home.

Upon landing, Bigelow ignored *Capitaine* Thenault's orders that he seek immediate treatment for his wounds and continued to fly escort missions. By the second week of September his wounds had become

Left: Edmund Genet, Stephen Bigelow, Mrs. Rene Simon, and Edward Parsons at the Simon Villa at Pau celebrating Thanksgiving, 28 November 1916. Mrs. Simon was the American wife of Lt. Rene Simon, chief acrobatic instructor at Pau. Above: Sergent Stephen Bigelow in the cockpit of his SPAD scout preparing for a patrol, Senard, 1917.

dangerously infected, and he was escorted to Auxiliary Hospital No. 6 where he lay seriously ill for some weeks.

On 14 September 1917, shortly after his arrival at the hospital, *Sergent* Bigelow was officially cited in French Army orders and was decorated with the *Croix de Guerre*, with Star for his heroic actions in protecting the allied bomber. During his convalescense he learned that his application had been approved for transfer as instructor to the U.S. Army Air Service. In January of 1918, while still a patient in hospital, he was honorably discharged from the French Army following his classification as physically unfit for further active service.

As with many young Americans who had sought adventure and excitement in France, Stephen Bigelow had arrived home, his nerves taut, his health broken. The vibrancy of his combat experiences had made his options in civilian life appear pale by comparison. At age 24 he had been forced to confront the awful reality that whatever he might do following his service with the *Escadrille Lafayette* when compared with the immediacy of war, with the intimacy he had found with the squadron, must seem insignificant and anti-climatic.

In 1921 Bigelow married Anne Meredith. They divorced in 1923. He married again in 1932 to Mary L. McDevitt. In January of 1936 his only child, Stephanie Sohier was born. Shortly after her birth, he and his second wife separated. A listlessness seemed to permeate him as he survived on a family remittance. His health continued to decline. Chronic tuberculosis, stimulated by alcohol, ravaged his body. He died in Boston on 6 January 1939 at age 44.

CHARLES BLAKE

Charles Raymond Blake was born on 8 November 1897 at Westerly, Rhode Island, the son of James Thomas Blake, and Annie (Fletcher) Blake of Derby, England. Blake graduated from Westerly High School in 1916.

Blake served in the *American Ambulance Field Service* with *SSU8* from January to June 1917 in the Verdun and Champaigne sectors. On 4 June he enlisted in France's *Service Aeronautique*. From 19 June 1917 to 8 March 1918 he attended the aviation schools at Avord, Tours, and the *G.D.E.* He received his *brevet* on the Caudron on 27 October 1917. From 11 March to 18 April 1918 *Caporal* Blake flew at the Front with *Escadrille BR. 29.*

On 17 March Blake was commissioned a First Lieutenant in the USAS at Chaumont, France and was attached to his former French squadron where he served from 18 April to 3 September 1918. While in the Air Service, Lt. Blake flew more than 100 hours over enemy lines, completing 37 bombardments of enemy territory. He was officially credited with the destruction of one enemy aircraft. He participated in the Chateau Thierry Defensive, the Aisne, Oise and Montdidier Offensives, and in the bombardment of the Vosges and lower Rhine region.

From 6 September to 24 December 1918 Blake served as instructor at the 7th A.I.C., Clermont-Ferrand. He returned to the U.S. in February 1918, and after eight months in hospital he was discharged at Fort Adams, Rhode Island on 20 October 1919. For his World War service, Lieutenant Blake was awarded the U.S. Distinguished Service Cross for extraordinary heroism, and France's *Croix de Guerre* for action against five enemy Fokkers on 9 August 1918. His DSC citation describes the action:

"Lt. Blake, with Second Lt. Earle W. Porter, observer, while on a reconnaissance expedition at low altitude far beyond the enemy lines, was attacked by five German battle planes. His observer was wounded at the beginning of the combat, but he maneuvered his plane so skillfully that the observer was able to shoot down one of their adversaries. By more skillful maneuvering he enabled his observer to fight off the remaining planes and returned safely to friendly territory."

After his discharge Blake attended the Rhode Island State College from October 1919 to February 1920. He married Clarissa Parsons Holland of New Hampshire on 2 June 1920. They eventually had four children: Roger Holland Blake, Nancy Fletcher Blake, Barbara Parsons Blake, and Charles Rowe Blake.

Blake attended the NYU College of Engineering from September 1921 to May 1924. He studied at the Federal Bureau of Investigation and at the National Police Academy in Washington, DC and became a State Policeman in Texas. Blake remained in the U.S. Army Air Corps Reserve and served as a Major in WW II. Blake died at Walter Reed Army Hospital on 9 May 1958 from a cerebral hemorrhage at age 60. He is buried at Arlington Cemetery.

ARTHUR BLUETHENTHAL

Arthur Bluethenthal was born on 1 November 1891 in Wilmington, North Carolina, the son of Leopold Bluethenthal, president of Bluethenthal Dry Goods Company, and Johanna (Dannenbaum) Bluethenthal. He was educated at Phillips Exeter Academy, graduating in 1909, and at Princeton University, Class of 1913. While at Princeton, he played center on the school's football team.

Following graduation, Bluethenthal was line coach at the University of North Carolina, Chapel Hill, for the 1913–14 football season. He then went to New York City to work as a tobacco broker for the Tobacco Products Corporation. While in New York, he also worked briefly for his father's Bluethenthal Dry Goods Company.

On 6 May 1916 "Bluie" left his father's business to join the *American Ambulance Field Service*. He could not sign up for the full six months because of his fall contract to coach the Princeton football team. But once he had arrived in France, Bluethenthal obtained a release from his contract with Princeton.

He was attached to Section Three of the *American Ambulance Field Service*, serving at the Battle of Verdun with the French 129th Infantry Division. He was awarded France's *Croix de Guerre*, with Star for his conspicuous bravery around Verdun. In October of 1916 he was transferred to Solonika, Greece and served in the Balkins with the 57th Infantry Division, Armee d'Orient, until 11 May 1917.

On 7 June 1917 Bluethenthal enlisted in France's *Service Aeronautique*. From 9 June to 15 March 1918 he trained at Avord, Pau, and the *G.D.E.* He received his *brevet* on the Caudron on 22 September 1917.

Following a four-week leave in Wilmington, North Carolina, *Sergent* Bluethenthal returned to France and was attached to *Escadrille Br. 227* on 17 May 1918. Toward the end of the month his official transfer came through to join the U.S. Naval Aviation, but *Sergent* Bluethenthal refused to leave his squadron. He continued to pilot his observation plane until 5 June 1918, the day he was killed while directing artillery fire over Maignelay, in the region of Amiens. He was posthumously awarded France's *Croix de Guerre*, with Palm, and the *Medaille Militaire*. His remains were buried at Esquennoy, Oise, north of Breteuil.

Sergent Bluethenthal's body was exhumed in 1921 and reinterred in the Oakdale Cemetery at his home at Wilmington, North Carolina. On 30 May 1928 the Wilmington airport was named Bluethenthal Field, in his honor. Bluethenthal's memorabilia is now a part of the New Hanover County Museum and is on display at the New Hanover County Airport, Wilmington.

PIERRE BOAL

Pierre de Lagarde Boal was born on 29 September 1895 in Thonon-les-Bains, France, and attended the St. Paul's School in Concord, New Hampshire, Class of 1915. He enlisted in the French Army in August of 1914 and served at the Front in the First French Regiment of Cuirassiers until he transferred to the *Service Aeronautique* on 24 May 1916. Boal attended the aviation school at Buc from 5 June 1916 to 1 February 1917. Unable to master the Bleriot monoplane, he was proposed for *radiation*, but instead of accepting his discharge from the French Air Service, Boal remained at Buc at his own request, acting as interpreter for his American comrades.

In January of 1917, when the Bleriot School was moved from Buc to Avord, Boal was granted a leave to America where he continued to give "invaluable cooperation" to the Corps' executive committee in Paris. Boal was commissioned a Captain in the U.S. Air Service on 27 May 1917 and received his *brevet* on the Curtiss trainer at the U.S. Aviation School at San Antonio, Texas. After serving for several months in the U.S., he was ordered back to *Plessis Bellville*, France where he was placed in charge of all American aviators temporarily assigned to French units at the Front. He also continued to work closely with Major Gros as Aviation Liaison Officer between the French and American Air Services. He was later made commander of the 400th Air Squadron overseas.

Boal married Jeanne de Menthon in 1919 and entered the U.S. Foreign Service that same year. During his twenty-eight-year career with the State Department he served in various capacities in Mexico City, Mexico; Belgrade, Yugoslavia; Warsaw, Poland; Berne, Switzerland; Lima, Peru; Ottawa, Canada; Geneva, Switzerland; Boliva and Nicaragua. In 1929 he served as the acting secretary general of a commission credited with averting war between Boliva and Paraguay. He also served as chief of the Division of Western European Affairs at the State Department, and as an adviser to the Geneva Disarmament Commission in 1930. Late in 1945 he was appointed to the Emergency Advisory Committee for Political Defense in Montevideo, Uruguay, with the rank of ambassador.

The honors Boal received included a Purple Heart for service in World War One, the Lafayette Flying Corps ribbon, France's *Croix de Guerre*, the *Legion of Honor*, the Order of St. Sava of Yugoslavia, and Peru's Commander Order of the Sun. In addition to being a member of the Early Birds, he was a member of the Metropolitan Club of Washington, D.C., the Long Island Aviation Country Club, and the Officers' Club of the 28th Division of Pennsylvania.

Once retired, Boal lived mostly in France, but also maintained a residence in Boalsburg, PA. He died on 24 May 1966 at the American Hospital of Paris, France following a stroke at age 70. He was survived by his widow, Mrs. Pierre Boal of Paris, France; a daughter, Mrs. Blair Lee III of Silver Spring, Maryland; a second daughter, Mrs. Gaston d'Harcourt of Le Havre, France; and ten grandchildren.

ELLISON BOGGS

Ellison Converse Boggs was born in Coconut Grove, Florida on 14 May 1899, the son of Adam Alexander Boggs and Emma (Converse) Boggs. He attended St. Luke's School, Wayne, Pennsylvania. On 4 August 1917 he joined France's *Service Aeronautique*. At age 17, Boggs was the last, and one of the youngest Americans to be accepted into the Lafayette Flying Corps.

From 5 August 1917 to 18 April 1918 Boggs attended the aviation schools at Avord, Tours, Pau, Cazeaux, and the *G.D.E.* He received his *brevet* on the Caudron on 23 October 1917. Illness prevented him from going to the Front until 21 April when he was assigned to *Escadrille SPA 81*. *Sergent* Boggs flew with this Squadron until the November 11 Armistice, providing excellent service. He was awarded France's *Croix de Guerre*, with Star for the part he played in the destruction of an enemy balloon on 15 July 1918.

Following World War One Boggs spent some time in Los Angeles where he was involved in aviation. He relocated to Miami Beach, Florida and married Edna Eckley. He became Boat Inspector of Miami Beach and served in that capacity until his death on 8 November 1940 at age 41.

WILLIAM BOOTH

William Vernon Booth, Jr. was born in Chicago, Illinois on 8 October 1889, the son of William Vernon Booth, President of Booth Fisheries Company, and Helen (Lester) Booth, the sister of Mrs. Philip D. Armour. He attended the Fay School, Southborough, Massachusetts, before entering St. Marks School in 1903. While at St. Marks young Booth was considered a good scholar and an excellent athlete, playing on the football and hockey teams, and captaining the baseball team.

Booth entered Harvard College with the Class of 1913 then attended the New York Law School. After receiving his law degree, he entered the law firm of Platt and Field.

Upon the U.S. entry into World War One, Booth immediately applied for entry into U.S. Aviation but was rejected for being under weight and height. On 19 May 1917 he sailed from New York for France to join the Lafayette Flying Corps. On 3 June 1917 he enlisted in the French Foreign Legion. From 10 June 1917 to 8 January 1918 he attended the aviation schools at Avord, Pau, and the *G.D.E.*, receiving his *brevet* on the Caudron on 26 October 1917. On 10 January 1918 "Rat" Booth was assigned at the Front with *Escadrille SPAD 96*. He flew continually with that squadron with the exception of a ten-day leave which he was granted for bringing down and enemy aircraft. He used the opportunity to marry Ethel Forgan, the daughter of David R. Forgan, a Chicago financier, on 27 April in Paris. Following a wedding trip to Cannes, Booth returned to the Front on 14 May while Mrs. Booth returned to her work at a Y.M.C.A. canteen.

Sergent Booth flew with *SPAD 96* until 25 June 1918, the day he was wounded in aerial combat near Longpont. On that date Booth was on patrol when he and four comrades engaged in violent combat with 18 Fokkers. Greatly outnumbered, Booth desperately tried to extricate himself from the melee. But his SPAD fighter was struck with an incendiary bullet which set his machine on fire. At the same instant an enemy round struck and shattered his right leg.

In great pain, *Sgt* Booth nevertheless kept his head and put his fighter into a steep dive. A 6,000-foot plunge extinguished the fire. But Booth's engine had stopped, and he made a forced landing near Longpont in the center of No Man's Land.

Fire from the German trenches poured into Booth's downed fighter Despite severe wounds and burns, he crawled from his machine and set it afire to keep it out of enemy hands. He then crawled through No Man's Land until rescued by French troops.

Booth was evacuated to the Scottish Woman's Hospital at Royaumont where his right leg was amputated in an attempt to save his life. On 4 July he was decorated with France's *Medaille Militaire* and the *Croix de Guerre*, with Palm. On 10 July 1918, with his wife at his side, the 29-year-old succumbed to his wounds and was buried at Royaumont-Asnieres-sur-Oise. *Sergent* Booth was posthumously decorated with the *Legion d'Honneur*.

CLARENCE BOSWORTH

Clarence Marsh Bosworth was born in Brandon, Vermont on 5 September 1894, the son of Birney Boardman Bosworth, a manufacturer, and Phoebe (Marsh) Bosworth. Bosworth attended the University of Vermont and graduated from that institution in 1917. In May of 1917 he applied for and was endorsed for admission into the Lafayette Flying Corps by James McElhone, head of the Corps's New York City office.

On 9 June Bosworth sailed from New York for France aboard the *Espagne*. On 5 July 1917 he enlisted in France's *Service Aeronautique* in Paris. He received his *brevet* on the Sopwith on 26 December 1917.

On 19 February 1918 Bosworth returned to the United States on one month's leave of absence. He returned to France on 16 October 1918 and was released from the *Service Aeronautique*.

EDGAR BOULIGNY

Edgar Jean Bouligny was born in New Orleans, Louisiana on 3 January 1888, the son of Jean Edgard Bouligny, a printer, and Marie Lucie (Delery) Bouligny. He was a descendant of Dominique Bouligny, a soldier who commanded a regiment of militia in the Louisiana Territory. Following the Louisiana Purchase, Dominique became an American citizen and eventually a member of the U.S. Senate (1824–1829).

In 1890, when Edgar was two years of age, his father moved his family and printing business to Mexico City. That same year he accidentally shot his shop manager and spent five months in a Mexican jail. Upon his release he returned to New Orleans where he died in 1903.

Edgar Bouligny's adventurous career began two years later when, at age seventeen, he was shanghaied to China. He later served in the U.S. Army from 1906-1912, then mined gold in Nevada. On 6 August 1914 he was among the first group of Americans to enlist in the French Foreign Legion. He was also the first American in service to France to be wounded in combat. During his nearly three years in the Foreign Legion he would be wounded three more times. He was awarded France's *Croix de Guerre*, with Star during the Battle of Champagne.

On 1 May 1917 *Sergent* Bouligny left the trenches and entered France's *Service Aeronautique*. From 7 June 1917 to 30 March 1918 he attended the aviation schools at Etampes and Chateauroux, receiving his *brevet* on the Farman on 13 July 1917.

On 24 April 1918 he was assigned to the Front with *Escadrille Nieuport 501* of the Army of the Orient, a combat and reconnaissance squadron. *Sergent* Bouligny flew Farmans and Nieuports with that squadron on the Serbian and Albanian Fronts until 14 June 1918. On 24 October 1918 he was commissioned a 2nd Lieutenant in the U.S. Air Service, a rank far beneath his worth and experience after having served more than four years in France.

While he was in France, Bouligny met Odile Hubeau, a photographer for *La Guerre Aerienne*, when he came to her office with pictures he had taken at the Front. The two became romantically involved and formed a business partnership.

Lt. Bouligny was discharged in the United States shortly after the War. He returned home to New Orleans where his fine war record was recognized in a triumphant reception. In January 1919 Bouligny returned to France to revisit the battlefields. He was reunited with Odile Hubeau, and they married in Sevres, north of Paris, the village of her birth. Odile taught him photography, a career they both pursued in Paris for several years. Bouligny then returned to the United States, bringing Odile with him. It was then that Odile learned that Bouligny was already married. Bouligny divorced his American wife, and he and Odile set out on a photographic odyssey in a mule-drawn covered wagon from New Orleans to Long Beach, California. He then bought an old Ford automobile and drove it to Jersey City, New Jersey.

Bouligny's restless spirit again took him back to France. He found the change too great there, and boarded a freighter at Marseilles, working his way around the Mediterranean and Africa. Upon his return to New Orleans, Edgar and Odile took an automobile trip around the States, into Canada, and Mexico, taking photos along the way which they sent to travel journals for publication.

The intensity of Bouligny's war service was such that he could not settle into a normal life. Constantly restless and explosive in temperament, he often flew into rages and beat Odile. His family attributed his personality change to shell shock from the war, and advised Odile to leave him. He was killed in New Orleans by his 33-year-old wife on 18 May 1931 in their apartment at 409 Bourbon St. Mrs. Bouligny shot her husband in self-defense following a severe beating from him. Bouligny was 43 years of age at the time of his death. He is buried at the Chalmette National Cemetery, New Orleans, Louisiana.

ALGERNON BOYESEN

Algernon Kene Boyesen was born in New York City, New York on 13 October 1880, the son of Hjalmar Hjorth Boyesen, a Norwegian-born author and professor of German Literature, and Elizabeth (Kene) Boyesen. In 1899, Algernon enrolled in Columbia University where his father had once chaired the German language department. He left that institution as a non-graduate of the Class of 1903.

In 1903 Boyesen married a young Pleasantville, New York woman named Adelaide. From this union two daughters were born. Boyesen took an apartment in New York City and began a career as a dramatist.

Boyesen was living in France in October 1913 when his version of the play, "Manon Lescaut"was rejected by the New York theater as being too offensive for the American public. In August 1915 Adelaide Boyesen filed for divorce from Algernon Boyesen at White Plains, New York. They had been residing in Hendon, England when she had returned alone to their Westbury, New York country home. Boyesen remained in England where he was alleged to have become a member of the Royal Flying Corps, stationed near London.

In November of 1915 Boyesen left England for France "to join the flying corps." On 15 January 1916 Boyesen enlisted in France's *Service Aeronautique*. On 8 March 1916 he was released because of inaptitude.

In 1925 Boyesen's brother, Hjalmar, an overseas attorney with the William Nelson Cromwell New York law firm, was chosen as one of the committee that was formed to erect the Lafayette Flying Corps monument near Paris. He played a major role in the development and construction of the monument which was dedicated in 1928.

In 1932 Algernon Boyesen was living at Bois de Boulogne, Biarritz, France.

Lester Brady

Lester Strayer Brady was born in Lock Haven, Pennsylvania on 5 August 1892, the son of Harry Ellsworth Brady, a dentist, and Emma (Strayer) Brady. He attended local schools in Lock Haven. During his senior year he contracted a mild case of pulmonary tuberculosis which responded quickly to bed rest at home. Following his rapid recovery Brady spent two years at the Pennsylvania State University studying forestry before entering Cornell University. He graduated from that institution in 1913 with a degree in economics.

In 1916, when it appeared that America would shortly enter the European War, Brady enrolled at the U.S. Officer training camp at Plattsburg, New York. He followed this training with private flying lessons at the Curtiss School of Aviation at Newport News, Virginia and at Mineola, Long Island, New York.

In April of 1917 Brady sailed for France aboard the *S.S. Chicago*. On 28 May 1917 he enlisted in France's *Service Aeronautique* and until 23 February 1918 he attended the aviation schools at Avord, Juvisy, Pau, and the *G.D.E.* He received his *brevet* on the Caudron on 6 November 1917. From 23 February to 13 April 1918 *Caporal* Brady served at the Front with *Escadrille SPAD 26*. On 16 April he was commissioned a Second Lieutenant in U.S. Aviation and was placed on duty in Paris until 7 June 1918. On that date he became a ferry pilot at the American Acceptance Park, Orly. He served with the First Observation Group from 16 July to 21 August 1918. From 30 August to the Armistice he served with the 135th Aero Observation Squadron, the 27th Pursuit Squadron, and the 1101st Aero Squadron, USAS.

Upon his return to the United States, it was discovered that Lieutenant Brady's tuberculosis had reactivated, and he began a recuperation at the U.S. Army General Hospital at Oteen, North Carolina and later at the Fitzsimons General Hospital in Denver, Colorado

On 31 July 1920 Brady was discharged as cured from the Colorado hospital and re-entered university. He completed four years of post-graduate work while attending eleven colleges and universities, filling all the requisites for his doctorate, but he never submitted his thesis. He became a life insurance salesman in New York City. In 1926 he married Catherine McElroy. One son, Douglas, was born of this union.

At the outbreak of World War Two Brady sought to regain a commission in the Air Force but was rejected because he was rated 100% disabled from his previous bouts with tuberculosis in spite of having been free of the disease for many years. However, his interest in aviation remained keen, and he flew often as a civilian pilot.

In the late 1940s Brady turned his energies to real estate development and was quite successful in this endeavor. He died of pneumonia at the Leech Farm Veterans Hospital near Pittsburgh, Pennsylvania on 30 December 1968 at age 76.

RAY BRIDGMAN

ESCADRILLE LAFAYETTE

Ray Claflin Bridgman, the 30th volunteer of the *Escadrille Lafayette*, was born on 31 May 1895 in Lake Forest, Illinois, the son of Walter Ray Bridgman, a professor of literature at Lake Forest College, and Leoline (Waterman) Bridgman, a writer.

Bridgman entered Yale University in 1913. He majored in History, and was admitted to the academic honorary society, *Phi Beta Kappa*, in his junior year. In the spring of 1916, while still in his junior year, the twenty-year-old Bridgman chose to leave school in order to enlist in a Red Cross relief organization in Europe on behalf of the Allied cause. An ardent pacifist, he held no illusions concerning the glories of war. In fact, he abhorred everything about war. Yet in spite of his strong beliefs, he served in France courageously for nearly two and one-half years.

Bridgman sailed for the British Isles aboard the U.S. merchant ship *St. Louis*. On 12 June 1916 he reached London then left for Paris. While in Paris he changed his mind about enlisting in the Red Cross and chose instead to "fight for France," which he considered "a privilege and a high honor."

On 24 July 1916 Bridgman enlisted in French aviation. He attended the flying schools at Buc and Juvisy, receiving his *brevet militaire* No. 5009 on the Caudron on 5 December 1916. He entered Nieuport instruction on 10 December at Avord and completed his instruction on 14 January then attended gunnery and acrobacy school at Cazeaux and at Pau.

On 27 March 1917 *Caporal* Bridgman arrived at the *G.D.E.* at Plessis-Belleville. On 13 April he joined *N.49* at the Front and flew combat patrols until 27 April. On 1 May he joined the *Escadrille Lafayette* at Ham.

N.124 had seen a period of intense aerial activity since its relocation to Ham on 7 April in support of the Nivelle offensive. In the month preceding Bridgman's arrival, *Sergent* Edmond Genet, *Sergent* Ronald Hoskier, and Hoskier's French gunner *Caporal* Jean Dressy had all been killed in combat. In May alone the *Lafayette* pilots would fly more than 350 patrols. Bridgman would fly his share of them. He made note of this in letters home to his parents. In a *7 May* letter to his mother, he wrote:

"Brilliant days of sunlight. A great attack is being made just South of us. Most of our patrols are made there for the present. There are fights every day. The danger from the special cannon (antiaircraft) against aeroplanes is almost as great as the German ships themselves."

The following day Bridgman wrote how he had been "just south of St. Quentin. There have been very many fights."

On 23 May the squadron's executive officer, French Lieutenant deLaage, was the next pilot to die. Bridgman noted this in a *28 May* letter home.

"The escadrille has suffered a terrible loss in the tragic death of our French lieutenant, Lt. de Laage. He crashed to the ground near our field, and died in a few moments. I was a witness to the accident, and was with him when he died. I admired and loved him. He was by far the noblest Frenchman I have known, and I believe almost the noblest man."

Sergent Bridgman received his new SPAD S.VII fighter in time to make the Squadron's 3 June flight to the airfield at Chaudun. In a *6 June* letter to his parents he explained the purpose of the move was "to be closer in touch with the fierce sector existing around Laon." He added that "the days are so long now that sometimes our escadrille goes out at 3:30 a.m. and the last patrol does not return until 9 p.m."

On 17 July *Spa.124* was moved to Saint Pol-sur-Mer, near Dunkirk, to support the forthcoming 3rd Ypres Offensive. Instead, the pilots were given a brief respite from the rigors of combat flying, as Bridgman indicated in a *27 July* letter.

"We are now in a sector where we can see England from the air. We are in tents and enjoy the sea bathing. Our American flag, floating above the escadrille, serves as a good target for German air ships and we are shot at by long distance guns from both land and sea."

On 11 August the stay at Dunkirk ended abruptly, the date the Squadron was moved to Senard. *Sergent* Bridgman made note of this in a *16 August 1917* letter:

"Flew down from Dunkerque the other day, and are now stationed a few miles from Verdun. Dunkerque was by far the finest place we have been. We had a part in the preparation for the recent Flanders offensive. But terrible storms hindered the proceedings a great deal. Here we are encamped near the Forest of Argonne."

Subsequent letters mention "intense activity over the Verdun area," and *Sergent* Bridgman tells of how he was attacked by six German planes and had to "race for French lines." Good weather made patrols a two and a three-a-day occurrence.

In his letters home to his parents Bridgman continually spoke out against the destruction of the war and against "men of every nation who are willfully and blindly continuing this destruction . . . I loathe death," Bridgman wrote. "I loathe this wild slaughter where nations enlist on the side of humanity's worst enemy, death, annihilation."

American trainees at the Buc School of Aviation, 30 October 1916. (Top row, l.r.) Carl Dolan, Ray Bridgman, Arch Johnston, Pierre Boal, Isadore Court, Herman Chatkoff, Kenneth Marr. (Seated, l.r.) Arthur Aten, W.K. Appleton, Lief Barklay, Guy Magley.

In a *15 September* letter Bridgman again explained his position to his father, who let his son know that he thought his letters sounded unpatriotic. After this date Bridgman no longer expressed his convictions with emotion and stuck strictly to relating events.

"I am grieved to have written you that last letter, a letter of misunderstanding and of a sort of baffled hope. It cut a bit deep that you should speak of my letters as 'dolorous wails.' Better for me to keep silent than to speak in such a fashion at this time when hope and faith in the future are so desperately needed. . . . Unless a new world is coming, life will not amount to much, and I would gladly die to help even a bit to bring a new world into being. So I am not worrying much about death, though I curse whoever in a high place is so lavishly and devilishly sacrificing the millions of youth of my generation. Many of those who rave most about the Kaiser are no more men and far less intelligent than he is. I inflict all this upon you, because you have some sympathy with my point of view."

"As far as the air fighting goes, the most exciting times have been here in this region. Fights are very numerous. The other day I got 'shot up,' as they say. In company with one French lieutenant and a sergeant, I attacked four German biplaces. I dove for almost a

Captain Ray Bridgman (center) commanding officer of the U.S. 22nd Pursuit Squadron, beside his SPAD S.VII scout with Lieutenant Davis (l) and Lieutenant Little, two of his pilots, October 1918.

mile, firing continually toward the end. Then I turned back toward the French lines. I had gone a bit too far, and the Germans got some good shots at me. Two bullets hit my machine, but no harm resulted. Anti-aircraft burst right underneath me. When I reached home, I found that a piece of shrapnel had pierced a hole in underneath. I am getting thoroughly used to being 'shot at' by German aeroplanes."

Bridgman's good effort at Senard did not escape *Capitaine* Thenault's eye. He later noted the "gallant and excellent" work the young pilot had done "during this whole period, never sparing himself in the slightest."

The *Lafayette Escadrille* left Senard for Chaudun on 18 September to fly support for the 3rd Aisne Offensive. The fall air turned cold and *Sergent* Bridgman noted "it is quite bitter at high altitudes ... Autumn mists and storms are hindering our flying somewhat. Pitiless, unending war, deadening in its monotony. The thunder of the cannon is tremendous in our present region."

On 24 September *Sergent* Douglas MacMonagle was killed in combat with a bullet through the head while flying patrol over Triaucourt. By 30 October the fall storms had moderated and Bridgman described the squadron's subsequent activity.

"A great day it has been, brilliantly clear and bracing. There was much excitement in the air. The German anti-aircraft guns certainly improve daily in their accuracy.

"It has been a splendid victory up here. The French artillery was overwhelming, and the German trenches were battered into nothingness. Thick weather hindered us from flying much the first day of the attack, but afterwards we flew over the battle, and shot our machine guns into the German lines, at bridges and reserves, if we could see any of the roads."

"The escadrille has diminished greatly. One man taken prisoner and two killed since August. Two have gone to instruct, and one is detached for work with the American Expeditionary Force. We are all soon to be taken into the American Army. It is beautiful to feel and see the Croix de Guerre on your own breast. I only wish that I had done something to deserve it more. I am most glad to have the Croix de Guerre for your sake and Mother's."

The citation which accompanying *Sergent* Bridgman's War Cross, with Star, dated 29 October 1917 mentioned no particular engagement, but spoke of him as a "combat pilot, skillful, modest and conscientious; (who) has always fulfilled with greatest keenness the missions which have been entrusted to him."

Former Lafayette Flying Corps pilots while serving with the U.S. 103rd Aero Pursuit Squadron, June 1918. (l.r.) Major William Thaw, commanding officer, Captains Robert Soubiran and Ray Bridgman, First Lieutenants Christopher Ford and Gorman Larner.

Fourteen November 1917 found Bridgman returning to the Squadron following two days in Paris where he had taken the physical examination required for a commission in the American Air Service. He noted how "winter has indeed come down upon France. The guns thunder on and on, but our work is disastrously hindered by storms and blinding fog."

In a 2 December letter Bridgman wrote: "There has been considerable flying of late, in spite of stormy weather. Yesterday Lufbery and I had a great time together, playing in the mist with a lot of German aeroplanes and shells. He is a superb flyer, and it is a treat to watch him."

Following their release from the French Air Service on 1 December, prior to their taking their commissions in the U.S. Air Service, Bridgman and several other *Lafayette Escadrille* pilots were suspended in a state of military limbo. Bridgman was finally commissioned a Captain in the U.S. Air Service on 26 January 1918. The day following his commission Captain Bridgman noted "there are only six of us at present out here at the front."

Those six *Lafayette Escadrille* pilots formed the nucleus of the U.S. 103rd Aero Pursuit Squadron which came into being on 18 February 1918. Captain Bridgman served as a flight commander with the squadron. And with Major Thaw in charge as commanding officer, it soon became an active fighting unit. On 23 April Bridgman wrote home that "the squadron has many aerial combats daily, and many issue successfully."

On 6 June Captain Bridgman became attached to the U.S. 139th Aero Pursuit Squadron. On 14 June he wrote home: "The Lafayette squadron is being all broken up. The old pilots are sent to command and help run new American squadrons, and new pilots take their places at the Lafayette. I am down in the American sector with this new squadron. Captain Dudley Hill came with me. Luckily we have Spads to fly. Jim Hall was down here when taken prisoner. I miss him terribly."

In two subsequent letters, written while with the 139th, Bridgman noted morosely: "The war changes little. I have now flown over the whole front...The world is now a slaughter house. He who slaughters most is honored more than a poet or a Christ."

On 10 August Captain Bridgman received his first confirmed victory when he downed a Rumpler C near Xivray. On 15 August he became commanding officer of the U.S. 22nd Aero Pursuit Squadron and served in that capacity until the end of the war. He mentioned his appointment in a 1 September letter to his parents.

"Have I written you since I was given command of this new squadron? After ten days of preparation, we are operating. All my planes and pilots are here, and we hope to achieve high results."

On 12 September 1918 Captain Bridgman scored the first American aerial victory (his second victory) seven hours into the Saint Mihiel offensive. He barely escaped with his life, however, when the engine of his SPAD fighter was struck by ground fire, and he was forced to land on recently liberated enemy territory. His reconnaissance report described the action.

"At 6.55 attacked Hannoveraner at 800 M., over Bois le Pretres. Fired 150 shots at very close range from good position. E.A. went down straight through clouds as out of control. Lost sight E.A. in clouds. Confirmation requested.

"At 7.10 saw 5 enemy Rumplers in region of St. Marie at 1500 meters. They retreated far into their lines too quickly for me to attack.

"Motor was hit by bullet from ground, was flying at 200 M. Was obliged to land just beyond former German lines near Remenauville."

Captain Bridgman's 14 September reconnaissance report described continuing enemy activity:

"While waiting around Mars-la-Tour for Salmsons, our patrol saw 6 E.A. in region of Conflans. Tried to get into position to attack E.A. but they turned and went far away. At this moment patrol was attacked by a group of 12 enemy Fokkers. Patrol was completely broken up and numerous combats resulted."

On 18 September Bridgman wrote his father: "The air activity is very intense. The enemy had many planes in the air, and there was a great deal of fighting."

On 24 September Bridgman knocked down a D.F.W. C near Woel, and on 28 September he scored his fourth victory against a Fokker near Montfaucon. On 24 October he penned one of his final letters home:

"My squadron has done well in its brief career. We have had twenty official victories in about six weeks. A third of my pilots have been lost during this six weeks. It is heart-breaking to see one's friends disappear day after day. It seems a long century since I set sail for Europe. America is a far-off dream. I wonder what the future holds. The pall of war obscures everything with its darkness."

By the November Armistice Captain Bridgman had logged more than 600 hours over enemy lines. He had been twice cited in General Orders. The squadron he commanded, the U.S. 22nd Aero Pursuit Squadron, had flown 117 patrols, had fought 90 engagements, and had received official confirmation for 46 victories in less than three months at the Front.

Captain Bridgman's overseas service was best summed up by his friend, James Norman Hall in his history of the Lafayette Flying Corps when he said of his colleague: "He was one of the keenest pilots, one of the most aggressive fighters the Squadron ever had, and this despite the fact that he hated war with his whole soul. . . . His Spad was always a battle-scarred old bird, and if he happened to be flying a new machine, in a week's time wings and fuselage would be plastered over with patches of fabric."

Captain Bridgman returned home to Lake Forest, Illinois unannounced and unheralded on Christmas morning 1918. The local newspapermen had been searching for him, but in his usual quiet manner, he wished to bring no attention to himself.

In the spring of 1919 Bridgman completed the requirements for graduation from Yale, and on 12 June he and Dorothea Eaglesfield from Indianapolis, Indiana, were married in Indianapolis. In the summer of 1919, Bridgman joined the Chautauqua Tour, a traveling group of lecturers, musicians, and entertainers. Accompanied by his wife, he gave talks on "The Romance of Flying" in many midwestern cities and towns.

His first teaching assignment was at a college preparatory school for boys run by his brother, Donald Bridgman, at the H.F. Bar Ranch near Sheridan, Wyoming.

He then began graduate school training in European History at the University of California in Berkeley. Later he studied at the New School for Social Research in New York City. He was an instructor in European History at New York University for the next twenty-nine years until ill health forced his resignation. He taught faithfully and was considered a brilliant and inspiring teacher and scholar by many of his students. He participated very little in university life outside the classroom, and he never sought an advanced degree.

Not long after Bridgman's return to the United States he was contacted by James Norman Hall. Disillusioned by what he had seen of "Western civilization" during the war, Hall had wanted his friend to accompany him to Tahiti to live, but Bridgman had declined. The horrors of his Great War experience increasingly haunted Bridgman as did the faces of the young men he had shot down in combat. For this reason, he avoided any contact with his former squadron mates, or with anyone else who reminded him of those times.

In 1920 the Bridgmans had a son, David. Another son, James, was born in 1921, and a daughter Caroline in 1922. Bridgman would say nothing to his family about his war experience. In his private life and in his teaching of history, he preached constantly for peace and international cooperation. He was deeply disappointed when the United States Congress rejected membership in the League of Nations. He spent the rest of his life arguing against war.

Bridgman embraced humanism and universalism, lived an ascetic life, and shunned personal ambition and private property. He never owned a house. All through the years, he supported his family as best he could. Mrs. Bridgman tutored privately and taught French and music at the Staten Island Academy to supplement the family income.

Bridgman became more and more reclusive even within his own family. He spent his leisure time taking long walks by himself or listening to classical music or writing poetry in his study. He had two books privately published. His long poem, "Star of the Renaissance," was set to music and performed by a choral group in New York City. He also memorized hundreds of poems-from John Milton to Edna St. Vincent Millay and John Masefield - which his family never tired listening to. Despite his extreme shyness, he occasionally recited some of these poems in public lecture halls and on the radio in New York City. He loved beauty of all kinds and never stopped delighting in the ferry boat ride across New York Harbor from Staten Island where he had lived from 1922 on.

Although Bridgman was very impersonal toward his wife and children, he loved them. And they remained loving and loyal to him and understood the roots of his pain.

When World War II began, Bridgman became even more intense and sad that his and many others' anti-war efforts had not been successful. When his sons enlisted in the war, it was excruciatingly painful to him. It was almost impossible to hold any lingering illusions for "the new world" he had dreamed of for so long and yearned for ever since he had been a young combat pilot in Europe trying to help "save the world for Democracy." But then the United Nations was established in 1945, with the United States one of its founding members, and his hope was renewed once again.

In the fall of 1951, with this country embroiled in the Korean War, Bridgman developed phlebitis and pneumonia but would not seek medical help. Seriously ill, despondent, in his 56th year, on 9 November 1951 he slipped from the deck of the Upper New York ferry into the New York Bay and drowned. On 17 December 1951 he was buried in an unmarked grave in Silver Mount Cemetery on Staten Island. Many of his former students, upon learning of his death, wrote Bridgman's family with expressions of condolences and admiration for their former professor.

JASPER BROWN

Jasper Cornish Brown was born in Manhattan, New York on 21 February 1888, the son of John E. Brown, an educator, and Annabelle (Cornish) Brown. As a youth Jasper grew up in Harlem where his father was a prep-school principal. He attended Morris High School in Morristown, New Jersey while living on his maternal grandfather's nearby farm.

With the European War raging, Brown enlisted in the *American Ambulance Field Service* and sailed for France. Once there, on 19 June 1917 Brown enlisted in the Lafayette Flying Corps. From 20 June to 1 February 1918 he attended the aviation schools at Avord, Pau, and the *G.D.E.*, receiving his *brevet* on the Caudron on 6 November 1917.

On 29 March 1918 *Caporal* Brown was assigned at the Front with *Escadrille SPAD 67*, the same day he was commissioned a Second Lieutenant in the U.S. Air Service. Brown flew diligently with that Squadron, engaging in many combats, until the 11 November Armistice. He unofficially brought down two German biplanes north of Chalons, which were too far behind enemy lines to be confirmed.

On 22 October 1918 Brown was promoted to First Lieutenant. On 2 December 1918 he was awarded France's *Croix de Guerre*, with Palm for his numerous combats and diligent work with *SPAD 67*, having flown over the lines at the Somme, Champagne, Aisne, Verdun, and Marne fronts. Lieutenant Brown was discharged at Garden City, Long Island, New York on 1 April 1919.

Following World War One Brown worked as an insurance salesman in his uncle's Prudential Insurance firm. He married in 1939. His wife passed away in 1942.

Brown worked for the State of New Jersey during and following World War Two as an inspector for the Bureau of Weights & Measures. In the early 1950's he moved to Fort Worth, Texas where he again sold insurance until his retirement in 1954. He then moved to Venice, Florida shortly after his retirement and married Elizabeth Muzzey Larson in June of 1955. Jasper Cornish Brown died in Venice, Florida on 17 December 1966 at age 78.

Stafford Brown

Stafford Leighton Brown was born in Newton, Massachusetts on 25 October 1895, the son of George W. Brown, a leather merchant, and Eugenie (Stafford) Brown. He graduated from Newton High School in 1915. While there he was a prominent athlete and classmate of David Putnam and William Wellman. Brown entered Dartmouth College in the fall of 1915. While at Dartmouth Brown proved an all-around athlete and captained the tennis team. Brown transferred the following spring to the University of Vermont, Class of 1919.

In 1915 Brown attended the Plattsburg training camp. On 12 March 1917 he joined the American Ambulance Field Service at Boston and sailed for France aboard the *Rochambeau*. Brown served as an ambulance driver with Sections 17 & 19 for four months in the Verdun sector. On 21 July 1917 Brown enlisted in France's *Service Aeronautique*. On 12 September he was released for disciplinary reasons.

On 2 January 1918 Brown enlisted in the U.S. Enlisted Reserve Corps, at Paris. He was assigned to Detachment 1, Flying Cadets, 2d Aviation Instruction Center, Tours. On 4 May 1918 he transferred to the 3d Aviation Instruction Center, Issoudun. He was discharged on 12 June 1918 to accept a commission. He was appointed 2nd Lieutenant, Air Service on 20 May 1918, and was promoted to First Lieutenant on 22 July 1918.

Brown was plagued by flying accidents. On one occasion he was slightly injured and his passenger killed when the plane he was piloting overturned shortly after he had left the ground. On another occasion his plane caught fire while he was in the air, and he barely escaped with his life. Brown was attached to the Acceptance Park at Orley, ferrying machines to the Front. On 28 September 1918, while flying a SPAD fighter in the Nantes sector, he was killed by a fall as a result of a defect in his machine. He was 23 years of age at the time of his death. He was buried at Hargeville, Suresnes, France.

EVERETT BUCKLEY

Everett Timothy Buckley was born in Walnut, Illinois on 3 February 1893, the son of Timothy W. Buckley, a land overseer, and Maude (Jess) Buckley. As a youth Buckley attended schools in Kilbourne, Illinois. He attended the University of Illinois agricultural school from 1911-12 but left that institution before taking a degree to become an automobile racer.

On 6 January 1917 Buckley enlisted in France's *Service Aeronautique*. From 16 January to 30 July 1917 he attended the aviation schools at Buc, Pau, Avord, and the *G.D.E.* He received his *brevet* on the Bleriot on 2 January 1917. On 3 August 1917 *Caporal* Buckley was assigned to *Escadrille SPAD 65*, Lieutenant Nungesser's squadron, and flew with that squadron until 6 September 1917.

On 6 September, while on patrol with *SPAD 65*, *Sergent* Buckley was shot down in combat and landed near Dun-sur-Meuse. Following his capture by German civilians, Buckley was placed in a number of prison compounds in Germany and attempted several escapes. He was severely beaten for his escape attempts.

On 27 July 1918, on his fourth attempt, he managed to escape from a prison farm near Varingenstadt into Switzerland, the first American to do so. On 28 July he arrived in Berne and was returned the next day to the American embassy in Paris. Buckley sailed for the U.S. with the Andre Tardieu Mission as a lecturer, landing in New York City on 10 November 1918.

Following World War One Buckley moved to California to work as a truck driver. It was there that he married Ruth Harp. He died of a heart attack in San Francisco on 9 June 1945 at age 52.

THOMAS BUFFUM

Thomas Bradley Buffum was born in Florence, Italy on 26 October 1894, the son of Dr. Charles Thomas Buffum, a physician, and Elizabeth (Bradley) Buffum. As a young man Buffum attended the Sillig's Institute in Switzerland, and the Gunnery School, Washington, Connecticut before entering Harvard University. He took his A.B. degree from that institution in January of 1916 in order to join the *American Ambulance Field Service* in France.

Buffum drove ambulances with SSU 8 from April to October 1916 during the battle of Verdun. From October 1916 to June 1917 he served with the *Armee d' Orient* on the Salonica and Macedonian fronts. He returned to France after America entered the war and on 15 June 1917 he joined France's *Service Aeronautique*. From 27 June 1917 to 20 March 1918 he attended the aviation schools at Avord, Pau, and the *G.D.E.* He received his *brevet* on the Caudron on 31 October 1917.

On 24 March 1918 *Caporal* Buffum was assigned to the Front with *Escadrille SPAD 77*, flying with that squadron until 4 May 1918. On that date Buffum was shot down over Montdidier and made a prisoner of war. In early July 1918 Buffum and two British officers escaped from their German prison compound but were recaptured several days later. Buffum was finally released in December of 1918 and arrived in Paris on 10 January 1919. Following his discharge he returned to the United States in February of that same year.

Following the War Buffum decided to attend the Harvard Medical School. But having gone to Tahiti to visit Nordhoff and Hall, he returned too late to enter school. Instead, he went to Long Island, New York to work for a motion-picture company. He worked at that occupation for three years before spending a year cruising along the Atlantic Coast in a small yawl. While on a trip to the interior of British Guiana, he took some moving pictures which he later sold to the news reels.

From 1924–1930 Buffum taught French and science at the Morristown School, Morristown, New Jersey. He next worked for the Bethlehem Shipbuilding Corporation until the fall of 1932 when he returned to the Morristown School where he served as master until 1938. On 6 September 1937 he married Beatrice Bell, whose father was head of the B & O Railroad. They had one daughter, Susanna. Buffum spent the next four years teaching boys sailing at Orleans on Cape Cod while involving himself in ocean racing. From 1940 until his retirement in 1963 Buffum worked as a department planner at the Quincy Shipyards, Bethlehem Steel Corporation. Buffum died on 23 November 1970 In Rockville, Maryland at age 76.

EUGENE BULLARD

Eugene Jacques Bullard was born in Columbus, Georgia on 9 October 1894, the son of Octave Bullard, a black stevedore, and Joyakee Bullard, a Creek Indian.

Bullard left home at an early age and wandered the South working at odd jobs until age ten when he stowed away aboard the merchant freighter *Matherus* in an attempt to reach France. He had heard that the French treated Blacks with dignity, and his dream was to reach that country.

The *Matherus* docked at Aberdeen, and young Bullard spent several years in Scotland and in England working as a prize fighter. Bullard fought 42 professional bouts, his most memorable on 28 November 1913 when the nineteen-year-old boxer finally reached France to fight a ten-round match in Paris.

Shortly after the outbreak of World War One Bullard enlisted in the French Foreign Legion on 19 October 1914. He was assigned to the Third Marching Regiment, and fought heroically at Artois, Champagne, and Verdun with that unit and with the 170th French Regiment, receiving two wounds and France's *Medaille Militaire*, and the *Croix de Guerre*, with Star.

After recovering from the wounds he had received at Verdun, Bullard was ordered to the *Service Aeronautique* to serve as a gunner. On 15 November 1916 he requested and was granted the opportunity to become a pilot. From 30 November to 20 August 1917 he attended the aviation schools at Cazeaux, Tours, Avord, and the *G.D.E.*. He received his *brevet* No. 6259 on the Caudron on 20 July 1917, thus becoming history's first Black aviator.

Caporal Bullard served diligently at the Front with *SPAD 93* from 27 August to 13 September 1917, and with *SPAD 85* from 13 September to 11 November 1917. On that date, following a confrontation with a French officer, he was transferred back to the 170th French Infantry Regiment, and spent the remainder of the war in a service battalion.

Bullard was officially discharged on 24 October 1919. He left the military service bitter and angered at the prejudice he had confronted. He felt particular enmity toward Dr. Edmund Gros, the head of the Lafayette Flying Corps, feeling that Gros's prejudice had prevented his transfer to the U.S. Air Service and had ultimately led to his expulsion from the *Service Aeronautique*. It is evident that at least some of Bullard's accusations toward Gros were accurate. Gros had an

obvious dislike of Bullard and refused to recognize his fine war record by including him with a handful of Lafayette Flying Corps pilots who were denied the Lafayette Flying Corps certificate and commemorative ribbon. When Bullard applied for acceptance to U.S. Aviation, Dr. Gros sent the following in a letter to AEF Headquarters:

"Bullard, who is a former prize fighter, knocked out a French adjudant for which he was given 10 days prison. He was given 20 days more for illegally wearing the Fouragere to which he was only entitled to as a member of the Foreign Legion.

"This leads to his total radiation from the Aviation Section of the French Army and to his transfer to the ranks of the French Infantry.

"Under these conditions will you consider of course that he is morally unfit to form a part of the United States Army, and you can reject him on those grounds. At least one dark cloud is dispersed from our horizon."

Following World War One Bullard worked as a musician in various Paris nightclubs while continuing his career as a professional boxer. On 17 July 1922 he married Marcelle Eugenie Henriette in Paris. Two daughters and a son were born of this union. In the fall of 1924 Bullard opened the Le Grand Duc nightclub in Paris. He later opened a gymnasium. Mrs. Bullard died in 1936. Shortly after Germany attacked France, Bullard rejoined the French army. Just before France fell to the German onslaught, Bullard escaped to the United States, arriving in New York on 9 July 1940.

For a time Bullard worked as a travelling salesman. His last job was an elevator operator at Rockefeller Center. On 9 October 1959 France bestowed upon him *Chevalier* of the Legion of Honor. Bullard died at the Metropolitan Hospital in New York City of cancer on 22 October 1967 at age 67.

RICHARD BULLEN

Richard Nixon Bullen, was born in Kenosha, Wisconsin on 26 August 1890, the son of George W. Bullen, a Chicago malt magnate, and Mary Lenore (Nixon) Bullen. George Bullen was a direct descendant of Anne Boleyn, and at Lake Oconomowoc, Wisconsin, he built a castle fashioned after the Haver castle in England. The Haver castle was presented by Henry VIII to the Boleyn family in order that he might marry Anne Boleyn, whom he accused of adultery not long after their marriage then had her beheaded for the crime.

As a youth young Dick Bullen and his brother, "Gay," spent their summers between the "Minnewoc" estate, and White Bear Lake, Minnesota where they raced class A yachts. He and Gay were given a yearly allowance of $7,000. Dick attended the Lawrenceville School, Lawrenceville, New Jersey. In 1907 he spent nine months in Africa on safari. That fall he entered the University of Wisconsin at Madison. In 1911, six months before he was to receive his degree, Dick dropped out of school and married Ruth Taylor, an actress whom he had met while playing a part in the Musical comedy "Jeanette." Ruth Taylor Bullen died of a mysterious virus on 21 February 1914.

Dick Bullen was working in real estate in Miami, Florida, when he decided to enlist in the Lafayette Flying Corps in New York City. On 23 June 1917 he sailed for France aboard the *Rochambeau.*

Bullen enlisted in the *Service Aeronautique* on 6 July 1917. He attended the aviation schools at Pau and the *G.D.E.*, receiving his *brevet* on 23 September 1917. While *Caporal* Bullen was at the *G.D.E.* he was released from service on 7 January 1918 because of injuries he had received in a flying accident at Pau.

Caporal Dick Bullen returned to the United States and became involved in War Bond drives. While on one such bond drive he married Mary Kilgore Miller in New York.

Following World War One, Dick became engaged in business with his brother, Gay, in Morocco. He was on an African Safari when Gay died in a speed-boat accident on 16 June 1929. Following Gay's death, Dick returned to the United States. Not long after a daughter, Mary, was born. Dick later moved to Florida and involved himself in business with his brother, John. He died at Miami Beach, Florida on 14 July 1936 at age 45.

WILLIAM BULLEN

William Graham Bullen, was born in Kenosha, Wisconsin on 24 December 1895, the son of George W. Bullen, a Chicago malt magnate, and Mary Lenore (Nixon) Bullen. George Bullen was a direct descendant of Anne Boleyn, and at Lake Oconomowoc, Wisconsin, he built a castle fashioned after the Haver castle in England. The Haver castle was presented by Henry VIII to the Boleyn family in order that he might marry Anne Boleyn, whom he accused of adultery not long after their marriage then had her beheaded for the crime.

As a youth young "Gay" Bullen and his brother, Richard, spent their summers between the "Minnewoc" estate, and White Bear Lake, Minnesota where they raced class A yachts. He and Dick were given a yearly allowance of $7,000. "Gay" attended the Lawrenceville School, Lawrenceville, New Jersey. In 1913 he entered the University of Wisconsin at Madison, and competed in golf tournaments in the summer.

In 1917 Bullen entered Marquette University at Milwaukee as a student in the arts and science school. That same year he asked for and was granted the Dean's leave to go to France as an ambulance driver.

Once in France, Bullen first enlisted in the Artillery School at Fontainblau but transferred to the *Service Aeronautique* on 14 July 1917. From 27 July to 10 March 1917 he attended the aviation schools at Avord, Juvisy, Pau and the *G.D.E.* He received his *brevet* on the Caudron on 29 September 1917.

From 14 March to 17 April 1917 *Caporal* Bullen flew at the Front with *Escadrille SPAD 162*. On one patrol his fighter was hit by anti-aircraft fire, slightly wounding him.

Sergent Bullen left the *Service Aeronautique* to join U.S. Naval Aviation and was commissioned an Ensign on 3 November 1918. Following World War One, in 1925 he served in the Rif War in Morocco with the *Escadrille de la Garde Cherifienne*, a unit composed of American volunteers. As a reward for fighting for the Sultan of Morocco against Abd el Krim, Bullen was invited to live in the palace as one of the family where all his desires were catered to.

He later became engaged in business in Morocco with his brother, Richard, while making his home at Antibes, France. On 16 June 1929 Gay fell from a fast-moving motor boat off the coast of Juan Les Pins and was instantly killed when he was struck by the boat's propeller. He was 34 years of age at death.

PHILIP BUSH

Philip Nelson Bush was born in Roxbury, Massachusetts on 21 December 1894, the son of Arthur Richmond Bush, an executive with the General Electric Company at Schenectady, N.Y., and Mary (Cole) Bush.

In the fall of 1915 Philip Bush enrolled in the University of Pennsylvania's Engineering school. He left that institution to join the American Ambulance Field Service.

On 9 May 1917 Bush enlisted in France's *Service Aeronatique*. From 23 May to 13 January 1918 he attended the aviation schools at Avord, Pau, and the *G.D.E.* He received his *brevet* on the Caudron on 3 October 1917.

On 19 January 1918 *Caporal* Bush was assigned to the Front with *SPAD 73*, flying with that squadron to 2 May 1918. On 8 May 1918 *Sergent* Bush was commissioned a First Lieutenant in the U.S. Air Service and continued to fly with his French squadron until 21 July 1918. On that date he was ordered to duty at Paris, and served at Choisy-le-Roy, and at the American Acceptance Park, Orly until the 11 November Armistice. While with the French Air Service he was awarded France's *Croix de Guerre*, with Palm. He received his discharge on 21 January 1919.

Following World War One Bush was employed by the International General Electric Company in Manila, the Philippines, and in Shanghai, China. He later worked for the Ford Motor Company at Schenectady, New York. It was there that he married Betty Taylor. On 25 February 1942 he entered active service as a Major in the Army Specialists Reserves and was shortly appointed to Lieutenant-Colonel and assigned to Bolling Field, Washington, D.C. While he was assigned there, he accidentally shot himself while cleaning his pistol and died of his wounds on 28 October 1942 at age 47. He is buried at Arlington National Cemetery, Washington, D.C.

LOUIS BYERS

Louis Leslie Byers was born in Philadelphia, Pennsylvania on 21 July 1894, the son of Louis Temple Byers and Georgena (Work) Byers. As a young man Louis attended Central High School, Philadelphia and the Staunton Military Academy, Virginia. He also attended the University of Pennsylvania from 1911-13 and the University of Illinois. During summer vacation he took a one-month naval training cruise on the Battleship *Rhode Island*.

While Byers was at the University of Illinois, in March of 1917 he enlisted in the American Ambulance Field Service in Chicago. On 13 June 1917, soon after his arrival in Paris, Byers enlisted in France's *Service Aeronautique*. From 27 June 1917 to 10 July 1918 he attended the aviation schools at Avord, Pau, Cazeaux, and the *G.D.E.* He received his *brevet* on the Caudron on 5 December 1917.

From 13 July to 18 July 1918 *Caporal* Byers was attached at the Front to *Escadrille SPAD 38*. On that date while on a patrol in the Champaigne sector he ran out of fuel behind enemy lines and made a forced landing in the region of Marquises. *Caporal* Byers was captured and made a prisoner of war at Rastatt, Baden. While a prisoner, he edited the prison paper, "The Barbed Wireless."

On 22 October 1918 Byers and two other prisoners escaped captivity but were captured six days later in the Black Forest and returned to prison and placed in solitary confinement for two weeks. The Armistice brought about Byers's release on 3 December 1918.

Following World War One Byers became associated with the Abrasive Company of Philadelphia, remaining with that firm until August of 1930. On 31 January 1925 he married Adelaide Merrick. They had one child, Georgena Adelaide. He then became works manager for the Sterling Grinding Wheel Co, of Tiffin, Ohio. He died on 26 September 1946 at Westfield, Massachusetts at age 52.

ANDREW COURTNEY CAMPBELL

ESCADRILLE LAFAYETTE

Andrew Courtney Campbell, Jr., the 29th volunteer of the *Escadrille Lafayette*, was born in Kenilworth, Illinois on 19 November 1891, the son of Andrew Courtney Campbell, a livestock broker, and Cornelia (Morton) Campbell. Campbell's paternal grandfather, Benjamin H. Campbell, had been the director of the National Bank of Illinois and was the largest stockholder in one of that state's earliest railroads. His paternal aunt, Annie, was the wife of General Orville Elias Babcock, General Grant's personal aid and one of a handful of Union officers present at the surrender of General Lee at the McLean House, Appomattox Court House, Virginia.

In his youth, "Coty" Campbell attended Kenilworth's public schools. He attended the Mercersburg Academy, Mercersburg, Pa. from September 1909 to June 1910. He spent his summers at his parents' resort home at Whitehall, Michigan where his feats soon won him a daredevil's reputation among the junior members of the White Lake summer colony.

Campbell had first impressed his friends one day when storm-driven waves had been breaking over the pier upon which they stood just below the Sylvan Beach lighthouse. One young lady present had challenged that no man could survive the stormy waves. Prodded by her words, Campbell immediately jumped into the frothy waters, only to be miraculously rescued by a life-saving crew seconds before his demise.

Fast boats and speeding roadsters provided exciting entertainment for "Coty" Campbell and his friends. Campbell's closest friend, "Aleck" Bayne, served as shipmate aboard the "Lorelai" when they had won the Silver Loving Cup at the 1908 White Lake regatta. When the two were not racing across the lake, they were terrorizing the local motorists on the nearby parkway.

The Campbell/Bayne friendship survived beyond their White Lake summers, and later when "Coty" journeyed to France to enlist in the *Service Aeronautique*, "Aleck" had followed him into aviation. First Lieutenant James Alexander Bayne was killed at the Front on 8 May 1918 while flying with *SPA 81*.

"Coty" Campbell's first experience with aviation had come about on White Lake the day his launch had met a downed hydro-aeroplane floating off Camp Holly. The plane's pilot had hailed Campbell's launch and had asked for a tow. Campbell pulled alongside the hydro-aeroplane then climbed onto its floats to question the pilot, one Bill Johnson, about the workings of the aeroplane. His curiosity satisfied, Campbell agreed to tow the plane, but only if Johnson would teach him to fly it.

When Johnson had agreed to Campbell's proposal, "Coty" lashed a rope to the aeroplane and gunned the launch, picking up enough momentum for the plane to rise. It was soon under its own power and flying out of sight.

Campbell attended the University of Virginia, Class of 1914. He left that institution without taking a degree and developed a reputation as a playboy as he half-heartedly pursued a dancing career.

In the late spring of 1916 young Campbell returned to his parent's sixteen-room mansion in Kenilworth, Illinois. He had grown bored with life and had felt a

Sergent *Andrew Courtney Campbell beside his three-wing Nieuport 17C.1 fighter, Chaudun airfield, 7 July 1917. Campbell is obviously pleased with himself after having defied enourmous odds in bringing the crippled plane safely to ground.*

growing desire to participate in the European War. As always, he acted quickly, compulsively. It was mid-June when he told his parents of his intentions. He retired to his room and put everything in order then filled his grip with an extra set of clothes and descended the stairs to his weeping mother.

"Don't fret, Mother," he had said. "Everything will be all right. I am merely going to do in my small way for France what Lafayette and Rochambeau did for us."

Campbell picked up his flute and played several of his parent's favorite Scotch airs. He then kissed his mother goodbye, and hurried out the door with his father to the train station.

Campbell sailed for France in June of 1916 aboard the *LaFayette*. He enlisted in French Aviation on 20 July 1916. On 8 September he began pilot training at the Buc school of aviation. On 23 November 1916 he received his *brevet militaire* on the Caudron and by 10 April 1917 he had proceeded through the schools at Juvisy, Avord, Cazeaux, Pau, and had arrived at the G.D.E.

Campbell joined the *Escadrille Lafayette* at Ham on 15 April 1917, the day before fellow *Lafayette* pilot, *Caporal* Edmond Genet, was killed in combat north of Montescourt. Courtney, as Campbell was called by the Squadron's pilots, soon developed a dual reputation with the *Lafayette*. First and foremost he was recognized as a stalwart, dependable fighter pilot. But he also enjoyed casting himself in the role of "jester" and "wild man." And as one fellow pilot said of him: ". . . (his) whole life seemed devoted to giving other people thrills and to whom trouble gravitated as naturally as iron filings to a magnet."

Campbell led a charmed life, as he proved on 7 July 1917, the day the Squadron was ceremoniously presented with an American flag by Mrs. William Gibbs Macadoo, the daughter of President Woodrow Wilson. While in his Nieuport 17 C.1 "test-hopping" a new motor fifteen-hundred meters above the Squadron's aerodrome, Campbell had placed his fighter into a short power dive and had performed a snapping loop at the end of it. But at the top of his loop the stress against his left wing plane had begun snapping bracing wires. Hanging upside down, he watched wood and cable groan and give way as his lower left wing separated from the Nieuport's fuselage and fluttered into the forest below.

The odds that Campbell would ever bring the fighter safely to ground were one in one million. The immediate result should have been that the Nieuport would fall into a spin that would have peeled away the rest of the wings.

However, as Campbell exited the loop, he was able to put his ship into a "perfect counterbalance." This, he maintained as he placed the Nieuport, now a biplane on the starboard and a monoplane on the port, into a gentle glide while avoiding a fatal spiral. He volplaned his bus into a perfect landing on a beet field ten kilometers from Chaudun.

Campbell's "white-knuckle" experience would have unnerved a normal pilot. But Courtney viewed it as a joke, and he climbed into the back of the ambulance which had come to retrieve his remains, forcing its attendants to make several bar stops where he related his unbelievable experience over drink.

Campbell was not one to dwell on morbid alternatives to his incredible landing. In fact, within a half hour of his arrival at the Squadron's aerodrome, he was back in the sky zooming about in another bus. Campbell was cited and later awarded France's *Croix de Guerre*, with Star for safely landing his damaged fighter.

While Campbell was recognized by the Squadron's pilots as "a skillful and courageous fighter", his recklessness in the air was also well known and gave more than one patrolling *Lafayette* pilot "the leaping jitters."

While still a young playboy in civilian life, Campbell had relished tailgating the car ahead of him while zooming down the parkway in his Stutz Bearcat. Now Campbell, the pilot, found great sport in placing the nose of his bus with its whirring propeller dangerously close to the ship before him. And it was to Edwin Parson's terror and Courtney Campbell's delight that while Parsons was leading one of many patrols with Campbell, the wild young Scotsman would place his deadly prop not three feet from the tail of Parson's SPAD.

Parsons often found his attention directed not toward Boche aircraft but in trying to keep out of Campbell's way. But neither pleading nor physical threats could convince Campbell to desist his dangerous practice.

"I shook my fist at him," Parsons said, "and cursed back into the prop wash, but he only grinned and waved and threw kisses; it was maddening."

On 18 August 1917, the day that Harold Willis was shot down near Dun-Sur-Meuse, Courtney Campbell had a second incredible bit of luck. He, Willis, and a number of the other Squadron pilots had flown fighter escort for a bombing raid deep inside enemy territory. Approximately sixty aircraft, bombers and fighters, had taken part in the raid, all using the long, narrow Senard airfield for takeoff and landing.

Orders had been issued to all pilots that each pilot returning from the mission was to taxi to the end of the field rather than turn into the hangars lining both sides of the airstrip. This would allow greater space for incoming planes to land and would lessen the risk of a collision between aircraft.

But when Courtney Campbell brought his SPAD down on the airstrip, he elected to turn it into his hangar rather than follow orders and taxi to the far end of the strip despite the fact that a Sopwith bomber was dropping to earth right on his tail. Campbell swung the nose of his fighter toward the Squadron's hangar and began taxiing across traffic, directly in front of the fast-approaching Strutter. By then the bomber pilot had reduced his speed too far to attempt another takeoff. His only alternative was to bear straight ahead, directly into the side of Campbell's SPAD.

The bomber's huge prop chopped Campbell's wing to bits, stopping not a foot from the young American's head. The force of the impact sent the SPAD cartwheeling before it came to rest as a pile of kindling.

Campbell extricated himself from the wreckage and surveyed the rubble that had been his SPAD. He lit a cigarette and calmly strolled across the buzzing airstrip to his Squadron's hangar. *Capitaine* Thenault stood in its shadows barely containing his rage. Campbell knew he must use all his charm to escape his due punishment for deliberate disobedience. And not to the complete surprise of his fellow Squadron mates, that's exactly what he did. The roar from the engines of the incoming aircraft drowned out his words. But whatever Campbell said to Thenault must have worked. For he soon left his visibly calmer commander for the bar, that impish Campbell smile playing over his lips.

On 19 September 1917 Campbell had his third miraculous escape from death when he buried the wheels of his SPAD into the top wing of French Lieutenant Maison Rouge's bus. Campbell, returning as part of a patrol which Maison Rouge had led over the right bank of the Meuse, was evidently not satisfied with the excitement just provided by six German Albatroses and the subsequent violent combat which saw David Peterson down one of the enemy fighters.

Campbell had sought further stimulus as the patrol neared the Squadron's field. And for some unexplained reason he chose Lieutenant Maison Rouge as his latest source of entertainment. Breaking formation, Campbell placed his own SPAD dangerously above Maison Rouge's upper wing, refusing to move away in spite of the Lieutenant's violent gestures that he do so. The anger he saw in Maison Rouge's face merely egged Campbell on, and he began bouncing his fighter up and down, coming ever closer to Maison Rouge's upper wing. Then suddenly Campbell miscalculated. And before he could redress, he had sunk his tires through the skin and framework of the Lieutenant's wing.

Locked together like mating dragonflies, the two fighters made turns above the field as the Squadron's ambulances jockeyed for position near their inevitable crash site. Campbell chose the only one way to extricate himself from Maison Rouge's fighter. He gave his SPAD full throttle and reefed the stick back. His bus rose like a Phoenix off the top of the Lieutenant's SPAD, barely managing to leave Maison Rouge with his upper wing.

Maison Rouge landed immediately, his twitching face at once a portrait of fear and rage. But Campbell, oblivious to the emotions he had wrenched from the Frenchman and to the consternation he had caused his comrades on the ground, took his time in bringing his own bus down. First, he put on an acrobatic display which nearly eliminated the tops of the Squadron's hangars. Typically, Campbell then landed without apology, expecting approbation rather than reprimanded, his lips curved into a mischievous grin.

Eventually Campbell's ceaseless antics began to wear on the others. Two incidents put him in further disfavor with the men. Chouteau Johnson, who had served nearly a year with the Squadron prior to Campbell's April 1917 arrival, held special enmity toward the upstart pilot and particularly enjoyed badgering him during Squadron poker games. When Campbell had had enough, he challenged Johnson "Come outside and let's settle it." Minutes later Johnson staggered back inside, so badly beaten he had to be taken to hospital.

A few days later Campbell pulled a stunt which brought the French authorities down upon the Squadron. Before going out on patrol, the Lafayette pilots would test their machine guns as standard operating procedure. One day a complaint came in that a SPAD with an Indian head insignia had killed a French farmer who was mowing his field. The incident was narrowed down to Campbell who admitted being the culprit. But he claimed that, while preparing to test his gun, he had dived on a farmer to scare him but had accidentally touched his machine gun trigger, shooting the farmer off his machine. Because of Campbells claim that it was an accident, and there was no proof otherwise, the matter was dismissed. But one pilot who knew Campbell well believed that Campbell had killed the farmer intentionally. And he later recalled that Campbell had shown no remorse for his actions.

On 1 October 1917 Lady Luck refused to ride with Courtney. Perhaps it was because he had misplaced the talismans which he felt certain would protect him in the air. They were two pink garters, which he claimed he had acquired from a young virgin beneath the dark of the moon. He had worn these charms faithfully on his shirtsleeves beneath his tunic, convinced of their protective powers to defend him while on patrol.

Uneasy without his talismans, Courtney Campbell had left on patrol with Henry Jones over the Chemin-des-Dames, north of Soissons. Over the Ailette reservoir they had come upon four German Hanoveranner reconnaissance planes which were circling beneath them, directing artillery fire. With Campbell above his shoulder, Jones dove down through the heart of the formation and came up under one of the German planes, firing both machine guns.

As Jones watched his stream of bullets penetrate the German's fuselage, the machine gunner in the rear cockpit fired on him through a panel in the bottom of the plane. Three bracing wires were shot off Jones's right wing and his fuselage became a sieve punctured by thirty-eight bullets. With his fuel tank spewing petrol, Jones glided toward allied lines, his immediate concern whether his wing would stay on taking precedence over what Campbell's fate might be.

Sadly, Jones's view of Campbell just prior to the combat was the last sighting by any ally of the wild young American. Thus, the "jester" that had made his friends laugh and cringe, and the "skillful and courageous fighter" was lost forever to his Squadron mates, gone at age 26.

For two months after his disappearance Campbell was carried on the Squadron's rolls as "missing, presumed dead." Then in November a German pilot dropped a message over allied lines which reported his death in combat. An accompanying map showed his resting place by a marsh at Pargny where his bus had fallen.

Shortly after his disappearance, *Sergent* Campbell was again cited in French Orders and was awarded the *Metaille Militaire* and a palm to his *Croix de Guerre*. The translation of his citation reads:

"Andrew Courtney Campbell, sergeant-pilot with Squadron 124; American citizen who volunteered for the duration of the war; ever since his arrival in the squadron he has given astonishing proof of bravery and extraordinary sangfroid; October 1, 1917, fell in a badly damaged aeroplane within the enemy lines after a hard combat against two enemy aeroplanes."

The German aviation journal *Flugsport* stated in its 28 November 1917 issue under the heading "List of 149 Enemy Aircraft Downed in German Territory in October, 1917" that ". . . a French Spad single-seater, No. S4245, was downed and its occupant, Andrew Courtney Campbell, was dead." But not until 8 February 1918 was Campbell officially listed as dead by the French, and by War's end, his remains had not been recovered by the Allies.

Lafayette Flying Corpsman, Captain Edgar Guerard Hamilton, while directing the U.S. Investigating Committee to find graves of missing American aviators unaccounted for, allegedly came upon *Sergent* Campbell's remains in late October 1919 under barbed-wire entanglements in the marshy area near Pargny. According to Campbell's mother, Cornelia Campbell, the enemy had clearly marked her son's grave, and when the body was exhumed, the breast pocket of the tunic had contained a letter and a Christmas card from her to Courtney, and a snapshot of Campbell's father and Courtney's pet fox terrier which was taken on the lawn in front of their Kenilworth home.

Some of Campbell's Squadron mates, including Ted Parsons, believed that his remains were never recovered, and that Captain Hamilton had been trying to put the Campbell's minds at rest, to end their state of limbo regarding Courtney's status. The exhumed body was first buried in Ploissy cemetery on the Soissons road. It was later moved to the American Cemetery at Belleau Wood before it was finally transported in 1928 to the Lafayette Escadrille Memorial at St. Cloud.

Some four years later, in September of 1932, the world press carried an alleged amnesiac victim's claim that he was Andrew Courtney Campbell. The F.B.I. soon learned that the imposter was one George Reynolds, a Navy deserter from World War One. Reynold's claim prompted an indignant response from one H. Andres, a former German pilot, who said he had shot Campbell down near Filain. He said that while he was accompanying an artillery biplace from Laon to Chemin-des-Dames, he was attacked by a single seater. After he had shot the plane down, he had gone to the crash site where his victim's body lay, shot through the head. There, he learned via his military papers that he was an American, Andrew Courtney Campbell. Andres later conveyed this information to the French by dropping them the detailed sketch to the grave site.

HUGH CAMPBELL

Hugh Gordon "Scotty" Campbell was born in Harrison, New York on 11 January 1897, the son of Mr. & Mrs. Frank B. Campbell. He received his Bachelor of Arts degree from Hobart College, Geneva New York, Class of 1916. Shortly after graduation, on 29 July he enlisted in the *Norton Harjes Ambulance Service* and served with Section 5 in France.

On 27 June 1917 Campbell enlisted in France's *Service Aeronautique*. From 24 July 1917 to January of 1918 he attended the aviation schools at Avord, Pau, Cazeaux, and the *G.D.E.* He received his *brevet* on the Caudron on 3 December 1917.

Before going to the Front with French Aviation, *Caporal* Campbell enrolled as an Ensign and student Naval Aviator at Paris, France on 11 June 1918. On 29 July 1918 he became Naval Aviator HTA #1073 at Moutchic, France. He was attached as a chasse pilot to *Escadrille Saint-Pol (SPAD 313)* and was cited for bringing down a German aircraft. He was awarded France's *Legion of Honor* and *Croix de Guerre*, with two palms. He later served in the Northern Bombing Group at Dunkirk, at Brest, and after the Armistice, at Bay Shore, New York. He received the U.S. Navy Cross for service as a seaplane pilot in France. His citation is as follows:

"For distinguished service in the line of his profession as a pilot of airplanes, in which capacity he showed exceptional daring and fought several times within the enemy lines, attacking at low altitudes."

Following World War One Campbell worked for an aircraft fireproofing company. On 9 June 1920 he was stunting with a JN4 over the ocean near Atlantic City, New Jersey when he attempted an Immelman turn and landed upside down in the waters off Kentucky Avenue. He died instantly from a skull injury. He was 23 years of age at death. Following elaborate military services, he was buried at the family vault at Batavia, New York.

JOSEPH CARRERE

Joseph Maxwell Carrere, Jr. was born in New York City, New York in June 1898, the son of Mr. & Mrs. Joseph M. Carrere. He attended the Groff-Clark School, New York. He left school at the end of the fall term to sail for France and enter the European War. On 22 February 1917 Carrere enlisted in France's *Service Aeronautique*. On 10 July he was released because of inaptitude.

Carrere returned to the United States. On 26 October 1917 he enlisted at Fort Hancock, New Jersey in the 22nd Company, New York Coastal Defense Corps. Private Carrere was sent to France following the Armistice and served overseas from 4 January 1918 to 28 February 1919. He was honorably discharged on 14 March 1919.

THOMAS CASSADY

Thomas Gantz Cassady was born in Freedom, Indiana on 5 January 1896, the son of Otto Ezra Cassady, a farmer, and Edith (Gantz) Cassady. As a young man he attended nearby Spencer public schools before entering the University of Chicago in the fall of 1914.

In the winter of 1916–1917 Cassady left school for France to enroll in the American Field Service. He served six months with SSU 13 and was wounded on 25 May 1917 while removing and transporting wounded during the intense bombardment of the field hospital to which Section 13 was attached. For his heroic service, Cassady was awarded France's *Croix de Guerre*, with Palm.

On 10 July 1917 he enlisted in France's *Service Aeronautique*. From 3 August to 24 December 1917 he attended the aviation schools at Avord, Tours, Pau, and the *G.D.E.*, receiving his *brevet* on the Caudron on 6 October 1917.

On 26 December 1917 *Caporal* Cassady was attached at the Front to *Escadrille SPAD 157* and flew with that squadron until 16 February 1918. On 22 February 1918 *Sergent* Cassady was commissioned a First Lieutenant in the U.S. Air Service. With no opportunity to fly or fight in an American squadron, Cassady requested that he be reposted to the French. That same day he was assigned to *SPAD 163* and flew with that squadron until 8 September 1918. While with *163* he was credited with the official destruction of five enemy aircraft and had three unconfirmed victories. Fellow American, *Adjudant* James Connelly, also had five official and two unconfirmed victories with *SPAD 163*, making the two Americans the top aces of the Squadron. The Escadrille was later named "Escadrille des Connelly-Cassady" in their honor.

On 8 September 1918 Lieutenant Cassady joined the 28th Aero Squadron, USAS, 3rd Pursuit Group, as a flight commander. While with that Squadron Cassady was credited with the destruction of four enemy aircraft, including two Fokker D VII fighters.

Lieutenant Cassady flew in many major campaigns, including Champaign, 1917; Belfort-Vosges, 1917–

1918; Chateau Thierry, Second Marne; Champaign. 1918; St. Mihiel and Aisne-Meuse Argonne. For his outstanding war service Cassady was awarded France's *Croix de Guerre*, with three Palms and two Stars, the U.S. Distinguished Service Cross, with Oak leaf, for extraordinary heroism, and the Aero Club of America Medal. On 6 August 1918 he was created a *Chevalier* of the *Legion Of Honor*. The citation accompanying the award acknowledged Cassady's heroic devotion to France.

"A first lieutenant of the American Army who came to serve France at a moment when no military obligation required him to do so. He was seriously wounded while in the ambulance section where he had volunteered and was cited in the Order of the Army. Was then transferred to the air service and there demonstrated to all, chiefs and comrades, his high character, his qualities as a pilot, and his absolute heedlessness of danger."

On 18 March 1919 Lieutenant Cassady was promoted to Captain. He returned to the United States on 4 April 1919 and was discharged from the military on 22 April 1919.

Following his World War service Cassady became involved in the investment banking business in Minneapolis, Minnesota. While working there for Halsey-Stuart & Co. he met and married Edith Harrison in September 1923. In 1926 Cassady moved to Chicago, Illinois where he founded the investment firm of McGowen, Cassady and White. Cassady dissolved his firm near the outbreak of World War Two when the Republic of France, though the United States Government, again called for the assistance of their friend, Thomas Cassady.

On 6 December 1941, the eve of the Japanese attack on Pearl Harbor, Hawaii, Cassady was summoned to Washington, DC, by Frank Knox, Secretary of the Navy. He had been recommended by French military leaders, because of his intimate knowledge of the French people and their language, to work with the French Underground. Cassady accepted the dangerous assignment of serving as an undercover agent in Vichy, while serving ostensibly as Naval Air Attache under Admiral Leahy, American ambassador to the Vichy Government.

Commissioned a Lieutenant Commander in the U.S. Navy, Cassady reached Vichy in early 1942 and served as an OSS agent with orders to aid escaping allied pilots. While working in this capacity and dressed as a civilian, he made four clandestine trips into occupied France. Cassady continued his efforts until the Germans occupied Vichy in November 1942, when he was arrested by them and detained until February 1944.

Upon his release from captivity through a diplomatic exchange between Germany and Brazil, Cassady continued his OSS work in Algers where he helped to plan the invasion of Southern France. After the Allied liberation of Paris, Cassady was put in charge of all secret intelligence for France and Germany. He continued in this capacity until the War ended in Europe.

For his outstanding War service, Lieutenant Commander Cassady was awarded the U.S. Legion of Merit, and France made him Commander of the *Legion Of Honor*. Jean Belaird, former French Consul-General in Chicago, when asked if he knew Thomas Cassady, replied "Ah oui!—Tom Cassady; C'est le grand ami de la France."

Following World War II Cassady became a partner in the Farwell, Chapman & Co., Chicago investment firm. In 1953 he headed up the Chamberlain Manufacturing Co. of Waterloo, Iowa. When he retired from business, he avidly pursued his hobby of fly fishing and worked diligently for conservation issues. He died of cancer in Lake Forest, Illinois on 9 July 1972 at age 76.

OLIVER CHADWICK

Oliver Moulton Chadwick was born in Lowell, Massachusetts on 23 September 1888, the son of Austin Kilham Chadwick, President of the Lowell Five Cent Savings Bank, and Julia May (Moulton) Chadwick. Oliver attended Lowell public schools and Lowell High School, and the Phillips Exeter Academy where he was awarded the Yale cup for scholastic and athletic excellence.

Oliver Chadwick graduated from Harvard College in 1911, and from the Harvard Law School in 1914. A gifted athlete and goal tender on the Harvard hockey team, Chadwick coached the Harvard freshman track team and was the assistant coach for the varsity track team.

Later that same year he entered the legal firm of Stone and Webster. But from the beginning of the European War, Chadwick felt a desire to aid the Allies. He went to Canada to enlist in military service but was denied entry because he was an American citizen.

Chadwick returned to the United States and entered military training at the first and second Plattsburg camps in the summer of 1916. He put in his application for the French Foreign Legion, but before he could enlist he was sent to the Mexican Border for four months with Battery A, 1st Reg. Mass F.A., N.G.

Upon his separation from military service, Chadwick enrolled in the Curtiss Flying School at Newport News, Virginia. Once he had gained proficiency in the handling of aircraft, he sailed for France in late 1916.

On 17 January 1917 "Ollie" Chadwick enlisted in France's *Service Aeronautique*. From 23 January to 25 July 1917 he attended the aviation schools at Buc, Avord, Cazeaux, Pau and the *G.D.E.* He received his *brevet* on the Bleriot on 4 May 1917, and on 28 July he was assigned at the Front to *Escadrille SPAD 73*, Captain Guynemer's squadron.

On 14 August 1917 while on a volunteer patrol, *Caporal* Chadwick was killed in combat north of Bixschoote while coming to the aid of a British Sopwith aircraft under attack by three German Albatrosses. He was 29 years of age at death. Chadwick was posthumously awarded France's *Croix de Guerre*, with Star, and the Aero Club of America War Medal.

CYRUS CHAMBERLAIN

Cyrus Foss Chamberlain was born in Minneapolis, Minnesota on 28 February 1889, the son of Francis A. Chamberlain, a Minneapolis banker, and Frances (Foss) Chamberlain. Chamberlain attended Minneapolis public schools. He graduated from Princeton University, Class of 1910. He then returned to Minneapolis where he became a partner in the Marsh and McLennan insurance firm until the spring of 1917.

On 19 May 1917 Chamberlain sailed for France aboard the S.S. *Chicago* to enter the European War. He arrived at Bordeaux on 29 May 1917. On 3 June he enlisted in France's *Service Aeronautique* in Paris. From 6 June to 8 December 1917 he attended the aviation schools at Avord, Pau, and the *G.D.E.*. He received his *brevet* on the Caudron on 15 October 1917.

On 12 December *Caporal* Chamberlain was assigned at the Front to *Escadrille SPAD 85*. He flew with that squadron until 9 January 1918. On that date he was assigned to *SPAD 98* and flew combat patrols with that squadron until 13 June 1918 when he was killed in a mixup with 13 enemy fighters near La Ferte-Milon. He was 29 years of age at the time of his death.

Sergent Chamberlain was posthumously awarded France's *Croix de Guerre*, with Palm. On 10 July 1923 Wold-Chamberlain Field in Minneapolis was named in his honor. In 1928 his remains were removed from the military cemetery at Coulommiers to the Lafayette Flying Corps memorial near Paris.

CHARLES CHAPMAN

Charles Wesley "Carl" Chapman, Jr. was born in Dubuque, Iowa on 28 December 1895, the son of Charles Wesley Chapman, a lumber sales merchant, and Neva (Powers) Chapman. Carl attended West Waterloo, Iowa High School before enrolling in Amherst College, Amherst, Massachusetts, Class of 1918.

On 19 May 1917 Carl Chapman sailed for France aboard the SS *Mongolia* to join the American Field Service. He arrived in Bordeaux on 30 May and reported to the Ambulance Service headquarters. But when he learned he was to drive a cameon instead of an ambulance, Chapman decided to join the French aviation.

On 10 June 1917 Chapman enlisted in France's *Service Aeronautique*. From 16 June 1917 to February of 1918 he attended the aviation schools at Avord, Pau, Cazeaux, and the *G.D.E.* He received his *brevet* on the Caudron on 30 October 1917.

On 21 February 1918 *Caporal* "Pick" Chapman was commissioned a Second Lieutenant in the U.S. Aviation Section Signal Corps. He flew at the Front with the 94th Aero Squadron, USAS, from 3 March to 3 May 1918, the day he was killed in combat while on patrol in the Toul sector. His plane crashed within enemy territory. German troops buried Chapman's remains near the village of Remoncourt. He was 22 years of age at death.

Lieutenant Chapman was posthumously awarded France's *Croix de Guerre*, with Palm, and the U.S. Distinguished Service Cross for extraordinary heroism. His D.S.C. citation is as follows:

"When on patrol duty, he courageously attacked a group of four monoplanes and one biplane and succeeded in bringing one down before he himself was shot down in flames."

In 1928 Chapman's remains were transferred to the Lafayette Flying Corps Memorial near Paris.

VICTOR CHAPMAN

ESCADRILLE LAFAYETTE

"Victor Chapman was killed day before yesterday. . . . He was always running in and out and I used to tell him it did me good just to look at him. He was so big and strong, and gave out to everyone he met. These are sad days."

—Alice Weeks.
Paris, 25 June 1916.

Victor Emmanuel Chapman, a founding member of the *Escadrille Americaine*, was born in New York City on 17 April 1890, the son of John Jay Chapman, lawyer and litterateur, and Minna (Timmins) Chapman.

Victor's mother, Minna, was the oldest child of wealthy Bostonian George Henry Timmins.

Victor's father, John Jay Chapman, the great-great-grandson of John Jay, the first U.S. Chief Justice, was the son of the powerful and wealthy Henry Grafton Chapman, broker and president of the New York Stock Exchange. Victor's mother, Minna, was the oldest child of wealthy Bostonian George Henry Timmins.

Minna was devoutly Catholic, a part of her Italian heritage, and sought sanctuary "in prayer, religious books, observance, meditations." John Chapman, a devout member of the Episcopal Church of England, would later come to regard Minna's Catholic faith as "dangerous."

The Protestant religion young "Jack" Chapman confronted when he enrolled at St. Paul's School was austere and uncompromising. And he so passionately embraced its tenants that by the time he was thirteen both his masters and fellow students began to regard him as "queer." Ultimately young Chapman suffered a nervous disturbance, the first of several which would occur periodically throughout his life. His health collapsed. When he fell ill with pneumonia, Jack's parents had been forced to remove him from school and place him under private tutors to prepare his entry into Harvard.

While at Harvard, John Jay Chapman would not rank high scholastically but did at times demonstrate an unquestioned brilliance. He was popular with his fellow classmates and became a rebel of sorts.

Legionnaire *Victor Chapman standing beside his father, John Jay Chapman. His stepmother "Alce" Chanler Chapman is seated beside an unidentified woman. Amiens, France, November 1914.*

It was at Harvard that the faith of his youth completely collapsed, with Chapman even refusing to attend chapel. He had his degree withheld because of this. Ultimately, his B.A. was granted in 1884, and his M.A. in 1885. And following a year of travel in Europe, he returned to Boston to attend the Harvard Law School.

While in Boston to begin his course of study, Chapman had met and had fallen in love with the beautiful, fiery Minna Timmins. He pursued her with great vigor and determination, the beginning of a passionate courtship which would tragically witness Chapman's withdrawal from Law School in January of 1887.

One evening John Jay Chapman had escorted Minna to a social function at the Walter Cabot home in Brookline. He momentarily left Minna unattended and when he had returned to her side he had witnessed one Percival Lowell, a Boston bachelor, hovering over her and had misinterpreted the fellow's actions as being injurious to Minna. In a jealous rage, Chapman had immediately attacked Lowell with his walking stick and had severely bloodied him.

Chapman soon learned that Lowell was a member of Minna's dramatic club and had only spoken to her out of courtesy. Upon the realization of his error, Jack was horrified at his actions and was overwhelmed with remorse. Next morning, following a night of brooding, he had stoked the coal fire in his Cambridge apartment until it was burning brightly and as an act of expiation he had thrust his offending left hand "deep in the blaze" until the knuckles and finger bones were exposed and charred. Then he took a horsecar to the Massachusetts General Hospital where he was immediately anesthetized and his left hand was amputated.

John Jay Chapman was not granted a Harvard law degree. Through diligent study he was admitted to the New York bar in 1888. Then on 2 July 1889, following a stormy courtship, Chapman and Minna Timmins were married. Nine months later Victor was born.

From the moment of Victor's birth, mother and child became inseparable, as Victor grew to become both "idol and slave" to Minna and took from her for life her "old world Catholicism." His father would later write of Victor:

"Victor . . . while he had not the intellect of his mother and was an inchoate nature, there was from his infancy to his death something about him of silence, mystery, godhead.

"He continued to the end of his life to make the sign of the cross in saying the same prayers that she had taught him—which ended with the phrase— "and make me a big soldier of Jesus Christ who is the Lord and Light of the world." He folded his hands like a crusader as he said them. He was a part of the middle ages in this piety. His tiny trench-bible, which was full of pressed flowers and kodaks of his friends, was so much a miniature copy of his mother's bible that the little book seemed like the baby of the big one."

A second son, John Jay, was born to the Chapmans two years after Victor's birth. Five years later the birth of Conrad in New York tragically resulted in Minna's death in childbirth on 25 January 1897.

Following his acceptance to the Massachusetts bar in 1888, Chapman had practiced law for ten years and had involved himself in politics, writing two expositions on the dangers of the alliance between politics and business in the United States. He became a leader in the "good-government" movement and railed against the private citizen who supported the successors of William Tweed and New York's Tammany Hall. He formed the People's Municipal League in an unsuccessful attempt to defeat the Tammany candidate in the New York mayoralty race.

Chapman had never been a success at the practice of law. During his ten years at law he received no money, taking all cases without a fee. Chapman abandoned the legal profession shortly after the death of Minna. And upon the critical acceptance of his first book, *Emerson and Other Essays,* he worked toward building his growing reputation as one of the finest writers in the English language in the United States.

John Jay Chapman remained a widower approximately fifteen months before marrying Elizabeth Chanler on 23 April 1898. Chanler Astor Chapman, Victor's half-brother, was born of this union on 27 April 1901. Elizabeth Chanler had been the daughter of lawyer and congressman John Winthrop Chanler and Margaret Astor Ward Chanler. This great-granddaughter of John Jacob Astor had become one of the famous "Astor orphans" upon the death of her parents at a young age and had inherited one million dollars.

A political crusade Chapman had been waging in New York culminated in a severe dispute with Theodore Roosevelt. As in the past, the strain proved too much for Chapman's highly-wrought temperament and in May of 1901, one month after the birth of Chanler, he suffered a complete mental breakdown.

For two years Elizabeth Chapman nursed Jack at "Rokeby," the Chanler country estate at Barrytown, on the Hudson. When Chapman recovered his health and left his seclusion some two years later, he was transformed both physically and spiritually. Now wearing a beard, with the visage of a "fiery sage." he eschewed his earlier life of politics for a pursuit of spiritual truth.

Victor's loss of Minna at such a young age would produce in him the same "early maturity" it had brought to Minna with the sudden loss of her father. At the same time it created in him a friend to the suffering. For the short few years that Victor had known his mother, she would always remain a profound influence upon him. And for some time following her death, he would immerse himself in a brooding sorrow.

Victor soon developed a melancholy nature and a "fathomless humility." His father also noted that "There was a reminiscence of his mother's nature in Victor's friendship. He was always the leader, both leaning on and sweeping forward some subordinate nature who adored and followed."

Victor was at first resentful of his stepmother "Alce," but he eventually saw in her a kind and generous woman, and a sincere warmth developed between the two. However, Elizabeth Chapman could never fill the void in Victor's heart at the loss of Minna.

But with Victor so much his mother's child, a tension would always exist between himself and his father. John Jay Chapman later summed up their relationship in one sentence:

"Victor always regarded me with piety; but as for being nourished and fed by my ministrations, it was out of the question."

At times a simmering hatred that Victor held for his father would unexpectedly manifest itself. When it did the senior Chapman felt "as if I and the whole universe I lived in were intolerable to his soul."

Following the death of his mother, Minna, Victor turned his true affections toward his younger brother, Jay. A powerful bond developed between the two which ended tragically near Gratz in Austria when Victor was twelve and Jay was ten. The two had been playing together in the woods, Victor's favorite haunt, when Victor had momentarily left his brother beside a swollen river. He returned only in time to witness Jay's fall into the torrent where he soon drowned with Victor helpless to assist him.

Victor took the blame for his brother's death upon his young shoulders. In school he was dull and uninspired and only seemed to come alive in natural settings among woods and streams. Only danger, or the threat of it, brought him completely to life, and he constantly put himself in harm's way.

Unable to act to save his younger brother, Victor seemed to atone for this failure by participating in a series of life-saving episodes, a pattern he would follow unto death.

Following the death of Jay, Victor focused his attention on his youngest brother Conrad who worshipped his oldest brother. Of Victor, he would later write:

"Although I was seven years younger, I was very close to him, with his fine figure and gallant easy walk. He seemed not to know fear, and tried to teach others to overcome theirs. Once in Maine, when I was afraid of the icy-cold sea water, he got me out of bed early, pulled off my pajamas, and dropped me into the sea. Then hastily he pulled me out and rushed me to our kitchen, to give me a cup of hot milk."

Young Victor began his formal education at the Fay School, Southboro, Massachusetts. He spent two years there before enrolling in St. Paul's School, Concord, New Hampshire in September, 1903. He left that institution in 1907 to finish his schooling in Germany.

Chapman lived in Germany and for a time in France before returning to the United States to spend one year at the Stone School, Boston. In September 1909 he entered Harvard University and graduated from there with an A.B.in 1913.

Victor's father, John Jay Chapman, summed up this part of Victor's life in the following way:

"He had no aptitude for sports, none for books, none for music; but always a deep passion for color and scenery, and a real talent for all forms of decoration, which we hoped would lead him toward painting or architecture. His water-color sketches, done in 1913—14 in Paris, showed a great advance on earlier work; but the dreamer was still in his dream, —and art is concentration. His pleasure was in scenery. If you could place him in a position of danger and let him watch scenery, he was in heaven. I do not think he was ever completely happy in his life till the day he got his flying papers."

Chapman demonstrated a talent toward painting and architecture. And he sailed to Paris immediately following graduation to study under M. Gromort to prepare for admission to the Beaux Arts.

Now at the age of twenty-four, Victor's character had been formed. Chaste, without passion, clinging to his mother's Roman Catholic faith, he drifted aimlessly through life like a somnolent knight-errant groping for purpose and adventure. In August of 1914 the world would provide him with both. Then the slumbering young warrior rose suddenly to life.

Victor was in England where his parents had gone when the French army had been mobilized to fight Germany. When Victor had presented his father with his intentions to enlist for France, the senior Chapman had vehemently disapproved. But through the intervention of his step-mother, Alce, Victor was shortly on his way to Paris and by late September 1914 had enlisted as a private in the Third Marching Regiment of the First Foreign Regiment of the Foreign Legion.

Chapman involved himself heartily in this experience which he initially viewed as a "continuous picnic." And with his artist's eye, he saw his forthcoming service as "a splendid chance to view the country under all conditions."

Victor reached the trenches in the third week of December 1914 as a *mitrailleur* and was soon wounded in the right arm by a rifle bullet. However, he still maintained a youthful enthusiasm in spite of a new reality involving filth and dysentery.

In a letter to his father dated 10 February 1915, he described how he was "generally living a happy-go-lucky, hand-to-mouth existence." Then in an effort to put the senior Chapman's mind at ease, Victor went to extremes to downplay his experiences under enemy fire.

"The tap, tap of Boches' bullets on the face of my abri in the evening affects me about as much as the lap, lap of little waves against the side of my sail boat. The rumble of distant artillery passes unnoticed, and but mild curiosity is aroused by the chug-chug-chug of a machine gun as of a steam-motor boat rounding a bend."

In reading between the lines, the reality of his experience appeared in dark contrast. And while no attacks were launched by either side, Chapman's battalion was continually under fire and suffered numerous killed and wounded from barrages and enemy sniping.

For nearly three months Victor's battalion remained in the forward trench without *repos*. This established the record for all French units at the time. While there, he attempted to save the lives of several of his comrades lying wounded in no-mans land; and readily used his parent's wealth to provide the basic necessities for his comrades. He was soon affectionately known by them as "the American millionaire," particularly after he offered a battle surgeon 100,000 francs to save the life of a mortally wounded friend.

In a letter dated 17 March 1915 to his brother, Conrad, penned in the trenches at Frise, Victor momentarily let down his guard and wrote that "one has to "take them (the horrors) lightly for otherwise the life would be an unbearable nightmare."

Following a brief *repos*, Victor's unit was sent back to the trenches near Bas. The sector was quiet, with the enemy lines one-half mile off. On 4 July 1915 Chapman wrote to his step-mother, uncharacteristically complaining that he had served his term and had wasted ten months of his life neither helping the French nor injuring the Germans.

Some months earlier, while in the trenches at Amiens, at the suggestion of his father and uncle, Victor had written an inquiry to fellow American Norman Prince when he had learned that Prince was "getting up an American Aero Corps at Pau." He later wrote Mrs. Chapman that he wished he had not done so after reading a reply from another American, Elliot Cowdin, who was then assisting Prince.

"As for Aviation, I was too irresolute and inarticulate at Amiens to give vent to my feelings, and with the bunting that popular heroes are made of lying so thick, I did not like to show my true self. I have read Cowdin's letter; but it is perfectly obvious that I am not wanted and have been foisted on them by Uncle Willy and Papa. This is Prince's and Cowdin's show, they got it up. It is not for Americans in general. If I had been hauled out of the Legion in February I might have been a charter member of the Club, but I would not think of joining now."

Several days following his letter to Mrs. Chapman, Victor was sent to Montbeliard with his battalion to be absorbed by the decimated *2me de Marche*. On 1 August 1915 the news reached him as a surprise that he was to be transferred to French aviation. This had been secured through the influence and intervention of his father, John Jay Chapman, and with the assistance of Alce's two brothers then living in Paris, William Astor Chanler and Robert Chanler.

Victor concluded his eleven months of service in the Legion having seen one-half of his unit either killed or seriously wounded. Chapman's courage and cheerfulness had won him the admiration of both officers and enlisted men, and he departed their company "with many adieux and five fellows helping me on with my sack."

Victor reached Squadron *V.B.108* at Nancy on 8 August where he was assigned as a *mitrailleur-bombardier*. Americans Elliot Cowdin and Norman Prince were serving there as pilots. However, both men were in Paris at the time of Chapman's arrival lobbying their effort to form an American Escadrille.

On 20 August 1915 Victor flew on his first bombing run in a Voisin. On 24 August he flew a second mission, this time across the Rhine into Dilingen, Germany where he bombed a railyard.

On 9 September Norman Prince returned to the Squadron and told Victor that an American Escadrille should become a reality within a month. Meanwhile

The Escadrille Americaine *at Luxeuil 14 May 1916. Left to right: Kiffin Rockwell,* Captaine *Thenault, Norman Prince, Lt. deLaage, Elliot Cowdin, Bert Hall, James McConnell, Victor Chapman.*

Victor made application for aviation school and was accepted for pilot training through the aid of Prince on 22 September 1915.

Victor entered the School of Military Aviation at Avord on 26 September and initially trained on the Morane single-seater. At Avord he met the former American Legionnaire, Kiffin Rockwell, who would become Victor's closest friend. He received his *brevet militaire* on the Maurice Farman on 9 January 1916. Chapman continued his flight training at Avord, with fellow American Bert Hall serving as one of his instructors, finishing up at the R.G.A.(*Reserve General Aeronautique*) on 17 April 1916.

With uncle William Astor Chanler becoming one of the founding members of the Franco-American Flying Corps and with his other uncle Robert Chanler using his financial and political influence with the French government to aid Norman Prince, it became a certainty that Victor would be assigned to the newly-formed *Escadrille Americaine*, *N.124*. And so he was, at Luxeuil, on 20 April 1916.

As Victor's father later noted, his entry into the American *Escadrille* was the consummate moment in Victor's life. He was suddenly transformed into the knight he had sought to become, completely fulfilled and happy. On 13 May all that he had hoped for reached fruition when he flew as part of the Squadron's first patrol across enemy lines.

By the evening of 20 May 1916 the Squadron had completed its relocation to Behonne in the Verdun sector. A morning patrol on the 22nd initiated combat operations over German lines, part of a launched French counteroffensive. The pilots were to protect allied observation and artillery reglage machines.

On the 24th Victor Chapman flew as part of a patrol consisting of Kiffin Rockwell and William Thaw, and which was led by the Squadron's French commander, *Capitaine* Thenault. Thereupon, he began building his reputation as an intrepid fighter who attacked German aircraft whenever and wherever he found them, regardless of the circumstances.

On a "scouting expedition" along the length of the Verdun Sector, near Etien, the formation ran over the top of twelve German biplaces flying above their own lines. Thenault had hoped to avoid confrontation, feeling the enemy's numbers too great to provoke, when

Chapman broke out of the formation and dived to the attack. He fired upon and struck a Fokker which was seen to fall out of control. Soon a fierce combat raged from a height of 4000 meters to near ground level. The Americans, outnumbered three-to-one, fought heroically in spite of the odds. But the outcome of the action clearly favored the superior enemy forces as first Rockwell, then Chapman, and finally Thaw were each forced out of the fight from wounds when each man's aircraft was repeatedly struck by fire and heavily damaged.

Thaw's wound proved the most serious, a bullet in the left elbow which required hospitalization in Paris. For their actions that day, all three Americans were subsequently cited and decorated, with Chapman and Rockwell receiving promotion to *Sergent*.

Missions continued at a frantic pace with Chapman and Rockwell soon back in the air. On 17 June Chapman, in the company of several other American pilots, was patrolling along the right bank of the Meuse river when he spotted a formation of German aircraft on the opposite bank.

Chapman's decision to combat the enemy whenever and wherever he found them became an ungovernable passion. Without hesitation he disobeyed *Capitaine* Thenault's explicit orders not to cross the Meuse and dove to the attack, dragging the rest of the patrol into the maelstrom.

Engagements were fought without decision with Chapman's colleagues withdrawing singly from the fight. All returned to Behonne save Victor who refueled at Vadelaincourt then returned to the lines to attack two Aviatiks under the protection of three undetected Fokker fighters.

Chapman sent one of the enemy buses earthward in flames before the other pilots realized the attack was being pressed by a single Nieuport. Initially stunned by the boldness of the American's solo attack, the Germans quickly regrouped then swooped in on Chapman from all directions, raking his fuselage with intense fire.

Refusing to disengage from the fray, Chapman tipped back his Lewis machine gun to remove the empty drum and was fitting a full magazine onto the weapon when his Nieuport was again struck by enemy fire. One bullet severed his right aileron control rod then ricocheted into the cockpit where it sliced through Victor's helmet, deeply creasing his skull.

Bloodied, dazed, his machine gun empty, Chapman decided he had had enough combat that day. Feigning death, he allowed his Nieuport to spin earthward. Apparently the ruse worked, for the Germans abandoned him. Once out of harm's way, he gathered up the severed ends of the control rod in his powerful fists and reestablished control of his fighter by manipulating the joystick between his knees.

Victor Chapman being greeted by Norman Prince upon his return to the Behonne airfield on 17 June 1916 following aerial combat over Verdun which saw him wounded in the head.

Soon the Froidos airfield of *Escadrille* 67 appeared beneath him. He settled down for equipment repairs and to bandage his bleeding skull. Following lunch, he returned to Behonne, much to his friends' relief. Kiffin Rockwell, who had begun the patrol with Chapman, wrote a letter that same day to his brother, Paul, describing Victor's recklessness.

"Chapman has been a little too courageous and got me into one of the mess-ups because I couldn't stand back and see him get it alone. He was attacking all the time, without paying much attention. He did the same thing this morning and wouldn't come home when the rest of us did. The result was that he attacked one German, when a Fokker which we think was Boelke (the papers say he was killed but we don't believe it), got full on Chapman's back, shot his machine to pieces and wounded Chapman in the head."

Capitaine Thenault and Victor's American Squadron mates pled with him to go to hospital and seek treatment for his wound. When he refused, and insisted he remain active on the roster, they next suggested he take a short rest leave to Paris. When he rejected both suggestions, *Capitaine* Thenault bribed him by promising him a new and better machine, suggesting that he should at least rest until it arrived. Victor reluctantly acquiesced and for the next few days he hung around the aerodrome seeing to the readiness of his new fighter.

On 18 June 1916 *N.124* pilot Horace Clyde Balsley became the first American to be seriously wounded in aerial combat when he fell under attack from four enemy fighters while battling a German two-place east of Verdun. The bullet that smashed his right pelvic bone also perforated his intestines, and upon his rescue he was taken to the Vadelaincourt evacuation hospital.

Because of the severity of his intestinal wounds, Balsley's doctors restricted his liquid intake to only what he could suck from a wet bandage. He soon became dehydrated from a burning fever brought on by infection. His thirst grew unbearable. As he lay in bed, racked with fever, Victor Chapman's powerful figure loomed at his bedside. And somehow he knew that his friend "whom every one in our squad loved deeply" had come to his aid.

Chapman, upon learning that Balsley could not drink water, suggested to Balsley's doctor that he be allowed to suck the juice of oranges. The doctor agreed, but said none could be provided in the village. Undaunted, Chapman promised to return with oranges if he had to fly to Paris to fetch them.

What was thought impossible Chapman quickly accomplished, returning shortly to his suffering friend with a bag of oranges. While at the Front with the Legion he had accomplished a similar miracle when he had provided a cow in a few short hours for a *mitrailleur* in dire need of milk for his ulcerated stomach.

For successive days Chapman appeared at Balsley's side with a fresh bag of oranges. But on 23 June Victor failed to appear.

Elliot Cowdin had come in his place and had told Balsley Chapman's plane had broken down.

In reality, Victor had been in the air preparing to fly two bags of oranges to Balsley when he had seen Norman Prince, Raoul Lufbery, and *Capitaine* Thenault depart on patrol. Unable to avoid any opportunity which might provide an engagement with the enemy, Chapman had decided to join the patrol despite the fact he was not a participant, intending to visit Balsley upon its completion. Northeast of Douaumont in the Verdun Sector the regular patrol had run into five German aircraft. Following a brief combat, they retreated to French lines.

Unknown to his Squadron mates, at the onset of the combat Chapman had joined in, apparently feeling his friends were in jeopardy. When they had withdrawn without ever having seen Victor, Chapman, never willing to retreat, had stayed on to carry the fight to the enemy.

His fate had not been learned until some hours later when a French crew flying a Maurice Farman artillery regulating machine had returned to their Squadron late in the afternoon and had telephoned Thenault to report having witnessed from a distance a lone Nieuport attacking an enemy fighter. The brave pilot had been jumped by three other fighters but would not withdraw in spite of the odds until his plane was severely damaged and out of control.

As the Nieuport streaked earthward, it passed close to the French crew who could see Victor's hunched body, his head slumped over the side of the cockpit. Soon the Nieuport fell apart in the air then crashed to earth six kilometers inside enemy lines.

Back at Behonne, pilots and mechanics stood on the tarmac in disbelief still awaiting the return of their beloved Victor. They all remembered Bert Hall having said that Victor was always the last to return from patrols, usually "with his machine all shot to hell-struts and wires and braces cut off." But the settling darkness dashed their hope and pierced their hearts with fear. One by one each pilot left the tarmac until only Kiffin Rockwell remained. Then he too turned to go. He later would write of that time:

"As I left the field, I caught sight of Victor's mechanicien leaning against the end of our hangar. He was looking northward into the sky where his patron had vanished, and his face was very sad."

Although his body was not recovered, a funeral service for *Sergent* Victor Chapman was held in the American Church at Paris on 4 July 1916 as a part of the American Independence Day service. He was twenty-seven years old at the time of his death, the first American aviator to be killed in World War One. That same day Aristides Briand, France's Prime Minister, referred to Victor as "the living symbol of American idealism."

While the death of Victor dealt a crushing blow to all the members of the Squadron, most keenly, it tore at Kiffin Rockwell's heart, Victor's closest friend from their flight-school days and throughout their tenure with the *Escadrille*. The day after Victor's death, Kiffin wrote the following tribute to Victor in a letter to a friend in Paris.

"Victor had about the strongest character of any boy I have ever known. He was very frank, honest and never had let anything kill his ideals and had always lived a very clean life, never doing any harm, or believing wrong of anyone and giving out a lot of good to the people around him. We used to kid him a lot and tell him that a lot of his ideals were very foolish and all, yet it didn't affect him and he still believed these old ideals and died a glorious death. As we all have to die sometime, it isn't so bad, yet by living he could have accomplished a great deal in his life. Last night I went to bed, but I couldn't sleep thinking about him, especially as his bed was right beside mine."

Kiffin penned this tribute to Victor in a 23 June 1916 letter to his brother, Paul: "There is no question but that Victor had more courage than all the rest of us put together. We were all afraid that he would be killed, and I rooming with him had begged him every night to be more prudent. He would fight every Boche he saw, no matter where or what odds, and I am sure that he had wounded if not killed several. I have seen him twice right on top of a German, shooting hell out of him, but it was always in their lines and there being so much fighting here it is impossible to tell always when you bring down a machine. His head wound was not healed, yet he insisted on flying anyway, and wouldn't take a rest. The first time he was ever in an aeroplane he went as a passenger clear to Dillingham and dropped a bomb on the station there."

Following the war what was initially believed to be Victor Chapman's remains were exhumed near where he had fallen. However, the dental chart of the skull proved otherwise. Still, the remains were reinterred in the American Cemetery at Suresness under Chapman's name. Consequently, the remains in that grave at Suresness were not later removed to the *Lafayette Escadrille* Memorial near Paris. The crypt bearing his name today remains empty. For his war service, Chapman was decorated with France's *Medaille Militaire* and *Croix de Guerre*, with two Palms.

LOUIS CHARTON

Louis Charton was born in France of French parents. He later became a naturalized American citizen and lived in New York City. At the outbreak of World War One he was temporarily living in Toul, and he and his brother, John, enlisted in the French Foreign Legion in Paris in August of 1914.

On 25 November 1914 Charton entered the trenches with his brother near Craonne with the Second Regiment of the Foreign Legion. On 4 July 1916 John Charton was killed at Verdun. Louis was severely wounded on 10 July 1916 and left the Legion on 1 February 1917 to join the *Service Aeronautique*. He was awarded France's *Croix de Guerre*, with Star for his remarkable courage.

From 28 February to 20 August 1917, *Sergent* Charton attended the aviation schools at Chartres, Avord, Pau, and the *G.D.E.* He received his *brevet* on the Farman on 14 May 1917.

On 22 August 1917 Charton reached the Front with *SPAD 92* and flew with that squadron until 5 September 1917, the day he was shot down by anti-aircraft fire over enemy territory in the Verdun sector. Following his capture by the enemy he was interned in Switzerland for one year because of ill health. In spite of his written pleas to be released to France so that he might continue fighting the Germans, he did not regain his freedom until the November Armistice.

HERMAN CHATKOFF

Herman Lincoln Chatkoff was born on 12 February 1890 in New York City, New York, the son of Benjamin and Sarah Chatkoff. He attended New York City public schools, and for short time attended the Cooper Union School. After leaving school he found employment with several engineering and technical firms.

Having quarreled with his family. Chatkoff sailed for France and was working in a Paris garage washing automobiles at the outbreak of World War One. On 24 August 1914 he enlisted in the French Foreign Legion. He served in the trenches until 20 May 1916. On 24 May he enlisted in the *Service Aeronautique*. From 5 June 1916 to 20 April 1917 he attended the aviation schools at Buc, Chartres, Chateauroux, and the *G.D.E.* He received his *brevet* on the Caudron in September of 1916.

While awaiting assignment at the Front, *Sergent* Chatkoff requested his return to the trenches where he spent two more months in the front lines. On 25 April 1917 he was assigned to the Front with *Escadrille C.11* and flew with that squadron until 15 June 1917. He won France's *Croix de Guerre*, with two Stars for his courage in the air.

On 15 June 1917 *Sergent* Chatkoff had flown to Chaudun to visit his American friends of the *Escadrille Lafayette*. In the rear cockpit of his Caudron two-seater was Benjamin Woodworth, a Section One section chief of the American Field Service who had been serving in France since 1915.

When Chatkoff was about three-hundred feet over the airfield he began stunting the heavy plane to impress Woodworth and his friends on the ground. Chatkoff lost control of the Caudron and it hit the ground from an altitude of two-hundred feet and was smashed to pieces. Woodworth was killed outright while Chatkoff suffered many broken bones and a severe head injury, *Sergent Chatkoff* spent the next seven weeks in a coma. He came out of it with his mind permanently impaired.

Chatkoff was discharged in Paris on 31 January 1919. He paid his way back to the U.S. and entered school in Boston but could not continue because of his mental state. Constantly nervous and restless, Chatkoff began travelling around the United States working as a farm laborer. He was often found to be in a destitute condition and barely survived on a small pension which had been granted him by the Republic of France. Chatkoff could not qualify for U.S. Veteran's benefits since his service had been under the French flag. When the American Legion learned of his condition, they aided in the presentation of a special bill in the 71st Congress, "An Act For the relief of Herman Lincoln Chatkoff." The bill, which was approved on 5 February 1931, granted Chatkoff hospital treatment and compensation. After the bill had gone into effect Chatkoff spent many years in and out of Veterans Hospitals. He died at the Veterans Administration Hospital at Perry Point, Maryland on 10 April 1964 at age 75.

ROGER CLAPP

Roger Harvey Clapp was born in the United States, the son of Mr. & Mrs. Knight L. Clapp. He attended De Witte Clinton High School and later engaged in mining out West.

On 3 June 1917 Clapp enlisted in France's *Service Aeronautique*. From 13 June to 12 January 1918 he attended the aviation schools at Avord and the *G.D.E.* He received his *brevet* on the Caudron on 16 September 1917.

From 15 January to 28 February 1918 *Caporal* "Gonnie" Clapp was assigned to the Front to *Escadrille BR. 120*. While flying missions with that squadron, he proved to be a pilot of "absolute fearlessness." On 14 March 1918 he was commissioned a First Lieutenant in U.S. Aviation. Fellow Lafayette Comrade Henry Forster later recorded the following impressions of Clapp.

"There was a fellow with us from Waco, Texas—Roger Clapp by name—large, not attractive, but got along reasonably well. He was a great borrower. He may have overspent whatever he had. He kept careful account of it in a little brown notebook. I remember one day we were talking about Stuart Walcott, a rather religious sort of fellow who had been shot down almost as soon as he had reached the front. While we were talking, Roger Clapp came up.

"'What,' he said, 'Stu Walcott killed?' He then put his hand into his pocket, pulled out the little brown book, with lots of names in it, took out a pencil and drew a line through the name. That was certainly to him one way of paying the debt."

From 30 March to 15 May 1918 Lieutenant Clapp continued to fly with *BR. 120*. On 6 June he was assigned to the 96th Bombardment Squadron, USAS. He was the designer of the squadron's insignia.

Clapp flew with that squadron until 6 July 1918, the day he was killed while testing a new plane at Amaty. With Sergeant Dunn in the rear cockpit, Clapp was preparing to land when the machine stalled, side-slipped and crashed onto the airfield. It burst to flame as Clapp's comrades stood helplessly by, watching in horror as Clapp and Dunn were consumed in "the roaring crackling mass of wood and canvas."

CALEB COATSWORTH

Caleb James Coatsworth, Jr. was born in Buffalo, New York on 12 January 1890, the son of Mr. & Mrs. Caleb James Coatsworth. He attended Buffalo public schools prior to enrolling in St. John's College, Annapolis Maryland for six years. He attended the U.S. Naval Academy for 1 year. In 1916 the Coatsworth's family purchased and furnished an ambulance for the Norton-Harjes Ambulance Service. That same year Coatsworth sailed for France to drive the ambulance and served with Section 7.

On 20 February 1917 Coatsworth left the ambulance service for France's *Service Aeronautique*. From 24 February to 16 July 1917 he attended the aviation schools at Avord, Pau, and the *G.D.E.* He received his *brevet* on the Caudron on 25 April 1917. From 18 July 1917 to 20 March 1918 *Sergent* Coatsworth served at the Front with *Escadrille SPAD 80*. On 16 August 1917 he was slightly wounded in combat.

On 20 March 1918 Coatsworth enrolled in the U.S. Naval Reserve Flying Corps as an Ensign at Paris, France. He took his ground and flight training in France and became Naval Aviator (HTA)#1005 at Moutchic on 19 July 1918. He served as a combat pilot at Pauillac and Porto Corsini, Italy. For his World War One Service, Coatsworth was awarded France's *Croix de Guerre*, with Star (Ambulance service) and the Italian War Cross.

Following the Armistice Lieutenant Coatsworth returned to the United States. He chose to remain in the Navy and served at Bay Shore, New York, and at Chatham, Massachusetts aboard the USS *Shawmut*. While at Pensacola, Florida he qualified as Naval Aviator(LTA) on 12 January 1921. He next served aboard the USS *Texas* at San Diego, CA. He became a tester of naval aircraft and served with the Air Squadrons Battle Fleet, Pearl Harbor; aboard the USSs *Tennessee* and *Mississippi* at San Diego; the USS *Argonne* at Coco Solo; and aboard the USS *Vestal*. While he was stationed in California he met and married Olga Randolph Lowell. From this union two sons, James and Henry, were born.

Lieutenant-Commander Coatsworth died of a heart attack on 25 September 1939 near Avalon, California while aboard the yawl *Aquilla* off Santa Catalina Island. At the time of his death, at age 49, Lieutenant-Commander Coatsworth was on the staff of the repair ship, *Vestal*.

EDWARD COLLIER

Edward M. Collier was born in Bass Rocks, Massachusetts, on 27 March 1895, the son of Mrs. Margrete Collier. He attended Phillips Exeter Academy, Exeter, New Hampshire, Class of 1907.

Collier enlisted in France's *Service Aeronautique* on 14 October 1916. On 25 October 1917 he was released because of ill health. Following World War One he lived in Paris.

PHELPS COLLINS

Phelps "Eddie" Collins was born in Alpena, Michigan in 1894, the son of Richard H. Collins, manufacturer of the Alpena Flyer automobile, and Charlotte (Phelps) Collins. Mrs. Collins died in 1900 and Phelp's father remarried in 1902. Phelps Collins attended Alpena schools, and the University of Idaho at Moscow, Idaho. He left that institution to work as a builder and constructor.

On 17 May 1917 Collins enlisted in France's *Service Aeronautique* in Paris. From 22 May to 1 September 1917 he attended the aviation schools at Avord, Pau, and the *G.D.E.* He received his *brevet* on the Caudron on 28 July 1917.

From 2 September to 18 September 1917 *Sergent* Collins served at the Front with *Escadrille SPAD 313*. From 19 September 1917 to 7 January 1918 he served with *Escadrille SPAD 103*. From 7 January to 18 February 1918 he flew with *Escadrille SPAD 124*.

On 9 January 1918 Collins was commissioned a Captain in the U.S. Air Service Signal Reserve Corps. From 18 February to 12 March 1918 Captain Collins served with the 103 Aero Squadron, USAS. A tireless pilot, it is believed that Collins's extreme devotion to duty resulted in his death on 12 March while patrolling near Chateau-Thierry. Fatigue evidently overtook him. He fainted in the air and crashed to his death while out on patrol eight kilometers south of Charly-sur-Marne. He was buried at the Montfrenet Cemetery near Cuperly, Marne. His remains were later moved to the Meuse-Argonne American Cemetery, Romagne-sous-Montfaucen, Meuse. Collins was posthumously awarded France's *Croix de Guerre*, with two Palms for his outstanding war service. He was credited with the official destruction of one enemy aircraft.

In 1928 Collins's remains were interred in the Lafayette Flying Corps Memorial near Paris. On 27 August 1931 Michigan's first state airport was opened at Alpena. It was named Phelps Collins Airport in honor of Captain Collins.

JAMES CONNELLY

James Alexander Connelly, Jr. was born in Philadelphia, Pennsylvania on 15 June 1894, the son of Mr. James A. Connelly, a textile merchant, and Minnie (Riddle) Connelly. Connelly attended Philadelphia public schools before entering the Episcopal Academy at Merion on 5 April 1906. He left that institution in 1912 to enter into business.

Adjudant James Connelly (rt.) with Capitaine Georges Madon before Madon's SPAD fighter.

In the spring of 1917 Connelly left a banking position in Philadelphia and sailed for France to join the Lafayette Flying Corps. On 15 June 1917 he enlisted in France's *Service Aeronautique*. From 20 June to 12 January 1918 he attended the aviation schools at Avord, Pau, Cazeaux, and the *G.D.E.* He received his *brevet* No. 9711 on the Caudron on 3 November 1917.

From 15 January to 27 June 1918 *Sergent* Connelly flew at the Front with *Escadrille SPAD 157*. While with that squadron he was credited with the destruction of an enemy balloon and an aircraft. From 27 June until the 11 November Armistice *Adjudant* Connelly flew with *SPAD 163*. Connelly was a brave and heroic pilot. During his service in the Lafayette Flying Corps he was credited with the destruction of two balloons and five enemy aircraft. For his splendid service, *Adjudant* Connelly was awarded France's *Medaille Militaire*, and *Croix de Guerre*, with six palms. He was also awarded the U.S. Distinguished Service Cross for extraordinary heroism for his heroic action near Suippes, France on 6 September 1918. His DSC citation states: "An American pilot serving with the French Army, Sergeant-pilot Connelly attacked a formation of 12 enemy planes (type Fokker), shooting down the flight commander and forcing the remainder to seek safety. He continued with the unequal combat until his ammunition was exhausted."

Following World War One Connelly founded the coal company of James A. Connelly, Jr. He married Elizabeth Atterbury, daughter of W.W. Atterbury, president of the Pennsylvania Railroad. From this union two sons were born: James Alexander Connelly, III, and W. Wallace Atterbury Connelly. Eventually the Connellys divorced and in 1935 James married Pauline Arey, the former wife of Joseph Ricker Collins, also a World War One flyer. James Connelly died after an extended illness in a Bryn Mawr, Pennsylvania hospital on 1 February 1944 at age 49.

ALAN COOK

Alan Augustus Cook was born in Rochester, New York on 17 August 1896, the son of Dr. Robert G. Cook, a neurologist and physician-in-charge, Brigham Hall Hospital, Canandaigua, N.Y., and Mary Belle (Strong) Cook. Cook graduated from Phillips Academy, Andover, Massachusetts, class of 1914, and Harvard University, A.B. degree, 1918, where he majored in chemistry under the auspices of James Bryant Conant, future president of Harvard, with whom he published chemical discoveries. Desirous of entering the European war, Alan Cook left Harvard at the end of the 1917 spring term and sailed for France on 30 June 1917. On 21 July he enlisted in France's *Service Aeronautique*. From 31 July to 16 December 1917 he attended the aviation schools at Avord, Tours, Pau, and the *G.D.E.* Cook received his *brevet* on the Caudron on 20 October 1917.

From 20 December 1917 to 20 July 1918 *Caporal* Cook served at the Front with *Escadrille SPAD 157*. From 20 July to the November 1918 Armistice he flew with *Escadrille SPAD 163*. Cook attained the rank of *Adjudant* and was credited with the 11 August 1918 destruction of one enemy aircraft. He was awarded France's *Croix de Guerre*, with Palm for his courageous victory in "a very severe fight." After the Armistice, he was discharged from the French Army and returned to the United States in February, 1919.

After completing his degree requirements at Harvard, he began his business career in 1919 in Rochester's optical industry. He married Katharine Elmendorf Shove of Canandaigua, N.Y. on 6 October 1919, and their union resulted in one daughter and two sons. His flight career inspired one son to serve as a WW II 8th AF B-17 pilot, a grandson to take flying lessons, and a great-grandson to currently undergo pilot training.

While working at Wollensak Optical Co. of Rochester, Cook completed courses in mathematics and optics sufficient to qualify him as an optical engineer. He moved to Eastman Kodak Co. as a production supervisor from 1926 through 1929, but then elected to return to a career designing lenses, shutters and other equipment for the Bausch & Lomb Optical Co. In 1950 he rejoined Wollensak Optical Co., as head of their optics engineering department until his retirement in the 1960s. His work became respected by many notables not only in the field of optics but by such legends as Edward Land, of Polaroid fame, Prof. Link whose firm produced the Link Trainer, Colonel Goddard, America's rocket pioneer and many Hollywood movie directors.

Alan A. Cook died in Rochester, New York on 17 April 1973 at age 76.

LINN COOKSON

Linn Palmer Cookson was born in Carlinville, Illinois on 11 July 1896, the son of Percy Cookson, a dry-good merchant, and Stella (Palmer)Cookson. Linn Cookson attended the Carlinville High School and the University of Illinois at Urbana, Illinois. He was student at that institution when he sailed for France to join the Lafayette Flying Corps.

On 19 June 1917 "Doc" Cookson enlisted in the *Service Aeronautique*. From 27 June 1917 to February 1918 he attended the aviation schools at Avord, Tours, Pau, and the *G.D.E.* He received his *brevet* on the Caudron on 27 October 1917.

Caporal Cookson's service in French Aviation was marked by a series of illnesses. On 26 January 1918 he was commissioned a Second Lieutenant in the U.S. Air Service, but sickness again delayed his arrival at the Front. He served briefly as the engineering officer for the 94th Aero Squadron, USAS, until he was invalided home to the United States on 1 June 1918. He returned to Carlinville, Illinois and shortly thereafter had an operation for appendicitis. Upon his release from the hospital he began his recuperation at the home of his parents. Complications from his operation set in, and he died there on 16 September 1918 at age 24. He is buried at the Carlinville, Illinois Cemetery.

RUSSELL COREY

Russell Bracken Corey was born in New York City, New York on 25 June 1894, the son Mr. & Mrs. Edward Corey. He attended New York public schools, and schools in France. In 1914 he enrolled in Yale University (Sheffield Scientific School), Class of 1918.

On 21 July 1917 Corey enlisted in France's *Service Aeronautique*. From 1 August 1917 to February 1918 he attended the aviation schools at Avord, Tours and the *G.D.E.* While taking his training, he also acted as interpreter for his American colleagues. He received his *brevet* on the Caudron on 27 October 1917. While at the *G.D.E.* he was injured in an airplane accident over allied lines.

On 28 February 1918 *Sergent* Corey enlisted as an Ensign in the Naval Reserve Flying Corps (Naval Aviator HTA 553). He served as Assistant *Chef Pilote* at Montchic, France. In June of 1918 he was promoted to Lieutenant J.G. and in October to Lieutenant. Lieutenant Corey served with the North Sea Patrol and with the Northern Bombing Group on the staff or Sir Alexander Godley, Commander of 22nd British Army. Following the Armistice he returned to the United States and served at the Chatam Naval Air Station, Cape Cod, Massachusetts. He retired from active duty on 10 May 1919, and received an honorable discharge as a Lieutenant on 17 February 1922.

Following his discharge from the Navy, Corey began working in the real estate and insurance divisions of the Douglas Elliman Company. In 1924 he married Angela Caffrey. One son, Edward, was born of this union.

Shortly after the outbreak of World War II Corey re-entered the Navy as a Lieutenant. He served on the staff of the Fourth Fleet at the Eastern Sea Frontier, and on COMFAIR at San Diego, CA, and on the Gulf Sea Frontier and at Newport. He also served as Naval Liaison Officer with the Brazilian Air Forces from January 1943 to April of 1945. He retired at the rank of Commander on 29 December 1945. He was awarded the Bronze Star Medal for his World War II service.

Following World War II Corey returned to his duties as executive for a mortgage brokerage firm in New York City. He died of a heart attack at his home in Kensington, North Hempstead, Long Island, New York on 3 May 1954 at age 59.

EDWARD CORSI

Edward Charles Corsi was born in Italy on 10 November 1897, the son of Mr. & Mrs. M. Corsi. Edward Corsi attended Brooklyn High School. In October 1916 he sailed for France and enlisted in the Norton Harjes Ambulance Service. He served as a volunteer ambulance driver at the Front for six months.

On 15 May 1917 Corsi enlisted in France's *Service Aeronautique*. From 8 June 1917 to 14 May 1918 he attended the aviation schools at Avord, Pau, Cazeaux, and the *G.D.E.* He received his *brevet* on the Bleriot on 30 October 1917. Corsi served as an instructor before joining *Escadrille SPAD 77* at the Front on 30 May 1918.

On 14 September 1918 Caporal Corsi and Lt. Maurice Boyau, top ace of *SPA 77*, shared a victory by destroying an observation balloon near Etraye. Two days later Corsi, Boyau, and Aspirant Cessieux attacked a balloon near Merville. Corsi later described the action:

"We attacked a balloon and as we broke away were attacked by a patrol of Fokkers. A dog-fight ensued. I saw one German go down. As they broke off I saw a Spad on my right headed home. I flew with him back to the field. It was Cessieux. He had a bad leg wound and just about made it. My plane had been hit many times. Boyau never returned and though an extensive search was made, his body was never found." Corsi was twice wounded in the engagement. Following his recovery from his wounds, Corsi returned to his squadron and had several unconfirmed victories. He was awarded France's *Croix de Guerre*, with Palm and Star.

Sergent Corsi flew with *SPA 77* until the November 1918 Armistice. He continued to serve with his squadron in the Army of Occupation in Germersheim, Germany, flying captured German aircraft back to Paris.

In September 1919, while Corsi was in Paris, he met with Merian Cooper, former pilot with the 20th Aero Squadron, USAS, who convinced Corsi to fly with him as a volunteer for Poland against the Bolsheviks. Corsi joined Cooper's Kosciuszko Squadron as a Captain and flight commander. He served with the squadron until the Soviet enemy was driven from Polish soil, and an armistice was declared between the belligerents on 18 October 1920. For his valiant service, Captain Corsi was awarded Poland's *Virtuti Militare*, and Cross of Valor with two bars.

Sergent Edward Corsi on motorcycle while with SPA 77.

Following his Polish service Corsi returned to New York City and engaged in the aviation business, then in the insurance business. As a result of his service in the Kosciuszko Squadron, he wrote a book titled *Poland: Land of the White Eagle*, New York: Wyndham Press, 1933. Corsi died in New York City on 24 February 1971 at age 73. He was survived by his wife, Pearl, and daughter.

JOHN COTTON

John Rowell Cotton was born in Chicago, Illinois on 12 January 1898, the son of Dr. Alfred Cleveland Cotton, a Chicago physician. John Cotton attended Chicago public schools and Amherst College, Amherst, Massachusetts, class of 1918. While in his junior year he left that institution in the spring of 1917 to join the American Ambulance Field Service.

Cotton left the United States with the Amherst unit of the American Ambulance Field Service and sailed for France. He served in Paris for one week before transferring to aviation.

On 10 June 1917 Cotton enlisted in France's *Service Aeronautique*. From 16 June 1917 to 10 January 1918 he attended the aviation schools at Avord, Sacy, and the *G.D.E.* He received his *brevet* on the Caudron on 30 November 1917.

From 15 January to 17 June 1918 *Caporal* Cotton served at the Front with *Escadrille Breguet 120*. On that date he was commissioned a First Lieutenant in the U.S. Air Service and continued to serve with his French Squadron as a flight commander into September of 1918. For his outstanding service as a day-bombardment pilot, he was awarded France's *Croix de Guerre*, with two Stars.

From September to the 11 November Armistice Lieutenant Cotton was in charge of formation flying and aerial combat at the 7th Aviation Instruction Center, Clermont-Ferrand. Following the Armistice Lieutenant Cotton served with the 120th Aero Squadron, USAS, and the 90th Aero Squadron, USAS.

On 17 February 1919 Cotton was honorably discharged and returned to Chicago in 1920 and worked there until 1928. He married Marion Gould in that city in 1920. Two sons, Henricus and John, and a daughter, Ustan, were born of that union.

In 1929 Cotton moved to Jackson Heights, New York. He worked as a business analyst for his brother-in-law's firm. He was involved in business consulting until his death of pneumonia in Jackson Heights, New York, on 16 June 1933 at age 35.

ISADORE COURT

Isadore Newton Court was born in Newark, New Jersey, on 7 December 1888. He attended New York schools and Columbia University. He left school to pursue a career in architecture. On 2 June 1916 Court sailed from New York for France aboard the *Chicago* to take a position with his uncle's Paris-based architectural firm.

Shortly after arriving in France Court joined the French Foreign Legion. On 17 July 1916 Court enlisted in France's *Service Aeronautique*. On 15 February 1917 he was released because of inaptitude and entered the Automobile Service of the French Army.

On 18 May 1918, at Paris, Court enlisted in the U.S. Army in Company B, 302 Battalion, Transportation Corps. On 26 August he was assigned to Headquarters 1102 Replacement Squadron. He later served with Headquarters 1104 Replacement Squadron until his honorable discharge on 8 April 1919.

ELLIOT COWDIN
ESCADRILLE LAFAYETTE

Elliot Christopher Cowdin, a founding member of the *Escadrille Americaine*, was born on 3 March 1886 at Far Rockaway, Long Island, New York, the son of ribbon manufacturer and sportsman John Elliot Cowdin and Gertrude (Cheever) Cowdin. Elliot graduated from St. Paul's School in 1905 and from Harvard College in 1907. In November of 1914 he enlisted with the American Ambulance Service in Paris and served with an ambulance section attached to the 1st Division Cavalry of the Belgian Army. He was honorably released from the ambulance section in early February of 1915. That same month he enlisted in the French Army, Aviation Section, and began his training at Pau on 9 March. He received his *brevet militaire* on the Voisin on 29 April and completed his training on 1 May 1915.

On 3 May Cowdin was assigned to *Escadrille V.B.108*, 103d Bombing Group which was stationed at Nancy. There, he served as a bomber pilot in the Battle of Artois. Norman Prince also served as pilot in that squadron, with Victor Chapman serving as gunner-bombardier.

Cowdin had run into Norman Prince in February of 1915 in Paris and had enthusiastically embraced his plan to gather the Americans serving in the French Air Service into one all-American squadron. Cowdin supported and advanced this cause and helped to bring the *Escadrille Americaine* to fruition approximately one year later.

On 9 July 1915 Cowdin was cited in orders and was subsequently awarded France's *Croix de Guerre*, with Star for a 26 June engagement with two German aircraft. During that combat an enemy Taube was "seemingly seriously hit." According to the citation Cowdin's own plane had been seriously damaged in the melee and his flying helmet had been struck with bullets.

Cowdin left *V.B.108* for Paris on 15 August to begin training as a pursuit pilot. That same month he was promoted to *Sergent*. Upon completion of his training on the Nieuport, he was assigned to *Escadrille N.38* on 30 September and flew in the Battle of Champagne.

Sergent Elliot Cowdin beside his Nieuport 11C.1 fighter.

Ten days later, on 10 November, he left *N.38* for *Escadrille Nieuport 49*, joining that squadron on 12 November.

In December of 1915 Cowdin and fellow countrymen Norman Prince and William Thaw were granted a month's permission to America. They arrived in New York on 23 December aboard the SS *Rotterdam*. Their furlough, though intended as "a short rest", furthered the cause for an American squadron when the Government of France realized the interest the three American volunteers had created during their visit to the United States. The pilots became the focus of the eastern press who doggedly pursued them in spite of their reluctance to speak of their duties. Both sides of the European War seized upon the press coverage: the Allied supporters citing the flyers' example in a plea for America's intervention, while German sympathizers condemned their participation as a violation of U.S. neutrality and demanded their internment.

Upon their return to France, the three Americans' desire to form an all-American squadron was enhanced as the French Government realized that just such an organization would be useful in galvanizing America's sympathy for their cause. A few short months later, on 15 April 1916, through the combined efforts of Cowdin, Prince, Thaw, and others, the mens' dream became a reality.

In the interim Elliot Cowdin returned to *Nieuport 49*, flying with that squadron until 15 January 1916, taking part in the Battle of Hartmannsweillerkopf. In February he joined *Nieuport 65* and flew with William Thaw during the Battle of Verdun. On 18 April he was cited in orders and received a Palm to his War Cross for his diligent pursuit of enemy aircraft into their lines. On 20 April Cowdin was again cited and became the first American recipient of France's *Medaille Militaire*. He also received a second Palm to his *Croix de Guerre* for attacking twelve German planes and destroying one Fokker aircraft.

Ten days later, on 28 April 1916 Cowdin joined Victor Chapman, Bert Hall, James McConnell, Norman Prince, Kiffin Rockwell, and William Thaw at Luxeuil where the newly formed *Escadrille Americaine* (*N.124)* was stationed. He served there with the Squadron and at Verdun for just under two months until 25 June 1915 when he was forced to depart *N.124* for what was termed "heart trouble."

Elliot Cowdin's greatest contribution in World War One was his role in the formation of the *Escadrille Americaine*. While thrice cited as an "excellent, brave, devoted" pilot, some of his American comrades remained unconvinced by his combat record. Paul Ayres Rockwell, the Squadron's official historian, later noted that, "Some of (Cowdin's) fellow pilots I knew said he always played up to his Commanding Officers and obtained citations for work he had not done and victories he had not gained, by buying champagne for his Captains. They told me most of his flying was done in bars."

The fact that Cowdin served briefly with five French escadrilles added credence to the charge that during his tenure in the *Service Aeronautique* he had been a trouble maker given to brawling, and that he had had a penchant for taking extended or unauthorized leaves. One such "unauthorized leave" ultimately brought about his departure from the *Escadrille Americaine* when he was brought back to Bar-le-Duc charged with desertion. But because Squadron Commander Thenault did not want to bring adverse publicity to the Squadron, he downplayed the charge and allowed Cowdin to resign from *N.124*.

Later Hall & Nordhoff, when writing their Lafayette Flying Corps history, in order to protect Cowdin's reputation and that of the *Lafayette*, had listed his departure for reasons of "ill health." Regardless of the reasons, Cowdin's departure from the *Escadrille* was not mourned by his squadron mates.

Sergent Cowdin spent the months of August and September ostensibly in hospital. From October of 1916 to January of 1917 he became attached to the Royal Flying Corps in the capacity of testing and delivering French pursuit planes to the British Army. In January of 1917 he was deemed physically unfit for further flying and was discharged from the French Army.

Cowdin returned to the United States, and on 18 June 1918 he was commissioned a Major in the U.S. Air Service, Bureau of Aircraft Production. The fact that he was able to pass the military physical prior to receiving his commission casts further doubt on his alleged heart condition. Four days after joining the U.S. Air Service Major Cowdin returned to France to begin a series of inspection tours for the Air Service on the American, French, English, and Italian fronts. He returned home following the Armistice on 23 December and was discharged on 28 December 1918.

Following World War One Elliot Cowdin never married nor appeared to follow any business career. There is no indication that he pursued anything but polo, and that half-heartedly, as he settled for the life of a remittance man kept at bay by a healthy family allowance. He died in retirement at age 46 of bronchial pneumonia following a bout with influenza at Palm Beach, Florida on 6 January 1933. His New York *Times* obituary, while making note of his war record, only briefly mentioned his subsequent life, stating that he had been a "former New York Sportsman" then went on to list the clubs he had belonged to. Elliot Cowdin is buried at Hewlett, Long Island, New York.

AUSTEN CREHORE

Austen Ballard Crehore was born in Hackensack, New Jersey on 9 January 1893, the son of William Williams Crehore, a civil engineer, and Anna (Ballard) Crehore. Crehore attended Hackensack schools. When the United States entered World War One in March of 1917, he tried to enter the U.S Air Service and U.S. Naval Aviation. When his application was rejected by the American Examining Board because of a hearing problem, Crehore sailed for France to join French aviation.

On 16 July 1917 Crehore enlisted in France's *Service Aeronautique*. From 21 July to 28 November 1917 he attended the aviation schools at Avord, Tours, Pau, and the *G.D.E.* He received his *brevet* on the Caudron on 30 September 1917.

William Wellman, who trained with Crehore, wrote of his friend's determination in his book *Go, Get 'Em!*.

"The man who, perhaps, typified the general spirit of the school more than any other was Austin Crehore, of New York, who was with me throughout the training at Avord, Pau and at Plessis Belleville later. He was far from robust physically, and all the time was suffering intermittently from chronic appendicitis. Again and again one of the other men and I had to help him home to the barracks from the flying field, half carrying him; undress him and put him to bed, he was suffering so. But he stuck like a bulldog, reached the front and, during each of his first three weeks of fighting, brought down a Boche. Then, and not until then, did he feel that he had earned a rest, and he went to the hospital and had his troublesome appendix out."

From 1 December 1917 to the November 1918 Armistice *Sergent* Crehore served at the Front with *Escadrille SPAD 94*. Crehore was credited with the destruction of two enemy aircraft and was awarded France's *Croix de Guerre*, with two Palms for his fine war record. Crehore was later awarded France's *Legion d'Honneur*.

On 26 August 1919, upon his return from France, Crehore entered the New York-Toronto First International Air Race. Leaving Mineola in an Ansaldo S.V.A., he crashed into a tree at Albany on the first leg of the flight. He was nearly killed in the accident and spent three months in an Albany hospital.

Crehore gave up aviation and started the Crehore & Richardson insurance and brokerage agency. On 14 January 1924 he married Katherine Dennis in New York City. They had three children, two sons and a daughter. Crehore eventually became president of the Eifert, French and Company insurance company of New York. He retired in 1960. Crehore also served as president of the Lafayette Flying Corps Association and was the director of the American Society of the French Legion of Honor. He died in a Scotch Plains, New Jersey nursing home on 20 August 1962 at age 69. He is buried at Fairview Cemetery in Westfield, New Jersey.

ARTHUR CUNNINGHAM

Arthur Lawrence Cunningham was born in Everett, Massachusetts on 15 December 1895, the son of John F. Cunningham, a shoe manufacturer, and Mary Elizabeth (Ryan) Cunningham. He attended Medford High School and entered Harvard College, class of 1918(19).

Cunningham left Harvard and sailed for France on 26 May 1917 as part of the Harvard Ambulance Unit. He spent one month in Paris with this organization before joining French aviation. On 7 July 1917 Cunningham enlisted in France's *Service Aeronautique*. From 15 July 1917 to February 1918 he attended the aviation schools at Avord, Pau, Cazeaux, and the *G.D.E.* He received his *brevet* on the Caudron on 1 December 1917.

On 24 July 1918 *Caporal* Cunningham was commissioned a Second Lieutenant in the U.S. Air Service. From 4 March to I October 1918 he served with the 94th Aero Squadron, USAS, as pilot and later as operations officer. On 1 August 1918 Cunningham was promoted to First Lieutenant. From 1 October to the November 11 Armistice he served as Operations Officer for the First Pursuit Group.

Following World War One Lieutenant Cunningham taught English to foreign wives of American officers at the University of Grenoble, France. He married Adele Bilger of Grenoble, France, and they had one son, John.

Lieutenant Cunningham returned to the United States on 3 August 1919 and was discharged the following day. He became an advertizing manager for the Johns-Manville Company and was also involved in home construction. He died in a nursing home in Boston Massachusetts on 10 February 1972 at age 76.

FRAZIER CURTIS

Frazier Curtis was born in Boston, Massachusetts on 18 July 1876, the son of Greely Stevenson Curtis and Harriet (Appleton) Curtis. His great-grandfather had been one of the organizers of the "Boston Tea Party" and his father had commanded the 1st Massachusetts Cavalry in the Civil War.

Frazier Curtis attended Mrs. Shaw's and J.P. Hopkinson's private schools, and graduated (cum Laude) from Harvard College, c'98 (A.M. '99). In 1909 he married Gladys Margaret Raper of England.

In September of 1914 at the outbreak of World War One, Curtis sailed for England to enlist in the Royal Flying Corps. He was rejected because of his nationality and age and returned to the U.S., hoping to return to Europe with a Burgess-Dunne seaplane. While furthering his aviation skills at Marblehead, Massachusetts he had met Norman Prince, and the two had discussed the possibility of forming and American Escadrille to fight for France.

Upon the completion of his training at Marblehead, Curtis again returned to England but was rejected a second time by the British Service. In February of 1915 he went to Paris and he and Prince began to put in motion the organization of the *Escadrille Americaine*.

On 2 March 1915 Curtis entered France's *Service Aeronautique* and that same day began his aviation training at Pau. He continued with his training at Avord where he had two serious accidents and spent one week in hospital. Curtis's nerves were strained, and he was given a 45-day sick-leave. At the end of this furlough he applied for transfer from bombing to *chasse* training. On 8 August 1915 the French War Office, instead of granting his request, discharged him as unfit for further flying. Curtis spent the next four months in England recuperating. He then returned to the U.S. and in March of 1916 he organized the Harvard Flying Corps which trained students to eventually become pilots in the U.S. Air Service.

Curtis's nerves again broke down and he was forced to give up his work and go to California to recuperate. For a time he worked for the Government Forestry Service. He committed suicide in La Jolla, California on 26 January 1940 at age 63.

ALVAN CUSHMAN

Alvan Alexander Cushman was born in Sioux Falls, South Dakota on 19 May 1897, the son of George Alvan Cushman, and Gertrude Maria (Peterson) Cushman. He attended the Phillips Academy, Andover, Massachusetts, Class of 1916, before enrolling in the Lafayette College, Easton, Pennsylvania, Class of 1920. He left there after one year, in April of 1917, to enlist in the American Ambulance Field Service.

On 22 June 1917 Cushman enlisted in France's *Service Aeronautique*. From 29 June to February of 1918 he attended the aviation schools at Avord, Pau, and the *G.D.E.* He received his *brevet* on the Caudron on 4 December 1918.

Caporal Cushman enlisted in the U.S. Naval Aviation Reserve Force as an Ensign. He was sent to England and was seriously injured on 23 May 1918 while flying a B.E. 2 bombing machine. Cushman suffered a broken thigh which kept him in hospital for five months. When he had recovered from his injuries, he became a hydroaeroplane pilot and was stationed at Bolsena, Italy until the end of the War.

Following World War One Cushman worked with the American Red Cross in Italy before joining the Balkin Red Cross Commission which was headquartered in Belgrade, Servia. Upon his return to the United States he worked for the General Motors Acceptance Corporation in Pittsburgh, Pennsylvania, in New York City, New York, and in Paris, France. Cushman married Gertrude Harrison. The couple had three children. He worked for Crysler Corporation in South America before becoming a representative for Silva Incorporated, an export/import firm, in Cali, Columbia. He died there on 13 February 1948 at age 50.

PHILIP DAVIS

Philip Washburn Davis was born in West Newton, Massachusetts on 10 March 1888, the son of Samuel Warren Davis, a high school Latin teacher, and Maria E. (Washburn) Davis. Davis attended Newton High School before enrolling in Harvard College, A.B. 1908. Davis was an outstanding athlete and set records in track and in tennis. His pursuit of tennis continued following college, and he won several local tournaments.

Upon leaving school Davis entered the Boston office of the securities firm of Lee, Higginson & Co. He remained there two years before forming the investment securities firm of Chamberlain and Davis.

Prior to World War One Davis had served in the 1st Corps of Cadets, Massachusetts Volunteer Militia. In May of 1917 he sailed for France to join the Norton-Harjes Ambulance Service. He arrived in Paris on 2 June.

On 5 June 1917 Davis enlisted in France's *Service Aeronautique*. From 15 June to February 1918 Davis attended the aviation schools at Avord, Pau, and Cazeaux. He received his *brevet* on the Caudron on 28 October 1917. *Caporal* Davis was transferred to the U.S. Air Service and was commissioned a Second Lieutenant on 6 January 1918. He was assigned to the 94th Aero Squadron, USAS on 1 April 1918 and flew combat patrols with that Squadron in the Toul Sector until 2 June 1918.

On that day Lieutenant Davis, with pilots Campbell, Cunningham, and Eastman, was ordered to escort a British bombing squadron across the lines. The American patrol crossed into German territory at 18,000 feet. Before it could link up with the British bombers the patrol was jumped by a flight of six German machines.

Following a fierce combat, Davis linked up with Cunningham. Davis's fighter was some 50 yards from Cunningham's fighter, and Cunningham watched in horror as a "tiny flame spurted out of Davis's machine just behind the pilot's seat, and began to lick its way around the fuselage." Davis dived toward earth, and as Cunningham followed, horrified, Davis's fighter burst into flames. It crashed into a meadow at the edge of a wood just behind German lines.

Weeks later the 94th pilots learned that Davis's body had been taken from the plane and buried. During the aerial melee, he had been struck in the leg, and an incendiary bullet had lodged in his plane, finally setting it afire. He was 30 years of age at the time of his death. Lt. Davis was buried near EurlyWald (Richecourt), north of Seicheprey, where his sister had purchased the ground containing his grave.

GEORGE DOCK

George Dock, Jr. was born in Ann Arbor, Michigan on 20 January 1895, the son of Dr. George Dock, a physician, and Laura (McLeMore) Dock. When Dr. Dock moved his medical practice to St. Louis, Missouri, George attended the public schools there. He enrolled in Dartmouth College, Class of 1916, and took an early degree in order to participate in the European War.

On 11 May 1916 Dock enlisted in the American Ambulance Field Service in Paris. He served with SSU 2 for 1 year and 1 month and was awarded France's *Croix de Guerre*, with Star for his heroic and distinguished service.

On 10 June 1917 Dock enlisted in France's *Service Aeronautique*. From 20 June 1917 to 15 March 1918 Dock attended the aviation schools at Avord, Pau, Cazeaux, and the *G.D.E.* He received his *brevet* on the Caudron on 14 November 1917.

From 18 March 1918 to the 11 November Armistice *Sergent* Dock served at the Front with *Escadrille SPAD 12*. He flew numerous combat patrols deep inside enemy lines, attacking German troops and munitions. He was awarded a second *Croix de Guerre*, with Star while with *SPAD 12* for his exemplary service under the most hazardous of conditions.

When *Sergent* Dock was discharged from the *Service Aeronautique* he returned to the United States, to his family home in St. Louis. He had suffered a detached retina while in service and had complications with it. Following his recuperation, he became involved in advertizing before relocating to New York to work as an executive for the investment banking firms of William R. Compton Company, and Halsey, Stuart and Co. He also served as director of the Financial Advertizing Association. He later worked for the New York office of the Doremus & Co. financial advertizing agency.

He married Elizabeth Halsey and later retired to Norwich, Vermont where he was an active member of the National Audubon Society. He died on 21 October 1971 in Boston, Massachusetts at age 76.

CARL DOLAN

ESCADRILLE LAFAYETTE

Charles Heave Dolan, Jr. the 31st volunteer of the *Escadrille Lafayette*, was born on 29 January 1895 in Boston, Massachusetts, the son of Charles H. Dolan, an Irish Boston politician, and Anne (Mainwaring) Dolan. Dolan grew up around Boston and graduated from Mechanics Arts High School. His first interest in aviation began in September of 1910 after watching Claude Grahame-White win a race around the Boston Lighthouse in a Bleriot monoplane.

Dolan studied electrical engineering for one year at the Lowell Technical Institute which was associated with the Massachusetts Institute of Technology. It was there, in 1913, that he built aviation wind tunnel models to test Eiffel's theories on wing efficiency. He also made frequent visits to the nearby Burgess Flying School.

In 1914 Dolan wearied of school. Desiring adventure, he left Lowell Institute and took a cattle boat to England, working off his passage by tending to a cargo of 700 wild Canadian horses.

From 1914–15 Dolan worked for the firm of Thompson-Houston and S. Smith & Sons, Cricklewood, England, where he served as final inspector for the first English-made aeroplane magneto. From 1915–1916 he worked in Paris as an installation engineer, placing autopilots in F-40 Farman pushers for the Sperry Gyroscope Company, Ltd. of London. In this capacity, he helped Lawrence Sperry's aircraft win France's $25,000 Safety of Air prize.

While in Paris Dolan met several *Escadrille Americaine* pilots at Henry's Bar, at that time the popular hangout for American aviators on leave. In the course of their conversation the young Americans learned of Dolan's engineering abilities and felt he would be a genuine asset to their squadron. They suggested he enter into French aviation, and they promised to make every effort to see that he joined their ranks.

However, Dolan still had a commitment with Sperry. And when he was assigned to a job in Russia, he chose instead to join the French Foreign Legion, the first step toward becoming an aviator.

He enlisted on 11 August 1916. He immediately detached to the *Service Aeronautique* and on 30 August he was sent to Bleriot school at Buc near Versailles, receiving his *brevet militaire* on the Bleriot on 11 March 1917. He next journeyed to the school at Avord for Nieuport instruction. From Avord he went to the Pau school near the Basses Pyrenees, for acrobacy and combat training. Because of an immediate need for pilots at the Front, the French had Dolan bypass the aerial gunnery school at Cazeaux. He was sent instead directly to the *G.D.E.* at Plessis-Belleville near Paris where he remained in reserve for approximately ten days.

Sergent Carl Dolan in his SPAD S.VII fighter.

Dolan joined the *Lafayette Escadrille* at Ham on 12 May 1917. He soon became affectionately known as "Carl" to his squadron mates and before long he was piloting a SPAD with his monogrammed insignia on its sides.

The veterans of the Squadron took an immediate liking to Dolan and recognized in him a solid addition to their roster. They soon put to use his engineering abilities which they had first become aware of that previous summer at Harry's bar. They also grew to admire Dolan for the self-control he demonstrated at all times. The teetotaler in a squadron of hard drinkers, Dolan, however, was never intolerant of those who indulged. As Edwin Parsons later said of Dolan:

"There was one who had never and has never had a drink in his life up to this time, to my knowledge, and yet did his work and enjoyed the parties and had as much fun as anybody else. And that was Carl Dolan. He deserves a good big pat on the back for his abstinence."

Dolan also realized that, unlike himself, many of his squadron mates could not fly without the calming effects of alcohol. For this reason he always kept a bottle stashed in his footlocker as a bracer for the men heading out on patrol and as a relaxer for the pilots coming in.

He also did much to eliminate their anxieties and to built their confidence in their machines by overseeing the French mechanics and by convincing them to apply his own thoroughness to their work. Meanwhile, *Sergent* Dolan flew his share of patrols and engaged in a number of combats.

His most difficult task followed a predawn patrol flown by his friend, Douglas MacMonagle in the early morning hours of 23 September 1917. Dolan had spent most of a 48-hour Paris leave tending to the inebriated MacMonagle, and was guiding him back to their barracks when his friend had suddenly lurched into *Capitaine* Thenault's quarters and had rudely awakened their commanding officer. As a punitive measure, Thenault had ordered MacMonagle to join the patrol which had been assigned to Raoul Lufbery and Robert Rockwell, and which was set to begin in one hour.

With time for the mission drawing near, Dolan used strong coffee in an attempt to revive his friend as he wrestled him into his flight suit. He offered to fly the mission for MacMonagle, but his friend had refused his offer.

Raoul Lufbery and "Doc" Rockwell were the first to take off followed by MacMonagle in his wobbling SPAD. The pilots formed a V formation at 15,000 feet. MacMonagle took a back position in the V and drifted aimlessly about the sky.

Lufbery soon spotted a formation of eight Albatros fighters 1500 feet beneath them. He signaled for Rockwell and MacMonagle to climb steeply to get above and behind the enemy. But MacMonagle failed to follow. Seeing this, Rockwell peeled away in a desperate diving attempt to head him off. But he reached him too late. In horror, he watched his comrade fly head on into the enemy formation. A German bullet pierced MacMonagle's temple, killing him instantly. His SPAD nosed into a three-mile power dive that ended in the woods near Triaucourt.

Dolan was ordered by *Capitaine* Thenault to retrieve his friend's body from the French lines and bring it back to the Squadron for burial. He then faced the burden of meeting MacMonagle's widowed mother at the Senard train station where she had previously arranged to meet her son. Together the next day they attended MacMonagle's funeral at the military cemetery at Triaucourt.

Dolan flew with the *Lafayette* Squadron until 18 February 1918 when it evolved into the U.S. 103rd Aero Pursuit Squadron. Although he had been recommended for a Captaincy, on 14 January 1918 he was commissioned a First Lieutenant in the U.S. Air Service, a rank beneath his experience and ability.

Dolan joined the 103rd roster and served dutifully with that famous squadron as pilot, engineer officer, and flight commander. He saw action on several fronts

Major David Peterson, First Lieutenant Carl Dolan, and First Lieutenant Henry Jones at Carlstrom Field, Arcadia, Florida, November 1918. Peterson would tragically die four months later in a plane crash while testing an experimental D.9 aircraft on 16 March 1919 over Florida.

and took part in the Saint Mihiel and Meuse Argonne offensives. On 12 October 1918 he was officially credited with the destruction of a Fokker fighter in combat near Fontaines-Haraumont at 10:55 o'clock at 1000 meters altitude. On 16 October 1918 he was ordered to the United States to report to the Office of the Director of Military Aeronautics, Washington, D.C. to serve in an advisory capacity on pursuit work.

Around the end of October Dolan returned to the United States with Major David Peterson aboard a transport which docked at Norfolk, Virginia. Together they went to Washington, D.C. and were given the option of going to Carlstrom Field, Florida or to Rockwell Field, California. Both chose to serve in Florida, and following a week's leave they arrived at their duty station at Carlstrom and Door Fields, Arcadia, Florida. At Carlstrom Field Dolan was put on the Control Board and was also made Engineer Officer in charge of planes and equipment.

Following the death of his friend, Major Peterson, who was killed in a flying accident in Florida on 16 March 1919, Dolan was transferred to Washington, D.C. to the Office of Director of Air Service, Information Group from 1919-1920. There, he served under Colonel Horace Hickam as head of the Library Division. Dolan resigned from the military in 1920. He was awarded France's *Croix de Guerre* and *Legion of Honor* for his World War service.

In May of 1921 Dolan was selected by General Billy Mitchell and Colonel Hickam as American Air Advisor to China. His function was to help the Chinese develop a military air arm. While en route to China, in Honolulu, he met Miss Ramona Frances Morgan. Dolan courted her on the ship's voyage to China. They were married shortly after their arrival in Peking, on 10 July 1921.

At the time of Dolan's arrival in Peking, the city was politically up for grabs and was being fought over by three warlords. Dolan, in the company of another American pilot, four British pilots, and one French pilot, managed to establish the first flying school in China that same year. This group also flew the first airmail flights from Peking to Petahoa from 1921–1923, and established China's first airline, the Peking-Tientsin Air Line. In reality, the airline was an offshoot of the fledgling military air force which Dolan had previously helped to establish, its pilots taken from the fifty or so he had managed to train.

At Nanyuan, ten miles from Peking, the Chinese established an airfield for this new group of pilots on a mile-long military parade ground. Nearby they erected a number of French-made canvas hangars which housed sixty World War One single-engine Avro biplanes for primary training use. Secondary training was conducted in forty FE-2B pusher-propeller aircraft. The advanced class flew twenty-five Vickers Vimy's.

While in China, in 1924, Dolan received three decorations from the Chinese Government, including the Sixth Degree of the Striped Tiger, this for saving the life of U.S. Secretary of the Navy Edwin Denby and various Chinese officials by safely crash- landing the Vickers Vimy he had been piloting near the Great Wall following the loss of one of its engines. Later, when Dolan helped to design China's first airmail stamp, he worked both the Great Wall and a Vickers Vimy into the design.

Dolan returned to Hawaii from China in 1925 and was appointed by Governor Farrington to the Territorial Air Board. He was made chairman of the Committee on Design and Development of Airports. In this capacity, Dolan designed John Rogers Field (now Honolulu International Airport). While the Dolans were in Honolulu, on 16 March 1927 Charles H. Dolan III was born.

In July of 1929 Dolan conducted an inter-island air route survey which indicated that Inter-Island Stream Navigation Company should enter the airline business. His favorable report resulted in the formation of Inter-Island Airways, Ltd., the predecessor of Hawaiian Air.

In 1930, in Atlanta, Georgia, he made a survey for Sperry Company of its recently purchased Pitcairn Aviation, Inc. and its subsidiaries. In 1931 Dolan helped Pitcairn evolve into Eastern Air Transport Inc. of Atlanta. This would later become Eastern Airlines. He was the company operations manager from 1931–1933. He then served as vice president in charge of operations from 1933–1934. He served as an aero valuation engineer for the Air Mail Division, U.S. Post Office Department, at I.C.C. and other hearings from 1934–1935. He then became the assistant technical advisor for the U.S. Senate Air Safety Commission in 1936.

From 1936–1940 Dolan served as vice-president and American manager of the Intercontinent Corporation. He next became president of Wissahickon Tool Works Inc. from 1940–1941. In October of 1941 he became the general manager of the Aero-Hydraulic Division, Chicago Pneumatic Tool Company.

From 1942–1945 Dolan was President of Commonwealth Aircraft Company, Kansas City, Missouri, the second largest manufacturer of combat gliders in World War Two. He was not recalled to active duty due to this important defense assignment, although he held a reserve commission as a Colonel in the Army Air Force. From 1946–1947 he was co-owner of Allied Engineering Company in New Haven, Connecticut.

In 1947–1948 he was executive director of the Air Safety Board which was set up by the Interstate and Foreign Commerce Committee of the U.S. Senate. He was for all practical purposes the U.S. Senate Aeronautical Advisor during the 80th Congress, primarily on air safety.

Dolan was recalled to service for the Korean Conflict with the rank of Colonel to serve as assistant to the Chief of Maintenance of the Air Force at the headquarters of the Air Material Command at Wright-Patterson Air Force Base. Following this assignment he became U.S. Air Force representative at Sperry Corporation, Long Island.

Until his retirement, Dolan served as a trouble shooter and consultant for Sperry Products, Inc. and for the Bulova Research and Development Labs, Inc.

Dolan entered semi-retirement at Lanikai, Hawaii. His aviation career had been composed of many "firsts." He installed the first insulated berths in aircraft; wrote the first airlines operational manual; and installed some of the first instrument landing systems. He also helped to write the original charter for the Civil Aeronautics Administration.

Throughout his postwar life Carl Dolan remained a staunch defender of the *Lafayette* Squadron's name, guarding it against the false membership of "ringers," and protecting its history against all manner of assault. This, he did with a vengeance.

And frequently he came to the aid of his less fortunate comrades who had been brutalized by their war experiences to the degree to which they had lost control of their lives. He often extended them loans, knowing they could never repay him, and personally assisted in the burial of two of them.

As one of the healthiest of the surviving *Lafayette* pilots, Dolan was often called upon to represent the Squadron in various ceremonies. He was at Luxeuil, France in July 1967 when a monument to the *Escadrille* was unveiled. Then in 1976 Dolan was the guest of the French Air Force at the 60th reunion of the *Lafayette Escadrille.* In November of 1981, he was invited with America's surviving World War One aces by France's President Francois Mitterand for special observances to commemorate the 63rd anniversary of the Armistice, and for a special reunion of all the World War One surviving fighter aces. Dolan, in frail health, made the trip to France as the *Lafayette Escadrille's* last surviving representative.

While Dolan viewed the trip as a great honor, it proved a considerable physical strain upon him. Prior to his departure from Hawaii he discovered he had cancer (lymphoma) and felt that he might not survive the journey. But Dolan decided to go ahead anyway. He probably contracted pneumonia when he visited the Lafayette Memorial near Paris on 10 November 1981. It was a chilly gray day and he was not properly dressed for such weather. He returned home to Honolulu, Hawaii where he died at age 86 on 31 December 1981, the last survivor of the 38 *Lafayette Escadrille* pilots.

ROBERT DONZE

Robert Louis Donze was born in Santa Barbara, California on 11 February 1895, the son of Eugene and Hortense (Chaffin) Donze. He attended Grammar and High School in Santa Barbara from 1900–1911. That same year he began work as a Marconi wireless operator.

Donze left California for Paris to join the Lafayette Flying Corps. On 7 November 1916 he enlisted in France's *Service Aeronautique*. From 8 November to 10 May 1917 he attended the aviation schools at Buc, Avord, Pau, and the *G.D.E.* He received his *brevet* on the Bleriot on 19 March 1917.

From 20 May to 15 June 1917 *Caporal* Donze served at the Front with *Escadrille Nieuport 93*. On that date he was seriously injured while test-hopping a new SPAD fighter.

When *Sergent* Donze's engine had failed him, he had attempted a forced landing. His wheels hit a trench obscured by grass, throwing Donze from the plane. As the SPAD's tail catapulted forward, it nearly severed his left arm.

Donze spent the next three months recuperating in hospitals in France and in England. While in England he met a nurse, Gladys McKensie-Penny of South Devonshire. They were married in England on 25 September 1917 and eventually had two sons.

Following a three-month recuperation *Sergent* Donze returned to the Front with *Nieuport 314*. In March of 1918 he was commissioned a First Lieutenant in the U.S. Air Service. He was assigned as test pilot at Le Bourget and was later appointed a flight commander and operations officer, American Acceptance Park No.1, Orly, France, in command of the fixed-motor division and the receiving division. France awarded Lieutenant Donze the *Croix de Guerre*, with Star, for his war service at the Front.

Following World War I Donze entered Cambridge University. He and his wife separated, and in 1923 he returned alone to the United States. He engaged in the radio business and worked as an insurance adjuster in Santa Barbara, California. During 1928–1929 he and a partner formed the Chadbourne-Donze Air Service, Inc. of Santa Barbara. Donze was also active in the Santa Barbara Aero Club. Donze later became involved in real estate development for the Bishop Realty of Santa Barbara. He died on 17 February 1970 in Los Angeles, California at age 75.

JAMES DOOLITTLE

ESCADRILLE LAFAYETTE

James Ralph Doolittle, the 37th volunteer of the *Escadrille Lafayette*, was born in Chicago, Illinois on 6 January 1894, the son of San Francisco publisher James Reuben Doolittle and Frances (Golsen) Doolittle. His paternal great-grandfather had been James Rood Doolittle, U.S. Senator from Wisconsin from 1857 to 1869, a close personal friend and adviser to President Abraham Lincoln.

Following preparatory school, Doolittle enrolled in Columbia University. He left there to join the Norton-Harjes Ambulance Service and served with that unit for a few weeks before entering French Aviation.

Doolittle enlisted in the *Service Aeronautique* on 16 October 1916. He entered the aviation school at Buc on 21 December and received his *brevet militaire* on the Caudron on 22 March 1917. He progressed through the aviation schools at Avord and Pau, arriving at the *G.D.E.* around the end of April to sharpen his flying skills while awaiting assignment to the Front.

On 2 May 1917 Doolittle was injured while at the *G.D.E.* at Plessis-Belleville when he wing-slipped a SPAD into the ground from an altitude of 700 feet. He broke his ankle and received severe facial cuts to the right side of his cheek and eye. He spent the next eight weeks in hospital recuperating.

Doolittle joined the *Escadrille Lafayette* at Chaudun on 2 July 1917 as a replacement for the injured James Norman Hall.

He was still learning the duties of patrol from the veteran pilots when the squadron was ordered to Saint Pol-Sur-Mer where it was to take part in the British Flanders offensive.

The pilots were plagued by thick cloud cover on the 17 July 1917 move from Chaudun to Dunkirk. Doolittle and several other pilots lost their bearing. When Doolittle came out of the clouds, he found himself directly above an unidentified airfield. He descended with the intention of landing to ask directions when his SPAD came under intense ground machine gun fire. He soon realized he was over a German airfield as he spotted the black crosses on the buses being hastily rolled out of hangers.

Sergent James Ralph Doolittle, the right side of his face disfigured from an airplane crash at the G.D.E., 2 May 1917.

Doolittle soared into the clouds for refuge. He exited them above a British observation balloon near Arras under attack by a German fighter. He rushed to the rescue of a single British Nieuport doing battle with the enemy and was immediately jumped by two other enemy fighters on opposite sides of his SPAD. He was struck twice in the leg and his motor was pierced by several bullets.

Doolittle turned his bus toward allied lines and began a long descending glide while being hotly pursued by the two German aircraft. As he approached the Brit-

Sergent Doolittle beside his SPAD S.VII fighter at Chaudun, July 1917.

ish lines he was again wounded, this time by British antiaircraft shrapnel as their batteries attempted to drive the enemy fighters away.

Doolittle dropped behind the British lines and landed in a plowed field where his machine overturned, reopening his old face wound. He was immediately taken to a nearby British hospital where he spent several weeks recuperating before he was granted release from the *Service Aeronautique* in order to return to America. For his action that day, *Caporal* Doolittle was cited in British dispatches and was later decorated with the British Military Medal. He was also awarded France's *Croix de Guerre*, with Palm.

Doolittle returned to the United States in the fall of 1917 and volunteered his services as a civilian instructor with the U.S. Air Service. He served at Mineola, Long Island, at San Antonio, at Gerstner Field, Louisiana, and at Dayton, Ohio. While acting as chief tester at Dayton, he was injured in another fall. Following his recovery, he was sent to Kenilworth Flying Field, Buffalo, New York.

While in Buffalo Doolittle met Miss Mary Louise Thomas of that city. They became engaged and set their marriage date for 1 August 1918.

On the morning of 26 July, Doolittle as pilot and Lawrence Dunham as observer left the Kenilworth Field in an experimental Curtiss aircraft powered by a Liberty motor. Doolittle reached an altitude of 300 feet and attempted to turn the plane. He turned too short with his motor throttled down and lost control of the machine. It crashed to the ground at 10:20 a.m., instantly killing Dunham and fatally injuring Doolittle who died in hospital one hour after the accident. He was 24 years old at the time of his death.

Doolittle's father ordered that his son's remains be cremated, and his ashes were then placed at the Albert Herter estate, the "Creeks", at East Hampton, Long Island, New York. In June of 1928 Doolittle's ashes were transferred to a crypt in the Lafayette Escadrille Memorial near Garches, France.

DENNIS DOWD

Dennis Dowd III was born in New York City on 22 April 1887, the son of Dennis P. Dowd, a real estate man, and Katherine (Oreily) Dowd, who died ten month after Dennis's birth. Dennis Dowd grew up on the Dowd homestead at Sea Cliff, Long Island. Following his graduation from prep school, he enrolled in Georgetown University, Washington, D.C., and received his A.B. degree in 1908. He then enrolled in New York City's Columbia University law school, class of 1911. He left that institution before receiving his degree and practised law first in Birmingham, Alabama then in Brooklyn, New York.

While Dowd had been a student at Georgetown University, he had become engaged to a young woman. Several years later, when he finished his education by clerking in a New York law office, he had asked his fiance to marry him, but she had put him off, thinking more about a "good time." When she refused to give him a definite answer, Dowd had written her this ultimatum.

"I am returning to New York. If I do not hear from you within a week, that our wedding will take place in October, I will sail for France, enlist in the Foreign Legion, and probably be killed in the war."

Dowd's fiance sent him a non-committal reply from Washington, D.C. A few days later she received a letter from Dowd, stating that he had sailed for France on 10 August 1914 aboard the *Rochambeau* to join the Foreign Legion. His fiance sent him a wire promising to marry him if he would return to the United States. Dowd replied that it was too late, that he was already in the French Army.

On 27 August 1914 Dowd enlisted in the French Foreign Legion and was among the first American volunteers to serve France in the Great War. He was assigned to the 9th Squad of the Second Marching Regiment of the French Foreign Legion. During the Battle of Champagne in September of 1915, he transferred to the 170th Line Infantry Regiment. On 8 October he was wounded in the right hand by shrapnel and was sent to hospital.

While Dowd was recuperating, he met Mademoiselle Paulette Parent de Saint-Glin, a Paris debutante who had adopted Dowd as a "godson." They became engaged and planned to marry after the war.

Dowd spent the autumn and winter in the hospital. While convalescing, he applied for French aviation. He learned of his acceptance about the time he was ordered to return to the Front. He asked his army surgeon to extend his leave for two weeks so that his transfer would take effect. The surgeon agreed and ordered Dowd sent to the physio-therapeutic institute at Epinal. But he incorrectly applied the psycho-therapy stamp to Dowd's papers and Dennis was sent to the mental ward at Epinel. It was several days before Dowd could convince the doctor-in-charge that he was, in fact, sane.

Once free of the mental asylum, on 28 March 1916 Dowd enlisted in the *Service Aeronautique*. From 13 April to 12 August 1916 he attended aviation school at Buc.

On 12 August 1916 Dowd was killed when his airplane fell and smashed into the training field. His monituer speculated that he had fainted at a high altitude. He was the second American airman to be killed in France, and the first American student-pilot to die in training. He was 29 years of age at the time of his death. Dowd's remains are interred in the St. Germaine cemetery.

MEREDITH DOWD

Meredith Loveland Dowd was born in Orange, New Jersey on 23 July 1895, the son of Col. Heman Dowd, a banker, and Mary (Loveland) Dowd. He was educated at The Newark Academy, Newark, New Jersey, and at the Asheville School, North Carolina. In 1914 he entered Princeton University, Class of 1918, where he was a member of the varsity football and baseball teams. In 1916 Dowd attended the military camp at Plattsburg, New York.

"Larry" Dowd left Princeton University in the fall of 1916 to join the American Ambulance Field Service. He served with Section One from 11 November 1916 until 3 May 1917.

On 14 May 1917 Dowd enlisted in France's *Service Aeronautique*. From 20 June to 29 December 1917 he attended the aviation schools at Avord, Pau, and the *G.D.E.* He received his *brevet* on the Caudron on 7 November 1917.

From 1 January to 6 February 1918 *Caporal* Dowd served at the Front with *Escadrille Nieuport 152*. On 8 April 1918 he was commissioned a Second Lieutenant in the U.S. Air Service. From 30 April to 17 July 1918 he was attached to the 471st French Squadron (Paris Defense Squadron) at Le Bourget. From 17 July to 29 August 1918 he served at the American Acceptance Park, Orly.

On 29 August 1918 Lieutenant Dowd was assigned to the Front to the 147th Aero Squadron, USAS. He flew with that squadron until 26 October 1918. On that date he was killed in combat near Dannevoux. His widow, Mary Brack Dowd, of England, was presented the U.S. Distinguished Service Cross which was posthumously awarded to Lieutenant Dowd for extraordinary heroism. His citation describes the action that led to his death.

"For extraordinary heroism in action near Dannevoux, France, October 26, 1918. Having been unable to overtake and join a patrol, Lieutenant Dowd alone encountered four German planes, which he daringly attacked. He fought with most wonderful skill and bravery, diving into a formation and sending one of the enemy machines to the earth. In the course of the combat his machine was disabled and crashed to earth, killing him in the fall."

Lieutenant Dowd was buried at the American Cemetery, Romagne-sous-Montfaucon, Meuse. His remains were later removed to the Lafayette Flying Corp Memorial near St. Cloud.

SIDNEY DREW

Sidney Rankin Drew, Jr. was born in New York City, New York on 19 September 1891, the son of Sidney Drew, a stage and screen actor, and Gladys (Rankin) Drew, an author. Sidney Drew attended Mount Pleasant Military Academy and graduated from the Cutler School, New York. He then pursued a career as a motion picture actor, starring in the romantic comedy, *A Girl Named Philippa*. He also wrote short stories and a one-act play entitled *What the Moon Saw*. He gave up his acting and writing career to become a motion picture writer and director.

In May of 1917 Drew joined the American Ambulance Field Service and sailed for France. Upon his arrival in Paris, when he discovered he was to drive ammunition trucks rather than ambulances, he applied for French aviation. On 9 June 1917 he enlisted in the *Service Aeronautique*. From 12 June to 22 March 1918 he attended the aviation schools at Avord, Pau, Cazeaux, and the *G.D.E.* He received his *brevet* on the Caudron on 17 November, 1917.

On 25 March 1918 *Corporal* Drew was assigned to *Escadrille SPAD 31*. He flew with that squadron until 19 May 1918, the day he was killed in aerial combat with five Albatross scouts over the town of Arvilliers, northeast of Montdidier. He was 26 years of age at the time of his death. The Germans buried his remains and placed a cross at the head of his grave that was constructed from the struts of his wrecked fighter. His body was later moved to Montdidier, and in 1928, to the Lafayette Flying Corps Memorial outside Paris.

JOHN DREXEL

ESCADRILLE LAFAYETTE

John Armstrong Drexel, the 32nd volunteer of the *Escadrille Lafayette*, was born in Philadelphia, Pennsylvania on 24 October 1891, the son of wealthy inheritor "Colonel" Anthony J. Drexel, Jr. and Margarita (Armstrong) Drexel. John Armstrong's grandfather had been a partner with J.P. Morgan and had founded the Drexel Institute in Philadelphia.

Shortly after Colonel Drexel had inherited his portion of his father's $30,000,000 fortune, he had chosen to live in Europe where he maintained elaborate residences in England and in France. He was an intimate friend of King Edward VII whom he entertained on a lavish scale at his London residence on Grosvenor Square. Drexel was also on friendly terms with his distant relative, William II of Germany when he was Kaiser.

The Colonel returned to his Philadelphia home approximately once a year for two weeks. It was there that he had once served in the Philadelphia City Cavalry and had helped to break the steel strike at Homestead, Pennsylvania as aide-de-camp to Governor Pattison.

It came as a complete surprise to the other trustees of the estate when Anthony Joseph II announced his withdrawal from Drexel and Company, Drexel Morgan & Company of New York, and Drexel, Harjes & Company of Paris. A spokesman announced the twenty-nine-year-old scion's reasons:

"Colonel Drexel leaves the firm because he does not care to assume the cares and responsibilities which are attached to the business. He is a young man who is fond of life as it is found in society. His pleasures would have to be curtailed immeasurably were he to continue closely identified with the business his father established, and he does not care for the confinement and close application to which he would be subjected. He prefers to be free footed and will withdraw."

Unlike his father and his grandfather before him, young Colonel Drexel chose to enjoy himself rather than to shoulder the responsibilities of birth. Thus, he relinquished his rights to the Drexel Estate to the other trustees and set out with his young wife for the court-life of London.

It was through the generous terms of his grandfather's will that John Armstrong Drexel, at the age of two, became one of this country's youngest millionaires. As was the family custom, John Armstrong received his early education at his father's residences through governesses and tutors. However, the Colonel, having broken the tradition, also seemed reluctant to raise his sons in the Spartan manner of his grandfather which saw descendants quickly became productive members of the House of Drexel.

Both John Armstrong and his older brother, Anthony Joseph III were sent to Eton College, England, where they soon earned reputations as millionaire playboys. John Armstrong attended Eton from 25 January 1905 to 3 April 1908 when he departed the institution for an intended career in "business."

While in England John Armstrong Drexel developed a sporting interest in aviation and soon lost all desire for business or school. There, in 1909, he paid pioneer British aviator Claude Grahame-White a substantial sum to teach him how to fly. Drexel soon held the Aero Club of America FIA Certificate No.8 and a Pilot-Aviator Certificate with the Royal Aero Club of the United Kingdom.

From unlimited funds bestowed upon him by his father, Drexel was able to purchase a Bleriot aircraft and was soon flying it at Pau, France.

In 1909 young Drexel began to promote aviation in England by financing a flying school near Beaulieu in co-operation with W.E. McArdle, of Bournemouth. In July of that same year Drexel competed for a cash prize by making a sea-flight from the aerodrome at Bournemouth, England, to the Needles, Isle of Wright, and back, a distance of twenty-one miles. Drexel finished in second place behind Frenchman Leon Morane.

Following the Bournemouth meeting, Drexel, with H. Delacombe as observer, flew his Bleriot from the Bournemouth aerodrome to the aerodrome of his Drexel-McArdle aviation school in New Forest. Then on 10 August 1910 he set the world altitude record during an aviation meet at Lanark, Scotland. "Chips", as he was now known through the press, left the ground in a Bleriot at 7 o'clock in the evening and attained a world's altitude record by rising to 6,604 feet. Several weeks previously he had set a new English record one Sunday evening in New Forest when he reached a height of 1,000 feet, topping the record of Louis Paulham who had done 977 feet at Sandown Park on 6 November 1909.

In 1909 American Glenn Curtiss had won the Gordon Bennett speed trophy at Reims, France. Because of his victory, the trophy was to be defended in October of 1910 at Long Island, New York, at the Belmont Park racetrack. Among the Americans chosen to defend the trophy were Glenn Curtiss, the Wright brothers, and four other Americans, including John Armstrong "Chips" Drexel. France entered three flyers on its team, and Great Britain, three, including Drexel's partner, McArdle, and his former instructor, Claude Grahame-White, who had first learned to fly at the Bleriot school at Pau in 1909.

On 22 October 1910 the Belmont meet began in a driving rainstorm. By the 24th the sky had cleared and a contest soon developed into an attempt to set a new altitude record. During the Rheims meet the previous year, the best altitude attained was 508 feet. During the first attempt at Belmont, by the French team, an astonishing height of 5,615 feet was reached. John Drexel, in his 50-horsepower Bleriot monoplane, soon topped that by hitting 7,105 feet, a new American altitude record. This was immediately broken by another American, Ralph Johnstone, who, on the last day of the meet, set a new world record in a 35-horsepower Wright "roadster" by reaching a height of 9,714 feet.

Competition began on 29 October for the Gordon Bennet International speed trophy and the $5,000 purse. Chips Drexel competed until a strong wind proved too much for his 50-horsepower Bleriot, and he had to withdraw from the competition after completing only seven of the required twenty laps. Britisher, Grahame-White took the competition and the prize money in a new Bleriot powered by a 100-horsepower Gnome rotary engine.

Drexel returned to Philadelphia upon completion of the Belmont meet. Above that city, on 24 November, he succeeded in breaking Ralph Johnstone's world record by attaining an altitude of 9970 feet in his Bleriot. However, the record was shortly disavowed because of the International Aeronautic Federation rule which stated that an official record for altitude must exceed the previous record by at least 100 meters (328 feet).

On 6 December 1910 Drexel left New York for England. He promised to return the following year to make another attempt at the record. However, two months later he announced that he was giving up aviation events. But on 12 May 1911 Drexel entered an exhibition flight near London and smashed his machine immediately after becoming airborne. He was able to walk away with cuts and bruises, but from pressure exerted by his parents, his sport flying tapered off considerably following the accident.

Shortly after the outbreak of World War One, John Armstrong's father was able to secure his son a position as a chauffeur for British Field Marshal Sir John French. He served in this capacity until he was invalided home on 11 December 1914 to his mother's London residence. The press called his "indisposition ... not serious."

In the fall of 1916, while at his father's Paris residence, John Armstrong decided to enlist in French avia-

The first American Airplane Racing Team. From left to right: John Armstrong Drexel of Philadelphia, Walter R. Brookins of Dayton, Ohio, and Charles K. Hamilton of New Britain, Connecticut.

tion. He joined the *Service Aeronautique* on 27 October 1916.

Drexel was sent to the Bleriot school at Buc but soon proved his mastery of flying to his moniteurs and was passed on without delay to the advanced flying course.

Drexel's air of detachment and his predictable absent-mindedness were the characteristics frequently recalled by his American colleagues in flying school. They would observe with stifled humor his walks to his training craft where he would invariably realize he had forgotten some article of his flying gear or his barograph for his altitude test.

Drexel received his *brevet militaire* on the Bleriot on 6 March 1917. Following acrobacy and gunnery school at Pau, he reached the *Escadrille Lafayette* at Ham on 10 May 1917.

While at Ham and at Chaudun, *Caporal* Drexel served dutifully with the Squadron, flying combat patrols nearly every day of his tenure there. But in his off-duty hours, Drexel was looked upon as a stuffed shirt. With barely contained mirth his fellow pilots listened to his cultivated Eton accent and observed his ongoing ritual as he readied himself for bed each night.

Upon shedding his uniform, before climbing into his cot, Drexel would first don an ankle-length flannel nightgown then put on a tasseled night-cap before tying up his moustache with a silk string which he knotted behind his head, this, to keep his moustache in place while he slept. Viewed as somewhat effeminate by his rough-hewn American comrades, he was never a popular figure with the Squadron during his short tenure there.

Henry Jones, who roomed with Drexel during his stay at Chaudun, remembered that Drexel had always had a valet and seemed lost without one. Jones had to tell Drexel what clothes to wear.

Through his father's influence, Drexel left the Squadron on 15 June, thirty-six days after his arrival there. He was shortly reassigned in a liaison capacity between the French and the U.S. Air Service staff at Paris where an office had recently been opened on the boulevard Hausmann.

Caporal Drexel remained in this capacity until he was commissioned a Major in the U.S. Air Service some two months later, on 14 August 1917, the first *Escadrille* pilot to receive such a commission.

This quantum promotion caused some resentment among Drexel's *Lafayette* colleagues since they viewed it as having been bestowed out of preference, and from his father's influence, rather than for merit.

At the time of his commission, Drexel's father, Colonel Anthony J. Drexel had been serving as a civilian interpreter to Major General A.W. Brewster, Inspector General of the A.E.F. From there, he could better judge the pulse of France's War effort and watch over his considerable real estate holdings and the Drexel interests in the Morgan-Harjes Paris bank which was then helping the beleaguered Republic with loans for munitions to keep the enemy at bay.

Following his commission, Major John Armstrong Drexel served in executive capacities in France and in England. In September of 1917 he was designated American Aeronautical Officer in the British Isles by Colonel R.C. Bolling. In mid-May of 1918 he was sent to the United States, to Washington, D.C. where he served as liaison officer at the American Air Department with the British Air Mission until the November 1918 Armistice. He concluded his war service at the rank of Lieutenant-Colonel. The British Government decorated him with the Order of St. Michael and St. George.

Following World War One, John Armstrong represented Drexel-Morgan banking interests. He became a partner in an affiliate firm, the Bond House of William P. Bonebright & Co. of London.

On 8 May 1926 John Armstrong made front-page copy in the *New York Times*. This time it was for his actions during the British general labor strike which was then tying up the English continent. Drexel helped to break the strike when on 7 May he had arrived at the London Station in his limousine and had handed his derby and overcoat to his chauffeur before donning a pair of engineer overalls. Emphasizing to the press that he had made mechanics his hobby, he climbed aboard the "Flying Scotsman", the prize Edinburgh train, and drove it out of the London station for Edinburgh amidst the cheers of volunteer workers. The *Times* article noted his past service in the *Lafayette Escadrille* and further stated that "once on a flight over the German ammunition depot at Bruges he dropped a ton of bombs."

Drexel remained involved in banking throughout his professional career. In the summer of 1934 he was elected director of the Anglo-South American Bank, LTD., and the British Bank of South America. Later that same year his father, Colonel Anthony J. Drexel, died of uremic poisoning on 14 December 1934 at the Philadelphia Ambassador Hotel. The press made little note of the Colonel's business acumen, only that he had entertained European royalty on a lavish scale and had been a keen yachtsman and owner of the *Vanduara*, the most handsome craft in European waters.

To his distress, John Armstrong Drexel soon learned that his father had left the bulk of his depleted estate to his long-time Parisian friend and her son. Under the terms of the Colonel's will, Madame Helene Barth, "a handsome middle-aged Frenchwoman," was to receive $25,000 in cash plus a life income from two trust funds totalling $500,000. John Armstrong received the sum of $5,000.

Colonel Joseph Drexel as he appeared in World War One while serving without rank as interpreter to Major General A.W. Brewster, inspector general of the A.E.F.

Having learned from his father's experiences, John Armstrong eschewed marriage. He developed a long-standing relationship with Lady Muriel Agnes Stuart Willoughby and maintained a working farm, at Bates Farm, England, where he raised thoroughbreds. He died at the Grosvenor Clinic, Ashford, County of Kent, England at age 66 of a heart attack on 4 March 1958. Lady Willoughby and Evelyn Curzon, another English woman, were the recipients of Drexel's estate. The complete Anglophile, it is in England where he is buried today.

NATHANIEL DUFFY

Nathaniel Edmund "Nat" Duffy was born in Buffalo, New York on 6 June 1891, the son of Mr. & Mrs. Edward E. Duffy. As a youth, Duffy attended parochial and private schools before enrolling in Princeton University. Duffy left that institution before taking a degree. In 1916–1917 Duffy served for nine months on the Mexican Border with Troop "I" of the New York National Guard. When he was mustered out in March of 1917 Duffy decided to enter the European War. The following month he enlisted in the American Ambulance Field Service and sailed for France.

Upon his arrival in Paris, Duffy sought a transfer to French aviation. On 24 May 1917 he enlisted in the *Service Aeronautique*. From 12 June to 23 April 1918 he attended the aviation schools at Avord, Pau, Cazeaux, and the *G.D.E.* He received his *brevet* on the Caudron on 14 November 1917.

On 25 April 1918 *Caporal* Duffy was assigned to *Escadrille SPAD 96, Group de Combat 16*. He flew many ground-strafing patrols and destroyed two enemy observation balloons. On 16 August 1918, while strafing German troop concentrations, he was shot down by enemy ground fire. Duffy was struck in the hip by machine gun fire but still managed to crash-land his SPAD just within French lines. His fighter overturned and he suffered severe facial lacerations and a broken nose. As plastic surgeons worked to put his face together, his features were permanently altered. Because of his injuries, Duffy was unable to continue flying. In December of 1918 he sought and was granted his release from the *Service Aeronautique*. For his exemplary war service, *Sergent* Duffy was awarded France's *Croix de Guerre*, with two Palms.

Demobilized in June 1918, Duffy returned to Buffalo, New York. There he met and married Mildred Pearson of that city. From this union a son and daughter were born.

Duffy worked at various jobs including clerk in the Hotel Statler until the opening of the Buffalo Airport on 1 February 1927. At that time he was appointed Superintendent of Service for that facility. A year later, when the current director was killed in an airplane crash Duffy succeeded him and held the post of director until his retirement in 1960.

Duffy served for 33 years as director of the Buffalo Airport, from February 1927, when it officially opened, until his retirement in July 1960. Duffy was in on the initial phase of the airport's construction and ran the overall operation of the facility. He died in Buffalo, New York, on 16 January 1972 at age 80.

WILLIAM DUGAN

ESCADRILLE LAFAYETTE

"There is no romance or other appeal in this war. It is a grim, sordid, merciless shambles with engines of destruction that revolt and disgust every decent sense and instinct a man has."

William Edward Dugan, Jr., the 27th volunteer of the *Escadrille Lafayette*, was born in Patchogue, Long Island, New York on 21 January 1890, the son of Rochester shoe manufacturer, William Edward Dugan and Sally (Hudson) Dugan. When William was three years old his mother died of tuberculosis. Upon her death, he transferred his affections to his maternal grandmother, Mrs. Martha Hudson, the widow of sea Captain Charles Hudson. Mrs. Hudson with her young children, including Billy's mother, had accompanied the Captain on his vessel on numerous voyages across the Atlantic and to the far east. Undoubtedly, in relating her adventures to young Billy, she had stimulated his imagination and had created in him a desire for adventure. Dugan's father later married Mary Brownell, cousin of suffragette, Susan Brownell Anthony, and sister of Frank Brownell, the developer of George Eastman's first light-proof camera.

Following preparatory school, Dugan spent one year at the University of Rochester before entering the Massachusetts Institute Of Technology, Class 1912. He left M.I.T. the spring of 1909 to take an engineering position with the construction firm then erecting the state education building in Albany, New York.

While in Albany, Dugan and two young friends quit their jobs and left for Panama on a tramp fruit steamer to witness the construction of the Canal. Dugan jumped ship in Nicaragua and hid in the jungles until the steamer had sailed then presented himself for a job at the office of the United Fruit Company.

Dugan was hired as assistant overseer of a banana plantation in Costa Rica. He remained there for three years until war seemed imminent in Europe, and he "saw a chance to embark on a really great adventure."

Soldat William Dugan as an American volunteer in the French Foreign Legion, December 1914.

In late August, 1914 he resigned his position and caught a fruit steamer for Bristol, England.

Dugan landed in London following an eighteen-day passage. He secured a passport to France from the American embassy, caught a boat from Southampton to Havre, and took the train to Paris.

On 19 September 1914 Billy Dugan enlisted at the French Foreign Legion recruiting office at the *Hotel des Invalides*, thus becoming one of eight future *Lafayette Escadrille* pilots to first serve France in the trenches. Eventually he would participate in every major battle from December of 1914 to May of 1916, including the early stages of Verdun. He was assigned to the Third Battalion of the *3me Regiment de Marche de 1er Etranger.*

Following training he left Paris with his unit on 28 November. A seven-day forced march to the Front followed which resulted in "90 per cent of the men becoming sick from poor food or the lack of it. Many were sent to hospitals, while those that stayed were much reduced in spirits and stamina." As Dugan's regiment neared the Front, seven days of useless maneuvers followed which further sapped the mens' strength and enthusiasm.

On 20 December Dugan's 3d Regiment was moved to the Somme to relieve the 2nd Regiment. Both friend and foe prepared for winter's onset. The legionnaires took casualties from German snipers and grenades while upgrading their trenches.

By 1 January 1915 the Regiment had made good progress. But a torrential cloudburst washed all their work away and filled the trenches with knee-deep "cold, liquid mud." Many Legionnaires fell ill from exposure. Rifles clogged and wouldn't shoot as parapets washed away. There was no place to rest.

When the men were withdrawn from the forward trenches, they worked from "daylight to dark" on the second line. Their mud-caked uniforms were filthy and lice infested. They hadn't bathed in a month. Morale ceased to exist.

On 19 January 1915 Dugan's unit returned to the trenches. Desperation pervaded the ranks. But spirits lifted as the weather turned cold and clear. In Dugan's sector, the enemy stood just 25 yards off. His Battalion was briefly withdrawn as French 75 cannons strafed the German trenches. The enemy retaliated with hand bombs, causing the death of several Legionnaires.

About this time William Dugan saw his first aeroplanes which were spotting for artillery. Fascinated, he watched "the evolutions of the graceful machines" and knew the sky was where he wanted to be.

By February of 1915 artillery exchanges between the belligerents had intensified. Dugan's unit was temporarily withdrawn to Cappy where his regiment was reformed from four to two battalions and sent forward to a new sector. Here, the chalk-stone soil made the trenches more habitable.

In this new sector Dugan joined company with future *Lafayette Escadrille* pilot, Victor Chapman, who had already spent two months there. Dugan remembered that Chapman "was a great one to prowl about in the dark and had close scrapes with German patrols and stray bullets."

By mid-March activity had intensified as the Germans exploded underground mines and hit the Legionnaires' trenches with *minenwerfers*. On the 17th Dugan's Regiment was relieved and left the sector it had held since mid-December 1914.

The Regiment was moved to a new sector near Dompierre. Spring brought warm weather and increased hostility. Fusilades, bombs, and trench mortar rounds were exchanged daily, and exploding mines became commonplace in the soft soil. Casualties greatly increased. As many as 450 enemy shells fell into Dugan's sector in one two-hour period.

Around 1 April, following the explosion of a French mine which wiped out 200 yards of German trenches, Dugan's unit was relieved to Cappy. It was here that he first met another future *Lafayette Escadrille* pilot, Edmond Genet. Dugan and his fellow Legionnaires were given a two-week repos and were issued new uniforms and equipment. His 3d Regiment joined the 2d Regiment since the 3d had to be eliminated due to losses, and what remained of three battalions were combined into two.

From 11 May to 28 June 1915 Dugan's Regiment occupied the Tilloloy sector. Most casualties resulted from the German cannon some 400 meters distant. Following a forty-eight hour Paris leave, Dugan's regiment embarked on trains for a journey to Montbeliard, near the Swiss border, in the Champagne sector. The remains of his regiment was combined with the decimated remains of five others to form the First Regiment. Dugan soon discovered that most of the troops he was joined to were from the Moroccan Division, many hardened criminals from France's African Legion. To his dismay, the officers often proved a worse sort than the men. However, Dugan's battalion had the good fortune to be assigned a captain of Swiss nationality named Junod who proved firm, but just. It was during this time that Dugan's friend, Victor Chapman, departed for French aviation.

William Dugan (far left) with fellow American Legionnaires shortly after their transfer to the 170th Line Infantry, October 1915. Paul Pavelka is seated top row, 2nd left. Another future Lafayette Flying Corpsman is Marius Rocle, front row, center. Note Dugan's swollen and blackened eye, the result of a shrapnel wound while fighting in the "Bois de Sabots."

On 16 September 1915 Dugan's Legionnaires again entrained to St. Hilaire. As they climbed down from the cattle cars, they could see the sky ablaze with exploding shells, their ears filled with a dull continuous roar. They marched several hours to a pine wood. A great offensive had been planned for the next morning, a combined French-British assault. Dugan's unit marched into place and settled into a small wood. At daybreak the French batteries opened up.

All day long Dugan sat in the woods listening to the heaviest barrage of the war. Overhead, many artillery aircraft circled. The men moved to another wood and pitched tents and were first issued steel helmets along with new uniforms. They waited for four days as the barrage continued, a "solid roar of thousands of guns."

At dawn on 25 September Dugan dropped into the trenches with Edmond Genet. At ten o'clock the first French wave went over. Dugan's second wave was next to go and immediately came under withering fire from shrapnel, gas, and bullets. Chaos and confusion reigned as troop congestion retarded movement. Wounded were trampled under staggering soldiers gasping for breath through inadequate gas masks. Charging French cavalry fell into the trenches, trampling their own troops with their horses. Dugan's wave finally reached the German trenches some 800 yards distant and routed any enemy cowering in dugouts. There he met a disabled German, his intestines seeping through a bayonet wound.

The attackers caught their breath as French 75's opened with a tremendous barrage. They then advanced through shellfire to a small wood and halted for the night. The sky was filled with gun flashes and rifle fire.

At dawn Dugan's regiment advanced to the edge of the "Bois de Sabots." It was here that the Germans, with their many *mitrailleuses*, had stopped the French offensive. A German plane appeared overhead and almost at once a screaming storm of shrapnel rained down on the legionnaires. Dugan cowered in a shell hole through a hellish six-hour barrage.

"Friends and comrades were blown to pieces; we could see their remains thrown into the air. One chap I knew was hit square by a shell and annihilated. Another had both legs blown off. The concussions of the explosions were stunning. Soon we were troubled by gas. Several small shell pieces hit my steel helmet and a minute portion struck me in the left eye, causing it to swell and bleed profusely. Rocks and debris rained on us, and in misery we glanced over our small parapets hoping for orders to move."

Finally Dugan's battalion commander waved his men to retreat. They broke and ran and "trampled to pulp" those who fell. They did not stop running until they had reached a small wood before the French batteries. Those Legionnaires who had fled without their rifles and ammunition were sent back to the maelstrom to retrieve them.

The next day Dugan's legionnaires were ordered back to the same wood to endure another furious barrage. Men covered their ears and opened their mouths to save their eardrums. Communication was impossible. The concussions from the explosions struck their faces and caused them tremendous headaches.

Then on 28 September Dugan's Battalion was given an order which had already expended two other battalions. Capture the German machine gun fort in the "Bois de Sabots."

The enemy could see their every move. The Legionnaires were enfiladed on all sides by machine gun fire, and they fell like leaves. Dugan was one of a handful of men to reach the enemy wire. There they lay, under thundering fire for six hours.

Dugan clawed a hole to hide in. He used discarded equipment to construct a small parapet.

"As darkness came it began to rain and I took the chance of stepping over to one of the numerous dead and taking his tent cover to keep myself dry. Machine guns spat a leaden stream. Men all about me were crying out and dying, and bullet after bullet slapped into my little parapet. The screams of the wounded were heart rending. Men swore and cursed and begged for mercy and water, and called on their mothers and their God. And then, in a drizzling rain, as German star shells lit up the scene and dropped in our midst, I fell asleep."

Hours later Dugan was kicked awake and ordered back to the rear. There, the meager survivors wolfed down rations from the soup kitchen which had been prepared for three times their number. Then he fell into a deep sleep. He had survived the first awful day of the Battle of Champagne.

The next day Dugan ran into Paul Pavelka, another Legionnaire destined to fly for the *Lafayette Escadrille*. They spent the day dodging shell fire. A little later they ran into Edmond Genet. On 6 October 1915 they were sent back into the line.

While moving forward through an old German communication trench Dugan's section came under heavy shell fire. The section before his own was completely obliterated and Dugan had thought his eardrums had burst from the concussion of the explosions. Upon reaching their trench, the Legionnaires immediately came under fire from shrapnel, high explosives, and trench torpedoes. This lasted for four days. Their parapet was blown in, burying the men. They dug each other out.

After four days of this hell, Dugan's regiment was relieved. The next day, the twenty American Legionnaires were transferred to the 170th French Regiment in answer to a request they had made two months earlier. On the 13th of October they found their new regiment seventeen kilometers behind the lines. Upon joining the 170th the Americans were immediately sent with it back to the Front, to Bois 28, in Champagne. It was there that American Legionnaire, Robert Soubiran, was wounded. He, like Dugan, was also destined to fly in the *Lafayette Escadrille*. For eight days Dugan and his comrades cowered under constant bombardment then were sent to the rear for repos.

Winter was soon to arrive. The French offensive had failed. Dugan's regiment was ordered to establish a solid, permanent line. In mid-November he and Pavelka were granted a four-day permission to Paris, Dugan's first leave in twelve months at the Front.

While in Paris Dugan explored the possibility of getting into aviation. He returned to his regiment at Chalons. Soon after rejoining it he fell and nearly severed a finger.

New Year's Day 1916 was spent on the march. The regiment stopped in Givray, in the Argonne. Pavelka left for aviation and Dugan and Marius Rocle, another American who had been with him for many months, continued to petition for aviation. Dugan wrote to Victor Chapman, then in flying school, and Chapman did all he could to secure Dugan's transfer.

Dugan was granted a second permission to Paris. Chapman met him there, with William Thaw and Norman Prince. They went together to the Ministry of War and to the Franco-American Committee and the paperwork was set in motion which would eventually effect Dugan's transfer to aviation some months later. Dugan returned to Givry to his regiment which had hunkered in for a bitter winter.

By mid-February Dugan's regiment was again on the march, this time in the direction of Verdun. A tremendous German bombardment was taking place, and a long-threatened enemy offensive was soon expected.

On 26 February Dugan left San Andre on a twelve-hour march through snow. Near Verdun the night sky was ablaze from cannon fire. Dugan's regiment took refuge in a small fort west of the city. Next day they crossed the Meuse river and reached a wood, the "Bois d'Hospice." They were soon sent forward as reserves toward Douaumont, but after a march through knee-

Sergent William Dugan with Soda at Saint Juste, April 1917.

(r.l.) Raoul Lufbery, William Dugan, Dudley Hill, Chouteau Johnson, Stephen Bigelow, and Didier Masson, Chaudun, July 1917.

deep mud, they were ordered back to the wood. Soon, they were ordered to Fort Vaux.

Dugan journeyed through a nightmare to reach Vaux. He later wrote of it:

"I shall always see it. Dead horses, wrecked kitchens and trucks and guns and dead stretcher bearers with their stretchers and their charges still on them; arms and heads and legs and man meat and horse meat all one great sickening trail all the way to Vaux; shrapnel and shell breaking in and around our column; wounded screaming and all of us with fear in our hearts."

The German batteries caught Dugan's regiment in an open field before Fort Vaux. Shells burst in their midst, killing many. In the darkness of night the men dug in.

On 2 March 1916 German airplanes appeared overhead. Almost immediately a tremendous barrage hit the regiment. Dugan's close friend, Jack Janz, an American, was hit. That same day Janz had predicted his own death by cannon fire. Dugan shook Janz's shoulder. "Then I saw his face. There was no face left. It had been blown off, and Jack was dead."

The bombardment was relentless, the heaviest Dugan had yet endured. The incoming shells from the German heavy guns roared like express trains. They came by the thousands, leaving craters forty feet in diameter. Crouched in a ditch in numbing ice water, he prayed to be spared, then to be killed, "anything to end the bombardment." The earth rumbled and heaved and vomited itself skyward in great clumps of earth, rocks, and great trees which showered down as kindling. Then the cannon stopped and the enemy attacked, intending to go through Dugan's regiment on to Verdun.

Dugan frantically fired his rifle and prayed for a barrage into the German ranks from the French 75s. The batteries soon opened up, each cannon belching twenty shells a minute. The attacking Germans faltered, broke, and fled.

Nightfall brought a comparative quiet to the battlefield. Dugan and Rocle used the lull to bury their friend, Jack Janz. Fort Vaux still stood, "surrounded by pieces of wood, guns and men."

Dugan's regiment was withdrawn to the rear for a twelve-hour rest. On the way he slaked his thirst from a scummy pool with a foot sticking out of it. Heading in, they were hit by gas shells.

Following a brief repos, the men were ordered back to the nightmarish wood, the "Bois d'Hospice." They were to attack and take the village of Douaumont.

As soon as they had left their shelter for the attack across a plain, the Germans opened on them with a heavy shrapnel barrage which pursued them to the village of Fleury. Under cover of night they reached the trench leading to Douaumont. The trench was under bombardment and packed with men. Confusion reigned.

As Dugan and his friends stood ankle-deep in icy water, shrapnel shredded the air. Next morning five German planes spotted Dugan's trench. With him were Americans Rocle and David King. All day until darkness the enemy aircraft directed artillery fire down on Dugan's regiment. King named the spot "Death Valley."

Dugan and Rocle hadn't eaten in two days. When they were given soup, they vomited. They took water and vomited. They had been too long without both. Moving toward Douaumont, the men fell many times on the ice. When they halted they were ordered to dig a new front line fifty yards before the Germans in the village of Douaumont.

At dawn on 7 March German riflemen opened up on Dugan's regiment. In spite of the night's digging the trench stood just waist deep and would not hold all the men as they crouched for cover. Many were killed. Night found the men sick and fatigued. Morale had ebbed to its lowest point. Fortunately, their trench line formed a salient in the German positions, and the enemy could not shell them without hitting their own troops.

Dugan's regiment was relieved at nightfall and was returned to the "Bois d'Hospice." The wood was now "stumps of trees, smashed up equipment and human limbs." The regiment left the haunted woods for the rear, sick from hunger and exhaustion. Dugan had colic, could not eat, and had lost twenty-five pounds. All his comrades were in similar condition, or worse. Yet they were put on a three-day forced march to Vaucouleurs.

Upon their arrival there, Field Marshal Jofre reviewed the brigade. Most of the men were too sick to stand. Soon men fell deathly sick and died in convulsions from the effects of gas. Dugan was given ten-days exemption from service to regain his strength.

Troops were brought in to rebuild the depleted ranks. The men stayed in repose for a month then on 11 April they were ordered back to Verdun. Easter Sunday, 23 April 1916 was spent on the march to the Front. Dugan was exhausted and sick at heart at the loss of so many of his friends. In a letter home he wrote this to his parents.

Sergent William Dugan beside his SPAD S.VII fighter, Chaudun, July 1917.

"After Verdun my morale is rather low, and I dread very much the next one. I am now credited with three campaigns and am beginning a fourth. There are few in the Army who have been out so long without intermission and I have frankly had more than enough."

Dugan's regiment reached the woods near Verdun on 24 April. They rested there until dark, then skirted Verdun and halted in a suburb. They set off again and reached Fort Souville. Soft sticky mud pulled at their limbs and clogged their rifles. Enemy shelling fell relentlessly. The desolate landscape had been pounded level and dead were strewn everywhere. They fell into position, relieving the men already there.

Dugan's regiment, only thirty meters from the enemy lines, fell under both German and French bombardment. A battery of French 155s opened up and dropped their rounds short, killing or wounding half of Dugan's section.

A French offensive was planned to begin at noon on 1 May 1916. A two-hour bombardment was to precede it to soften the enemy lines. Because of the proximity of the Germans, all French troops were to be withdrawn from the forward trench with the exception of two men from each squad who were to endure the shelling and "prevent a reaction by the enemy." Dugan was one of the two chosen from his squad. He knew it was a death sentence.

Lieutenant William Dugan at home in Rochester, New York, in the spring of 1919. (l.r.) Frank, William, Charles, and Harold. William, Sr. stands behind his sons on the steps of their home.

At noon the trenches were cleared with the exception of Dugan and a handful of others. Then the bombardment began. He found a small hole to crouch into as the trench around him was smashed and obliterated. Shell fire rained down upon him from friend and foe alike, and his head ached horribly from the concussion of the shells.

After two hours the bombardment ceased. Miraculously, Dugan had been spared. The sections of the trench that had not been obliterated soon filled with men. Whistles blew, and Dugan went over with the attackers. Stumbling over churned-up ground, they made their way to the German trenches. Just as Dugan reached them, an enemy hand bomb exploded near him, wounding him in the face and hip.

Dugan's company fell into the German trenches, occupying them. Dugan fought bravely and took several prisoners for which he would later be awarded France's *Croix de Guerre*, with Star and promoted to *Sergent*. But the enemy immediately bombarded with *minenwerfers* then counter-attacked. Dugan's troops were driven back to their own trench. However, the rest of the division had held its position, and the attack was considered a success.

Meanwhile Dugan's company came under fierce enemy attack. They fired into the German ranks until their rifle barrels were "sizzling hot." Dugan spotted a large Prussian trooper repairing his parapet. He fired several times at him and missed. The German seized his gun and fired at Dugan. A sharp pain exploded in his left arm, and it dropped useless at his side. Abandoning his rifle, he clambered out of his trench toward a dressing station. He jumped into a large shell crater. Two Germans cowered inside. He motioned toward his arm, and they applied a tourniquet and placed his arm in a sling.

He and another American, Charles Hoffecker, who had been wounded severely in the backs of both thighs, tumbled into the French communication trench. A German bombardment pounded around them until dark.

Dugan and Hoffecker were eventually evacuated to a small town back of Verdun. At a nearby church their wounds were dressed and bandaged and they were given injections against lockjaw. Here, the two separated, with Hoffecker going to the south of France where he died of his wounds a few weeks later, while Dugan went to Bar-le-Duc, to the hospital for broken limbs.

The enemy bullet had passed through his wrist and had cracked a bone. A piece of grenade was taken from his hip. On 10 May 1916 he was placed on a hospital train for Saint Etienne.

On a number of occasions while at the Front Dugan had requested transfer into the *Escadrille Americaine*. But his commanding officers would not believe that such a squadron would ever become a reality and thus had refused to assist him in his efforts. It was only after he had been wounded and was at the Saint Etienne auxiliary hospital that he was able to see his dream realized. His father had been pulling some Paris strings, and word had reached Dr. Gros of Dugan's predicament. Gros visited Dugan in hospital, and by the time Dugan had recuperated from his wounds, Gros had arranged his transfer into the *Service Aeronatique*.

During June Dugan received orders that he had finally been transferred to aviation and was ordered to report to the school at Buc when he had finished his convalescence. A month later he left Saint Etienne, and after a week's leave in Paris, he began his career as an aviator.

Sergent Dugan enlisted in French Aviation in early July of 1916. On the 9th of that month he arrived at the school at Buc. On 20 December 1916 he received his *brevet militaire* on the Bleriot. He then went through the schools at Juvisy, Avord, Cazeaux, and Pau before arriving at the G.D.E. pilot pool. He joined the *Escadrille Lafayette* at Saint Just on 30 March 1917.

The diminutive "Duke" Dugan soon joined his former Legionnaire comrades, Edmond Genet and Robert Soubiran, on combat patrols. As in the trenches, he quickly proved his worth as a fighter. One comrade, Edwin Parsons, later wrote of him:

"Bill and I were very close friends and I greatly admired him for his bravery not only in the air but in the Foreign Legion. We fought side by side in the air and he was always a most dependable stalwart in a fight. On the ground, he was a most congenial companion. No hardship, no crisis seemed to affect his constant cheerfulness and good humor."

Dugan's closest escape while in the *Escadrille* came in April of 1917 when, on a morning patrol, he was attacked by two German Albatros fighters deep inside enemy territory. They disabled his aircraft, and chased him to allied lines where Dugan managed to elude them and land on a British airfield.

While he was overseas, Billy's mother had asked his childhood sweetheart, Romona Adams, to write to him. When he had come home on leave, they were married on 8 September 1917 at Bronxville, New York. When Dugan returned to France, he took the new Mrs. Dugan with him.

On 11 January 1918 *Sergent* Dugan was commissioned a First Lieutenant in the U.S. Air Service. He joined the U.S. 103rd Aero Pursuit Squadron on 18 February 1918 and flew as a pilot with that squadron until 31 May 1918 when he was transferred to duty as Dispatching Officer at the American Aviation Acceptance Park, Orly Seine. While at Orley, Lieutenant Dugan was made officer in charge of repair and testing, and performed the final inspection of French aircraft before they were turned over to the American Air Service.

For his fifty months of constant service to France, Lieutenant Dugan was awarded France's *Legion d'Honneur*. Following the signing of the Armistice and the end of his overseas service, Lieutenant Dugan and his wife returned to the United States. He resigned his commission in March of 1919 at the rank of Captain. He then returned with Romona to Costa Rica where he was made a farm manager for United Fruit Company before eventually taking charge of twenty-nine plantations.

While Dugan was in Costa Rica, in early January 1921, Romona gave birth to a son, William, III. Romona fell ill shortly after and died unexpectedly on 11 January. Billy wanted Romona to be buried near her family. Leaving his infant child in the care of a white nurse, he accompanied Romona's body to New York.

The day following Romona's funeral on 25 January, Billy received a telegram at his father's home informing him that his infant son was gravely ill. He hurried back to Costa Rica and for a time the child seemed to rally. Dugan planned to send him to relatives in New York in hope that he would recover. But before he could do so, his son died in his arms on 29 April, 1921. William again returned to the States and buried his son beside Romona on 9 May.

For a brief time following the death of his family, Billy returned to his father's home in Rochester. He had lost his purpose in life, and rejected the offer of a vice-presidency with United Fruit Company. But when that firm asked him to explore new territory for banana plantations, he saw in the offer a distraction from his grief.

After taking the company yacht to Nicaragua and to Panama, he then headed far up river in a canoe and would trek deep into the jungles. For two years he explored the territory south of the Panama Canal. While he was there, he debunked the white-Indian theory put forth by one Professor Marsh, an explorer, proving instead that the inhabitants there were albino Indians.

He continued with his work, reporting on areas suitable for United Fruit's new banana plantations. He made maps for prospective railroads and ports.

By late summer of 1924 he had completed his work in the jungles of Central America. He fell ill with what was thought to be malaria and was sent home by his firm to New York to recuperate.

On 3 September 1924, shortly after arriving at the Patchogue, Long Island home of his maternal grandmother, Mrs. Martha Hudson, Billy fell ill and was rushed to the Patchogue hospital. He died there at age 34 on 4 September 1924 of septicemia. He was buried on 7 September in Woodlawn Cemetery in the Bronx, New York, beside his wife and infant son.

The senior William Dugan had not heard of his son's return to New York, and only learned of his death when a reporter arrived at his Rochester home and showed him a copy of the local paper. He was deeply grieved at his loss. That same year his shoe manufacturing concern, the Dugan and Hudson Company, would fail due to a lengthy labor strike.

Left: William Dugan in 1920, shortly after resigning his commission from the U.S. Air Service. Right: Romona Adams Dugan, William's childhood sweetheart who died unexpectedly after giving birth to Willaim's son, on 11 January 1921.

LOWELL DULON

Lowell Richards Dulon was born in New York City, New York on 19 April 1893, the son of Ruold and Mabel (Richards) Dulon. Dulon attended the University of Wisconsin, Madison, Class of 1915. He left school before completing his degree and worked as a teacher.

Shortly after the United States entered World War One, Dulon sailed for France and enlisted in France's *Service Aeronautique* on 15 June 1917. On 21 October he was released because of inaptitude.

Following his release from French aviation, Dulon joined the AEF and was assigned to U.S. Artillery. He was commissioned a First Lieutenant in the U.S. Army and was assigned to the Coast Artillery Corps.

SHERBURNE EATON

Sherburne Eaton was born in Woburn, Massachusetts on 28 June 1897, the son of Charles and E. (Sherburne) Eaton. He was educated at Cambridge Latin School, 1915, and Harvard College, Class of 1919. While at Harvard he participated in freshman and varsity football.

Eaton left Harvard and sailed for France where he enlisted in the Foreign Legion (Infantry) on 2 March 1917. On 25 March Eaton enlisted in France's *Service Aeronautique* in Paris. He was released on 1 September 1917 because of inaptitude following his second smashup in training.

On 27 October 1917 Eaton was invalided out of the *Service Aeronautique* because of the injuries he had received in training. Eaton was then detached on special service to the U.S. Army, and in April of 1918 he was attached to the 1st Aero Squadron, USAS, as intelligence officer and interpreter.

Plagued by his service injuries, Eaton returned to the United States to lecture on the second and third Liberty Loans and on the Red Cross drive with the "Men from the Front" Section, U.S. Committee on Public Information. He then acted as inspector of aeroplanes and served on the "Safety Committee." On 23 August 1918 he entered the Field Artillery Central Officers' Training School, Camp Zachary Taylor, Kentucky. On 28 November 1918 he was honorably discharged.

Following World War One Eaton worked as a bond salesman. He also worked for Colonial Western and Canadian Colonial Airways, Inc. as an assistant to the president. He resigned to become a representative for the United Fruit Company. In 1948 he purchased and began operating a dairy farm in Kortright, New York. He married Florence Tyler of Oneonta in 1953. Eaton retired from business in 1960. He passed away in Sidney, New York on 16 July 1965 at age 68.

STUART EDGAR

Stuart Emmet Edgar was born in Nutley, New Jersey in 1890, the son of Howard Edgar, a cattle rancher, and Louise Griffin (Smith) Edgar. Stuart Edgar attended Newark Academy, Newark, New Jersey before enrolling in Cornell University, Class of 1913. Edgar remained at Cornell for two years in the law course before leaving that institution in 1911 to work as a reporter on the staff of the New York *Evening Sun*.

Edgar left his job on the *Sun* to enter the European War. From 1916 to May of 1917 he served as an ambulance driver with Section 7 of the Norton-Harjes Ambulance Service.

On 9 May 1917 Edgar joined France's *Service Aeronautique*. From 17 May to 8 December 1917 he attended the aviation schools at Avord, Pau, and the *G.D.E.* He received his *brevet* on the Caudron on 23 September 1917.

On 11 December 1917 *Caporal* Edgar was assigned to the Front with *Escadrille Nieuport 158*. He flew with that squadron until 28 March 1918. On 1 March 1918 Edgar was commissioned a First Lieutenant in the U.S. ASSRC. From 4 April to 30 May 1918 he served at the American Acceptance Park at Orly.

On 30 May Edgar was assigned to the 103 Aero Squadron, USAS and flew with that squadron until 17 August 1918. On that date he left the aviation field to make a patrol, when at a height of 400 feet his engine stopped dead. His machine lost speed and spun into the ground, instantly killing Lieutenant Edgar. He was 28 years of age at the time of his death. He was buried at the American Cemetery at Bruley, near Toul. His remains were later removed to the St. Mihiel Cemetery, and on 1 June 1928 to the Lafayette Flying Corps Monument, St. Cloud.

Donald Eldredge

Donald Herbert Eldredge was born in Rockford, Illinois on 5 November 1896, the son of Herbert and Daisy (Ames) Eldredge. At the age of two he moved with his family to South Bend, Indiana where his father served as president for the George Wyman and Co. department store. While attending South Bend public schools, Donald worked as a newsboy, and a cash boy for his father's department store. He attended the Culver Military Academy Summer Naval School at Culver, Indiana in 1914.

Eldredge left South Bend to attend Dartmouth College, Hanover, New Hampshire, Class of 1919. He was a sophomore at Dartmouth when he and other students formed the first Dartmouth ambulance unit on 15 April 1917. He and his classmates sailed for France on 6 May 1917 aboard the S.S. *Espagne* to serve in the American Ambulance Field Service.

On 13 June 1917 Eldredge joined France's *Service Aeronautique*. From 20 June 1917 to 20 February 1918 he attended the aviation schools at Avord, Pau, and the *G.D.E.* He received his *brevet* on the Caudron on 3 December 1917.

From 24 February to 10 June 1917 *Caporal* Eldredge was assigned at the Front to *Escadrille SPAD 76*. On 17 June 1918 *Sergent* Eldredge was commissioned a Second Lieutenant in the U.S. Air Service. On that date he was again attached to his old French squadron and flew with *SPAD 76* beyond the 11 November Armistice to 1 December 1918 when he was ordered back to the United States. Eldredge was awarded France's *Croix de Guerre*, with Palm for his excellent and heroic service at the Front. He was credited with the destruction of one enemy aircraft.

On 17 February 1919 Eldredge returned from overseas on the S.S. *Rotterdam*. On 20 February he was discharged at Garden City, New York.

Donald Eldredge returned to South Bend, Indiana and rejoined the George Wyman & Co where he quickly rose to vice-president and sales-promotion manager. He later became president of the company, and was one of South Bend's most influential business and civic leaders. He also helped to establish South Bend's first airport.

Eldredge married Elsie Lippincott in South Bend. The couple had three sons and a daughter. Donald Eldredge died suddenly of a heart attack at age 47 while on a business trip to New York City on 17 January 1944.

CHESTER ELLIOTT

Chester Arthur Elliott was born in Peninsula, Ohio on 18 February 1897, the son of Arthur Monroe Elliott, a tire company executive, and Ethel (Mott) Elliott. Elliott attended public schools before enrolling in Harvard University on a music scholarship, Class of 1915-17,19-20. While at Harvard, he studied pre-med and chemistry. He played the piano at social functions to earn spending money.

Elliott left Harvard to join the American Ambulance Field Service in France. From 1 February to 1 July 1917 he served at the Front with SSU 9. Elliott was gassed while hauling wounded in his ambulance.

On 2 August 1917 Elliott enlisted in France's *Service Aeronautique*. He received his *brevet* on 22 October 1918. On 1 February 1918 he was released at the *G.D.E.* at his own request in order to take a commission in the U.S. Military. On 26 August 1918 Elliott was commissioned a Second Lieutenant in the U.S. Air Service.

Lieutenant Elliott remained in France and studied chemistry, math, and French at the Soborrne. On 1 July 1919 he returned to the United States. He was discharged on 10 July. That fall he returned to Harvard University to finish his education. He again left school before completing his studies to join Chevron of California in their aviation sales department. He spent a year in South American with the United Fruit Company. He married Ethel Ann Myers in 1922 and from this union two daughters were born. Shortly thereafter, he returned to California and was employed by Standard Oil and by Phillips Petroleum.

During World War Two Elliott served in a civilian capacity as an aviation petroleum specialist. In 1944 he signed on with Exxon of Standard Oil. He was involved in aviation all the time he was working for the oil industry. He saw to it that aviators such as Charles Lindbergh and Amelia Earhart had fuel at their refueling stops. Elliott retired from business in 1960. He died in Canoga Park, California on 29 January 1981 at age 83.

DINSMORE ELY

Dinsmore Ely was born in Chicago, Illinois on 16 May 1894, the son of Dr. James Owen Ely, a physician, and Emma (Dinsmore) Ely. As a young man Ely attended the New Trier High School, and the University, and Chicago Latin Schools. In the fall of 1914 he entered the Massachusetts Institute of Technology, Class of 1918. Ely attended the Institute through 1917, studying architecture.

In May of 1917 Ely volunteered for service with the Tech unit of the American Field Service. On 25 June he sailed from New York on the *Rochambeau* and arrived in France on 4 July. At the American Field Service office in Paris he discovered that ambulance drivers were no longer in demand, and he was to be reassigned to drive a camion. Ely objected to this and immediately resigned from the Field Service.

On 13 July 1917 Ely enlisted in France's *Service Aeronautique*. From 20 July 1917 to 20 February 1918 he attended the aviation schools at Avord, Tours, Pau, Cazeaux, and the *G.D.E.* He received his *brevet* on the Caudron on 25 October 1917.

On 24 February 1918 *Caporal* Ely was assigned to the Front with *Escadrille SPAD 102*. He flew ground-strafing missions with that squadron until 1 April 1918, seeing service in the Toul sector. On 6 April 1918 he was commissioned a Second Lieutenant in the US Air Service. He was reassigned, at his own request, to his French Escadrille.

On 18 April 1918, while flying a SPAD to rejoin his unit at Montdidier, he met with an accident near Villa-Coublay and was taken to Military Hospital No. 1 in Versailles with a fractured skull. He died three days later without regaining consciousness.

This phrase from his last letter to his parents, "It is an investment, not a loss, when a man dies for his country," was used throughout the United States in aid of the Liberty Loan campaign. One-million dollars was subscribed as a memorial to Lieutenant Ely. He was buried at the Des Gonards' Cemetery, Versailles. His remains were later moved to the Lafayette Flying Corps Memorial near Paris.

JOHN ENDICOTT

John Endicott was born in Newtonville, Massachusetts on 3 February 1894, the son of George Gardner Endicott, and Emily (Cunningham) Endicott. He prepared at the Noble and Greenough School, Boston, and received an A.B. degree from Harvard University in 1915. Following graduation he entered Harvard Law School but left there before taking a degree to serve in the American Red Cross. On 5 June 1917 Endicott enlisted in France's *Service Aeronautique*. On 12 September he was released for disciplinary reasons.

On 15 October 1917 Endicott returned to the United States. On 12 December 1917 he enrolled as a seaman 2d class US Naval Reserve Force. On 4 March 1918 he was promoted to chief quartermaster, assigned to Massachusetts Institute of Technology. On 3 June he was appointed ensign, and served as instructor, Naval Aviation Detachment. On 10 December 1918 he was released from active duty.

Following World War One Endicott returned to Harvard Law School for one semester. He left that institution in 1922 and was employed by the Ford Motor Company in Detroit, Michigan. John Endicott moved to New York City in 1930. He died there on 2 November 1931 at age 36. He was survived by his widow, Muriel.

ROBERT EOFF

Robert Grimshaw Eoff was born in Christiansburg, Virginia on 1 January 1893, the son of Edward John Eoff, a newspaperman. He attended the Virginia Polytechnic Institute, Class of 1918, where he studied electrical engineering. In March of 1917 Eoff joined the American Ambulance Field Service. Upon reaching Paris, he was assigned to S.S.U. 18 at Vadelaincourt in April 1917 and served at Verdun.

On 14 June 1917 Eoff enlisted in France's *Service Aeronautique*. From 2 August 1917 to 20 January 1918 he attended the aviation schools at Avord, Tours, Pau, and the *G.D.E.*. He received his *brevet* #9535 on the Caudron on 27 October 1917.

On 24 January 1918 *Caporal* Eoff was assigned to the Front with Escadrille *Nieuport 157*. He flew with that Squadron until 27 March 1918, and was awarded France's *Croix de Guerre*, with Star for his 22 March 1918 destruction of an enemy aircraft.

On 11 June 1918 *Sergent* Eoff was commissioned a Second Lieutenant in the U.S. Air Service. From 29 June to 23 August 1918 he was assigned to the American Acceptance Park, Orly. On 23 August he was assigned to the 95th Aero Squadron, USAS. He served with the 95th until 5 November 1918 when he joined the 185th Aero Squadron, USAS, where he remained until the November Armistice. He then was transferred to Air Service Headquarters, Paris, where he remained for one year.

Upon his return to the United States in late 1919, Eoff finished the requirements for his degree then took a position with Edison Lamp Works. In 1928 he joined the R.C.A. Corporation, becoming a warehouse manager in Los Angeles. He married Dorace Thon there on 4 June 1939.

Eoff retired from R.C.A. in 1952. He then became a radio and television distributor for R.C.A. in Hawaii and retired in 1962 to California. Eoff died of pneumonia on 20 January 1989 in Laguna Beach, California at age 93.

EDWIN FAIRCHILD

Edwin Bradley Fairchild was in Manila, the Philippines in 1892, the son of George H. Fairchild, a sugar processing entrepreneur. Edwin Fairchild spent most of his youth in the Philippines were his father developed a process to make granulated sugar from cane.

Fairchild attended private schools in Manila. In 1909 his father sent him to school in Dresden, Germany where he was still studying at the outbreak of World War One. One day, while chatting with another student, a German citizen accused him of being a British spy and attacked him. After being briefly detained under suspicion of being a spy, Fairchild left Germany. He returned to the United States and entered Cornell University, New York, Class of 1919.

In early 1917 Fairchild left school and served four months in France with the AFS in TMU 526. Desiring to serve more actively, on 27 June 1917 Fairchild enlisted in France's *Service Aeronautique*. From 2 August 1917 to 14 January 1918 he attended the aviation schools at Avord, Tours, Pau, and the *G.D.E.*, receiving his *brevet* on the Caudron on 29 September 1917.

On 16 January 1918 *Caporal* Fairchild was assigned at the Front to *Escadrille* SPAD 159 and flew with that Squadron until the 11 November Armistice. Fairchild was a brave and skillful pilot. He fought numerous combats and was credited with the destruction of one enemy aircraft and one observation balloon. Fairchild was wounded and shot down on 22 October 1918 in the Argonne Forest. However, he continued to fly with his squadron and was promoted to the rank of *Adjudant*, and was awarded France's *Croix de Guerre*, with two Palms and two Stars. He was discharged on 19 December 1918.

Following World War I Fairchild returned to Manila to work on his father's plantation. In 1929 he married Mary Edith Phillips, an Australian, in Manila against his father's wishes, and was disinherited by his father because of it. From 1930 until 1941, when the Japanese invaded the islands, Fairchild ran his own food processing plants outside of Manila. The Japanese destroyed his plants and interned the Fairchilds in the Santo Tomas camp. They were freed when U.S. troops retook Manila.

On 29 March 1945 the Fairchilds were evacuated from Manila on the troopship *John Lykes* by the U.S. Army and brought to Los Angeles on 3 May. The Fairchilds continued to reside in California after World War Two. While Edwin had been in prison he had developed a process which made coconuts digestible to all people, since it was the only food available in any quantity near the prison. He patented the process in California. He was also an inventer with dozens of other patents to his name. He later worked in the food processing lab of the Anderson Pea Soup Company north of San Francisco. He died of cancer in Palo Alto, California on 7 July 1979 at age 76.

CLARENCE FAITH

Clarence Henry Hoggson Faith was born in New Haven, Connecticut on 9 February 1893, the son of Charles Burton Faith, a bookkeeper, and Lillie May (Hoggson) Faith.

Faith was a student at Tufts College Medical School, Medford, Massachusetts, when he decided to enter the European war as an ambulance driver. Faith left school following the fall 1916 semester to serve in the American Ambulance Field Service. He spent six months in the field with SSU 12 and was awarded France's *Croix de Guerre*, with Star for his courageous rescue of wounded French soldiers under bombardment.

On 25 May 1917 Faith enlisted in France's *Service Aeronautique*. From 17 July 1917 to 9 May 1918 he attended the aviation schools at Avord, Juvisy, Pau, and the *G.D.E.*, receiving his *brevet* on the Caudron on 28 October 1917.

On 6 May 1918 Faith was commissioned a Second Lieutenant in the U.S. Air Service. From 13 May to 9 June 1918 he served as a ferry pilot at the American Acceptance Park, Orly. On 9 June 1918 Faith was assigned to the 103 Aero Squadron, USAS and flew combat patrols with that squadron until the 11 November 1918 Armistice.

Lieutenant Faith returned to the United States in 1919. He served at Mitchell Field, New York until his discharge on 27 August 1919.

That same year Faith relocated to Philadelphia, Pennsylvania and began work as a self-employed consulting statistician conducting investigations and reports on investment matters. Faith entered the service again on 16 September 1941. He rose to the rank of Colonel and was Assistant Operations Office of the Concentrated Command, Office of the Assistant Chief of Air Staff, Headquarters, Army Air Force. From February of 1943 to April of 1944 he served in Cairo, Egypt.

Following World War Two Faith returned to his consulting business. He married Goldie Basch, a professor of Social Work at the University of Pennsylvania. Faith died in Philadelphia of heart failure on 26 February 1962 at age 69. He is buried at Arlington National Cemetery.

CEDRIC FAUNTLEORY

Cedric Erroll Fauntleroy was born on a farm near Church Hill, Mississippi on 22 November 1892, the son of Mr. & Mrs. W.J. Fauntleroy. As a young man he attended the Natchez Seminary but quit that school at age 15 and left Mississippi for Texas to work as a cowboy, a dish washer, and a chuck wagon cook.

Fauntleroy was working as a race car mechanic on the fair circuit when he decided to sell all that he owned to pay his way overseas to join the Lafayette Flying Corps. On 13 July 1917 he enlisted in France's *Service Aeronautique*. From 20 July to 2 November 1917 he attended the aviation schools at Avord, Tours, and Pau. He received his *brevet* on the Caudron on 29 September 1917. Because of his exceptional skill as a pilot and his innate mechanical ability, *Caporal* Fauntleroy was sent to Issoudun by the French Air Service to assist their American allies who were in urgent need of a pilot with his knowledge and skill.

On 5 February 1918 Fauntleroy was commissioned a First Lieutenant in the USAS and appointed *Chef Receptionnaire* for the American Aviation Instruction Center at Issoudun. From 4 April to 14 October 1918 he was officer in charge of the Acceptance and Inspection Division, A.A.P., Orly Field. While serving in this capacity, he supervised the acceptance of over two-thousand French planes, and made thirty-five hundred test flights without a serious accident. Fauntleroy also served as an aviation instructor and was credited with teaching Eddie Rickenbacker how to fly.

On 10 October 1918 Fauntleroy was promoted to Captain. On 25 October 1918 he was assigned to the U.S. 94th Aero Pursuit Squadron and flew with that unit until the November Armistice. Captain Fauntleroy was recommended for the U.S. Distinguished Service Medal for his fine record with the U.S. Air Service as tester of aircraft. Fauntleroy was promoted to Major on 19 March 1919 and on 20 March he became the Commanding Officer of the U.S. 138th Aero Observation Squadron and served in that capacity until 1 May 1919.

Following his release from the U.S. Air Service, while Fauntleroy was in Paris he made contact with Merian C. Cooper, formerly of the 20th Aero Squadron, USAS. This meeting resulted in the founding of the American Kosciuszko Squadron which was assembled to protect Poland from Soviet aggression. Fauntleroy was given the rank of lieutenant-colonel with initial command of the squadron since he had outranked Cooper in the U.S.A.S.

Fauntleroy returned to the United States to recruit pilots for the squadron. While in Chicago, in March of 1919, he married Dorothy Louise Bigelow in that city.

Fauntleroy served Poland diligently in its war against the Bolshevist regime. So successful was Fauntleroy's command, that he was placed in charge of three Polish Air Force squadrons in addition to commanding the Kosciuszko Squadron. He served in this capacity until the Soviet enemy was driven from Polish soil and an armistice was declared between the belligerents on 18 October 1920.

For his outstanding war service to Poland Fauntleroy was promoted to General and was awarded the *Virtuti Militari* and the War Cross with 3 bars, the General Haller Medal, and several other decorations.

Some time after the Colonel's return to the United States, two daughters, Joan and Jacquelyn, were born to the Fauntleroys. Aviation, which played such a prominent role in Fauntleroy's career, would also prove a source of great tragedy for him. Mrs. Fauntleroy was killed in a plane crash on 27 May 1937. The Colonel's son-in-law was killed in an air accident in 1946. And he lost his daughter, Joan in a plane crash on 13 November 1947 while she was working as a stewardess with Western Airlines.

During World War Two Fauntleroy had owned and operated two parachute manufacturing plants for the U.S. Government in Manti, Utah, and in San Diego, California. During other periods in his career, Fauntleroy worked as a bank vice president, a rancher, a campaign manager for Wendell Wilkie's 1940 campaign, and a race track executive. Following the end of the War he retired to Tuscon, Arizona and bred and trained thoroughbred and quarter horses. He remained involved in racing circles until his death of cancer on 27 November 1963 at age 71. He is buried in Church Hill, Mississippi in the Presbyterian Burial Grounds.

IAN FERGUSON

Ian Fearchar Ferguson was born in New York City, New York on 10 May 1896, the son of Farquhar Ferguson, a physician, and Juliana (Armour) Ferguson. He attended Harstroms School and was a student there when he decided to join the Lafayette Flying Corps.

On 3 June 1917 Ferguson enlisted in France's *Service Aeronautique*. From 10 June 1917 to 8 January 1918 he attended the aviation schools at Avord, Pau, and the *G.D.E.*, receiving his *brevet* on the Caudron on 17 October 1917.

On 10 January 1918 *Caporal* Ferguson was assigned to *Escadrille* SPAD 96 and flew with that squadron until the 11 November Armistice. *Sergent* Ferguson was a skillful and heroic pilot. He flew many patrols and was awarded France's *Croix de Guerre*, with Palm for his 12 April 1918 combat with a "Fokker triplace."

Following World War One Ferguson lived on a family remittance in Greenwich, Connecticut. On 9 May 1925 he married Ruth Cozzens Amory. He later lived for five years on a boat which he had built in Oxford, Maryland. He died of cancer in Easton, Maryland on 30 January 1972 at age 75. He is entombed at the Armour family Mausoleum, Woodlawn Cemetery, New York.

JOSEPH FLYNN

Joseph Flynn was born in Philadelphia, Pennsylvania, the son of Mr. & Mrs. John Flynn. On 22 May 1917 Flynn enlisted in France's *Service Aeronautique*. On 3 August 1917 he was released because of dishonorable conduct.

One Joseph T. Flynn trained as a pilot at Foggia, Italy and graduated from that school on 18 March 1918. On 22 October 1918 Lieutenant Flynn was assigned as a pilot to the 141st Aero Squadron, USAS, and served with that unit until the 11 November Armistice. It cannot be verified if this is the same Joseph Flynn who had previously enlisted in the Lafayette Flying Corps.

Christopher Ford

ESCADRILLE LAFAYETTE

Christopher William Ford, the 38th volunteer of the *Escadrille Lafayette*, was born in New York City on 2 October 1892 the son of Christ Ford and Anna (Gensch) Ford. Christ Ford, an Irish laborer, had immigrated to the United States during the Irish potato famine. Anna Ford had immigrated from Germany.

Christopher was orphaned as a young child and grew up in the care of an aunt, Mrs. Harriett Cluff, on New York's lower east side. In order to survive in an impoverished environment, he had to begin work at an early age peddling newspapers and was thus afforded few educational opportunities. However, in early adulthood he took extension courses through George Washington University to advance his education.

Prior to World War One Christopher had worked as an auditor for an accounting firm then as a reporter for the *Wall Street Journal*. He had an early interest in aviation and eventually saved the funds to journey to San Antonio, Texas to take flying lessons at the Marjorie Stinson Flying School. Ford arrived there in the spring of 1916 and learned to fly on a Model B Wright "pusher" and was awarded Aero Club of América Aviator Certificate No. 462.

In March of 1917, shortly after the United States declared War on Germany, Ford joined the American Red Cross in order to gain passage to France. On 14 April 1917 he left New York aboard the *S.S. Chicago*, arriving in Bordeaux on the 23rd. Ford enlisted in French Aviation via the Foreign Legion in Paris on 26 April 1917. He enrolled in the aviation school at Avord on 17 May 1917 and received his *brevet militaire* (#6926) there on the Caudron on 9 September 1917. From 10 October to 4 November 1917 he was at the G.D.E. at Plessis-Belleville. On 5 November 1917 he became the 38th and last member of the *Escadrille Lafayette* when he joined the Squadron at Chaudun.

A number of the American pilots serving with the *Lafayette* Squadron were well-educated and from prominent families. Because Ford had come from a deprived background, he never felt accepted by the other pilots. This created a desire within him to prove himself to them, and he often volunteered for extra combat missions. On one such flight he froze his face. On another, he seriously injured his hand. And through his dedication and enthusiasm he soon won the other pilots over, and before long "Chris" Ford became one of the most popular men in the *Lafayette*.

Major Christopher Ford shortly after his return to the United States on 12 July 1919.

Ford had an exemplary war record and was a fine combat pilot in spite of a mixup in his paperwork which sent him to the G.D.E pilot pool without having taken the acrobacy course at Pau. He arrived at the Front with only nine hours of Nieuport training.

Once at Chaudun, Ford immediately began flying patrols and daily proved he could stay alive despite his

Chris Ford aboard a Model B Wright "pusher" at the Marjorie Stinson Flying School, San Antonio, Texas, May1 1916.

handicap as he and his squadron mates confronted increased enemy fighter concentrations. On one of his initial patrols Ford noted:

"I didn't know what had happened on my first spin. I was at a high altitude, and, after trying about every combination possible on the controls, the ship finally levelled off. Back at the field I asked some of the fliers what I had done and the spin was explained to me."

While *Sergent* Ford flew with *Spa.124* he continued to learn his craft. As his confidence grew he had French tri-color lighting bolts painted on the wings and fuselage of his SPAD S.VII fighter which were visible to any aggressor at a thousand meters. Ford survived his tenure with the *Lafayette* Squadron to become a proficient pilot and when the Squadron evolved into the U.S. 103rd Aero Pursuit Squadron on 18 February 1918, he entered its ranks as a First Lieutenant in the U.S. Air Service. He was soon made a flight commander.

While with the 103rd Lieutenant Ford was credited with the destruction of two enemy aircraft, the first, a single-seater fighter near Rheims on 27 March 1918. His second victory came on 21 May 1918 when he downed an Albatros fighter near Armentieres. Ford was awarded the U.S. Distinguished Service Cross and France's *Croix de Guerre*, with Palm and Star for his actions. His D.S.C. citation, presented in 1919, portrays his valor:

"For repeated acts of extraordinary heroism in action near Rheims, France, March 27, 1918, and near Armentieres, France, May 21 1918. Near Rheims on March 27, Captain Ford, while on patrol duty with two other pilots, led his formation in an attack on eight enemy planes. After twenty minutes of fighting, the American formation shot down three German machines, of which one was destroyed by this officer. Near Armentieres, on May 21, he again led a patrol of six planes in attacking twenty enemy aircraft. The attack resulted in ten individual combats. Captain Ford shot down one hostile plane and with his patrol routed the others."

Ford's citation does not tell the full story of his heroism. During the 27 March encounter he, Major William Thaw, and Captain James Norman Hall attacked the German eight-plane formation. Lieutenant Ford's gun had jammed at the onset of the battle. Although helpless to defend himself, he nevertheless remained above his two comrades and repeatedly dived to their aid, and for half an hour managed to drive enemy fighters off his beleaguered comrades.

On 30 July 1918 Ford was transferred to the U.S. 213th Aero Pursuit Squadron and on 21 August 1918 near Etain he was credited with his third victory. Lieutenant Ford rejoined the 103rd on 1 October 1918 and flew with that squadron in Champagne, in Flanders, at Saint-Mihiel, and in the Argonne Offensive.

After 314 hours over the Front, on 15 October 1918 while Ford was leading a combat patrol, his SPAD S.XIII engine was struck by ground machine gun fire and he was forced down near Remonville, in the Argonne Forest, within enemy lines.

While a prisoner of war Lieutenant Ford was promoted to Captain on 6 November 1918. On 20 November, when the Armistice had not effected his release, he and other prisoners escaped from the Ger-

Captain Ford beside his SPAD S.VII fighter at La Noblette while with the U.S. 103rd Aero Squadron, March 1918.

Above: Captain Ford with other Allied POWs at the Villingen prison camp in Germany, October 1918. Right: Miss Marie Antoinette Vannemann (later Mrs. Christopher Ford) as a nurse serving at the Mt. Sernet French hospital, summer of 1918.

man prison camp at Villingen, Germany. Following a one-hundred-kilometer trek through the Black Forest, and which saw him swim the Rhine river, Ford rejoined the Allies at Colmar, France.

On 27 November Captain Ford rejoined the 103rd Pursuit Squadron at Foucacourt. On 6 January 1919 he was transferred back to the 213th Pursuit Squadron as commanding officer. On 3 May 1919 when the 213th was demobilized Captain Ford was transferred to the 110th Aero Replacement Squadron. For his exemplary war service, Captain Ford was promoted to Major on 14 May. On 29 May Major Ford was transferred to the 1st Replacement Depot at Saint Aignan and on 15 June 1919 he was made commanding officer of the Le Mans Provisional Battalion No. 263. On 5 July he was ordered back to the United States and he arrived at Boston harbor aboard the *S.S. President Grant* on 12 July 1919.

When Ford had been serving on the Champagne front he had seriously burned his hands while trying to extinguish a barracks fire. At the hospital nearby his squadron Ford had met Miss Marie Antoinette Vannemann, a nurse from Johns Hopkins Hospital who had come to France in 1914 to serve with the French Hospital Corps. From a prominent Philadelphia family, she too had been orphaned as a child but had been brought up by other family members in a culturally refined atmosphere.

Miss Vannemann was awarded France's *Legion d'Honneur* "for extreme devotion and care of wounded Allied soldiers." She was also awarded France's Medal of Recognition No. 8843 "for remaining on duty at post while serving in the French hospitals at Mt. Sernet and Estree St. Denlee during violent bombardment by the enemy."

Ford had begun courting Miss Vannemann during his stay at the hospital. She was soon won over by his strikingly handsome good looks, his engaging manner, and his infectious laugh. She went on to serve with the Red Cross in Italy and in Greece while Major Ford continued with his overseas duties.

The wedding of Major Christopher Ford and Marie Antoinette Vannemann at Brynn Mawr, Pennsylvania, 28 July 1920. Ford's former Lafayette Escadrille comrade, 1st Lieutenant Carl Dolan, served as Ford's best man.

Approximately one year after their return to the United States, Major Ford and Miss Vannemann were married at Bryn Mawr, Pennsylvania on 28 July 1920. Fellow *Lafayette* pilot Lieutenant Carl Dolan acted as best man at their wedding. Major and Mrs. Ford eventually had four children: Christopher, Jr., Marion, Mary Elizabeth, and Richard.

Upon his return to the United States from France, because of his fine war record, the U.S. Army asked Major Ford to remain in service. He served at Mitchell Field, New York from 12 July to 26 August 1919 at which time he entered the hospital at Cooperstown, New York. Upon his discharge from the hospital one month later Ford was assigned to the Air Service Liquidation Division at Washington, D.C. On 20 April 1920 he was made officer in charge of Flying and Training at Carlstrom Field, Florida and remained in that capacity until 25 July 1922 when he entered the Army Air Corps Tactical School at Langley Field, Virginia.

Following a two-year stint as squadron commander at Langley Field Major Ford served a five-year hitch as executive officer for the 1st Corps Area, Boston, Massachusetts. While flying a mission over Swampscott, Massachusetts he nearly lost his life when his aircraft developed engine trouble at 1,000 feet and he and his mechanic were forced into the icy waters off Jeffer's point. Ford and his mechanic managed to cling to one wing of the aircraft, the only portion of the plane left unsubmerged, until they were rescued by a Swampscott fisherman and taken in for medical treatment.

On 1 September 1930 Major Ford began serving a two year stint as flying instructor and squadron commander at the Air Corps Training Center, Brooks Field,

San Antonio, Texas. It was there that Mrs. Ford succumbed to complications following the Caesarian birth of her last child, Richard, on 7 May 1931.

Major Ford went on to attend the Command and General Staff School at Fort Leavenworth, Kansas and at Lono, Arkansas. On 2 October 1934 he was made commanding officer of the 3rd Pursuit Squadron at Clark Field, the Philippines. He remained in that capacity until March of 1937 when he was ordered to Cleveland, Ohio, there to become aviation instructor for the aviation section of the 37th Division, Ohio National Guard. On 16 November 1940 Ford was promoted to Lieutenant Colonel and was ordered to Chandler Field, Atlanta, Georgia to serve as aviation instructor for the Organized Reserves and was also made Commanding Officer of the 4th Corps Air Detachment. While on duty in Atlanta, he met and married Ruth Carter of Vinings, Georgia on 15 October 1941.

In late 1941 the military decided to remove 385 officers from flying status. Colonel Ford, with more than 10,000 air hours, was one of the pilots selected. Rather than serve on a non-flying basis, he chose retirement from the Army Air Corps and did so on 31 March 1942.

Colonel Ford was deeply hurt by his treatment at the hands of the Air Corps. For a time he served as the assistant chief inspector for the Consolidated Vultee aircraft plant at Miami. Always the vigorous outdoorsman who enjoyed hunting and fishing, and who kept himself in top shape, Ford, who had never been seriously ill, was shocked to learn in 1944 that he had terminal cancer. After one year of courageously battling the disease, he succumbed to the illness and passed away at age 52 on 9 April 1945 in the Atlanta Veteran's Hospital. On 12 April Colonel Ford was buried at Arlington National Cemetery with full military honors.

TOD FORD

Tod Ford, Jr. was born in Youngstown, Ohio on 19 May 1887, the son of David Tod Ford and Carrie (Arms) Ford. Ford's parents moved to California while Tod was a child. Ford attended California public schools, and became a friend of Charles Nordhoff. Following high school, he attended Leland Stanford University where he received a law degree. Ford briefly practiced law in California before joining the U.S. diplomatic service in France. Following the U.S. entry into the European War, Ford resigned from the diplomatic service to become a combatant.

On 21 May 1917 Ford enlisted in France's *Service Aeronautique*. In August 1917 he was removed from flying training because of ill health. He returned to the United States as a member of the French Mission. He was released from French Service and enlisted in the U.S. Air Service on 26 December 1917. Ford was commissioned a First Lieutenant in the U.S. Air Service and returned to France for technical and liaison duties. He was discharged on 19 December 1918.

Following World War One Ford became a freelance writer and newspaper columnist, doing book reviews for California papers. He married Anne Perkins, and from this union a daughter was born.

During World War Two Ford worked as an optical engineer at a bombsight manufacturing firm. Ford died in Santa Barbara, California on 21 June 1954 at age 67.

HENRY FORSTER

Henry "Harry" Forster was born in New York City, New York on 21 March 1889, the son of Frederick Prentiss Forster, a lawyer and real estate developer, and Edith (Allen) Forster. Henry Forster was educated at Groton School, Groton, Massachusetts, and at Harvard College, 1911, AB in History & Government. While at Harvard Forster was a member of the varsity rowing crew.

Prior to World War One Henry Forster worked as a bond broker while serving for two and one-half years in Squadron A, Cavalry, New York National Guard, and six months in the 2d Battalion, New York Naval Militia (Aviation), where he gained some flying experience.

Forster sailed for France on the *Rochambeau* and on 13 June 1917 he enlisted in France's *Service Aeronautique*. From 20 June 1917 to 31 January 1918 he attended the aviation schools at Avord, Pau, and the *G.D.E.*, receiving his *brevet* (#9686) on the Caudron on 31 October 1917.

From 1 February to 2 August 1918 *Sergent* Forster flew with Escadrilles *C 74*, *SPAD 102*, *BR 224*, and *SPAD 15* on the fronts at Ypres, Amiens, Oise, Aisne, and Chateau-Thierry. While in French service he had a total of 45 hours of combat patrols over the lines.

On 19 July 1918 *Sergent* Forster transferred to the U.S. Naval Reserve Force as an Ensign. He served at the Dunkirk Station flying Italian Henriot seaplanes from 8 August to 24 August 1918 before attending the Bombing School at Boscombe Down, Salisbury, England from 7 September to 22 October 1918. He was then ordered to St. Inglevert, France for service with the Northern Bombing Group.

On 9 January 1919 Ensign Forster returned to the United States on the *SS Calamares*. He was promoted to Lieutenant J.G. that same month and resigned from the Naval Reserve Flying Corps on 5 February 1920

Following his World War Service, Henry Forster became a real estate broker and management executive in New York City. He married Helena Fish at Garrison-on-Hudson, New York on 28 August 1920. The couple had four sons and one daughter.

For 50 years Henry Forster continued to be active in real estate in New York State. For many years he served as president of Brown, Harris, Stevens, Inc., a large New York Firm. He died on 11 September 1989 in New York City, New York at age 100. Mr. Forster is believed to have been the last surviving member of the Lafayette Flying Corps.

ERIC FOWLER

Eric Anderson Fowler was born in Quogue, Long Island, New York on 24 July 1895, one of fourteen children of Anderson Fowler, a meat-packing executive, and Emily (Arthur) Fowler. Eric was educated at St. Bernard's School, New York; Hill School, Pottsdown, Pennsylvania; and Princeton University, Class of 1919.

Eric Fowler joined the American Ambulance Field Service on 6 August 1916. He was attached to Section Four until 9 June 1917, the day he enlisted in France's *Service Aeronautique*. From 20 June to 27 November 1917 he attended the aviation schools at Tours, Avord, and Pau. He received his *brevet* on the Caudron on 21 October 1917.

Following his preliminary training, Fowler went on to the school at Pau, Bassess Pyrenees, for his acrobatic training. On 27 November 1917, the day he was to graduate from that school, he was in the air at 700 feet with forty other planes when his training craft was seen to dive out of control then nose into the ground behind a hanger. *Caporal* Fowler was killed instantly. He was 22 years old at the time of his death. It was believed he had fainted in the air, since his controls were found intact and no parts were missing from his machine. Charles Chapman, an American comrade who was present at the time of his death, wrote this of the incident.

"He (Fowler) worked all morning doing acrobatics, then went to the next class without eating and fainted in his machine."

Eric Fowler's funeral was held at the Episcopal Church at Pau. Two of his sisters who were serving in the Ambulance Corps were in attendance, as was his brother, Major Harold Fowler, who was in charge of the American Aviation Squadrons assigned to the British Expeditionary Forces. Eric Fowler was buried at Pau on 28 November 1917 with full military honors. His remains were later removed to the Lafayette Flying Corps Monument, St. Cloud.

WILLIAM FREY

William Frey was born in New York City, New York on 15 February 1892. He was educated in New York schools then joined the Rhineland Import Paper Company as a salesman and buyer.

On 1 April 1916 Frey sailed for Europe aboard the *Rochambeau,* ostensibly to conduct overseas business for his employer, but on 22 April 1916 Frey enlisted in France's *Service Aeronautique*. He received his *brevet* on 7 September and on 23 November he attended the acrobatics school at Pau.

Before being assigned to the Front, Frey was granted a permission to the United States. Frey did not return to France and was considered a deserter by the French Army.

EDMOND GENET

ESCADRILLE LAFAYETTE

Edmond Charles Clinton Genet, the 20th volunteer of the *Escadrille Lafayette*, was born on 9 November 1896 in Ossining, New York, the son of Albert Rivers Genet, a lawyer, and Martha Rodman (Fox) Genet. Genet was the great-great-grandson of Edmond Charles "Citizen" Genet, the French Republic's first Minister to the United States in 1792, and Cornelia Tappan Clinton Genet, daughter of New York's Governor De Witt Clinton, whom "Citizen" Genet married in 1794.

Young Edmond Genet received his education in public and private schools and at Mount Pleasant Academy in New York. Early on he took an interest in seafaring and hoped to eventually become a U.S. Naval officer.

On 25 October 1912 when Edmond was sixteen years old his father passed away from a lingering illness. Though Edmond was still in school, he helped to support his mother by working at the Chilmark Farm Diary at Ossining.

In December of 1913 when Genet was in his seventeenth year, he left Ossining for the Naval Training Station at Newport, Rhode Island to enlist as an ordinary seaman. Mr. V. Everit Macy, owner of the Chilmark Dairy, had been impressed by Edmond's diligence at work and had learned during their conversations of Genet's dream of attending the U.S. Naval Academy at Annapolis. Macy was able to use his influence to secure for Edmond the opportunity to take the entrance exams for the Naval Academy. However, in February of 1914 Genet failed the mathematics portion of the exams which caused his rejection as an appointee.

In his letters to his mother and friends, Genet painted a self-portrait of an intense, brooding youth who was extremely self-critical. He often called himself a "failure" and "the black sheep of the family." However, upon failing his exams, he concluded his letter to his mother on an upbeat note, stating that "perhaps some day in the distant future I will turn out to be a white one and be something worthwhile."

Caporal Edmond Genet in Paris, 4 September 1916.

On 10 March 1914 Seaman Genet was stationed aboard the *U.S.S. Constellation* at Newport, Rhode Island. He wrote home of a powerful desire to be out of the United States until he could "make good in something if that will ever be possible." And he indicated that if he did not put out to sea with his vessel but remained on land, he intended to "jump ship" and say "good-by to the Navy."

Edmond Genet on 20 January 1914, the day he set sail for France.

However, Genet managed to stick it out, and his spirits soared when he learned that he would soon be sailing for Mexican waters where hostilities had broken out between the United States and counterrevolutionary general Victoriano Huerta. On 23 April 1914 Genet sailed to Vera Cruz, Mexico aboard the *U.S.S. Georgia* as part of the Atlantic Fleet.

Edmond had volunteered for service with the landing-forces, but by the time he touched Mexican soil on 30 May the fighting between the Americans and General Huerta's forces had already passed. He spent the months of June and July aboard ship just off Vera Cruz, performing menial tasks. He also earned three days in the brig for losing his temper at a Chief Petty Officer.

On 7 August 1914, following three months at Vera Cruz, Genet sailed with the *U.S.S. Georgia* for Port Au Prince, Haiti. When off duty, Edmond continued to study for the spring entrance exams for Annapolis in hope that he might again be allowed to compete.

In October 1914 the *U.S.S. Georgia* docked at Hampton Roads, Virginia. By mid-November the ship had reached the Navy Yard at Charlestown, Massachusetts and Genet had rapidly lost interest in the forthcoming Annapolis exams. His curiosity and attention had been drawn toward the growing European conflict. He felt that the United States would eventually be a part of that war, and for the first time his letters expressed a recurring thought, "some hopes of dying for my country."

23 November 1914 found young Genet having just begun his eighteenth year and anxiously looking forward to the *Georgia's* scheduled "Canal-Pacific trip to the Exposition." And he had again resumed his study for the Annapolis exams. He had also planned to attend night school in Boston until the ship's scheduled departure on 15 February.

But on 14 December Edmond attended services at Boston's St. Paul's Cathedral and was deeply moved by a sermon on "the present war and the relation of American Christians to it." Stirred by the sermon, and further prompted by his French heritage, these combined with the fact that his childhood sweetheart, Gertrude, has been on "a pleasure-trip to Europe" since early fall, caused the impetuous youth to quickly alter his plans.

Genet had been granted leave from 29 December 1914 until 8 January 1915. But after spending the holidays with his mother, Edmond jumped ship and headed for Washington, D.C. instead of returning to the Charlestown Navy Yard. There he secured a passport from the French consul, giving his age as twenty-one and his spurious purpose for going to France-to make inquiries concerning a family estate.

Armed with a cross about his neck and a copy of the New Testament in his pack, the young crusader was about to enter a holy war of which he knew nothing. With the enthusiasm of a boy scout setting out on a camping trip, not even his wildest nightmares contained the horror the war would soon bring to him. Still he had one mission from which he would not be deterred: a self-fulfilling pursuit of death.

On 14 January Genet wrote his mother a letter from the Athens Hotel in New York, again calling himself "the black sheep of the family". He went on to tell her that "I never expect to come back-death seems nearer to me than any possible chances of going through the horrible ghastly conflict which is carousing over Europe without meeting death."

Prior to writing his mother, Edmond had gone to the French Consul to have his passport visaed then had purchased passage on the *S.S. Rochambeau* to Havre, France.

When Genet's ship was delayed, although subject to arrest, he did little to conceal himself and in fact made careless appearances in public including visits to theaters. Edmond sailed for France on 20 January 1915. One of his last letters from the Athens Hotel in New York was to Jeannette Halstead, an actress friend in Boston. With high emotion he wrote her that he was sailing for France "to throw away my life, if need be, in the war." He went on to say:

"There appear to be very, very few possibilities of my escaping through the conflict alive. It looks to be a long one and sooner or later I expect to have to give up my life on the battle-field. I care nothing about that. Death to me is but the beginning of another life-better and sweeter. I do not fear it."

Genet concluded by saying: "Good-bye to you all. You may be sure that I shall never forget you even while I may be prodding Germans with my bayonet or being stuck myself."

On 29 January Edmond arrived in Havre feeling upbeat. He no longer wrote of the prospects of his death, but of the end of the war and the subsequent "openings and opportunities." He told his mother that "Most people go to America to win their fortune. I am going to Europe and win it if there is such a feat possible-God helping me."

On 3 February 1915 Edmond Genet enlisted in the Foreign Legion at Paris as a common *soldat*. He left that same day for the Legion's barracks at Lyons for a period of training. One of the first friends he made while in Paris was fellow American Norman Prince who had also come over on the *Rochambeau* and was soon to enlist in the *Service Aeronautique*. He had stirred Genet's interest in aviation which would grow with Edmond's service in the trenches.

By 20 February Genet had been moved twenty-five miles from Lyons to the *Legion Etrangere* training camp of Valbonne to begin drilling and target practice. On 20 March *Soldat* Genet left Lyons on a troop train for the Front in the North of France. He was now a member of the 3d Marching Regiment, 1st Bataillon, 4th Company. Five days later, on 25 March, he entered the trenches at Cappy and immediately came under fire from German riflemen and artillery some 400 meters away. On 1 June following a particularly fierce barrage he wrote Jeannette Halstead that if he were back home in his old High School, he would currently be graduating with "the Class of '15."

By the middle of the month a torrential rain filled the trenches with mud and water. As Genet and his comrades tried to clear out the trench many were killed or wounded.

Soldat Genet with the Foreign Legion at Compeigne, April 1916.

On 22 September 1915 Genet entered the Battle of Champagne with fellow American William Dugan. In a 11 October letter to his brother, Rivers, he described the horror he had encountered there.

"As you will already know from my recent letter to Mother I have been through a considerable amount of horror and real war these past three weeks. I'm not out of it by any means either . . . I certainly have been through enough fire lately to last and satisfy any one for a lifetime, but there's lots more coming I know. How I came through alive and unhurt is more than I can tell. So many of us were cleaned out in the attack—a matter of a brief ten minutes. Anybody who has the least bit of heart trouble or nervous prostration sure wants to keep out of a modern war. It's no place for an invalid."

Genet's desertion from the *Georgia* and how to resolve it became a frequent topic of his letters home. He desperately wanted to clear his name, and his mother has been making an effort to assist him through the aid of a Navy Chaplain.

In a letter to Jeannette Halstead he further described what he had been through at Champagne:

"In an attack we made on Sept. 28th, out of our company of 250 there are not quite 60 left. . . . The rest were either killed or wounded by the murderous mitrailleuse and shell-fire we faced. . . . That same night there were but thirty-two of us able to be collected from two companies (500 men) who were able to assist in carrying back behind the lines some of the badly wounded.

"On the first day of the actual fighting, the 25th, I believe I had the closest shave that I've had yet. We were advancing after the retreating Germans who had been driven from their entrenchments and they were firing big shells into the advancing ranks. Our section happened to get exposed to view on top of a ridge and immediately they hurled a huge 210 shell at us. Somehow we felt it coming—heard it too, and we all just dropped flat on our stomachs in the soft mud. . . . The huge shell burst a couple of yards behind me and killed two of my squad and wounded four or five others. Since that day I love such shells only at a great, great distance.

"There are many little bits of interesting things to say and describe about the sights of a battlefield after the battle and trenches which have been stormed and taken, but really it is best that you simply imagine such sights for they are too morbid and ghastly to write about. You'd have horrible nightmares for nights and nights after.

Top: Soldat Genet (2nd rt.) with the 1st Battalion, 4th Company, 16 Escoriade, Verdun, Janaury 1916. Bottom: Eleve-Pilote Genet at the Buc School of Aviation standing by a Caudron, Janaury 1916.

"Besides all the strain of being continually under fire we were soaked by rain almost all the time and slept in mud and water every night. You can imagine how we felt and looked after ten or twelve days of such life."

Genet's regiment remained in the line under constant bombardment until 17 October. Only then was the Moroccan Division, to which he had recently been assigned, relieved by the 170th French Infantry Regiment.

On 22 December 1915, following a period of repose, Genet's Battalion was marched back to the Front through drenching rain and snow. Now at Verdun, the men endured a month of miserable weather before they were withdrawn from the line on 27 January 1916 for a month's maneuvers. Genet mentioned in a letter to his brother that of the 48,000 men who had volunteered for the Legion since the beginning of the war, only 5,000 remained fit for service, the rest having been killed or wounded.

By 23 February Genet had been sent to the trenches near Compeigne where he remained for six days under snow blizzards and sporadic enemy bombardments. Edmond experienced periods of repose and front-line

Above: Gertrude Talmage, Edmond Genet's unrequited love. Right: Caporal Genet and Sergent Hinkle passing time at Sainte-Juste, March 1917.

service until he was granted Paris leave on 20 April and left the front-line trenches.

He arrived in Paris on 21 April 1916, Good Friday. He spent Easter at the apartment of his friend and former Legion comrade, Doctor David Wheeler. Later Genet had attended religious services and had received Communion at the American Church of the Holy Trinity. A deeply religious youth, he possessed an odd blend of religious heritage bestowed upon him by both a Quaker and Catholic ancestry. And he never missed the opportunity of a religious service regardless of denomination. "Church is Church," he had written, "anywhere and in any denomination."

While in Paris Edmond had gone to see Doctor Edmund Gros, head of the Franco-American Flying Corps, about his delayed permission to enter French Aviation. Gros promised to present a new demand for Genet and to work hard for his transfer.

But by the 28th of April Edmond had returned to the Front. He remained there until 30 May 1916. The previous day, through the aid of Doctor Gros and Paul Rockwell, his request for transfer to Aviation, for which he had applied and had passed a physical the previous October, was finally granted to him.

On 1 June, following sixteen months service with the Legion, Genet arrived at the aviation depot at Dijon to pick up his equipment. He immediately left for the Buc aviation school to begin his flight training.

On 20 June Genet wrote a brooding letter to his mother regarding his desertion from the U.S. Navy. He felt that he had deceived his friends by not letting them know the conditions of his departure. His agony was further exacerbated by press stories about his current experiences in which the reporters had been led to believe that he had "quit the Navy O.K." He longed to remove the "blot" from his record. Should he meet death in France, he lamented, "it would be much easier to meet if I knew I was O.K. with my own loved country."

On 16 July while in Paris Genet first "confessed" his desertion from the Navy to his friend, Major Frank Parker, U.S. Observer with the French Armies, and sought his advise. Parker promised to help him clear his record and assured him that he would rehabilitate himself by his fine service in France. Genet had also volunteered to become Parker's observer within the aviation school and promised to help secure whatever information he could about the French Aviation Corps.

On 5 August 1916 the weather at Buc was too breezy to permit flying. Edmond sought and was granted Paris leave and lunched with Paul Rockwell. When he returned to the school that afternoon he was overjoyed to find a letter from his "darling Gertrude." For some time his numerous letters to her had gone unanswered, and young Genet had been in torment.

Genet's "Gertrude" was twenty-year-old Gertrude Talmage, granddaughter of the Reverend Thomas De Witt Talmage, the famed Presbyterian preacher of the late 1800's and pastor of the immense Brooklyn Tabernacle where thousands flocked to hear his eloquence and bold denunciation of evil. Gertrude had been born in Pittsburgh in 1895 to the Reverend Frank De Witt Talmage, also a Presbyterian minister, and Gertrude (Barlow) Talmage, formerly of Ossining. Rev. Talmage had a church in Pittsburg. Upon his untimely death at age 42, Mrs. Talmage had returned to Ossining to be near her brother, George, who ran the local hardware store.

The Barlow's had been friends of the Genets. Consequently, Edmond had met and had fallen in love with Gertrude. But since Edmund had left home at age sixteen, what must have been a brief, chaste relationship, Genet had built up far beyond reality. In a previous diary entry he had called her "the very image of love and gentleness and sweetness to me. God grant that I may get back to her and have her for my own dear wife some day."

What Genet could not know, and what Gertrude had lacked the heart to tell him, was that she had met young Roy Minich, a Congregationalist minister, and the two were about to become engaged. Instead, she let Genet believe that both his letters to her and her's to him had been mysteriously "lost in the mail."

Gertrude's 5 August letter was the last Genet would receive from her. He would later lament: "Day after day, week after week, its the same unbearable waiting for word from her."

Upon his return to Buc, on 10 August Genet had just left the ground in a 50 h.p. Bleirot trainer when his motor had quit. He crashed to earth from 50 meters and demolished the plane, and in the process injured his left hip and shoulder. He spent two days in the school hospital. The following evening American Dennis Dowd, Genet's former Legion comrade, was killed in training when the Caudron he was flying fell from 500 metres.

Caporal Genet received his *brevet militaire* on the Caudron at Buc on 3 September 1916. He graduated top of the class which included Americans Steven Bigelow and Edwin Parsons. Through Major Parker's direct intervention he was able to bypass the school at Avord and go directly to the Nieuport School at Pau. Following Paris leave, on 11 September Genet arrived at Pau to begin Nieuport instruction which he completed in eighteen days.

On 2 October Genet arrived at the Cazeaux gunnery school. He completed that course on the 26th then returned to Pau on 1 November for combat and acrobatic training. On 10 December, having completed his course at Pau, he left for Paris on a ten-day leave with Bigelow and Parsons. They had been ordered to report to the aviation reserve depot at Plessis-Belleville on the 20th. Genet spent Christmas in Paris with friends and New Years day at the G.D.E.

On 19 January 1917 *Caporal* Genet joined the *Lafayette* Squadron at Cachy in the Somme sector and was followed a few days later by his comrades Parsons and Bigelow. The American pilots at the Squadron were at first deceived by Genet's diminutive, cherubic appearance, by his "round, innocent, pink-cheeked face," and had thought at first a boy had entered their mist. But they soon realized their mistake once Genet had put on his new dress uniform with the French *fourragere* on his left shoulder, indicative of the citation bestowed upon his unit for its heroic action at Champagne.

Genet was gratified at having reached the Front, and his diligence to duty and cheerful mien made him an immediate favorite with the other pilots who soon tagged him "Smiler." But to the puzzlement of his American comrades Edmond later revealed a dark side to his personality and would often become sullen and morose. Not men to pry into a comrade's affairs, the

Caporal Genet preparing to fly a patrol from Sainte-Juste, February 1917.

pilots could not realize that the cause of his despondency was the burden of his "desertion" and "darling Gertrude's" unrequited love.

Due to inclement weather, little flying could be accomplished, and Genet found further time to brood. "Isn't it funny," he wrote in a 20 January letter to his mother, "that all the good stories about me seem to be getting around and the others are keeping dark. Some day those will out and then there'll be a mess of affairs."

On 27 January the Squadron moved to Saint-Juste. There, the Americans faced some of the coldest conditions yet encountered in France. It was so frigid in the "underground trench" where the pilots were billeted that the ink in Genet's fountain pen froze while writing his mother a letter.

The following morning Genet and the other pilots worked hard at constructing their new barracks. On the 29th Edmond made his first patrol over the lines in a Nieuport handed down from William Thaw and came under antiaircraft fire on the Somme River. On 2 February, while flying a morning patrol in his 110 h.p. Nieuport, the oil froze in his machine and his engine burned up. Six days later he nearly froze his face while flying in the bitter cold. This was unusual for French winters which were usually damp and rainy with only occasional wet snows.

When not flying, Genet had used his artistic skills to decorate the barracks lounging room where the pilots huddled when not sleeping. The Americans had covered the walls and ceiling with corrugated cardboard strips to hold in the warmth. On its smooth outer surface Genet had sketched and painted in startling scenes of aerial combat between French and German adversaries. On other panels he has done the Squadron insignia, a fierce Indian head, and had made pencil drawings of pretty Parisian girls.

At first the American pilots had appreciated the vibrant imagery splashed across the drab cardboard. But soon Genet's scenes only served as painful reminders of hard combat and fallen comrades.

On the morning of 15 February Genet fought his first aerial combat with two enemy "Valfiche" two-seaters deep inside their lines over Roye and drove both adversaries earthward before he was driven off by enemy antiaircraft fire.

On 1 March a German observation plane flew into allied lines. Genet ran to his fighter to answer the alert but was buffeted by a side wind which upended his Nieuport as he taxied for takeoff. The plane was demolished, but Genet escaped the wreckage without injury. On the 4th Genet took off into thick mist, got lost, and was forced down to refuel at three different airfields. He finally located his aerodrome some four hours and 450 kilometers later, and he landed exhausted and nerve strained. On the 6th he suffered frostbite on his nose and cheek following five hours in extreme chill over the lines. The skin on his face peeled away and he covered the open sores with vaseline.

On 16 March the French and British forces launched an allied offensive against the region between Roye and Soissons. As part of the operation, the Squadron began flying many patrols. On that first day of the offensive Genet was ordered to attack enemy observation balloons.

At dawn the following day a zeppelin alert reached the Squadron. Genet, with incendiary ammunition in his machine gun, was ordered skyward to intercept and attack it. But his flying instructions proved incorrect and he became lost in the clouds and was forced to land for lack of fuel at another aerodrome.

The next day, a Sunday morning, Genet flew two reconnaissance patrols over the new territory taken during the offensive. He witnessed a countryside left barren by the retreating enemy as they toppled trees and blew up roads, railroads and dams to impede the allied offensive.

On 19 March 1917 Genet left on a morning patrol with James McConnell across enemy lines. Soon after entering German territory they were enveloped by clouds and mist. When they flew into clear sky north of Ham they suddenly encountered two German two-seaters. Genet immediately closed with the nearest biplace, leaving McConnell to battle his wingman. That was the last Genet saw of his friend as Edmond was soon in a combat for his life.

As Genet closed on the German bus the two adversaries began circling each other. The enemy gunner fired his machine guns at Genet and immediately severed his main upper wing support. A guiding rod was shot in two and his upper wing was pierced with bullets. A ricocheting bullet fragment struck Genet in the left cheek, stunning him. Undeterred, he again closed with the German plane and fired at it until the two nearly collided. Then Genet dived to look for McConnell and immediately came under antiaircraft fire. Still, he continued to search for McConnell, circling the area for fifteen minutes in hope that his friend might appear.

Finally, in fear of losing his upper wing, and with his cheek pained and bleeding, Genet returned to the

Squadron's aerodrome, hoping McConnell had returned ahead of him.

His heart sank when he discovered that McConnell was still out. When he later learned that his friend had been killed, Genet's grief turned to guilt. He began second-guessing his actions on the patrol and felt he should have done more. Sick with grief at this painful blow, he vowed to avenge McConnell "if it costs me my own life."

On 27 March another crushing blow was dealt young Genet. On that day he received a letter from his mother in which she informed him that his beloved Gertrude has become engaged. Of this unrequited love, he replied:

"The simple truth itself is unbearable; can you half-way imagine what it means to me to hate to realize that she has been receiving my letters all these months and has just permitted me to keep on and on without telling me directly and instantly of her engagement? What particular matter does it make after all, though, when one comes straight down to it? Time is capable of healing over all wounds even if the scar remains forever—just as it is doing with the gash in my cheek made last week by that bullet, only that is infinitesimal in comparison to the size of this wound."

Despairing at the loss of McConnell and Gertrude, and obsessed with expunging the blot of desertion from his record, Edmond again wrote of death, telling his mother that if the war didn't end soon, there was little chance he would return to her alive. He went on to say that since the discovery of McConnell's remains beside his badly smashed Nieuport he had "chased two Boche machines, but could get up to neither, but I'll get one yet and more than one, or be dropped myself, to avenge poor Mac. I've already been told I was reckless in the air over the lines, but after this I vow I'll be more than reckless, come what may."

On 4 April Genet rejoiced when he learned that President Wilson had sought to declare war on Germany. But in his diary entry for that date he recorded a foreboding for his own fate.

"Somehow I've given away completely this evening. I feel sure there is something very serious going to happen to me very soon. It doesn't seem any less than Death itself. I've never had such a feeling or been so saddened since coming over to battle for this glorious France."

On 6 April 1917 the United States declared War on Germany. By 8 April the squadron had completed its move to Ham in the Somme sector to keep up with the rapidly advancing front. While there, Genet learned that he and McConnell had been cited for their 19 March actions.

On 15 April Genet wrote home that he would soon receive France's *Croix de Guerre* and his belated promotion to *Sergent*. He also learned that the bi-place he had battled the day McConnell had been killed had been disabled by him and forced to land on French soil, resulting in the capture of its crew.

Whenever possible, Genet had engaged the Germans since McConnell's death. Of the enemy he told his mother, they "are becoming much more active along here now and fights occur practically daily."

Genet awoke on the morning of 16 April feeling ill. For several days he had complained of fatigue and nausea, undoubtedly precipitated by the physical and emotional trauma he had suffered. Nevertheless, he made his scheduled 7 a.m. patrol with Walter Lovell and Thomas Hewitt. North of Saint Quentin their flight came under heavy antiaircraft fire and Genet's quick maneuvering to throw off the enemy gunners made him so sick he was forced to return to camp at 8:15 a.m. Telling the other pilots his "stomach was very upset", he went to bed.

Shortly after noon Lovell and Genet were to fly another patrol but Lovell noticed that Genet appeared tired and advised him not to fly. Genet acquiesced and Lovell set out without him.

But on 2:30 that same afternoon Genet left on patrol with Raoul Lufbery over the lines between Saint Quentin and La Fere. Previously, Willis Haviland had seen that Genet was ill and with his own fighter disabled, had offered to fly the patrol in Genet's place. But Genet had refused Haviland's gesture, and a half-hour later, with Lufbery leading, they reached Moy and came under enemy antiaircraft fire while flying at low altitude on account of the clouds.

Lufbery, glancing back, saw three shells burst approximately one-hundred yards behind Genet's plane. Genet suddenly turned his bus around and headed back toward French lines. Lufbery also turned and followed Genet for four minutes to make sure he was all right before he himself returned to the lines to finish the patrol.

Poilu in the trenches watched Genet's machine appear overhead. Then suddenly at a height of 1,400 yards, with its engine fully revved, it fell into a vicious corkscrew dive. A wing tore away and the Nieuport crashed on the road with tremendous force not three-hundred meters from the site when James McConnell had fallen.

The speed with which Genet's fighter had hit the ground made it impossible to pinpoint the cause of his death. His Nieuport had dug a five-foot deep hole in the hard packed road, and as Edwin Parsons later noted:" Every bone in his body was broken and his features were completely gone." It was assumed that Genet had either fallen ill and had fainted, or had been wounded by a shell fragment which rendered him unconscious.

When Lufbery returned to the aerodrome, he was much surprised that Genet had not returned ahead of him. Moments later a call came in with the news that Genet had fallen north of Montescourt, five kilometers within allied lines. Lufbery, Walter Lovell, and others immediately took off for the crash site in the Squadron's light truck to pick up the body. Upon viewing Genet's smashed Nieuport, Lovell wrote later:

"I have never seen so complete a wreck, and I have seen numbers. He had fallen with the motor in full speed in the middle of the road, which proves that the German shell had killed him or rendered him unconscious . . . I have lost a very dear friend, and a courageous comrade of combat: the squadron has lost one of the most conscientious pilots that it has ever had or ever will have."

In a simple ceremony under light snow Genet was buried in the little military cemetery at Ham. In compliance with his final written request, his body had been wrapped in the flag of France and his casket covered with the flags of France and the United States. Squadron Commander Thenault, in eulogizing young Genet, said this of him:

"He at once gave promise, as before, of a brilliant career. The only solicitude of his chiefs was to teach him moderation. With his juvenile ardor he was always ready to fly, whether to attack an avion or burn a drachen.

"He was young, and he seemed even younger. He was our Benjamin, and we cherished him as in a family one prefers the youngest, the weakest. But his heart was not weak, as on many occasions he caused us to see." At the conclusion of Thenault's eulogy the sun broke through the snow-laden sky and for an instant its rays "illuminated the bier like a benediction from heaven."

Of the death of Genet, Edwin Parsons would later write, "the Escadrille lost one of its most sincere and conscientious pilots. He gave his life for his ideals."

Dead at age twenty, Edmond Charles Clinton Genet had become America's first son to die following the U.S. declaration of war. He was posthumously awarded the French *Croix de Guerre*, with two Palms for his 19 March and 16 April actions. On 26 September 1922 he was posthumously awarded the *Medaille Militaire*.

Shortly after the end of World War One, Secretary of the Navy Josephus Daniels cleared Genet's name of the technical charge of desertion when he had it expunged from his record. He then notified Edmond's mother that her son's offense of desertion had been nullified by his conduct while serving France. And that his Navy record from that point on "shall be considered in every respect an honorable one." In 1928 Genet was reinterred at the *Lafayette Escadrille* Memorial in Villeneuve l'Etang near Paris.

WILLIAM GIBSON

William Wallace Gibson was born in Savannah, Georgia on 11 November 1894, the son of Carleton Bartlett Gibson, an educator, and Martha (Newcomb) Gibson. He attended the Troy Academy, and the Rensselaer Polytechnic Institute, Troy, New York, Class of 1917. He left that institution in the middle of his sophomore year to enter business.

Gibson was working in Bridgeport, Connecticut for the Remington Arms Company when he decided to enter the European War. On 13 July 1917 Gibson enlisted in France's *Service Aeronautique*. On 13 September 1917 he was released because of inaptitude.

Gibson received a commission as a Second Lieutenant in the U.S. ASSRC and received his *brevet* at Tours. On 3 February 1918 Lt. Gibson was assigned at the Front to the 94th Aero Squadron, USAS. On 16 April 1918 he was assigned to the 1101 Aero Squadron, USAS where he served until the 11 November Armistice. It is believed that Gibson died in 1957.

Left to right: William Bullen, Gibson, John Endicott and William McGinn.

JOSEPH GILL

Joseph Francis Gill was born in Indianapolis, Indiana on 6 October 1897, the son of Mr. & Mrs. J.P. Gill. He attended local schools there and the Shortridge Middle School, Indianapolis, before enlisting in the American Ambulance Field Service.

In the Spring of 1917 Gill sailed for France and enlisted in France's *Service Aeronautique* on 10 June 1917. From 13 June 1917 to February 1918 he attended the aviation schools at Avord, Pau, and the *G.D.E.*, receiving his *brevet* on the Caudron on 4 December 1917.

Caporal Gill was commissioned a Second Lieutenant in the U.S. ASSRC on 22 March 1918. On 8 June 1918 he was assigned to *SPA 471*, the Defense of Paris Squadron, at Le Bourget, and served with that Squadron until 17 July 1918. On that date he was assigned to the American Acceptance Park, Orly until 28 August 1918 when he was injured in an accident which prevented him from further service until after the end of the War.

Following World War One, Joseph Gill attended Stanford University, Stanford California. He received an undergraduate law degree in 1927 and a Phd in law in 1928. He became a practicing attorney in California. He married and later divorced. In 1948 Gill received a Master's Degree in speech and drama from Stanford University.

He succumbed to cancer on 17 April 1950 at the Palo Alto Veterans Hospital at age 52. He is buried at Golden Gate National Cemetery, San Bruno, California.

Lieutenant Joseph Gill, standing (3 l.) with other members of SPA 471, Le Bourget, France, July 1918. Fellow LFC pilot George Willard is standing (2 r.).

WILLIAM GILMORE

William Smith Gilmore was born in Cincinnati, Ohio on 12 May 1891, the son of V. Gilmore. Gilmore attended Boston's Massachusetts Institute of Technology, Class of 1913, where he studied architecture for two years.

In the fall of 1916 Gilmore sailed from the United States to France to serve in the American Ambulance Field Service. He spent some time in Italy before joining the Field Service on 16 January 1917. He served in the Verdun sector as an Ambulance driver with SSU 12. On 9 July 1917 Gilmore joined France's *Service Aeronautique*. He resigned to join U.S. Naval Aviation.

On 9 May 1918 Gilmore enrolled as a Landsman For Quartermaster at Brest, France. He received ground school and primary flight training at Moutchic, France. On 26 September Ensign Gilmore became Naval Aviator #1504. He was assigned as a seaplane pilot at Porto Corsini, Italy, and took part in bombing raids against the Austrian aerodrome at Pola, Yugoslavia. Following the Armistice, he returned to the United States and served at the Naval seaplane training station at Bay Shore, Long Island, New York. He was discharged as an Ensign on 28 November 1921.

Following World War One Gilmore worked as an architect. He retired to Bermuda. Gilmore passed away in Dobbs Ferry, New York on 3 July 1972 at age 81.

CLARENCE GLOVER

Clarence Merritt Glover was born in New York City, New York on 20 May 1894, the son of Benjamin M. Glover and Louise (Grossmiew) Glover. He attended New York public schools before enrolling in Yale University. He left that school without taking a degree and entered into business in New York. Desiring to enter the European conflict, Glover joined the Norton-Harjes Ambulance Service in 1916 and served in France with that unit for six months. He was expelled from the Ambulance Service for misconduct.

On 9 June 1917 Glover enlisted in France's *Service Aeronautique*. He immediately left for the United States where he was charged with posing as a Lafayette Escadrille pilot by a flying-school colleague who wrote an expose letter which appeared in a 20 August 1917 New York newspaper accusing him of being an imposter.

Glover had returned to France in September of 1917. He denied his colleague's charges in a 22 October 1917 letter to the New York *Times* which he had mailed from the Avord aviation school on 16 September 1917. Glover stated in the letter:

"I have been in France one year and during that time have been connected with the Norton-Harjes ambulance corps, being with them six months. On 10 May I received permission of 21 days to America on account of an injury to my arm, received while in the Ambulance service."

Glover denied that he had referred to himself as a Lafayette man but admitted that perhaps his friends had. From 16 September 1917 to April 1918 he attended the aviation schools at Avord, Pau, Cazeaux, and the *G.D.E.*. He received his *brevet* on the Caudron on 19 December 1917. He was the last Lafayette Flying Corpsman to be trained in the French schools.

On 1 July 1918 Glover was assigned at the Front to *Escadrille SPA 78* and flew with that Squadron until the Armistice. He finished his service as a *Sergent*.

Following World War One Glover returned to the United States. However, he shortly returned to France where he became involved in business in Paris. In 1940 he redeemed himself in the eyes of his comrades when he launched a campaign to raise money to restore the Lafayette Flying Corps monument near St. Cloud which had fallen into severe disrepair. Glover died in Paris in the summer of 1971 at age 77.

REGINALD GOURARD

Reginald Gerald Gourard was born in New York City, New York, on 31 March 1897. A foundling at birth, he was taken into the care of Aimie Crocker Gourard in the second week of his infancy and was raised by her.

Miss Gourard owned property in Paris, and while she was living in France, Reginald attended the Anglo Saxon School in Paris. He and Miss Gourard were residing in New York City in December of 1916 when Reginald applied for a passport to France to attend to business affairs for his stepmother.

Gourard sailed for France aboard the *Rochambeau* on 30 December 1916. On 11 February 1917 he enlisted in France's *Service Aeronautique*. On 4 May 1917 he was released because of inaptitude.

CHARLES GREY

Charles Gossage Grey was born in Chicago, Illinois on 20 June 1894, the son of Mr. & Mrs. W.C. Grey. Charles attended Chicago schools before enrolling in Columbia University, New York, New York.

In 1916 Grey was working as a reporter for a Chicago paper when his editor assigned him to interview a recruiter of ambulance drivers to serve with the French Army. Grey finished his assignment then joined the American Ambulance Service. Grey served with SSU 4 for 5 months, carrying wounded from the Front.

On 17 June 1917 Grey enlisted in France's *Service Aeronautique*. From 19 July to 24 November 1917 he attended the aviation schools at Avord, Juvisy, Pau and the *G.D.E.*, receiving his *brevet* on the Caudron on 26 September 1917.

On 26 November 1917 *Caporal* Grey was assigned at the Front to *Escadrille SPA 93* and flew with that squadron until 13 March 1918. On 21 March 1917 *Sergent* Grey was commissioned a First Lieutenant in the U.S. Air Service and that same day was assigned as a Nieuport test pilot at Le Bourget, and American Acceptance Park, Orly field.

On 1 August 1918 Lieutenant Grey was assigned to the 213th Aero Squadron, USAS as a flight commander. From 2 September to 3 November 1918 he flew in combat operations from Vaucouleurs, Bar-le-Duc, and Foucaucourt. Grey was credited with the destruction of five enemy aircraft and was awarded the U.S. Distinguished Service Cross for extraordinary heroism. His citation, dated 10 December 1918, describes the action:

"For extraordinary heroism in action near Montmedy, France, 4 November 1918. While leading a patrol of three machines, Captain Grey observed a formation of our bombing planes hard-pressed by twelve of the enemy. He attacked the leading enemy machine without hesitation, thereby attracting the enemy's fire and allowing the bombing machines to escape undamaged." Grey was promoted to Captain on 6 November 1918. He was discharged in France in 1919.

Following World War One Grey lived in Paris with his French wife and worked in the US Diplomatic Corps. He was struck down by an automobile while walking near his home in Palm Beach, Florida, and died at age 92 on 6 March 1987.

NORMAN GRIEB

Henry Norman Grieb was born in Philadelphia, Pennsylvania on 2 July 1895, the son of William Gottlieb Grieb, Founder and President of the Ajax Tire & Rubber Company, Inc. of New York City, and Sarah (Gesemyer) Grieb.

"Shorty" Grieb was a graduate of William Penn Charter School, Philadelphia, of Phillips Academy, Exeter, New Hampshire, and of the Hotchkiss School, Lakeville, Connecticut.

While Grieb was a student at Yale, Class of 1918, he served in the R.O.T.C. In the spring of 1917 he enrolled in the Yale Ambulance Unit which was being organized for service in France. On 26 May 1917 Grieb sailed for France with fifty-eight other Yale students aboard the S.S. *Rochambeau*.

Once in France, on 13 June 1917 Grieb enlisted in the *Service Aeronautique*. From 19 June to 28 August 1917 he attended the aviation school at Avord.

On 1 July 1917 *Caporal* Grieb was taken with serious injuries to the aviation camp hospital. The cause of his injuries vary. One source stated that he "suffered fractured ribs and a contusion of the lungs" as a result of an airplane accident. Another source traced his injuries to "a serious wound caused by the accidental discharge of a revolver." A third source said he sustained his injures when he was run over by an automobile. A fourth sources stated Grieb "had been given his aviator's license and was making a trial flight when his motor died and a forced landing was necessary. While working on the motor an automobile crashed into the aeroplane and it was found at the hospital that Grieb had suffered fractured ribs and contusions of lungs." As a result of his injuries, this source continued, he developed pneumonia and died. A fifth source, an American comrade, claimed he had contracted venereal disease from "the d'Avord girls" and after several weeks in hospital, had ended his life with an overdose of morphine.

Whatever the cause of his injuries, Grieb was removed to the French base hospital at Bourges, Cher, where he remained in the convalescence ward of the hospital until his death on 26 August 1917. He was 22 years of age at the time of his death. He was buried the next day with full military honors in the old cemetery at Bourges. The French Government posthumously awarded *Soldat* Grieb the *Medaille Militaire*. In 1928 his remains were transferred to the Lafayette Flying Corps Memorial, St. Cloud.

JAMES GRIER

James Murray Grier was born in Philadelphia, Pennsylvania on 15 October 1896, the son of Frank G. and F. May Grier. James Grier attended Philadelphia schools before enrolling in Temple University in Philadelphia. He left that institution in 1917 to join the American Ambulance Service.

When Grier arrived in France he decided to join French aviation. On 1 August 1917 he enlisted in the *Service Aeronautique*. From 2 August to 10 April 1918 he trained at the aviation schools at Avord, Tours, Pau, and the *G.D.E.* On 6 January 1918 he received his *brevet* on the Caudron.

On 10 April 1918 *Caporal* Grier enrolled in the Naval Reserve Flying Corps as an Ensign at Paris. On 19 July 1918 he became Naval Aviator #1006 (HTA) at Lake Bolsena, Italy. He served at Moutchic, and at Porto Corsini under Lieutenant Willis Haviland.

Following his World War Service, Ensign Grier remained in the Navy and served at Bay Shore, Hampton Roades, and at Langley Field; also aboard the USSs *Oklahoma* and *Arizona*, and later at Mitchell Field.

On 6 September 1920 LTJG Grier was on duty at Mitchell Field, Long Island, New York. He was flying a JN-4 #38070 above the National tennis championship matches at Forest Hills, New York, with Sgt. J.P. Saxe, who was photographing the competition from the rear cockpit. Grier was flying the plane at an altitude of 500 feet at reduced air speed in order that Saxe might take the best possible shots. He turned into the wind shortly after passing the grandstand. His plane fell into a spin and dove vertically to earth, killing Saxe and mortally injuring Grier. Lieutenant Grier died at 5 pm on 6 September 1920 in St. John's Hospital, Long Island City, New York. He was 23 years of age at the time of his death. He was buried in Philadelphia, Pennsylvania.

DAVID GUEST

David Porter Guest was born in Richmond, Virginia on 25 November 1895, the son of Commodore John Guest, U.S. Navy, and Lucille (Powers) Guest. Porter was educated at the Episcopal High School, Alexandria, VA, the Philips Exeter Academy, Exeter, New Hampshire, and at Yale University, Class of 1918, where he studied mechanical engineering.

While a student at Yale, Guest took a primary aviation course at the Curtiss Flying School at Newport News, Virginia. Following the United States' entry into the European War, on 2 June 1917 he sailed for France to join the conflict.

On 6 July 1917 Guest enlisted in France's *Service Aeronautique*. He received his *brevet* on 17 October 1917. On 30 December 1917 Guest was released upon his own request in order to join the U.S. Air Service. Guest returned to the United States. He enlisted in the Air Service and was commissioned a Second Lieutenant on 4 March 1918. Guest served as an instructor at Door, Carlstrom, and Gerstner Fields. On 15 September 1918 he was commissioned a First Lieutenant. He was discharged on 15 January 1919.

Following World War One Guest worked as a bond salesman for two years. He returned to school and took courses in corporate finance and commercial law then began working as an accountant for a number of firms. In 1926 he married Eleanor Putnam Lakin of Scarsdale, New York in New York City. From this union two children were born.

During World War Two Guest served in the U.S. Army Ordnance Legal Division as a consultant on contract negotiations. In 1945 he became a negotiator for the U.S. Maritime Commission until his retirement. He died in an Alexandria, Virginia hospital on 4 July 1967 at age 71.

ANDRE GUNDELACH

The majority of the Lafayette Flying Corpsmen who served at the Front during World War One flew or fought in *chasse*, or fighter squadrons. However, thirty Americans served in bombing and observation squadrons as pilots and observer/gunners. Occupying the seat of a cumbersome bombing aircraft was more dangerous than flying a single-seater fighter, and eight of the thirty men who chose to serve in bombers and observation craft were killed at the Front.

American chasse pilot George Dock, who flew with SPAD 12 and 31, paid this tribute in his unpublished memoirs to his colleagues who flew in the bombing and observation planes.

"Nothing surpassed the courage and tenacity of the bombers. At all hours of the day they crossed the lines in large patrols, and flew far into German territory in spite of heavy fire from Archy and numerous formations of Fokkers that were eager to attack these cumbersome, slow-moving machines. Regardless of their serious losses they fought their way to their objectives, never turning back until their missions had been accomplished.

"It was inspiring to see them flying steadily on, never swerving from their course to avoid storms of anti-aircraft shells, while Fokkers darted like ill-tempered gnats around them, trying to break up their formation by throwing them into panic. The exploits of an "ace" who brings down his two or three dozen enemies are well worthy of renown, but in respect of heroism and grit, to say nothing of their absolute military value, the day bombers were the peer of any aviators in the French Army.

"Pursuit pilots could bide their time and take advantage of an opponent's weakness or surprise, crossing the lines then returning into their own territory whenever things were unfavorable. But bombing pilots who went far into enemy country often fought against greatly superior numbers of combat machines, risked much more than any chasse pilot a forced landing beyond the Hun lines, and ran the gauntlet of Archy fire which was more demoralizing than even the attacks of Fokkers. The high quality of the bombers was due largely to the bravery of the squadron commanders, who always led the patrols on their raids and insisted on strict discipline."

Andre Gundelach, the son of a Chicago, Illinois, cabinet maker, epitomized the best of the young Americans who served as bombing and observation pilots. Andre was born in Chicago on 16 January 1895, the son of August Gundelach and Alice (Burkhart) Gundelach. Andre attended Chicago parochial schools. His mother died while he was still a youth, and in 1914 Gundelach enlisted in the U.S. Navy and served as a

boatswain's mate. While aboard ship with the Asiatic Fleet, he made a round-the-world cruise.

Of French ancestry, (his mother had been born in France of English and Belgian parents), in late 1916 Andre asked for and received an honorable discharge from the Navy in order to join French aviation. In February of 1917 Gundelach sailed from New York to Bordeaux, France. On 20 March 1917 he enlisted in France's *Service Aeronautique*. From 24 March to 10 July 1917 he attended the aviation schools at Avord, Pau, and the *G.D.E.* He received his *brevet* on the Bleriot on 31 May 1917.

On 12 July 1917 *Caporal* Gundelach was assigned at the Front to *Escadrille SPA 95* and flew with that squadron until 8 September. While with *SPA 95* he was awarded the French *Croix de Guerre*, with Palm for his destruction of an enemy aircraft. On 24 September he requested and was granted a transfer to *Escadrille Sop. 111*, a bombing squadron, and flew with that squadron until 21 December 1917.

On 9 November 1917 Gundelach had been commissioned a First Lieutenant in the U.S. Air Service. From 3 January to 18 May 1918 he served as an instructor at the 7th Aviation Instruction Center, Clermont-Ferrand.

On 23 May 1918 Lieutenant Gundelach was assigned to the 96th Aero Squadron, USAS, a Day Bombardment Squadron. A fellow pilot, Lt. Charles Codman later penned these observations of Gundelach. "Gundelach obviously knew what he was about. That keen blue eye, the rare but infectious smile that lit up the bony face under its thatch of blond hair, were reassuring, and his manner of handling a Breguet as if it were a pursuit plane inspired confidence."

Lt. Gundelach flew with the 96th Aero Squadron as a flight commander until 12 September 1918. On that day he was killed in combat in the Saint Mihiel Sector.

Gundelach was posthumously awarded the U.S. Distinguished Service Cross for extraordinary heroism. His citation describes the action that resulted in his death.

"For extraordinary heroism in action near Buxieres, France, September 12, 1918. Lieutenant Gundelach with Second Lieutenant Pennington H. Way, Observer, volunteered for a hazardous mission to bomb concentrations of enemy troops. They successfully bombed their objective, but while returning were attacked by eight enemy planes. Their plane was brought down in flames and both officers were killed."

The total details of the mission are not included in the citation, a mission which had begun on the opening day of the St. Mihiel offensive. Lieutenant Gundelach had been ordered by the squadron Commander on a bombing run against troop concentrations on what the squadron log called "the worst flying day in many months." That morning Lt. Codman had observed Gundelach gazing upward into the storm. Gundelach told the Headquarters officer "We shall have to wait until it clears."

"Your orders are to leave for Buxieres with your formation at ten o'clock," the officer demanded as he checked his watch.

"And throw away eight planes? No formation can get through this, far less get back."

Gundelach stalked off the field and took his objections to Major James Dunsworth, the C.O. Moments later the grimfaced Gundelach appeared from the operations building and scanned the lowering clouds. He stalked across the airfield to his Breguet bomber and he and his observer climbed on board.

The airfield was being lashed by squalls of rain as Gundelach and Lt. Way headed into the gale with a full load of thirty-two bombs beneath the sagging wings. Near the end of the airfield, a violent gust of wind helped to lift the laboring bomber above the chimney of the Y.M.C.A. hut. An instant later it disappeared into the storm.

Gundelach and Way managed to drop their load on the target. But upon their return their Breguet plane, No. 4495, was attacked by eight enemy aircraft and crashed in flames near Commercy. Way was burned to death in the wreckage. Gundelach escaped the intense flames by jumping from his cockpit. His body was found several days later some distance from the crash site.

DAVID GUY

David Wade Guy was born in St. Louis, Missouri, on 28 September 1897, the son of William E. Guy, and Katherine (Lemoine) Guy. Guy attended the Morristown School before entering Princeton University, Class of 1919. However, he left that institution in his sophomore year in February of 1917 to serve in France with the American Ambulance Field Service. From 19 February to 25 July 1917 Guy served as an ambulance driver with S.S.U. 15 at Hill 304 and in the Mort Homme Sector.

Guy joined France's *Service Aeronautique* on 21 July 1917. From 1 August to 28 November 1917 he attended the flying schools at Avord, Tours, Pau, and the *G.D.E.* From 2 December 1917 to 1 January 1918 he served at the Front with *SPAD 155*. From 1 January to 1 June 1918 *Sergent* Guy flew with *SPAD 156*. He then served with *SPAD 38* from 1 June to 7 November 1918. While flying with *SPAD 38* in the Champagne Sector on 1 July, he brought down a German Rumpler observation plane for which he was later awarded France's *Croix de Guerre*, with Palm.

On 7 November *Sergent* Guy entered the U.S. Army Air Service as a First Lieutenant and was assigned to the 1st Aero Squadron, USAS. From 20 November 1918 to 5 January 1919 Lieutenant Guy served in the Army of Occupation at Treves, Germany. He returned to the United States and was discharged on 14 April 1919 at Mitchell Field, Long Island, New York.

Following World War One, David Guy entered Massachusetts Institute of Technology and received his degree in Chemical Engineering in 1922. He then returned to St. Louis and began work for the Monsanto Chemical Company as a chemical engineer. He married Miss Maury Howard Crawford in Paris on 16 July 1923. From this union two daughters, Valerie and Althea, were born.

Guy also worked for Ingersoll-Rand Corporation, and later, in the financial department of the Pennsylvania Insurance Company. He eventually founded and became president of the Wellington Foundation, Inc. a mutual fund company, and served as its director for many years. He died in a Byrn Mawr, Pennsylvania hospital on 28 May 1960 at age 62.

JAMES HALL

ESCADRILLE LAFAYETTE

James Norman Hall, the 34th volunteer of the *Escadrille Lafayette*, was born on 22 April 1887 in Colfax, Iowa, the son of Arthur Wright Hall and Ella Annette (Young) Hall. James Norman was one of five children born to the Halls in the small prairie community of Colfax (pop. 1749) which served as the trading hub for the surrounding farms. Arthur Hall was the bottler and shipper of *Colfax Mineral Water*.

Hall graduated from Colfax High School in June of 1904. He then began working full time as a clothing clerk in the Colfax Mercantile. At night he pursued a writing career and submitted his poetry to various national magazines such as the *Atlantic Monthly* and *Harpers*.

Shortly after starting full time at the Mercantile James was offered the opportunity to manage another mercantile in a neighboring community. But Hall turned the offer down and instead traveled to Grinnell, Iowa, some thirty-two miles from Colfax, where he enrolled in Grinnel College in the spring of 1906 as a member of the Class of 1910.

Hall put himself through Grinnell College and was awarded a Bachelor of Philosophy degree in 1910. Following graduation, he took a position in Boston with the Massachusetts Society for the Prevention of Cruelty to Children. He also took night classes at Harvard University to begin work toward a masters degree in English literature. Hall still held hopes of becoming a writer, and when time allowed he continued to struggle at the craft.

In April of 1914 Hall left Boston with the intention of becoming a sheepherder in Miles City, Montana at the suggestion of a friend who edited the newspaper there. However, he soon changed his mind and instead decided to go to England.

On 27 May Hall sailed for Liverpool aboard the steamship *Laconia* to make a pilgrimage to Nether Stowey, in Somerset, where Samuel Taylor Coleridge had lived while writing *The Ancient Mariner.* Hall had intended to spend the summer in England writing and bicycling through the countryside in search of inspiration. He later wrote of one such excursion through Wales following his failed attempt at visiting Joseph Conrad at his home. Hall had considered Conrad "the greatest artist of our times" and had made it to the author's dwelling only to quickly flee, lacking the courage to knock on his door. Hall had completed his journey on 8 August 1914:

"As I traveled north though the beautiful countryside in the direction of Wales, I had not the faintest premonition that I was enjoying with all England the last days of the Indian summer of an age that was all but gone. . . .

"It was in the village of Beddgelert that I walked, all unsuspecting, into my first news of World War I. I had arrived from Llangollen only the evening before and took lodgings for the night at an ancient inn called Llewellyn's Cottage, lighted by small leaded windowpanes in walls three feet thick. I well remember the events, such as they were, of that last evening I was ever to spend in a world at peace."

Ten days later, on 18 August 1914, Hall enlisted in London as a private in the British Army with the 9th Battalion, Royal Fusiliers. He represented himself as a Canadian citizen in order to be accepted into the British military forces which were not accepting Americans at that time. The following day, at the Hounslow Barracks, he was issued a British uniform and a Lee-Enfield rifle.

Hall's 9th Fusiliers became one of the battalions of the First Hundred Thousand. He trained with his unit for seven months at St. Martin's Plain, in Kent then received further instruction as a machine gunner at Aldershot before embarking from Plymouth for France in May of 1915 with the 12th Division.

Hall's military transport landed at LeHavre where he boarded a troop train for the Front and the trenches. He was soon promoted to lance corporal and put in charge of his machine gun section.

Near the end of September 1915 Hall's division was moved southward to take part in the Battle of Loos. Hall entered the trenches near the village of Hulluch and soon joined in the British assault against the German lines. Following the successful assault, Hall's C Company gun team took up quarters in a half-destroyed German surface dugout. Hall momentarily left the dugout to spell a soldier at the forward gun position when an enemy shell made a direct hit on the dugout, killing his seven comrades inside. Throughout the month of October Hall remained in the line enduring frequent attacks and counter-attacks and heavy artillery fire. Of that time, Hall would later write:

"For a full month we had no opportunity to remove any of our clothing. We were moved from one position to another through the wreckage of trenches where the tangled mass of telephone wires, seemingly gifted with a kind of malignant humor, caught in the piling-swivels of our rifles and coiled themselves around our feet. Trenches were repaired only to be destroyed while the work was still in progress. Twice we received orders that we were to attack at dawn; then the order was countermanded. There were many such orders and counterorders, alarms and excursions."

Upon the completion of the Battle of Loos, with winter at hand, Hall's division was withdrawn from the line to rest and to re-equip. About that time, in late November 1915, Hall had requested emergency leave home to America after receiving a letter from his mother that his father was critically ill with Parkinson's disease. Hall's commander had refused his request for leave. But upon learning that Hall was an American citizen who had enlisted illegally as a Canadian, he informed him he could be granted an immediate discharge. Then if he so wished, he could re-enlist after seeing his father.

Hall accepted the Colonel's proposal and was immediately discharged on 24 November 1915 "on the ground of being an American citizen." He set sail for the U.S. shortly before Christmas following one year and 99 days of service with the Royal Fusiliers, 172 of them in service in France with the British Expeditionary Force.

Shortly after Hall's ship docked at Boston he received a note from the editor of the *Atlantic Monthly* requesting that he call on him. At that time the magazine was associated with Houghton Mifflin Company. Ellery Sedgwick, the *Atlantic* editor, requested that Hall write a series of articles for his magazine about his experiences with the British Army. Hall accepted the proposal and shortly returned to Colfax, Iowa.

While at home he began working on the articles which would later become his book, *Kitchener's Mob*. While writing the articles Hall suffered pangs of guilt at not being in the trenches with his comrades. He planned to re-enlist in the B.E.F. but also felt a duty to finish his book, hoping to make the American public more sympathetic to the Allied effort against Germany and more aware of the suffering of the soldier in the trenches.

Five months after his discharge from the British Army Hall had turned his manuscript over to his publisher and was preparing to re-enlist, this time as a stretcher-bearer with the Royal Army Medical Corps. But while in Boston visiting Ellery Sedgwick, the *Atlantic Monthly* editor had requested that he do three articles on the newly-formed *Escadrille Americaine*. At that time Hall knew nothing of the squadron but immediately accepted the assignment. His interest in aviation had gone back to the fall of 1910 at Squantum, Massachusetts when Claude Grahame-White had flown his Bleriot monoplane around the Boston Lighthouse. Also witnessing that event was future *Lafayette Escadrille* pilot Carl Dolan.

Hall left Boston for London then went on to Paris where he met Paul Rockwell, then in charge

James Norman Hall while at the Buc School of Aviation, spring of 1917.

of the Paris bureau of the Chicago *Daily News*, and several of the *Lafayette* pilots, including Paul's brother, Kiffin. Rockwell introduced Hall to Dr. Edmund Gros who suggested he join the Franco-American Flying Corps instead of re-enlisting in the British Army.

Hall accepted Gros's suggestion and on 11 October 1916 he enlisted in Paris in the aviation section of the French Foreign Legion, L.M. No. 11921. About the same time his book, *Kitchener's Mob* was released to the public and met with some success.

Hall attended the French aviation schools at Buc, Avord, and the G.D.E. from 16 October 1916 to 14 June 1917, receiving his *brevet militaire* No. 6051 on the Caudron on 23 April 1917.

Caporal Hall joined the *Escadrille Lafayette* at Chaudun with fellow Americans Douglas MacMonagle and David Peterson on 16 June 1917 and almost immediately began flying combat patrols. On 26 June a group of U.S. Air Service officers had arrived at the Squadron for a tour of inspection. Group Commander Fequant ordered the pilots of the *Lafayette* Squadron to fly an exhibition patrol for the visiting Americans.

The sun had nearly set by the time the nine American pilots had prepared their planes for take-off. Hall had been one of the pilots chosen for the patrol but had difficulty in starting his SPAD and had left the airfield some minutes behind the other pilots. Once airborne, he hurried to catch up and was soon over the front lines which then stood just fifteen kilometers from the Squadron's aerodrome.

As Hall crossed into enemy airspace at 4,000 meters, he came under desultory antiaircraft fire but was heartened at spotting his comrades five kilometers inside German lines. Hall rushed to join the security of the seven-plane formation. But to his horror one of the group of fighters at his level opened fire upon him.

Hall instantly realized his mistake as the German bus poured tracer rounds into his SPAD. Hall banked sharply to avoid the deadly fire then felt a smashing, white-hot blow to his left shoulder. At the same instant his goggles where shot in two at the nosepiece and he was blinded by blood streaming down from his bullet-creased forehead.

Hall's SPAD fell into a corkscrew dive, its motor roaring at full revolution. The enemy pack hung on his tail, and another round pierced his flightsuit and creased his scrotum. Seeing Hall's SPAD falling out of control, the Germans broke off their pursuit. Hall's whole left side was paralyzed and he thought his left arm had been shot away. He placed his stick between his knees then throttled the SPAD's engine down with his right hand until the propeller stopped dead. Then Hall passed out.

When he came to he discovered he was lying on a stretcher in the Allied trenches in the Ravine of Ostel. His SPAD, a complete wreck, had pancaked squarely into a forward trench, shearing off both wings against the sandbagged parapets. Hall was rushed to a first-aid dressing station where his bloody combination and uniform were cut away. A bullet had entered his left shoulder from behind, missing the shoulder blade, but making a large hole where it had exited under the arm. Hall was soon sent to the American Ambulance hospital at Neuilly, near Paris, where he was decorated with France's *Medaille Militaire* and *Croix de Guerre*.

Hall used his convalescent time to begin a book titled *High Adventure* which dealt with his experiences in the French Flying Corps. In early September 1917 Hall left the Neuilly hospital. But by a mistake in orders, he was sent to *Spa 112* instead of back to the *Lafayette* Squadron. As he returned from Belfort to the Verdun sector, Hall's fuel pump quit while he was over the Vosges Mountains. He made a forced landing in a cow pasture on a "sugarloaf mountain" where he demolished his SPAD after striking a wire fence and apple trees. Miraculously, Hall escaped with only a broken nose, and he was ordered back to the G.D.E.

Sergent Hall soon wearied of Plessis-Belleville and "deserted to the front" in a battle-scarred Nieuport. On 3 October 1917 he arrived at the *Escadrille Lafayette*, now at Senard at the foot of the Argonne Forest. It was there that he first learned of the 24 September death in combat of his close friend and comrade, Douglas MacMonagle. Upon hearing the sad news, Hall had wept silently while swearing to avenge MacMonagle.

Hall flew a number of combat patrols through autumn and early winter. Anxious to avenge MacMonagle's death, he often returned from patrol with his fighter punctured by gun fire.

On New Year's Day 1918 Hall set out on patrol with William Thaw and others from the aerodrome at LaCheppe, in the Champagne sector. It was a bitterly cold morning, and the mechanics had had difficulty in getting the SPADs' engines warmed up. When the motors had reached the proper revolutions, the flight headed up into the clear, bitter sky.

Hall lost track of his comrades not long after becoming airborne. He spotted a speck a half-mile distant, and thinking it was Thaw, began to climb and overtake him. Hall soon realized his error as the black German crosses grew visible on the Albatros fighter's wings. Hall's unwary adversary had no sense of the closing danger and was entertaining himself by performing stunts.

At five hundred meters above the enemy Hall dove down. He closed to within one hundred meters then fired both guns, watching his tracers bounce off the Fokker's engine. The enemy bus soon fell out of control then disappeared into a cloudbank. When it momentarily reappeared Hall poured another 100 rounds into it before it flipped over and disappeared again into vapor.

Hall quickly noted his time and location, east of Brimont. Upon reaching his aerodrome he learned that his destruction of the enemy bus had been verified by Allied infantry who had seen the ship crash in no-man's land. Thus *Sergent* Hall received his first confirmed victory, and the very last for the *Escadrille*.

Captain Hall before his SPAD scout while with the U.S. 103rd Aero Pursuit Squadron, La Noblette, March 1918.

Hall was commissioned a Captain in the U.S. Air Service on 7 February 1918 and flew with the U.S. 103rd Aero Pursuit Squadron at La Noblette from its inception on February 18 until 29 March 1918. During his service there Captain Hall scored his second and third victories over enemy scouts, both on 27 March, the first over Reims and the second (which he shared with Major Thaw and Captain Christopher Ford) near Somme-Py. Hall had been leading the patrol and had led the attack against a German formation- five fighters and three two-seaters . The fight had lasted twenty minutes and had concluded in the destruction of at least three enemy aircraft. Hall, for his heroic role in this action, was awarded the U.S. Distinguished Service Cross and France's *Croix de Guerre*, with Palm. His French citation called him a "Pilot of great bravery who has engaged in numerous fights daily."

Capitaine Thenault, Hall's *Lafayette* commander, praised the American even further, in saying "What a splendid soldier he was! After having distinguished

Captain Hall in a German staff car shortly after his crash behind enemy lines. Hall's Nieuport 28 fighter had been knocked down by an unexploded antiaircraft round to its engine. Hall's bus slammed into the ground, breaking his nose and severely spraining both ankles.

himself in the British and French armies, he was to become once more conspicuous for gallantry in the American army. Few men have fought as real combatants in three armies and he must hold a unique record."

On 29 March Hall was assigned as a flight commander to the 94th Aero Pursuit Squadron, USAS, then at Epiez. His entry there was inauspicious and somewhat embarrassing as he somersaulted his fighter on the squadron's muddy field before a group of raw American pilots. On 29 April he scored his fourth confirmed victory which he shared with Lt. Edward Rickenbacker (his first) when they shot down an Pfalz fighter near St. Buassant.

On the morning of 7 May 1918 Captain Hall left the Toul aerodrome in the St. Mihiel sector with Lieutenants Rickenbacker and Green to challenge five enemy fighters in the air near Pont-a-Mousson. The 94th had been recently equipped with the Nieuport 28 fighter, and while the pilots immediately appreciated its power and maneuverability, one such fighter flown by Lieutenant Meissner had revealed an alarming defect. The fabric on its upper wing surface had split under pressure along the leading edge and had ripped back in great strips. Meissner had just managed to control the Nieuport and to bring it safely to ground. At the time of the incident it was thought that the defect had been unique to Meissner's fighter. Only later, to Hall's distress, was it learned that all Nieuport 28's had the same inherent weakness.

Hall's patrol located the enemy group above the village of Armaville, south of Metz. Holding an altitude advantage, the Americans dove to the attack. Hall had his foe squarely within his gunsights and was about to open fire when he heard an ominous ripping overhead. Instantly large pieces of shredded fabric began flapping in his slipstream. He immediately broke off the attack and turned toward allied lines.

Hall, remembering Meissner's experience, nursed his Nieuport homeward as light shown down through the upper skeletized wing plane. He throttled to the slowest possible speed to stay airborne, and yet keep his fighter in a straight-level descent. But as Hall lost altitude he came under fierce antiaircraft fire. Unable to take evasive action, his plane shook violently from a direct hit. He dropped out of control at 1000 meters.

Closing quickly to earth near Pagny-sur-Moselle, Hall reefed back on the stick. His landing gear sheared away as he slammed to ground and skidded across an open field. Almost immediately he was surrounded by German troops who lifted him from his fighter.

As Hall touched the ground shooting pain in his ankles told him that one or both had been broken. Also, for the second time he had broken his nose. A German artillery officer informed him that the mortal blow dealt his plane had been the result of a direct hit by a 37-millimeter round from an antiaircraft battery. Miraculously the round had failed to explode on contact but had pierced the engine and had stuck fast in one of the cylinders.

Hall was transported to the German war hospital at Jarny-Conflans. There he lay for six weeks in a ward while his ankles mended. He was then transported to the Bavarian town of Landshut and placed with other officer detainees. And while they were decently fed and well housed, their existence contained a deadly monotony within the high-walled prison yard.

With Hall a prisoner of war, his publisher, Houghton Mifflin, had released his book, *High Adventure*, in its uncompleted form. Hall had been unable to send the final two chapters on to his editor before being shot down.

At the signing of the Armistice Hall and two other American pilots were allowed to leave Landshut castle prison and make their way to the Swiss border. Without papers and still in their U.S. Air Service uniforms, they set out by train through chaotic Germany. They then took a steamer across Lake Constance to Romanshorn, Switzerland. From there they went to Berne and into American custody, arriving in Paris two days later.

Hall was assigned to temporary duty at the U.S. Aviation Headquarters in Paris. He sought and was granted permission from General Patrick, Chief of the Air Service, to make one last flight over the old front from the Channel coast to the Swiss border. Patrick could not have refused Hall's request, this pilot who had belatedly received his Distinguished Service Cross to add to the decorations bestowed upon him by France to include the *Croix de Guerre*, with five Palms, the *Medaille Militaire*, and the *Legion d'Honneur*. Hall made the journey in a 180-horsepower SPAD.

Upon completion of this flight Hall returned to Paris Headquarters where he was ordered by Dr. Gros, now a Lieutenant Colonel in the U.S. Air Service, to assemble and edit the histories of the pilots of the *Lafayette Escadrille* and the Lafayette Flying Corps. His associate in the project would be Lieutenant Charles Nordhoff, a Californian, and a former Lafayette Flying Corps pilot with *Spa 99*.

Charles Bernard Nordhoff as he appeared while serving with Escadrille N.99, January 1918.

Initially the relationship between Hall and Nordhoff had been far from amicable. Both were opposites in temperament. Hall was self-effacing and quickly won friends. Edwin Parsons, his Lafayette comrade, called him "one of the most loveable men in the world." He had an ability to get along with everyone, and was unquestionably the most popular pilot in the *Lafayette Escadrille*. In fact, his friend Paul Rockwell later wrote of him: "Jim Hall was one of the finest all-around persons I have ever known, and an honorable man in every way. He had just about everything, intelligence, courage, appearance, talents of many sorts, and was most modest and withdrawing."

Charles Nordhoff, on the other hand, while unquestionably brilliant, was aloof, distant, fastidious. And as one Lafayette colleague later said of him he was "a likeable fellow—at a safe distance."

When Captain Hall had first met Nordhoff, the Lieutenant had shook Hall's outstretched hand "in a coldly punctilious manner." The first thought to flash through Hall's mind was: "Lord! How am I to work with this man?"

Hall later recalled that he would never have imagined that in spite of this strained first meeting a firm bond to span 28 years would develop between the two

Charles Bernard Nordhoff and James Norman Hall at work in Hall's library at Papeete, Tahiti, circa 1930.

men both in life and in literature, from 1919 until Charles Nordhoff's death in Montecito, California in 1947. But initially all the two men seemed to share in common was a desire to visit the South Seas.

Hall and Nordhoff sailed to the United States aboard the SS *Mauretania* in February of 1919. Both men received their discharges at Garden City, Long Island then headed for Martha's Vineyard, Massachusetts to complete the Lafayette Flying Corps history that spring and summer. In the early autumn of 1919 they turned over the finished manuscript for the two-volume history of the Lafayette Flying Corps to its publisher, Houghton Mifflin Company.

Before leaving Boston they met with Mr. Thomas Wells, editor of *Harper's* Magazine, who contracted with them to do a series of articles about the South Seas. He gave them a generous advance to live on while they worked at their craft.

Hall had a previous commitment to fulfill. He had agreed to give a series of lectures across the country on World War One aviation. Nordhoff returned to his home in California to wait for him. Hall completed his obligation by the end of the year and in late January 1920 he and Nordhoff sailed to Tahiti.

The two authors collaborated on *Fairy Lands of the South Seas* which was published in 1921. Just prior to the book's publication, on 4 December 1920 Nordhoff had wed a nineteen-year-old Tahitian native girl, Vahine Tua Tearae Smidt, who would eventually bear him seven children. Nordhoff remained in the South Seas writing a series of youth books while Hall set out for Iceland to concentrate on his poetry. He had planned to finance his trip by writing a travel book on Iceland and was advanced $5,000 by Harpers to that end. Ultimately Hall lacked the inspiration to complete the project and wound up borrowing the funds to repay Harpers advance with interest. He returned to Tahiti in early 1924 dispirited and in debt.

Nordhoff had been overjoyed at seeing his friend again. Now married and following a disciplined regimen, Nordhoff had enjoyed some success at his craft. But even in Tahiti Hall found no inspiration to write and spent his time canoeing and hiking the lagoons and mountains of the Islands.

But two events eventually brought Hall back to his typewriter. The first was his marriage in 1925 to Sarah Teraiveia Winchester, the daughter of an English sea captain and a Tahitian mother. Despite the great differences in their backgrounds and ages—Hall was thirty-eight and "Lala" sixteen—it proved a loving marriage that lasted the remainder of Hall's lifetime and resulted in two offspring, a son Conrad, and a daughter Nancy.

The second event which curbed Hall's ennui and rekindled his inspiration came about through the generosity of his friend, Nordhoff. While Hall had been experiencing his dead calm, Nordhoff had continued to produce further editions of his successful juvenile series which had centered around his youthful hero, Charles Selden. Keenly aware that his friend was floundering, Nordhoff had suggested that he and Hall combine to write another "boys book" which would take Selden into World War One and the U.S. Air Service. Hall was more than pleased to again join forces with Nordhoff, and they began their collaboration on the project in 1927.

The book, titled *Falcons of France*, was marvelously well done and borrowed autobiographical material under the guise of fiction from both mens' experiences in the Lafayette Flying Corps. It was published by Little, Brown & Company in 1929.

James Norman Hall shortly before his death at age 64 on the island of Papeete, Tahiti, 6 July 1951.

Upon the completion of *Falcons of France* both authors had begun searching for another topic to collaborate on. In the winter of 1916 when Hall had been at Bleriot school at Versailles, during a period of unfavorable flying weather he had secured a short Paris leave. While browsing in Brentano's bookstore he had come upon Sir John Barrow's narrative *The Mutiny of the Bounty* which had been published in 1831. He had purchased the volume and had kept it as part of his library.

When Hall had first suggested to Nordhoff that they do a work of historical fiction involving the Bounty incident, Nordhoff had been less than enthusiastic. However, after reading Barrow's narrative, his enthusiasm soared, particularly when their combined research revealed that little had been written on the subject in subsequent years, and nothing of substance with regard to historical fiction.

The authors decided to write the work as a trilogy, and following the critical success of *Mutiny On The Bounty* in 1932, they followed up with *Men Against the Sea* and *Pitcairn's Island* in 1934.

Ironically, following the Bounty trilogy which had taken the two authors five years to complete, suddenly it was Hall who took charge for the duo as his writing became governed by a disciplined control. Now it was Nordhoff who was floundering as an artist as he journeyed down a path of unbridled excess which ultimately destroyed him as an artist and contributed to his death.

Nordhoff, in a series of liaisons with young Tahitiennes, had for some time been unfaithful to his wife, Vahine, the mother of their seven children. Out of revenge Vahine had taken Nordhoff's young chauffeur as her paramour. When Nordhoff had learned of it he had gone after Vahine's lover with a pistol. Unable to catch him, Nordhoff announced to his wife that he was "getting a divorce" and walked out of their house, never to return.

The *Hurricane* was the next novel undertaken by the Hall/Nordhoff partnership. But following this title the partnership no longer operated on an equal basis of mutual effort. The team's *Atlantic* editor, Edward Weeks, made note of this when he said "he (Nordhoff) deferred more and more of the work to Hall. Jim never mentioned this to me but I could tell from our correspondence that it was so."

Hall, always a true friend to Nordhoff and well remembering Nordhoff's kindness to him when he himself had floundered as a artist, continued to take on more and more responsibility for the team in spite of that fact that Hall's own son, Conrad, had fallen ill with chronic colitis and Hall himself from filariasis and slight elephantiasis. These illnesses forced Hall to sail with his son to the United States to seek treatment for Conrad and himself.

Meanwhile Nordhoff, back in Tahiti, paralyzed with despair at having lost his skills as a writer, continued to pursue a life of excess centered in sex and alcohol. He took a mistress, Teuria, and in three years time she bore him three sons which Nordhoff refused to legally recognize. After *Botany Bay* which *Atlantic* published in 1941, Edward Weeks, the team's editor stated that "the creative imagination which powered their novels was largely Jim's."

Nordhoff's contribution diminished to the degree to which *Men Without Country* was mostly the work of Hall, and their last title, *The High Barbaree*, published in 1945, was completely Hall's work. But regardless of the fact that those last works were dual in name but solo in effort, Hall insisted that Nordhoff share equally in whatever royalties the books might generate. While always considerate of his friend's past contributions, Hall also realized Nordhoff's name joined with his as authors guaranteed the sale of books.

Some time in 1940 Nordhoff had abandoned the South Seas for good. He relocated to Santa Barbara, California and officially divorced Vahine there. At a cocktail party he met Laura Grainger Whiley, an interior decorator, and they married on 12 June 1941. Nordhoff made several futile attempts at returning to his writing and even took on another partner. Still, he continued to drink heavily and grew more morose and dispirited that he had expended his talents and had no inspiration left within him.

It was fitting that when the end had come for Nordhoff, James Hall would be a guest at his Montecito home. Following dinner, the two close friends had visited and had reminisced late into the evening. On the morning of 11 April 1947, when Nordhoff had not gotten up for breakfast, Hall and Mrs. Nordhoff had entered his room. They found him propped up between pillows, a book in his hands, with the reading lamp still on. Charles Nordhoff had died of a heart attack.

James Norman Hall continued to write and saw two other books published after Nordhoff's death. In 1950, the year Grinnell College conferred upon him an honorary degree of Doctor of Literature, Hall began to experience the symptoms associated with chronic heart disease. He had nearly completed his autobiography when he was stricken by a heart attack on 6 July 1951 and died at age 64 on the island of Papeete, Tahiti. Following a moving funeral attended by friends, including several hundred native mourners, he was buried behind his home on a Tahitian hillside facing Matavi Bay where the *Bounty* had first dropped anchor.

WESTON HALL

ESCADRILLE LAFAYETTE

"He had his day and it had been a big one. He lived on it more or less for the rest of his life." Bill Henry. *Los Angeles Times*, Dec. 14, 1948.

Weston Birch Hall, a founding member of the *Escadrille Americaine*, was born on a farm near Higginsville, Missouri on 7 November 1885, the only child of George Hall and Georgia (Johnson) Hall. Bert's father, George, at age fourteen, had served as an orderly for Confederate cavalry Colonel Joe Shelby in Company F, 2nd Missouri Cavalry. Colonel Shelby had been one of a handful of Confederate commanders who had refused to surrender at the end of the War. Instead, he had taken a number of his troopers, including Corporal Hall, into Mexico with him to fight for Emperor Maximilian against Benito Juarez.

When Maximilian was captured and executed by Juarista troops in 1867, George Hall had returned to the United States with Colonel Shelby. He then worked for a time on steamboats on the river at New Orleans before moving in 1876 onto Colonel Shelby's old farm near Higginsville, Missouri. He also served as postmaster and depot agent at nearby Page City. It was there that he met and married the widow, Georgia (Johnson) Page. In November of 1885 Weston Birch, their only child, was born.

The stories of George Hall's exploits during the Civil War and in Mexico had been vividly related within his community. Both in Page City and in Higginsville he had been viewed as a local hero. His son Birch, or "Bert" as he later came to be known, grew up on stories of his father's military exploits. They made a lasting impression on his young mind. He idolized his father and hoped some day to emulate the glamorous life his father had led as a young cavalryman.

Bert worked as a farm hand from an early age and thus had little opportunity for a formal education. In his early teens he ran away from home and found employment as a section hand on the Rock Island Railroad. In the early 1900s he arrived in Kansas City. There, he signed on as chauffeur for Justice Casimir J. Welch, driving Welch's Moon motor car. Hall was one of the few individuals in Kansas City capable of handling the Moon's "curtain-pull" steering wheel, a trick he had learned on a Knox car and a 1-cylinder Cadillac.

In 1904 Bert left Kansas City and eventually arrived in Dallas, Texas. He signed on with the Forepaugh

American members of the Foreign Legion at Toulouse, September 1914. Front l to r: Carstairs, William Thaw, James Bach. Back: Casmeze, Sweeny, Casey, Bert Hall.

and Sells Brothers Circus where he was soon billed as "the human cannon ball" while being shot daily from a cannon. In the summer of 1907 Bert returned to Higginsville to sell to his father the land he had inherited from his mother upon her death in 1905. He evidently used part of the $450 from the sale for train fare to San Francisco where he remained at least through 1910.

Some time prior to World War One, Hall had signed on as a seaman on a merchant vessel and had eventually landed in France. At the outbreak of hostilities between Germany and France, Hall had been employed as a taxicab driver in Paris. On 21 August 1914 he was one of the first forty-three Americans to enlist in the French Foreign Legion at the Hotel des Invalides in Paris.

He and his American comrades left Paris for Rouen on 25 August. There, they received their uniforms then joined their regiment at Toulouse. For three weeks they performed field maneuvers, long marches, bayonet and rifle practice before beginning a sixty hour trip in boxcars to Camp de Mailly for more drilling and simulated attacks. They left for the Front on 30 September 1914.

Hall arrived at the Chemin des Dames Sector with James Bach, Kiffin and Paul Rockwell, Robert Soubiran, and William Thaw. There, he served dutifully and began to build his reputation as a yarn spinner, a reputation which would remain with him the rest of his life. One American comrade, David Wooster King, recorded his first impressions of Hall:

"We had a crude chimney in our dugout and were busy brewing chocolate. Bert, from Kentucky, ex-racing driver, and God knows what besides, was going all out. He finished, and there was a moment of awed silence. His sidekick broke the spell:

"'Bert, you know that's a damn lie.'"

"The answer was disarming."

"'Well, who the hell said it wasn't!'"

The sector the American Legionnaires occupied was relatively inactive during the first months of the war. Their duties consisted mostly of standing guard, or huddling in dugouts under artillery bombardment.

As Hall, William Thaw, and James Bach sat in their dugout sipping pinard and brewing chocolate, the conversation invariably led to aviation. Thaw would stir his comrades' imagination by relating his pre-war experiences along the French Riviera flying his Curtiss Model R Hydro aeroplane.

Characteristically Bert Hall would try to top Thaw's experiences, relating how he had first learned to fly in France in 1909 and how, four years later, he had become the first man to pilot a plane in combat while with the Turkish Army in the Balkan War. Hall would later contradict himself in print, saying he had first flown at the Buc aviation school. Nevertheless, all three men longed to be out of the dugout and into the air.

The aerodrome for *Escadrille D.6* sat thirty-two kilometers from their position. In October Thaw had secured permission from his senior officer to hike to the aerodrome to see if he could secure a transfer from the Foreign Legion to French aviation for Bach, Hall, and himself.

In November Thaw's visit paid its first dividends. Bach's orders came through for transfer to aviation, and he left on the 15th for Saint-Cyr to begin his training. Thaw, following a second visit to the *D.6* aerodrome, was assured that the transfer orders for Bert Hall and himself were on the way. Then on 24 December 1914 Thaw was ordered to join the roster of *Escadrille D.6* as a *soldat-mitrailleur*. Four days later, on 28 December, Bert Hall was ordered to join James Bach at Saint-Cyr. However, his arrival date was delayed when he contracted the mumps.

At the Buc aviation school Bert Hall demonstrated the bluff and bravura which characterized his adult life. Hall's Legionnaire comrade, David Wooster King, made note of the incident:

"Bill (Thaw) and Bert had gone to the aviation. Bill with his previous experience qualified at once as a pilot. Bert's debut was typical. He climbed into the machine, had the controls explained to him, and started off just as Bill had done. Full speed ahead, up, and then down with a crash. After they had extracted him from the debris the officer in charge questioned him.

"'What went wrong?'

"'I don't know.'

"'You don't know! Haven't you ever been in a plane before?'

"'No.'

"'What in God's holy name do you mean-starting off like that?'

"'Well, I thought I might be able to fly.'

"They decided he had enough nerve to be worth training."

Bert Hall published two books on his World War service: the first, titled *"En l'Air!* in 1918. The second title, *One Man's War*, was published in 1929. Historians have called both works unreliable fabrications. A comparison of the titles does raise troublesome questions and casts doubt upon Hall's credibility as a truthful chronicler. For he either omitted much pertinent information in his first volume, *"En l'Air!* which then appears in the later work, or the second volume, *One Man's War*, is merely an embellished version of the first title.

Bert Hall at the Behonne airfield following a patrol.

For example, the 1929 title, *One Man's War*, carries a lengthy description by Hall of his 1913 service in the Balkan War when he flew in the air force of Sultan Abdul Hamed of Turkey against Bulgaria. He describes how, for lack of pay, he switched to the Bulgarian side and fought his former employer. But when the Bulgarians also welshed on his wages, he intended to rejoin the Turks, but a lady spy turned him in. He was arrested and jailed but was saved from the firing squad at the last minute when his mechanic ransomed his life. Why, one wonders, did his earlier title, *En l' Air!* not mention his daring Balkan service?

Another glaring omission in *"En l'Air!* or perhaps a glaring addition in *One Man's War* pertains to Hall's Foreign Legion service. In his first title he rightly portrays his sector from late September to mid-December 1914 as having been relatively inactive with regard to combat with the enemy. And yet in *One Man's War* Hall graphically describes the death of his comrade "Corporal Weidman" as he and Weidman battled German troops in "No Man's Land" and Hall dispatched two towering Prussians with thrusts of his bayonet.

In truth, *Caporal* Weidemann, a full-blood German and sixteen-year Legion veteran, was killed during an enemy assault that also involved Kiffin Rockwell on 5 January 1915, nearly three weeks after Hall's departure for French aviation.

More questions arise when the two titles are juxtaposed during Hall's service with the *Escadrille Americaine*. One bothersome comparison involves a combat patrol which Hall indicated in both books as having begun before dawn over Forts Douamont, Vaux, and Hill 304. Featured prominently in both descriptions are Hall's Vickers machine gun (his plane would have been armed with a Lewis gun), his mechanic, Leon, and a citation.

"En l'Air!" has Hall dating this patrol by his diary: 22 June 1916. During the flight he made contact inside enemy lines with three German planes, a two-seater and two pursuit craft. He first closed with the bi-place and either wounded or killed the observer while the pilot fled. Hall's attention was suddenly diverted by two attacking enemy fighters. He fired on one and it burst to flame and eventually hit the ground. Hall then fought the other German until he expended his ammunition. He then returned to his aerodrome where he was welcomed by his mechanic, Leon, who informed him that his victory has been witnessed by a French observation balloon, and that Hall was in line for a citation and another Palm for his *Croix de Guerre*.

One Man's War records the same patrol. The time and location are the same, pre-dawn "Between Fort Douaumont, Fort Vaux and Hill 304." However, the date is now moved ahead to early July. During an ensuing combat between the three German planes—two pursuit planes and a bi-place—he first fought and downed one enemy fighter before taking on the two-seater. Hall again noted "the face and square head of the observer" as the two combatants rushed past each other, but at this point the similarity between the two versions ceases.

Hall gets on the tail of the bi-place and fires. But instead of wounding or killing the observer, they duel and Hall suddenly blacks out and feels "a terrible pain" in his mouth. A bullet from the observer's machine gun has pierced his right cheek and knocked out a number of his teeth before exiting the left side of his neck. He nearly faints for loss of blood, but manages to return to the squadron's aerodrome where he is greeted by Leon "yelling something about a citation" until he sees Hall's bloody face and begins screaming for help.

The obvious question is why did Hall not mention having been shot in the earlier book *"En l'Air!"*? Did such an event occur, or was Hall, by 1929, trying to squelch the story being circulated by Paul Ayres Rockwell, the *Lafayette Escadrille's* official historian, that Hall "had all of his teeth pulled out, one after another so he could get off of flying combat patrols."

As Hall continued with his story, he mentioned in both books that around mid-December 1916 he had left Paris for a journey that would take him into Russia. The purpose of this journey as put forth in both titles is vague and ill-defined. *"En l'Air!"* states that it was at the "urgent request from the military authorities in Russia and Roumania for some French aviators. They were needed on the east front, to show the Russians how we were playing the game."

Hall states in *One Man's War* that he went to Russia at the request of the French War Office "to help the Russians pep up their flying and also report back to Paris what is going on."

Almost as soon as Hall sailed for Russia, the inconsistencies between the two versions resurface. In *"En l'Air!"* he mentions how his ship was stopped by a German submarine. He goes on to say "the sea was too rough, however, to permit the men from the submarine to board our ship."

In *One Man's War* Hall states that the U-Boat pulled alongside and when the ship stopped "the German Commander came aboard and looked us over very carefully . . . The Germans searched the mail, took what they thought would be interesting and waved us good-bye."

In scrutinizing Hall's *En l'Air!* version of his eastern journey, these incidents can be sifted out:

· Hall brought down a Boche plane "on the Eastern front" for which he was subsequently decorated by the Czar with the Cross of St. George, the Order of St. Vladimir, and the Russian Legion d'Honneur.

· Hall then went to Roumania where he joined the Russo-Roumanian army and conducted an aerial bombing mission on 3 February 1917 against the Royal Palace at Sofia. A war council was being held there "by the Kaiser, Emperor of Austria, King of Bulgaria and the Sultan."

· Hall next returned to Russia where he noted conditions during the Revolution before escaping the country via the trans-Siberian railway "through Siberia, Manchuria, and Japan to San Francisco."

The expanded version of this journey as related in *One Man's War* puts forth more peripheral information. However, the highlights of the trip are these:

· On 20 January 1917 Hall downed a German two-seater while attached to the Russian Air Force at Menelkof.

· On 20 February at Galatz, Roumania, while attached to the Roumanian Army Air Corps, he shot down "two antique German Albatrosses" with his SPAD fighter.

· On 22 February he bombed the royal palace at Sofia where a meeting was being held by "the Kaiser, the Emperor of Austria, the King of Bulgaria, and the Sultan of Turkey." The following day he was decorated by King Ferdinand with the St. Stanislaus Medal and the Vertu Militaire.

· On 7 March he was decorated by the Czar.

· On 10 April he met U.S. Ambassador David R. Francis and became a frequent guest at his Petrograd home.

· Hall met Russian Revolutionary Leon Trotsky who offered him $5,000 then $10,000 for "a pear-shaped diamond" and his American passport.

· When Hall left Russia, he smuggled out platinum, gold, and jewels for a Czarist general's wife. When they reached safety in Yokohama, Japan she rewarded him with "$38,000.00 in American money."

What both versions of Hall's eastern journey lack is the specific information needed to corroborate his experiences: locations of aerodromes, names of squadron commanders, fighter unit designations, etc. Without this information, his story cannot stand unchallenged. Lieutenant John J. Niles, whose name appears as co-author of *One Man's War*, seemed to sense the dilemma and attempted to explain it away when he wrote this in his brief preface to the book.

"Lieut. Hall seldom carries a watch and has never owned a calendar, and the Russian part of this yarn was kept partly by the Russian calendar and partly by the European calendar, and the French parts of the story were apparently recorded by a method known as "Yesterday was the 9th, today is the 10th, and that makes tomorrow Friday."

Niles goes on to state: "Lieut. Hall and I have spent endless months in libraries trying to verify the events recorded in "One Man's War"; this research was carried on at the Public Libraries in Chicago, Illinois, Des Moines, Iowa, New York City, the Library of Congress, Washington, D.C. and the Bibliotheque Nationale in Paris."

But now the question arises, were Hall and Niles researching to verify events, or were they seeking the information needed to add a sense of realism to Hall's trip into Russia and Roumania? Unquestionably, the "yarn" certainly grew in the ten years that separated the publication of the two titles.

In *One Man's War* Hall sets the date of his landing in San Francisco as 21 July 1917. On 1 October he is in Washington, D.C. testifying before the Senate Committee on Aviation. He then rejects the offer of a Captaincy "in the Aviation Section of the United States Marines" before sailing for France on 6 January 1918 aboard the "Rochambeau" to rejoin his old squadron, *N.103*

James Norman Hall and Charles Nordhoff, in their two-volume history *The Lafayette Flying Corps*, state the following with regard to Bert Hall's service:

"In January, 1917, he (Hall) was granted permission to accompany the French Aviation Mission which was sent, at that time, to Roumania. He later asked for, and was granted permission to return to the United States, supposedly for the purpose of entering the United States Air Service. He remained in America until the close of the war."

Paul Ayres Rockwell, who had access to the Lafayette pilots' service record in France, said of Hall that he was listed as a deserter by the French *Service Aeronautique*.

But in *One Man's War* Bert Hall had himself landing back in France on 19 January 1918. He noted in his book that his old squadron "N-103, is somewhere out in the country nearby; I haven't found it yet." A few days later on 5 February 1918 he learned he was to do counterespionage work with "the Intelligence Section of the French Air Service." Hall asks one of the lieutenants at Headquarters why he was chosen for such an assignment, and one reason the lieutenant gives is "as a liar you're quite versatile."

Hall soon joins a nebulous squadron whose nationality seems to be British. He never indicates its designation nor location but merely calls it "the outfit." He makes numerous trips between the Front and Paris where he is constantly visiting with Raoul Lufbery. One interesting and revealing bit of dialogue from Hall's book has Lufbery asking Hall:

"Did you really go to Russia, or was that all hokum?" And I said, "Why, hell, no, I was there. Honest!" Luf said, "Well, it surely was a good yarn, and I like a good yarn, true or not. Still keeping the old diary, Bert?" I said, "Yes." And he said, "I'll bet that'll be a prime piece of hokum when it's finished." And I said, "Right!"

According to *One Man's War* Hall fought several aerial combats while on patrol. On 26 March he recorded that he had downed a "pfalz single-place" which was confirmed. By the end of the book he is making a motion picture for "Air Service GHQ" which is to be shown in the United States to boost recruiting. Hall concluded his tale when his "pen ran out of ink" sometime after 28 May 1918, following a series of "undated" diary entries.

The historian must ask, is there any value in Bert Hall's two books beyond the entertainment found in a well-told tale? The answer is yes, if one focuses on the period of Hall's service with the *Escadrille Americaine*. He often gives interesting insights into the character of his fellow pilots and into the conditions under which the squadron lived and fought.

An added challenge for the historian is to sift the facts from the fiction of Bert Hall's French aviation service. As nearly as can be established, the facts are these:

Hall appeared at the Aviation Training Headquarters at St. Cyr on 15 January 1915. Upon leaving the Foreign Legion he had come down with the mumps which had delayed his arrival there for three weeks. Several days after reaching St. Cyr he was reunited with James Bach, his Legion comrade. The two Americans then went to the Buc school together. They were joined there by William Thaw who had been serving as a *soldat-mitrailleur* at the Front with *Escadrille D.6* On 15 March 1915 Thaw, a pre-war pilot, would become the first American to receive a French *brevet militaire*. Hall did not fare as well. He crashed and demolished his trainer against the side of a hangar.

Hall and Bach eventually went on to Pau where they joined Americans Elliot Cowdin, Frazier Curtis, and Norman Prince. Both Hall and Bach smashed up so many training machines that the school's authorities began to wonder if they were German spies. Both *breveted* on the Caudron, Bach on 4 July 1915, followed by Hall on 19 August 1915. Hall joined Bach at the Front in early September following several weeks at the General Reserve pilot pool near Paris. Bach had been serving with *MS.38* at Chalons-sur-Meuse in the Champagne sector.

For his first month *Caporal* Hall flew reconnaissance missions in a Morane Saulnier. The squadron then was supplied with two-seater Nieuports. *MS.38* became *N.38*, and Hall and Bach became two of France's first twenty Nieuport pilots.

Hall continued to fly with *N.38* until 1 January 1916 when he was returned to Avord for pursuit training. Once Hall had completed the course, he served as an instructor there. Victor Chapman was one of Hall's pupils. Then on 28 April 1916 *Sergent* Hall joined six other Americans at Luxeuil les Bains to become part of the *Escadrille Americaine*.

On 1 May *Sergent* Bert Hall was promoted to *Adjudant*. The Squadron flew its first regular patrol on 13 May, fighter support for the 4th French Bombardment Group. On 18 May the Squadron scored its first victory when Kiffin Rockwell downed an L.V.G. reconnaissance plane between Mulhouse and Hartsmannweillerkopf.

On 19 May the Squadron moved to the Verdun sector. The hangars were located at Behonne near Bar-le-Duc. The pilots lived in a stone villa between the flying field and the town.

The Squadron flew its first sortie from the Behonne airfield on 21 May. The following day *Adjudant* Hall took off on his second patrol of the day with Norman Prince, Kiffin Rockwell, and William Thaw. He became separated from the other pilots and eventually engaged a German aviatik at 12,000 feet over Malancourt.

Hall fought the two-seater down to 3000 feet with the German observer firing tracer rounds into Hall's right wing and to within a foot of his cockpit. Hall dived under the enemy's tail then nosed up and fired two long bursts from his Lewis gun into the belly of the bi-place. The enemy bus came apart and fell in pieces near the French forward trenches. Hall's victory was confirmed before he reached the Behonne hangars. On 18 June he was awarded France's *Medaille Militaire* and *Croix de Guerre*, with Palm for this 22 May victory.

Hall and Kiffin Rockwell fought an inconclusive air battle on 21 July. James McConnell related the combat in a letter to Mrs Alice Weeks the following day: "Kiffin has as usual been doing fine work. Yesterday he and Hall were together when he dove on a Bosche. Kiffin didn't see the Germans higher up. Down he came, a Bosche followed, Hall dove on the Bosche and two more dove on Hall. It was a parade downward, everyone shooting at once. Hall got ten holes in his plane and Kiffin and he got away by hiding in the clouds."

Bert Hall's second confirmed victory came on 23 July on a mid-afternoon patrol when he jumped a Fokker fighter "decorated up like a new saloon." The enemy single seater attacked and fired but dove under Hall's bus and surfaced before his Lewis gun. Hall quickly maneuvered onto the enemy's tail. When the German dived, he followed him down and closed the distance between them. Hall fired two short bursts. The Fokker lurched, rolled on its back, then spun 10,000 feet to earth where it crashed before the eyes of thousands of French soldiers between Vaux and Damloup. Hall's victory was quickly confirmed. However, his citation was late in coming, and Kiffin Rockwell complained of this to his brother, Paul, in a 31 July 1916 letter.

"I don't think . . . that a full report of my work has gone out of this office, and a number of times my report on a fight has been changed. The machine that Hall brought down X.(Norman Prince) did his best to prove that it wasn't brought down, and so far Bert hasn't even been proposed for a citation."

Hall's third victory came on 28 August 1916, northeast of Douaumont, while the Squadron was still at Bar-le-Duc. He dove out of the sun on an early morning patrol and struck a "Boche photographic machine." A running fight ensued with Hall battling the enemy observer. A burst from the German's guns stitched holes across Hall's upper wing before rounds from his own machine gun killed the enemy pilot and set the forward cockpit afire. The German bus hit the ground inside allied lines and was easily confirmed.

It was shortly after Hall's third victory that friction between Hall and the other American pilots became apparent. One historian has said that the fight had gone out of Hall following the death of Victor Chapman, and that he had sought extraordinary means to avoid flying missions. But yet, at the time the friction began,

Bert Hall and the Escadrille Americaine at Luxeuil for the funeral of Norman Prince, 17 October 1916. By the end of the month Hall would be transferred to another squadron because of friction between himself and his comrades. Front row, l to r: Hall, Lt. deLaage, Capitaine Thenault, William Thaw, Father Armonier, Chouteau Johnson. Back row: Laurence Rumsey, Paul Pavelka, Emil Marshall, Didier Masson, Dudley Hill, Robert Rockwell.

Hall's three confirmed victories were surpassed in the Squadron only by Raoul Lufbery's four, and even his harshest critic, Paul Ayres Rockwell, the Squadron's official historian, grudgingly admitted that Hall had done good work. This is attested to by the fact that Hall was awarded France's *Medaille Militaire* and *Croix de Guerre*, with two Palms since having joined the Squadron. So his conduct in the air should not have been the issue.

Among the reasons given for Hall's ultimate dismissal from the Squadron was that he was an habitual liar, a check kiter and forger (of Norman Prince's remittance checks), and a card sharp who cheated his fellow pilots in games of chance. Hall did state in *"En l'Air!"* that "In bad weather we play games, read and gamble. Mostly gamble. Poker and bridge are the two leading pastimes of that kind." And in *One Man's War* when the Squadron was at Bar-le-Duc he states "I had some luck earlier in the evening in a poker game. Hope I'm not bounced off before I get a chance to spend it."

Whether or not Bert Hall had cheated at cards, the opportunity for winning big certainly existed. Hall, the Missouri farm boy, older and more worldly than the young idealists surrounding him at the poker table, had never experienced the niceties of life. No family remittance had ever been his, no race horse, no gift of an airplane.

And when he had sat cupping his poker hand, a rough among diamonds, and had read without difficulty the giveaway expressions on the young faces about the card table, he well knew that Chapman and Cowdin and McConnell and Prince and Rumsey and Thaw were all scions of millionaires. The lambs certainly were there for the shearing.

In reading Kiffin Rockwell's war letters, it is easy to see that he and Hall had a friendly relationship. Hall, whom Rockwell affectionately called "Bertie" was a frequent patrol mate of Kiffin's. And despite his brother,

Paul's later negative opinions of Hall, it was evident that Kiffin both admired and respected Hall.

The facts contradict Paul Rockwell when he said that the fight left Hall after Victor Chapman's death. As Kiffin noted in his letters, following Chapman's death Hall continued to fight and engage the enemy and scored two of his confirmed victories.

It is also apparent when reading Kiffin's letters that Rockwell had a strong dislike for Norman Prince, and at the time of his death was making a concerted effort to have him "run out of the Escadrille." Bert Hall was a Rockwell ally. Prince had tried to deprive Hall of his 23 July victory. But when Rockwell was killed on 23 September 1916, evidently Prince rallied the other pilots against Hall and ultimately forced his expulsion from the *Escadrille*.

On 28 September 1916, five days after Kiffin Rockwell's death, Hall made a formal request for transfer to *Escadrille N.103*. However, he continued to fly patrols in spite of the shortage of available fighter aircraft. He shared *Capitaine* Thenault's Nieuport with Paul Pavelka when Thenault had taken permission to Biarritz on 6 October.

Bert Hall on his first trip into China, the summer of 1929.

By 18 October *N.124* had moved to the frigid hell at Cachy and on 29 October Hall's orders had arrived for transfer out of the *Escadrille Americaine* and into *N.103*. It is interesting to note that the last two sentences in Hall's diary entry for the 29th are these: "I think the Lafayette Escadrille is glad to get rid of me. I don't blame 'em." And yet on 29 October 1916, the date of Hall's diary note, the Squadron was still being called the *Escadrille Americaine*. The name *Lafayette Escadrille* still existed only in the mind of Dr. Edmund Gros.

Hall left *N.124* on 1 November 1916. It was a bitter parting, as recorded by Emil Marshall, a former Legionnaire from New York, and one of the nonflying personnel with the Squadron. He penned in his diary what he had witnessed that day how Hall had shaken his fist at the other American pilots upon leaving the Squadron and had shouted out "You'll hear from me yet."

Adjudant Hall joined the roster of *N.103* at Cachy on 16 November 1916. He received his fourth and final confirmed victory on 26 November and was awarded another Palm to his *Croix de Guerre* with the accompanying citation:

November 26, 1916. Bert Hall, Adjutant Pilot in Escadrille N.103.

"Clever, energetic, and courageous pilot, full of spirit. Daily attacking enemy planes at very short distance. On November 26, 1916, shot down a German plane at two hundred meters from our trenches. The following day, after a combat held quite near, returned with his machine hit by several shots, also a shot in his helmet. (Three additional citations.)"

Hall left *N.103* on 14 December 1916. Evidently he had received orders to report on 16 December to the under secretary of War in Paris. He had put in a request to accompany the French Aviation Mission to Roumania. He arrived in Paris on the 17th. and beyond this date what Hall's movements were is difficult to establish.

As James Norman Hall and Charles Nordhoff indicate in their *Lafayette Flying Corps* history, he was granted permission to go with the French Aviation Mission to Roumania in January of 1917. It is not certain when he then asked for and was granted permission to return to the United States to join the U.S. Air Service. However, he did sign his name to the Laurence Rumsey family guest book in Buffalo, New York on "September 10, 1917." Hall and Nordhoff said that "he remained in America until the close of the war."

However, Hall married Della Byers, an entertainer, in Paris on 12 April 1918. He evidently remained in France until sometime after the 11 November Armistice. Shortly after his return voyage to the United States Bert, Jr. was born in New York City on 26 February 1919.

Now out of uniform, Hall and his new family headed west to Hollywood where he produced and starred in a film titled *A Romance of the Air*. In 1920 he made another aviation film titled *Border Patrol*. He traveled the country with the films, appearing in person and lecturing before each screening. When Hall, Della, and their infant son reached Connecticut, Hall struck out alone. Bert failed to return. And Della divorced him in 1921.

Bert took a second wife, Helen, soon after the divorce was final. Two sons, Donald and Norman, were born before Hall separated from Helen and married a third wife, Genevieve. Weston, a fourth son, was born of that union.

From approximately 1921-1929 Hall was on the payroll with Twentieth Century Fox Studios in a variety of capacities: actor, writer, aviation consultant. He continued to travel the country lecturing on his war experiences to movie audiences prior to the screening of films. In early 1923 he appeared in Yokohama, Japan where he presented himself as "head of a film concern."

As a top-drawer bridge player -an art he had refined while with the *Escadrille Americaine* -Hall also supplemented his income by playing tournaments around the country.

Throughout most of his adult life, Bert Hall had the uncanny ability to manipulate the press. He used his association with the *Escadrille* as the magnet to draw attention to himself when it suited his purposes. But sometimes it was to his embarrassment and the chagrin of his former squadron mates. As during the 4 July 1928 dedication of the *Lafayette Escadrille* monument near Paris when three women claiming to be Mrs. Bert Hall had shown up to occupy the same reserved seat.

Bert Hall was not present at the dedication. Nor was there any mention of Hall's name in the official dedication booklet presented to the former pilots of the Lafayette Escadrille and the Lafayette Flying Corps. As a bit of revisionist history, at the urging of Dr. Edmund Gros and Paul Rockwell, who considered Hall a deserter from the *Service Aeronautique*, Hall's name was not included in the booklet's list of "Surviving Members" of the Lafayette Escadrille.

Elizabeth Chapline, the last Mrs. Bert Hall.

The release of *One Man's War* in 1929 regenerated interest in Hall. His story reached a wide audience when *Colliers* magazine excerpted portions of the book in three April 1929 issues.

The 10 May 1929 issue of the New York *Times* made note of a letter which Hall's co-author John J. Niles had received and which had been posted from Paris where Hall had been "on business." The *Times* quoted from the letter that Hall was now "General Cham, chief of Air Service with Nationalists Chinese at Canton." It also pointed out that Hall, "A 20th century American soldier of fortune....has been fighting in China since soon after the World War." The article concluded with mention of Hall's pre-war service in the "Balkan Wars."

Bert Hall, in fact, was in China in the summer of 1929, but not as "General Cham" nor as the commander-in-chief of the Nationalist air force. The press role Hall had assumed, undoubtedly as a publicity stunt to push sales of his book, was that of "General Chang Hui-chang," a Chinese warlord and Hall's employer.

China was a country divided into factions at the time of Hall's arrival there. Hall had signed on as aviation instructor and advisor to the faction then being led by one General Chang Hui-chang. But Hall had duped the Western press into believing that he himself was "General Chang."

While in his advisory capacity to General Chang, Hall had concocted a convoluted scheme whereby he

Bert Hall in Tientsin, North China, August 1933, where he had gone to sell German pistols and ammunition to a Chinese warlord.

hoped to make himself a small fortune through the sale of a squadron of nonexistent Douglas bombers to the Chinese General.

Hall set up a dummy corporation called the National Import Company. He then talked his namesake, General Chang, into a $100,000 advance which Hall was to apply toward the purchase of the Douglas bombers. Hall had General Chang place the funds into escrow in an American branch bank in Canton. He then had the money transferred to the main San Francisco branch in the name of an accomplice who promptly withdrew the funds while Hall beat a hasty retreat for the States.

At approximately the same time Hall's scheme was being consummated, the 27 April 1930 issue of the New York *Times* was calling Hall "The most outstanding aviator soldier of fortune" while comparing him to Lawrence of Arabia. The article further stated that Hall was "General Chan, Commander of air forces of Chinese Nationalist Army and an active figure in the military that have affected a large part of China."

As soon as Hall had landed in California in the summer of 1929 he was arrested and brought into custody in the Los Angeles County jail. Chinese officials representing General Chang had filed a $100,000 suit against Hall in San Francisco for failing to carry out the contract for the purchase of the Douglas airplanes. But General Chang eventually withdrew the charges upon the recovery of a portion of the advance when he realized how foolish he might appear for having been flimflammed by Hall.

Another person who felt she had been flimflammed by Hall was Madame Marcelle Hall of Paris. In early 1932 appeals had been made on her behalf to Hall's former squadron mates to use their influence to induce Hall to support what she claimed was their two offspring.

In January of that same year Hall had met Elizabeth Chapline, a vaudeville actress eighteen years his junior, at the Hotel Muchleback in Kansas City. Hall had been struck by her beauty and had immediately passed a note to her male companion saying so.

Hall had arrived back in Los Angeles flat broke, which did not deter him from checking into the exclusive Hotel Ambassador. By then he had gone through the small fortune he had extracted from China. Unpunished and unrepentant from his past escapade, he had decided to go to the well one more time.

Elizabeth Chapline's mother had lived in Los Angeles, which was Hall's business location at that time. When Beth had traveled there to visit her mother and pursue an acting career, Hall had vigorously courted her. The twenty-nine year old Beth was smitten by Hall, and in February of 1932, as Hall planned his return to China, Beth made the decision to accompany him.

The 5 February 1932 issue of the New York *Times* carried an article on Hall datelined Los Angeles. It noted that Hall was about to sail for China, for the "Sino-

"General Chang Hui-chang" on the Puget Sound lecture circuit, 1935.

Japanese fighting zone." Hall mentioned that his contact was one T.V. Soong, former Chinese Minister of Finance. He then stated: "I have no official connection with the Chinese Government-yet. But Soong will return soon to his former position, and then, who can tell?"

Once Hall had landed back in China he had begun his search for another gullible victim. He settled on one General Ho Chu-Kuo, who was Chief of Ordnance for his faction. Hall soon began to put a deal together with General Ho which involved the sale of German pistols and ammunition. Meanwhile he lived on cash advances paid to him by the General in Mexican currency.

Beth Chapline wrote a series of letters to her mother in California while she was in China with Bert. In them, she speaks of her love for Hall and of his pending business deals which she admits to know little about. On 29 April 1933 she wrote the following to her mother from "Tientsin, N. China."

"Really the way conditions are between Bert and me is most amazing. It hardly seems believable that after having been together better than a year, I know absolutely nothing about his finances that is not from him. I have guessed a great deal, but he is so proud that I don't want to embarrass him by probing too closely. I can't go into detail, but it's something big. He has been so good to me that I adore him and always shall. My love for Bert has intensified everything to an alarming extent."

On 27 June 1933 Beth again wrote from "Tientsin, N. China" that:

"Bert has done some very good business, and is held up now by his inability to get a huchao-a, a government permit for the landing of goods from foreign countries. Money is in the back awaiting delivery. Would be quite a nice profit. One thing I know is that Bert won't throw away and give away as he has always done when he gets his hands on some money again. At least I don't think he will."

In September of 1933 Hall sent Beth on to Japan. He then received a postdated check from General Ho for $9,755 which he cashed at a Tientsin bank. Then on 9 October he sailed off to Kobe aboard the *Choko Maru* to join Beth. But Hall was denied entry into Japan. F.P. Lockhart, the American consul general in Tientsin, when informed of Hall's scheme by the Chinese, had contacted the Japanese authorities and asked that they force Hall's return to Tientsin.

Lockhart had Hall placed under arrest as soon as the steamer docked in China. He was then transported to Shanghai to face charges at the United States Consular Court. Hall pled guilty on 1 November "to a charge of engaging in illegal firearms importation into China for sale to military forces." He also pled guilty to receiving money under false pretenses for converting the money he had obtained from General Ho to his own use.

On 11 November 1933 Hall was sentenced by the U.S. Court of Shanghai to serve two and one half years in Federal prison. Hall was extradited to San Francisco on 14 November. From there he was taken to McNeil Island Federal Penitentiary in Washington State to begin serving his sentence. The ever-faithful Beth, convinced of Bert's innocence, lived in a boarding house close to the penitentiary while he served his time.

Upon his release from prison, Hall quickly rebounded. Undaunted, he was almost immediately back on the lecture circuit making presentations before screenings while doing some free-lance writing on the

Bert Hall standing behind Shirley Temple at the Twentieth Century Box studios in Hollywood. Alice Faye and Robert Young are on his left.

side. He resurrected the role of General Chang which had once been his bread and butter. It would feed him again. Movie adds in Puget Sound papers challenged the reader to come to the theater and "meet in person General Chang, one of the most picturesque personalities in the world today...Bert Hall, famous war ace, soldier of fortune, world adventurer, High Mandarin of China."

But Hall eventually wearied of the role and left Seattle for Hollywood in the fall of 1936 to return to work for Twentieth Century Fox studios. He again took up the craft of technical advisor and a script writer.

Bert and Beth were married in Las Vegas on 25 June 1937. Some months later, he dropped her off at her sister's home in Los Angeles and kept on going. Beth was devastated. She filed for divorce. When Bert read about the action in the papers he rushed back to her-said he had left her as a result of financial problems and because he felt Beth would be better off without him. They were reconciled. Beth called off the divorce.

Bert's money woes were temporarily solved the following May when he won a windfall judgment against the Douglas Aircraft Company in Los Angeles Superior Court. The suit stemmed from his 1929 China trip and the sale of the Douglas aircraft. After twice bringing his case to court, Hall was able to convince Judge Henry M. Willis that he had engineered the sale of 20 Douglas airplanes to the Chinese government on 23 May 1931 for the total value of $440,000. Despite the fact that the planes were never delivered Hall was awarded a $40,000 judgment as a commission and reimbursement for expenses incurred. Douglas eventually settled with Hall out of court.

One historian called Bert Hall "a loveable old rogue." This terse character summary seemed to be the consensus of those who knew him, both male and female.

In spite of his shenanigans with the *Lafayette* Squadron, Hall apparently was a welcome post-war visitor to the homes of the former *Escadrille* pilots. They enjoyed his company and retained a genuine affection for him. William Thaw, in his post-war correspondence, always spoke well of Hall. But James Norman Hall, in a 6 September 1943 letter to Willis Haviland, mentioned how "Bert Hall was still going strong, and I guess he is the same old four-flusher as ever."

By late 1939 Hall had wearied of Hollywood and was looking for somewhere to invest the judgment he had won against Douglas. Sometime in 1940 he and Beth arrived in Dayton, Ohio. Around 1944 they moved to the tiny hamlet of Castalia and Bert made a career change which seemed totally out of character. It was there that "the world's number one soldier of fortune" started the Sturdy Toy Factory with local resident Harley Hoffman.

At Hall's request Hoffman designed a toy shaped like a German V-1 rocket. It had a cigar-shaped balsa fuselage and was fired by a catapult. Eventually 3,000 such toys were manufactured and sold to Toy Mart and Highbee's. Hall's firm also manufactured other toys of cast aluminum: boats and helicopters. Eventually the company would employ ten people and would also produce small portable air compressors and machinery.

Hall often chased rainbows across the roads of Ohio and Kentucky as he promoted his various enterprises. He was returning to Castalia from one such trip to Kentucky at 6 a.m. on the morning of 6 December 1948 and had been heading west on Route 6 two miles outside of Fremont, Ohio.

Suddenly Hall's automobile veered off the road and tore through a fence on the Charles Hudson farm before striking a utility pole. When local officials arrived at the scene and found Bert's lifeless body slumped on the floor, they had assumed from the shattered windshield and the bruise on his forehead that the sixty-three year old Hall had died of injuries. But after the coroner had made his examination, it was learned that Hall had died of a massive heart attack.

The following day Beth contacted Arlington National Cemetery requesting that her husband's remains be buried there. The Department of the Army immediately notified her that a crypt would be provided for Hall. At her husband's living request, she had his remains cremated in preparation for services to be held at Arlington at 3 p.m., December 14, 1948.

But in a sudden reversal the Department of the Army reneged. Arlington would not accept Hall's remains after all, evidently having discovered his felony conviction. Beth had Bert's ashes placed in a small black box. She next wrote Henri Bonnet, France's Ambassador in Washington, D.C. requesting her husband's ashes be placed in a crypt in the *Lafayette Escadrille* Monument near Paris. Bonnet responded that he would contact France's Minister of Foreign Affairs in Paris who would see to her request.

But several days later the Secretary to the Ambassador wrote Beth back suggesting she contact former Lafayette Flying Corps pilot Charles Grey who was then serving on the American Embassy staff is Paris. Grey turned Beth's request over to Paul Ayres Rockwell who immediately rejected it on the grounds that Hall was ineligible to be buried in the Monument's crypt since he had not died in the First World War.

However, this could not have been the true reason. For when William Thaw had died of pneumonia in Pittsburgh in 1934, Rockwell had raised no objection to William Nelson Cromwell's attempt to have Thaw's remains transferred from Pittsburgh's Allegheny Cemetery to the Lafayette Memorial. It was evident that the Lafayette survivors did not want "the loveable old rogue" buried in hallowed ground.

Following this rebuff, Beth turned Bert's ashes over to his friend, Will D. "Billy" Parker. She asked Parker, a test pilot and aeronautical engineer, and manager of Phillips Petroleum's aviation department, to make the proper arrangements to have Hall's ashes scattered over his home town of Higginsville, Missouri. On 20 January 1950 the final arrangements were made.

Parker and his co-pilot climbed into his twin-engine aircraft with the black box containing Hall's ashes. It was a cold, cloudless winter day over Missouri. As Parker later recorded the event:

"I took 'Uncle' Bert to Higginsville and flew around over that part of the country for about ten minutes while he disembarked." So, nearly a year and two months after his death, Bert Hall had found a resting place. But as he exited the black box into the crisp Missouri air, he did it in the usual Bert Hall fashion. The slip stream from the plane's prop blew some of the "old rogue's" remains back into the cockpit, irritating the co-pilot's eyes.

EDGAR HAMILTON

Edgar Guerard Hamilton was born in New Castle, Pennsylvania on 19 November 1891, the son of Louis Hamilton, an engineer, and Hariott Guerard (McComb) Hamilton. Following grammar and preparatory school, Hamilton worked out west for a time before deciding to go to Europe. He returned to the United States to enroll in the Carnegie Institute of Technology of Pittsburgh. He left that institution in February 1917 to join the American Ambulance Field Service.

On 27 February 1917 'Ham' Hamilton enlisted in France's *Service Aeronautique*. From 3 March to 25 July 1917 he attended the aviation school at Avord. He received his *brevet* on the Caudron on 12 June 1917.

Because Hamilton spoke fluent French, and had a sound knowledge of aviation techniques, he was assigned as an instructor and technical advisor for the American student-pilots. From 10 July to the 11 November 1918 Armistice, he served at various French airfields as a *moniteur* and technical instructor, finishing his service as a *Sous-Lieutenant* at Chateauroux.

Hamilton was disappointed that he was not assigned to the Front. However, he served diligently and was of great assistance to his American comrades who passed through the French flying schools.

Following World War One Dr. Gros assigned Hamilton, James Norman Hall, and Charles Nordhoff to write the history of the Lafayette Flying Corps.

He was also a member of the American Graves Registration Committee which helped to locate the graves of missing Lafayette Flying Corps pilots. He managed to locate the remains of all but six airmen.

Hamilton returned to the United States after the war and attended Harvard University from 1924–1925. He later earned a degree from Carnegie-Melon. He conceived the idea of gathering all the fallen Corpsmen in one monument near Paris. In 1928 he assisted in the transfer of their remains from various grave sites in Europe to the Lafayette Escadrille Memorial, St. Cloud.

Hamilton spent thirty years in the French Foreign Legion in Algeria and in Morocco building roads, tunnels, and bridges. He retired as a Legionnaire Major in 1946 following an incapacitating attack of typhus and returned to the United States to live.

In August 1958 Hamilton accidentally struck and seriously injured a young girl with his automobile in Connecticut. After transporting the girl and her mother to the local hospital, he left his address and insurance information then drove to New York City. The child died shortly after and Hamilton was arrested and jailed in the Tombs jail. Eventually Hamilton was released and given a suspended sentence. Hamilton immediately returned to France, then to Rabat, Morocco. He suffered a stroke and spent his last years in the French Foreign Legion home, unable to see, read, or write. He retained his U.S. citizenship. He died on 16 January 1984 at age 92 in the Home of the Aged, Rabat Morocco.

ROBERT HANFORD

Robert Marshall Hanford was born in Brooklyn, New York, on 11 March 1895, the son of Frederick Lewis Hanford. Robert was orphaned at an early age and was raised in New York City by his grandmother, Mrs. E. Marshall. He attended Georgetown University Law School, Washington, D.C., class of 1917. He left that institution before completing his course of study in 1915 to work as a salesman in Brooklyn.

On 4 April 1917 Hanford joined the American Volunteer Motor Ambulance Corps in New York City. On 14 April 1917 he sailed for France aboard the *Rochambeau* and served briefly in France with its Ambulance Corps.

On 24 May 1917 Hanford enlisted in France's *Service Aeronautique*. From 5 June to 15 October 1917 he attended the Aviation school at Avord. On that date *Soldat* Hanford was taking his test for his French *brevet*. While flying a Caudron from Avord to Chateauroux where he was attempting to land, he collided with a Farman bi-plane which had approached his aircraft from Hanford's blind spot. He was instantly killed as both aircraft came hurtling to earth. He was 22 years old at the time of his death.

JOHN HARRISON

John B. Harrison, Jr. was born in Bloomfield, New Jersey, the son of John Harrison. On 12 June 1917 he enlisted in France's *Service Aeronautique*. On 16 October 1917 he was released because of inaptitude. He later joined the British Royal Air Force.

WILLIS HAVILAND

ESCADRILLE LAFAYETTE

Willis Bradley Haviland, the 16th volunteer of the *Escadrille Americaine*, was born on 10 March 1890 in Minneapolis, Minnesota, the son of Dr. W.H. Haviland, a physician, and Grace (King) Haviland. He attended Kemper Military School at Boonville, Missouri from 1904-1905. He then enlisted in the U.S. Navy in 1907, serving until 1911. Following his discharge he enrolled in Iowa State College, Ames, Iowa, and studied electrical engineering for two years. In 1912 he enlisted in the Illinois National Guard and served until 1915.

Sergent Willis Haviland at Cachy with the lion cub, Whiskey.

Haviland sailed for France in the fall of 1915. He joined the American Ambulance Field Service and served with S.S.U.2 on the Pont-a-Mousson sector. He enlisted in French aviation via the Foreign Legion on 26 January 1916.

Haviland began his aviation schooling at Pau on 30 January. He was soon joined by Frederick Prince, Robert Rockwell, and Robert Soubiran. While he was at Pau, his mother, Mrs. Grace Haviland, arrived in France and joined him there. Not long after the pilots had completed their first triangle course, the school was temporarily shut down and they were ordered to Buc where they finished their altitude tests and triangles in one day. Mrs. Haviland left Pau when her son had gone on to Buc. She settled in Paris and would remain there until the end of the war.

On 20 May Haviland received his *brevet militaire* on the Caudron and returned to Pau for Nieuport instruction. Following gunnery and acrobatic school, he was sent to the *G.D.E.* pilot pool until 20 October 1916 when he was ordered to *N.124*.

Willis "Havvy" Haviland, Frederick Prince, and Robert Soubiran all joined the roster of the Squadron at Cachy in the Somme sector on 22 October. Four days earlier *N.124* had been transferred from Luxeuil to fly support for the Somme Offensive, and had become a part of French *Group de Combat 13*. Seven days prior to the pilots' arrival Frederick Prince's brother, Norman, had died of injuries he had sustained while attempting a landing after a flight protecting allied bombers striking Germany.

The Squadron's new home was no longer the comfortable villa of the past, but was "a portable barracks newly erected in a sea of mud." Damp mist seeped through its barn-like walls into the mens' bones. With no bedding and blankets, they were forced to sleep in their flying suits. For lack of cooking facilities, the pilots ate at the mess of the other fighting units stationed on the field.

But in time, hard work and some Paris purchases transformed the barracks into a habitable quarters which served as a barrier against the cold and mud. How-

Sergent Willis Haviland and his mechanic beside his Nieuport scout at Cachy, winter of 1916.

ever, thick mist blanketing the Somme frequently kept the pilots grounded, and Raoul Lufbery accounted for the Squadron's only two confirmed victories while it was there, on 27 December 1916 and on 24 January 1917.

On 26 January the *Escadrille Lafayette* was more than willing to depart the frozen terrain of Cachy for the Ravenal airfield near Saint-Juste-en-Chausse. The Nevelle offensive had been planned for the spring, and the pilots had been ordered to lie low until their participation in it.

In March the situation in the air changed dramatically. *Sergent* James McConnell, one of the original members of the Squadron, was killed in combat on 19 March 1917 near Jussy. On 7 April *Sergent* Haviland and the other pilots flew their planes to Ham in order to keep up with the rapidly advancing Allied front. 16 April saw the death of *Caporal* Genet south of St. Quentin.

Genet, who had flown an early morning patrol with Walter Lovell, was ill when the time had come for his afternoon flight. Haviland, with his own plane disabled, had offered to fly Genet's plane in his place. Genet had declined Haviland's offer and had flown to his death. Later, Haviland mentioned Genet's death in a letter to Mrs. Alice Weeks.

"Genet, a brave little rooster, was killed in an accident. I went over in the car to get his body and saw the machine. It had plunged to the ground full motor. You can imagine the wreck. He must have fainted, air sickness as he was with Lufbery who saw him turn across the French lines and start home."

The death of *Sergent* Ronald Hoskier closely followed Genet's death when he and his observer were shot down and killed on 23 April in a Morane Parasol above Saint Quentin.

On 26 April *Sergent* Haviland, while on patrol with Lieutenant William Thaw and *Sergent* Chouteau Johnson, scored his first and only confirmed victory when their patrol had run directly into a formation of Albatros fighters. In a combat of sudden confrontations the two formations had sought and fought each other as they broke out of the clouds into canyons and pools of blue. Johnson tumbled an Albatros scout and Haviland an enemy bi-place, as he noted in an April 27 letter to his mother.

"Does this good news please you? Yesterday evening at half past six I shot down a large two-seater German plane. It fell to the ground inside the enemy lines, southeast of Saint-Quentin, and was confirmed officially this morning by artillery observers. The Captain will propose me for the War Cross with palm today, I think. Yesterday Chouteau Johnson also shot down a machine, which has been confirmed; that makes six sure ones for our Escadrille in the past ten days.

Didier Masson (left), Willis Haviland, and Raoul Lufbery at Cachy, winter of 1916.

Lufbery got two and Lieutenant de Laage two last week as well as two doubtful ones.

"We are having a good bit of action about here of late and are making up as best we can for the loss of McConnell, Genet, and Hoskier."

On 17 May *Sergent* Haviland received France's *Croix de Guerre*, with Palm in recognition of this victory. 23 May found him in hospital from an ankle injury. While climbing out of his fighter, he had re-injured an ankle which he had sprained some years previously while playing basketball. While in hospital recovering, he had received what must have been one of the earliest airmail letters when "Bob Rockwell flew by my window yesterday afternoon, waving his hand and dropped a bundle of mail for me."

The 23rd was also the day the Squadron's executive officer, French Lieutenant deLaage de Meux, was killed when he crashed his new SPAD S.VII fighter into the Ham airfield shortly after takeoff.

On 3 June 1917 the spacious airfield at Chaudun became the Squadron's next base. The pilots left there on 17 July in their new SPADs for the airfield at St. Pol-Sur-Mer. The flight was made in the evening under poor weather conditions, with Haviland and David Peterson each smashing up their planes while attempting landings through mist and low clouds.

The airfield at Senard near the Argonne forest served as the next base for *Spa.124*. The Squadron settled there on 11 August to partake in a planned French offensive at Verdun. The weather was ideal for flying and many patrols were flown by Haviland and the other pilots.

On 18 September the Squadron returned to Chaudun. *Adjudant* Haviland, "the handsomest man in and the Beau Brummel of the outfit", did not go with the others, but left to join *Spa.102*. He arrived at its aerodrome on 1 October 1917 and remained with that squadron until 28 December 1917 when he enrolled as a Lieutenant (Senior Grade) in the U.S. Naval Reserve Force Center at Paris. On 1 February 1918, at Mountchic, he became Naval Aviator (HTA) #577. On that date he was made chief pilot at the U.S. Naval Air Station at Dunkirk, conducting submarine patrols until 25 March 1918 when he was reassigned to the 13th Squadron Royal Naval Air Service.

On 2 May 1918 Lieutenant Haviland was sent to the Lake Bolsena training station northwest of Rome to learn to pilot the Italian Macchi L-3, M-5, and M-8 flying boats. Once he had completed his training, he was ordered to erect and command the Naval Air Station at Porto Corsini, Italy, eight miles from Ravenna.

Lieutenant Willis Haviland standing near one of his squadron's Italian Macchi M-8s complete with Navy goat insignia.

On 23 July 35 officers and 330 enlisted men were placed under his command along with 35 Italian flying boats and 250 tons of supplies.

The purpose of the Porto Corsini station was to serve as a base to strike Austria-Hungary's main naval bases on the upper Adriatic Coast, at Trieste, and at Pola (today Pula, Yugoslavia). At the direction of his superior, Lieutenant Commander John Lansing Callan, he brought along some 'old hands' as flight commanders, choosing when he could those navy pilots who had served in the Lafayette Flying Corps: Ensigns Caleb James Coatsworth, Jr., James Murray Grier, Daniel Elliot Huger, Austin Gilette Parker, and Leonard Minor Reno.

Almost immediately Haviland found himself at odds with Callan, the Commander of U.S. Naval forces in Italy. Callan subscribed to the philosophy of a non-flying commanding officer, while Haviland, a veteran with more than 14 months of combat flying, felt he must lead by flying.

Haviland stood near the hangers and watched his pilots leave for an early-morning patrol over the Austrian lines. When they returned from the mission, their planes shot full of holes, he vowed to lead the next patrol.

Lieutenant Haviland, the "flying commander," preparing to lead a patrol out across the Adriatic.

He spearheaded a strike against the Austrian aerodrome at Pola, flying far in advance of his formation. When he spotted a squadron of Austrian fighters beneath him, he dove into the heart of their formation, his machine guns rattling. When his command caught up with him, they drove the enemy back to the Pola aerodrome. Later that week, Haviland became the first American to carry out a night bombardment raid on that enemy station.

The friction between Haviland and Callan reached a head by the end of August 1918 with Callan threatening to replace Haviland as commanding officer of the Porto Corsini station. For a time Haviland acquiesced to Callan's demands but soon began flying missions again. A staff inspector turned him in to headquarters in Paris and a court martial was proposed for Haviland's "direct disobedience of orders."

Haviland's defense was terse, but original, as stated in his reply to the charges: "I wasn't leading my men—I was too far ahead of them for that."

A look at the Porto Corsini station's record indicates Haviland's style of leadership paid great dividends. Not only did his squadron become the most heavily engaged U.S. Naval Aviation unit in World War I, it was also the most decorated.

One of its pilots. Charles Hazeltine Hammann, was awarded the U.S. Medal of Honor. Fourteen other pilots, including Haviland, were eventually awarded Navy Crosses. Lieutenant Haviland's Navy Cross citation reads:

"For distinguished and heroic service as pilot of a seaplane engaged in patrolling the waters of the War Zone, escorting and protecting troops and cargo ships, operating against enemy submarines, and bombing the enemy coast, showing at all times courage and a high spirit of duty."

Before the end of October 1918 the bombing stopped in Italy.

Shortly after the 4 November Armistice between Italy and Austria-Hungary, Vice Admiral H. T. Mayo, Commander-in-Chief of the U.S. Atlantic Fleet, visited Porto Corsini and issued the following commendation:

"I have been greatly gratified to notice the perfect condition of the station and of the planes and the discipline and smartness of the men. I am therefore glad to heartily commend Captain Willis B. Haviland and the officers and enlisted men of the station who have brought the station to the highest point of efficiency."

By the time Lieutenant Haviland had finished his World War service, he had become one of the most decorated members of the *Lafayette Escadrille*. His military awards would eventually include the U.S. Navy Cross, the French *Croix de Guerre*, with 2 Palms and one Star, the Belgian *Croix de Guerre*, with Palm, the Italian War Cross, and the Italian Military Medal. The following extraction from his citation for the Italian War Cross best sums up his overseas service.

"Lieutenant Willis B. Haviland ...made numerous flights of bombardment patrols, always demonstrating singular courage and high spirit of duty and was always an example to his subordinates."

Haviland's comrades and subordinates, both in the French and U.S. Naval Air Service, always spoke highly of him. Fellow *Lafayette Escadrille* pilot, Edwin Parsons, said of Haviland: "He was a very brave and fearless fighter and a fine companion." Another Lafayette comrade, Harold B. Willis, wrote the following of Haviland.

"I can see Willis-or 'Havvy' as we called him-a strong, good looking, clean courageous type if there ever was one! He did stand out! I do not recall his ever saying or doing a mean, unkind or cheap thing."

Daniel H. Keyser, a Naval seaman who served as Haviland's messenger at Porto Corsini, "stood in awe" of his commanding officer. He wrote this of him:

"He was one of the nicest persons I have ever known. I never heard him raise his voice. The (Naval) flyers were supposed to drop propaganda leaflets on Pola. The Austrians returned the visit next night with bombs. (Haviland) said 'to hell with the leaflets, we will take bombs next time.'

"The Austrians came over quite a few times but did no damage, for most of the bombs were duds. Captain Haviland never took cover during a raid. His orderly and I sat out the raids in a building used as a store

U.S. Naval pilots Ensign Coatsworth, Lieutenant Haviland, and Ensign Parker on leave in Venice, Italy, November 1918. Coatsworth and Parker were former Lafayette Flying Corps pilots: Coatsworth with SPAD 80, and Parker with SPAD 85 and 98.

house and we would see him strolling along outside as though he knew the bombs were duds."

Haviland served aboard the USS *Texas* following his Italian war service with former Lafayette Flying Corps pilot, Ensign William Carey Van Fleet, Jr.. Their assignment was to introduce the U.S. Fleet to the English method of piloting land fighters from the ship's deck. This concept of a plane leaving the deck of a battleship was first pioneered by Eugene Ely when he took off from the USS *Birmingham* on 14 November 1910. He also made the first landing, on the USS *Pennsylvania* on 18 January 1911.

In December of 1918, while in Caribbean waters, Haviland had persuaded Captain N. C. Twining of the *Texas* to permit construction of a 40-foot-long runway by lashing timbers across the No. 2 guns of the vessel's forward deck. He then had his plane, a Sopwith *Camel*, winched to the deck and its wheels bound by a holding bridle which would be released by riggers at Haviland's signal. Haviland climbed into the cockpit and revved and raced the plane's motor until it seemed to the nearby sailors that the prop blast and vibration would tear its fuselage apart.

Haviland signaled for the cables to be released. The straining aircraft roared down the runway, dropped precipitously toward the sea, then climbed into the sky. Haviland eventually landed the *Camel* at Conde Bluff, Guantanamo Bay, Cuba.

While serving as aviation officer for the USS *Texas* at Guantanamo Bay, Haviland was promoted to the grade of Lieutenant Commander for his "imminently satisfactory" service on 1 April 1919. That same year he married Mary Lucile Satterthwaite. The following year, a daughter, Barbara, was born.

Haviland later served aboard the USSs *Oklahoma* and *Arizona* at Guantanamo; and aboard the USS *Shawmut* at Anacostia. One of his final assignments was to train U.S. Naval pilots to fly land planes. He was honorably discharged from the Navy as a Lieutenant Commander on 27 December 1925.

A civilian again after ten years of continuous military service, Haviland became an auto sales executive.

Military biplane being winched onto the deck of the USS Texas from which Lieutenant Haviland would take off, thus introducing the U.S. Fleet to the English method of piloting land fighters from the decks of naval ships.

He then became manager of the Atlanta Airport, Atlanta, Georgia. He left that position to become an insurance executive in Columbus, Ohio. He next took charge of the Universal Aviation Schools which was headquartered in St. Paul, Minnesota. While serving in that capacity, he was appointed by the aeronautics branch of the Department of Commerce to advise and assist the government in the regulation of flying schools.

In 1935 Haviland became Western manager for Standard Brands, Inc.(Fleischmann Distilling Corporation) with headquarters in San Francisco. His territory included Alaska, Hawaii, Washington State, Oregon, and California.

When the USS *Oklahoma* was sunk by the Japanese at Pearl Harbor on 7 December 1941, Haviland sought reinstatement as a Naval Officer. He was accepted with rank of Commander and was made the supervising executive officer in charge of the construction of the new Naval Air Station designated for Whidbey Island, Washington. In November of 1943 Haviland assumed command of the Whidbey Island Naval Air Station which he ran with clock-work efficiency. His superiors noted this, and he was soon to be given a special assignment in the Central Pacific War Zone. He relinquished command of the Whidbey Island Station on 11 September 1944.

But Haviland fell ill before he could accept his new assignment and was instead sent to the U.S. Naval Hospital at Corona, California where x-rays revealed a lung tumor. He underwent surgery to have the tumor removed and died of post-operative pneumonia at the Naval Hospital at age 54 on 28 November 1944. Commander Haviland is buried at Arlington National Cemetery. He was survived by his widow, Lucile, and his daughter, Barbara.

JOHN HEILBUTH

John Raynolds Heilbuth was born in 1899, the son of Edouard Heilbuth. John attended Paris schools (Lycee Jason de Sailly) and was a seventeen-year-old student when he joined the American Ambulance Field Service on 29 May 1916.

Heilbuth served one year with SSU 2 and was awarded France's *Croix de Guerre* while with that unit. On 3 July 1917 Heilbuth enlisted in France's *Service Aeronautique*. He was released in August 1917 because of defective eyesight.

THOMAS HEWITT

ESCADRILLE LAFAYETTE

"How did we get so far away from ourselves? To where we now are from where we had hoped to be?"

Thomas Moses Hewitt, Jr., the 28th volunteer of the *Escadrille Lafayette*, was born in Westchester, New York, on 21 December 1894, the son of Thomas Hewitt and Sarah (Baum) Hewitt.

Following high school, Hewitt enrolled in the New York Nautical School. In early December 1915 he signed on as a seaman on a tramp steamer docked at Hoboken, New Jersey and which was destined for Manchester, England. Upon his discharge in England from the ship's crew, he attempted to enlist in the Royal Naval Air Service and the Royal Flying Corps but was turned down because he was not a British national. He then tried to enroll in the American Red Cross as an Ambulance driver, but since Hewitt hadn't reached the required minimum age of twenty-one, he was ordered to cable New York for his father's permission.

Hewitt refused to do so and instead crossed the Channel to France. Five days later, on his twenty-first birthday, he signed on with Section 7 of Richard Norton's Ambulance Service. He drove ambulances in the Champagne sector until early April of 1916. On 13 April he enlisted in the French Foreign Legion (L.M. No.11,826) then transferred to aviation.

From 3 July 1916 to 28 March 1917 Hewitt attended the aviation schools at Buc, Juvisy, Avord, Cazeaux, Pau, and the G.D.E. He received his *brevet militaire* on the Caudron on 24 December 1916. He had been impressive in training, both in acrobatics and marksmanship. However, his school record was not without blemish as he was twice punished for minor rules infractions. Still, his seeming aggressiveness and his impeccable appearance won over the officers from the *Escadrille Lafayette* who had come to the G.D.E. to select American replacements for their Squadron. Hewitt was chosen along with Americans Kenneth Marr and William Dugan, and the three joined *N.124* at Ravenal on 30 March 1917.

Edmond Genet, who had been in pilot training with Hewitt, became the first of the Squadron's pilots to express misgivings about him in his diary entry for 30 March:

"Three new fellows arrived from Plessis Belleville this afternoon to swell our already full personnel. They are Dugan, of Legion Fame, Kenneth Mar (sic) and Thomas (more intimately known as "Jerry") Hewitt. The first two are good fellows but Jerry is quite questionable. He will have to be tested out before I can feel confident in his "cran et sang froid."

Genet's misgivings were soon realized. Hewitt failed to fulfill the promise his confident manner and

Thomas Hewitt as an ambulance driver, January 1916

Sergent Thomas Hewitt preparing to leave on a patrol, Chaudun airfield, June 1917. Behind Hewitt are the Lafayette Squadron's new SPAD S.VII fighter planes. The center fighter has Kenneth Marr's personal insignia "M" painted on its side.

finely-tailored new black uniform had suggested. He had the misfortune of joining the Squadron during a time of intense combat activity, and his initial exposure to both enemy aircraft and explosive ground fire unnerved the novice pilot. As Hewitt later recalled:

"My first patrol was at the time of the St. Quentin-Aisne retreat. We were over a crack German battery. My only thought was keeping track of my comrades. I landed ten miles from my own field, and didn't know where I was."

Squadron Commander Thenault may have shared partial blame for Hewitt's timidity. As with all new pilots, he should have broken Hewitt in on patrols behind the lines, giving Hewitt the chance to familiarize himself with the sector. So it was little wonder, having faced such a violent initiation, that his confidence left him as quickly as the creases in his uniform which soon became rumpled and spattered with food. Hewitt's timidity and unwillingness to aggressively confront the enemy were soon noted and frowned upon by his American comrades who expected each pilot in the Squadron to shoulder his load. *Lafayette* pilot Edwin Parsons found Hewitt to be a contradiction as he later recalled in a taped interview.

"Thomas Hewitt, who didn't quite seem to fit, achieved the nickname 'Horrible.' It seems strange that a chap who made such a brilliant record in schools. Yet on any patrol which I led, Hewitt could always find some reason why he couldn't get beyond our balloon lines. He was not, even though he was a brilliant pilot, intended to be a wartime pilot and take his chances with the rest."

On 7 April 1917 the *Lafayette* Squadron was moved from Ravenal to Ham. Its two-month stay there saw a period of bitter fighting and death. As Hewitt saw Genet and Hoskier killed and viewed deLaage's crushed body in the wreckage of his SPAD, he shrank further from combat and frequently nosed his fighter home before the patrol was completed. Hewitt, himself, candidly recalled one instance.

"It was near Laon. Two Teuton craft, flying about 2,000 metres, were regulating artillery fire. My partner started down after them. The two machines started down and headed for Laon. It did not take long to see that the chase was useless, so we started for home. The Teuton gunners put barrage after barrage in front of us. I lost my partner in the wild run for home, but upon landing found that he had reached there about

The Lafayette pilots celebrating the U.S. entry into World War I. (l.r.) Dugan, Thenault, Hewitt, Thaw, deLaage, Lufbery, Marr, Parsons, Hinkle.

ten minutes before me. I was awfully glad to get out of it."

Before long Hewitt's pilot skills began to leave him as he demonstrated on the Squadron's 3 June move with Combat *Groupe 13* from Ham to Chaudun. The pilots of the *Lafayette* were ordered to ferry their new SPAD S.VII fighters to the Chaudun field. They were forewarned of the location of a deep irrigation ditch which ran the width of the new field. Red markers had been planted along the ditch to make it visible to the incoming pilots. Seventy-four pilots of *Groupe 13* brought their SPADs and Nieuports safely down on the field by carefully avoiding the ditch. Hewitt, the seventy-fifth, managed to roll his SPAD directly into it at such speed that he flipped and demolished his plane.

Hewitt, shaken but unhurt, threw off his safety harness and dropped to the ground as Squadron commander Thenault cursed him in two languages. Then as punishment for wrecking his SPAD, he ordered Hewitt back to Ravenal with orders to fly the new SPAD for Robert Soubiran, who was then on leave in Paris, back to Chaudun. The mistake was Thenault's. Hewitt managed to destroy Soubiran's SPAD in the same ditch not ten feet from where he had wrecked his own fighter. Thenault, enraged, immediately grounded Hewitt.

However, the possibility existed that on one of his flights to Chaudun, Hewitt's SPAD may have been struck by ground fire. As he later recalled: "You may never know when your machine is hit. Once, when I was coming from Ham to Chaudun, I felt something lift me. When I got to the ground my mechanic showed me a six-inch hole in the tail of my machine which I did not suspect was there."

When Hewitt was put back on flying status, he seemed unable to screw up the courage for combat patrols. On 17 September 1917 Captain Thenault removed Hewitt from the Squadron roster and ordered him back to the G.D.E. for retraining as a bomber pilot. Hewitt soon washed out of that school and requested his release from the French Air Service in order to return home to join his own country's army. By late October he had been released from the *Service Aeronautique* for reasons of "ill health" and by 1 November he was aboard ship sailing to the United States.

On 11 November 1917 Hewitt arrived in New York harbor with fellow *Lafayette* pilot Edwin Parsons who was heading home on leave. Hewitt told the press he intended to join the U.S. Air Service.

The February 1918 issue of *Air Travel Magazine* carried an article entitled AN AERIAL SOLDIER OF FORTUNE "Sergeant Hewitt's Own Story of Life in the Lafayette Escadrille." The bulk of the article focused on Hewitt's experiences as a student aviator. In spite of the prodding of two interviewers to relate his aerial combat experiences, it was to Hewitt's credit that he would not embellish on his *Lafayette* service and refused to succumb to his questioners' demands for the "vivid, personal details" of his "fights."

Hewitt apparently spent some months "recuperating" from his overseas service before he pursued his purpose for departure from French Aviation. When he finally did register for the military draft, he learned that he had forfeited his U.S. citizenship and had been classified as a "friendly alien." He next attempted but was denied admission into the U.S. Air Service.

The matter of Hewitt's citizenship was eventually resolved and on 25 October 1918 he enlisted in the U.S. Army in New York City. He served with the 30th Artillery, Coast Artillery Corps at Fort Wright, New York and with the 4th Company, Long Island Sound, Coast Artillery Corps until his honorable discharge as a private on 19 March 1919.

Following his military service Hewitt attended the National University Law School in Washington, D.C. from September 1922 to June 1925. During that period he married and had a son, Thomas Moses III in 1924. Hewitt received his Bachelor of Laws degree on 10 June 1925.

After graduation Hewitt briefly set up practice in Washington, D.C. In early 1926 he spent several months in Manila, the Philippines, promoting aviation. He returned to New York City in the spring, and he and his wife separated. That same year he was arrested and jailed for embezzlement. On 16 October he was taken from confinement by New York law enforcement authorities to Washington, D.C. to give a blood transfusion to his three-year-old son who was thought to be dying of anemia.

About the same time he was examined by the U.S. Veterans Bureau for mental disorder and was briefly placed in the psychopathic ward of Bellevue Hospital where he was "adjudged abnormal but not insane." On 10 September 1927 Hewitt was arrested for passing bad checks. His unlawful actions were attributed as the cause of his father's death.

Hewitt was apparently released from confinement before 12 December 1927. On that date he announced from Wallingford, Connecticut that he planned a world flight in memory of Raoul Lufbery and the *Lafayette Escadrille*. The Wallingford paper announced the following:

"A round-the-world flight to perpetuate the memory of America's Ace of Aces, Raoul Lufbery, and the Lafayette Escadrille, was announced here today by Thomas Moses Hewitt, Jr. of Bronx, NY, a former Lafayette Escadrille pilot...Wallingford Chamber of Commerce agreed to aid in sponsoring the trip. Hewitt proposes to make a globe circling flight and to arrive in Paris for the official dedication of the Memorial de L'Escadrille Lafayette on May 30, 1928."

Evidently Hewitt's plans failed to materialize. On 30 August 1928 he was arrested for passing two bogus checks totalling $1000 at the Hotel Biltmore in New York City. He was charged with grand larceny. Hewitt blamed his erratic behavior and numerous arrests on a "head injury" he claimed to have sustained during the war. He also claimed that he had been "cited for bravery" by the United States, Great Britain, and France" after having been shot down over Dunkirk.

Hewitt further claimed that he had delivered a squadron of planes to the Mexican government and had been superintendent for airway extensions for the Department of Commerce in the St. Louis and Chicago area. But at the time he had been in Washington, D.C. attending law school.

Hewitt was finally convicted on a petty larceny charge for having defrauded the Hotel Biltmore for $66.14. Hewitt, in a moment of truth, admitted "It is only when I am drinking that I do these things." In spite of a written plea for leniency by prominent Philadelphia attorney and former Lafayette colleague Charles Biddle, on 6 October Hewitt was given an indeterminate sentence at the state penitentiary.

While incarcerated, Hewitt attempted to use his association with the *Lafayette Escadrille* as a means for his possible release. He first approached Mr. William Nelson Cromwell, vice president of the "Memorial de l'Escadrille Lafayette." He next contacted officers of the *Escadrille Lafayette* Flying Corps Association, soliciting funds for an appeal. His actions caused great consternation among the membership, since as a former pilot of the *Lafayette Escadrille,* Hewitt's name appeared in all their publications.

On 18 July 1929 Hewitt sued out a writ of habeas corpus for himself stating that "he was not in a proper mental condition when he was tried" on the bad check charges because of "a war wound." The presiding judge ruled in opposition to upholding Hewitt's writ and ordered him back to the penitentiary.

Upon his release from prison in the fall of 1929, Hewitt returned to Washington, D.C. Over the next few years he was arrested a number of times for intoxication. He died on 24 May 1936 at age 41 in a rundown D.C. rooming house. The cause of death listed on his certificate was "acute congestive heart failure" which was precipitated by "acute alcoholism."

Following his death, Hewitt's body lay unclaimed in the District morgue. A friend of Hewitt tipped off representatives of the American Legion who in turn contacted a reporter at the Washington *Times* newspaper. An effort was launched through the press to locate his estranged wife to prevent Hewitt's body from being buried in Potter's Field.

Two days after Hewitt's death the *Times* ran a story under the headline "Legion Saves Hero Pauper's Burial." The story related how Edward Lewis, Commander of the American Legion's National Press Club Post, had remembered that Hewitt had once been a member of the *Lafayette Escadrille*. Lewis had then promised:

"This man will be buried in Arlington Cemetery, where he deserves to be buried. He will be accompanied to the grave by a guard of honor, a firing squad and a bugler to blow taps."

The following day's headlines trumpeted "Hero's Rites Await Lonely War Flyer." The article spoke of Hewitt's "many decorations" and listed his accomplishments as having raised "more than $4,000,000 for the Victory loan" and having been a frequent blood donor while "in the district jail on intoxication convictions."

The headline for Thursday, May 28, 1936 proclaimed "Hunt for Wife Holds up Rites of Flying Hero." The article mentioned how Hewitt's estranged wife still had not been located, and again noted that he was "a member of the famed Lafayette Escadrille and holder of numerous hero awards." His downfall was traced back "to a head injury received when his plane crashed during a forced landing near Pol, Belgium."

By Friday, May 29, 1936 the Washington *Times* had promoted Hewitt to the rank of a commissioned officer, stating "Capt. Hewitt Laid To Rest In Arlington." While in fact, the highest rank Hewitt had attained was that of *Sergent* in the French Air Service.

A followup article spoke of how the "Famed War Flier" was finally laid to rest in Arlington in the presence of his brother, his estranged wife, and his son who were there to "Pay Last Tribute To (the) Fallen Falcon." Also present and honoring the "war-time flying hero" were France's air attache to the United States and Hewitt's former French *moniteur*, Jean Auguste Maigret, then an employee in the Department of Agriculture.

LESLIE HICKSON

Leslie Matheson Hickson was born in Montreal, Quebec, Canada on 14 November 1892, the son of Richard Joseph Hickson and Emily (Matheson) Hickson. Leslie attended Erasmus Hall High School, Brooklyn, New York before enrolling in Amherst College, Amherst, Massachusetts, Class of 1914. Upon receiving his Bachelor of Arts degree, Hickson worked as an assistant advertizing copy writer in New York City. He resigned his position to attend Harvard University Graduate School of Arts and Sciences, 1915–1917.

Hickson left Harvard and joined the American Ambulance Field Service in New York on 25 June 1917 in order to secure passage to France. On 30 July 1917 he enlisted in France's *Service Aeronautique* at Paris. He received his *brevet* on 27 September 1917. On 11 February 1918 he was released at his own request following an airplane accident at Pau.

Following World War One Hickson returned to New York City where he was employed as treasurer of the Lotus Magazine Foundation. From 1929–1931 Hickson served as an advertizing manager for the Standard Oil Company in Pennsylvania and in New Jersey. From 1933–1936 he was sales promotion and advertizing manager for the Atlas Supply Co. On 4 August 1934 he married Lucile Rathbun in New York City. From 1937–1947 Hickson was advertizing director for the National Distillers Products Corporation of New York City. Because of his wife's illness, Hickson moved to

California but returned to New York when Mrs. Hickson died a few years later.

In New York Hickson re-entered the advertizing agency business but was plagued by serious illness and a broken hip.

Hickson died in his sleep of arteriosclerosis at his home at Mount Vernon, New York on 3 January 1968 at age 75.

DUDLEY HILL

ESCADRILLE LAFAYETTE

Dudley Lawrence Hill, the 12th volunteer of the *Escadrille Americaine*, was born on 6 October 1893 at Peekskill, New York, the son of Edward Finch Hill, founder of the Union Stove Works, and Jessee (Dyckman) Hill. Hill's youth was spent with his brother and sister between the large family home on Nelson Avenue in Peekskill and the summer estate on Fire Island, New York. He played football at Oakside High School and later again at the Peekskill Military Academy before graduating on 9 June 1914.

That fall semester Hill entered the Mechanical Engineering School at Cornell University but transferred in February of 1915 to New York University to take advantage of a special chemistry course offered there. He stuck it out just one month. Telling his friends he had to take part in some capacity in the European War, he requested and was granted a year's leave of absence from the University. Before the end of March he had set sail for France on *La Touraine*.

In Paris in late April 1915, Hill joined the American Ambulance Service, serving faithfully until the first of August as an attendant at the American Hospital at Nancy, and as an ambulance driver at the Front. In August his petition to join the French *Service Aeronautique* was granted. After joining the Foreign Legion Hill transferred to aviation, and from 25 September 1915 to 2 May 1916 he attended the aviation schools at Pau, Chateauroux, and at the G.D.E. On 17 March 1916 he received his *Brevet D'Aviateur Militaire No. 2999* in the Caudron. Following aviation school he served briefly as an instructor and tester of aircraft.

The fact that Hill made it through flight training was remarkable in light of his two physical handicaps. As a youth of fifteen he had become blinded in his left eye, the result of an ice hockey game on the frozen Hudson River. The following summer he had lost half of his hearing when he burst an ear drum while diving too deeply off the family sailing yacht at Long Island.

How Hill managed to obtain his pilot's license in spite of his handicaps was explained by *Adjudant* Edwin C. Parsons in an article in the Springfield *Republican* dated 24 March 1918:

"Dudley Hill of Peekskill, New York, was wholly blind in one eye, and yet by carefully learning all the letters on the card used in the eye test, he had been able to pass all of his examinations up to the school at Pau. When he took the physical test there, they put before him a card with different letters on it and his deficiency was found out. Never, declared the officers, could a one-eyed man be a successful pilot, and they decided at once to reject him. But while they were waiting for the order of dismissal, Hill went on training through the kindness of the instructors, who liked him, and before the order came he had taken his license as a pilot. Once he had it, there was no practical way to throw him out of the service and he became one of the most able and valiant members of the escadrille."

Caporal Dudley Hill at the G.D.E. pilot pool at Plessis-Belleville awaiting assignment to the Front, May 1916.

By the time *Capitaine* Georges Thenault and other representatives of *N.124* were ready to select another American pilot as they continued to build the *Escadrille's* complement, they settled on the one-eyed Dudley Hill. In retrospect, Thenault could not have made a better choice, for in Hill he soon discovered, in spite of his handicaps, a courageous, durable, unusually proficient pilot who logged more time (on the line ready to fly, or on patrol) than any other member of the *Escadrille*. It was as if Lady Luck was his co-pilot, for while so many of Hill's comrades were being killed or maimed, he entered a hospital just once, in mid-January 1917, while home on furlough in Peekskill for the removal of his appendix.

Hill's service with *N.124* began officially on 6 June 1916 and continued through 18 February 1918 when *Escadrille SPA.124* was incorporated into the U.S. Air Service as the U.S. 103rd Aero Pursuit Squadron. His nickname, "Dud," concealed an aggressive spirit. A cable from Paul Ayres Rockwell to the New York *Globe* dated 27 July 1916 noted Hill's capabilities.

"In a hot aerial fight over the Verdun battle fields Saturday Caporal Dudley Hill of Peekskill, New York, a volunteer pilot with the French aviation service, forced a German aeroplane to land precipitately on the shell-torn ground directly behind German lines at Ft. Thiaumont. For this feat he has been nominated for the grade of sergeant.

"Lieutenant (Alfred de Meux) de Laage, Sergent Kiffin Rockwell, and Caporal Hill were on patrol duty over the battle ground, flying at a great height, when they saw three German machines in the distance. They flew at once toward the hostile machines. Two of the German aeroplanes fled while Hill, who was in the lead, gave battle to the third. The German was evidently hard hit by the American's first volley, for he immediately dived to the ground, unable to make his regular landing place.....Since going to the Front, Hill had had several combats with enemy planes, and has proved himself a courageous fighter."

Self-effacing by nature, Hill was never one to draw attention to himself. As one of his comrades noted, ". . . Monosyllabic Dud has a long and enviable record of service which he never mentions, but is always generous in his praise of the records at the Front of other men."

Hill soon made *Sergent*. Before long he was senior-ranking American *Adjudant* in the absence of Lieutenant William Thaw and *Adjudant* Chouteau Johnson during their furlough to the United States. Perhaps Hill would have made more press copy if he had accommodated the American correspondents who frequented the *Escadrille*. But in the nineteen months he was with the *Escadrille Lafayette* he avoided the newshawks, preferring instead to bury his face in a journal, or to play chess with a squadron mate. Occasionally, however, a reporter would catch him. The 7 June 1917 *Philadelphia Record* dispatch declared from Paris:

"The Germans' latest trick is to paint their machines with the famous tri-color circle, long used by allied aviators as a distinguishing mark-- only the German tri-color is red, white and black and has a small iron cross in the middle. At a distance in the air, however, the colors are hard to distinguish and Sergeant Dudley Hill flew around a German ten minutes before he realized that it was not a French plane. The German, apparently appreciating the joke, did not fire, and just as Hill was about to attack, he dived away safely."

An *Associated Press* dispatch dated 19 August informed its readers how ". . . Aviator Dudley Hill of Peekskill, New York, had a narrow escape when two Germans attacked him with incendiary bullets. He escaped, however, with one bullet hole through the

Sergent Dudley Hill in Raoul Lufbery's Nieuport scout preparing to leave the Behonne airfield for a patrol, July, 1916. Before taking off, he will lock a 47-round ammunition drum onto the .303 caliber Lewis machine gun which is post-mounted onto the fighter's upper wing.

lower plane by outmaneuvering his opponents. He is believed to have brought down one German."

Hill's life with the *Escadrille* seemed a succession of "narrow escapes" both in the air and on the ground. There were bad landings which demolished aircraft, and once while tearing through a French village on a motorcycle, he lost control and crashed headlong through a window into a *madame's* kitchen. Unscathed, but chagrined by the breakage he had caused, Hill immediately wired home for a loan to pay the damage to the dwelling.

On 8 October 1917 *Adjudant* Hill was awarded France's *Croix de Guerre*, avec 'etoile (with Star). His citation reads: "American citizen engaged for the duration of the war. A good fighter pilot, dedicated to his duties. He participated in a number of engagements, particularly the last attack of Verdun with no consideration for his own safety. He showed everyone the best examples of bravery, spirit and hardiness. He particularly distinguished himself on the 18th of August while protecting a mission in which his plane was seriously damaged."

Near the end of his tenure with the *Escadrille Lafayette* Hill managed to rathole the funds to purchase a new French uniform. After shedding his old oil soaked service uniform for natty threads of horizon blue, he was asked and agreed to escort a colonel's wife to Paris for the day. By then, Whiskey and Soda, the squadron's two lion mascots, had been removed to the zoological gardens there, and Hill worked a visit to Whiskey's cage into the day's itinerary.

Once at the Paris zoo, Hill explained to Whiskey's keeper his relationship with the animal and managed to talk his way into the lion's cage. The instant the half-blind Whiskey spotted the half-blind aviator, the lion was so overjoyed at seeing his old squadron mate, he bounded up to him, planted his huge paws on Hill's shoulders in a welcoming embrace, and drove the unprepared Hill in his new uniform into the mire at the bottom of the cage. There, Whiskey further expressed his affection by mauling Hill with his teeth and paws, much to the delight of the gathering crowd who viewed the spectacle as an act staged for their benefit. They expressed their appreciation and delight through raucous laughter and applause.

Perhaps it was fortunate Hill was nearing the end of his tenure with the *Escadrille Lafayette*. His transfer to the U.S. Air Service would mean a new uniform of khaki as his horizon blue was relegated to the bottom of his footlocker.

On 17 November 1917 *Adjudant* Dudley Hill was made a Captain in the Signal Corps of the U.S. Air Service. Because of his defective eyesight, this required a waiver which had been personally issued by General John J. Pershing on 5 November. General Pershing recommended that "on account (of) three years flying with French army it is thought he would make a very useful officer in spite of marked vision defect."

On 18 February 1918 Captain Hill was assigned to the U.S. 103rd Aero Pursuit Squadron and flew combat missions until 1 June 1918. He then flew for two months (from 1 June to 1 August 1918) with the U.S.

139th Aero Pursuit Squadron. He commanded the U.S. 138th Aero Pursuit Squadron from 1 August to 1 November 1918. From that date to 1 December 1918 he was Commanding Officer of the 5th Pursuit Group.

On 21 March 1919 Captain Dudley Hill became one of only six *Lafayette Escadrille* pilots to be authorized by Field Marshal Petain to wear the French *fourragere* in the colors of the *Croix de Guerre* for having served both with *N.124* and the U.S. 103rd Aero Squadron when the squadron had received two citations in *l'Ordre de l'Armee*. The other recipients were Ray Bridgman, William Dugan, Henry Jones, Robert Rockwell, Robert Soubiran, and William Thaw. In that same month of March 1919 Captain Hill was discharged from the U.S. Air Service.

As must have been the case with most survivors of the *Escadrille Lafayette*, life after the Great War could only seem anticlimactic to Dudley Hill. His nearly twenty-eight months of numerous air sorties in the worst of conditions caused him to despise flying, and he never again took the controls of an airplane. Nor did he ever discuss, even with his closest of family, the bad times of his war experience. He chose only to mention the good memories, the humorous events, or those episodes which highlighted the unique qualities of his comrades.

Back home at Peekskill after the war he worked at the family's Union Stove Works until he was forced to give it up due to eye trouble irritated by dust in the molding rooms. On 28 January 1925 a grateful France awarded Hill the Cross of Knight (Chevalier) *de Legion d'Honneur* for his "splendid services during the War." In September of 1926 he began work for the Cuban Cane Corporation in Cuba, running and maintaining the sugar and molasses manufacturer's 167 miles of railroad.

At the onset of World War II Hill began work for a private firm in Canada which was contracting pilots there for the U.S. Army Air Corps. It was secret work, and Hill was paid a commission for each enlistment.

In 1941 he joined fellow Lafayette comrade Carl Dolan at the Wissahickon Tool Works. Dolan was president of that firm, and he found Hill a position as assistant to the manager of the Schulykill Manufacturing Company. That same year Hill married Evaline Craven in Norfolk, Virginia. They had a son, MacDonough, in 1942.

Following World War II Hill invested in plastics and became a purchaser for a firm producing farm implements in New York City. By then most of his *Lafayette* comrades had passed on: Bill Thaw of pneumonia; Chouteau Johnson (whose ashes Hill and Carl Dolan had scattered) of throat cancer. Then on 30 June 1951 Dudley Hill died of a heart attack at age 57 at his Peekskill home.

Captain Dudley Hill, USAS, beside his SPAD S.VII scout, spring of 1918.

The true mark of Hill's character had been written some thirty years before by one of his squadron mates. In James Norman Hall's words:

"He went to the Front when French combat squadrons were equipped with 15-meter Nieuport biplanes, with Lewis guns mounted on the top planes. From that time until his transfer to the American Air Service, Dudley served constantly as a pilot with N.124. He had flown with the Squadron on every sector of the Western Front. His experiences as a pilote de Chasse is as broad as his length of service indicates. His old French uniform of horizon blue had numberless baptisms of gas and burnt castor oil at a time when the entire American Air Force could have ridden comfortably in one Handley Page. And yet, to hear him tell of it—but who ever heard 'Dud' tell of anything in which he himself was concerned? Unless it was to speak of some headlong flight which he claims to have made from pursuing Germans."

EDWARD HINKLE

ESCADRILLE LAFAYETTE

Edward Foote Hinkle, the 24th volunteer of the *Escadrille Lafayette*, was born in Cincinnati, Ohio on 22 May 1876, the only child of wealthy manufacturer, William H. Hinkle, and Lucille (Foote) Hinkle. Hinkle's father had built the Ashland Iron and Steel Company at Ashland, Wisconsin, the largest charcoal and iron foundry in the world. His ancestors had founded the town of Hinklestown, Pennsylvania in 1709. His grandfather had published the McGuffey Reader.

Hinkle attended Andover Academy before receiving his A.B. in architecture from Yale University in 1899 and a B.S. degree from Cambridge University in 1901. Following his schooling, he then went to Paris, France to complete his study in architecture and design at the *Ecole de Beaux Arts*.

In 1905 he returned to the United States and that same year set up an architectural office opposite the Waldorf-Astoria Hotel in New York City. His projects included fine city homes and the Vanderbilt Riding Academy. He also submitted a design for the town of Gary, Indiana which was initially accepted but ultimately rejected by its planners.

Hinkle closed his office in 1910 to return to Paris for further study, and to serve as representative for several American companies including the Velsin Aeroplane Company. One of his early inventions was a helicopter.

Hinkle was in Bucharest, Roumania at the outbreak of World War One. He left that country for France as a freighter stowaway. He later recalled his experience:

"The travel situation got sticky, but an influential friend arranged for me to 'stow away' on a freighter carrying a cargo of flour to Constantinople. The Turks hadn't gone to war yet, but things were touch-and-go. I helped to repay my friend by delivering him a chart of the Bosporus mine field. I sketched it in the darkness while the freighter was escorted through by a German-commanded patrol boat."

Back in Paris, Hinkle spent the remainder of 1914 trying to wrangle his way into French aviation. His mother had been of French descent, and he had grown fond of France during his years spent there. He also had grown to hate anything German.

Sergent Edward Hinkle beside his SPAD S.VII fighter at Ham, June 1916.

"I hated the Germans even before the war. Twice I had near-duels with Germans. Thankfully, I bluffed my way out both times. A captain of the Prussian Guard challenged me once and I suggested we have it out with six-shooters. Only I thought it would be nice if we could continue shooting as long as we were able to, even if lying wounded on the ground, and until we ran out of bullets.

"'Oh, that's brutal!', the German said. The second time I wanted to use clubs. The German said it wasn't gentlemanly. I told him I was serious about it whenever I fought. My regard for France is why I joined the Escadrille."

On 20 July 1916 Hinkle joined the French Foreign Legion at Dijon. He then transferred into French aviation, was made a second-class *soldat* and issued a French uniform before being sent off to flying school.

Hinkle's first school was at Buc, near Versailles. His French *moniteurs* had thought him to old for aviation at age 40 and had tried to wash him out by demanding that he master the clipped-wing Bleriot Penguin in one day, a nearly impossible feat.

Hinkle arose at 5 a.m. the day of the test, feeling he had been ordered to complete an impossible task. By noon he was ready to concede defeat, but his comrades surrounded him and suggested he have a glass of red wine and go out to the field and try to pass. Hinkle remembered the incident:

"Now, the Penguin had a wheel under the tail section instead of a skid, and it was about as graceful as a cow on ice. The idea was to taxi the thing from one point on the field to another-in a straight line-and back. Well, I couldn't get the hang of it to save me. The tail-end would slew around on the wheel...sometimes it would tip over on a wing . . . everything but go straight.

"Lunch time came, and it looked like I was a wash-out, for sure. Then one of the men who'd had the course—I think it was Bob Soubiran-gave me the secret. Full throttle, zoom to the end of the field, whip her around, then full throttle back.

"I went back to the course after lunch, fortified with several cups of red wine. I took hold of the control stick and said to myself, 'Here goes nothing.'" By three o'clock that afternoon Hinkle had control of the unruly Penguin and was on to the next stage of his flight training.

Three months of additional flight training followed: primary school at Avord flying Bleriots before going on to Nieuports, the service planes. At Avord Hinkle received his *brevet militaire* on the Bleriot on 4 November 1916. Gunnery school followed at Cazeaux, near Bordeaux; then acrobatic school at Pau in the south of France, near the Pyrenees. While at Pau, Hinkle learned to kill his engine then to spiral down to land on a forty-foot-long area. Hinkle later related how that training had saved his life:

"While flying missions, I made seven forced landings. Sometimes in wheat fields and once in a woods. They had to take the wings off to get the plane out of there. But I never did damage a piece of equipment. I'm proud of the fact that during training and my time at the front, I never busted up a machine. There is no breakage on my record."

Edward Hinkle, Robert Soubiran, Frederick Prince, and Harold Willis in the forward Allied trenches attempting to confirm a Squadron victory.

From Pau Hinkle went to the G.D.E. pilot pool north of Paris at Plessis-Belleville where he honed his flying skills while awaiting an opening at the Front. He had proved that he was a capable flyer despite his age.

Hinkle requested assignment with the *Escadrille Lafayette*. But the French authorities felt him too old at age 40 for aviation and tried to deny him the squadron assignment. They had set the age limit for aviation at 28. But through the intervention of William Thaw and Doctor Edmund L. Gros, Hinkle was able to join the *Escadrille* at nearly 41 years of age on 1 March of 1917. Thaw, a fellow Yalee, used to visit Hinkle in his Paris apartment whenever he could secure a leave from his squadron.

Hinkle joined the Squadron at Ravenal where he was promptly given the moniker "Pop" by his younger

American comrades. Hinkle's memories of Ravenal were bleak.

"Ravenal was a wretched place. I remember the icy winds and the snow blowing across the flying field. We'd go out on patrols, stiff as corpses in the cockpits . . . the propwash chilling us to our bone marrow."

It was so cold in the air above the sector that Hinkle froze both ears while fellow pilot Edwin Parsons froze a toe inside three pairs of socks and a fur-lined flying boot and was unable to walk for a week.

"I was first at Saint-Juste, then Ham, and then Soissons. We moved from field to field. The second time we moved, the Germans bombed the field five hours after we got there, burning five hangars. The second night we could hear them but they could not find us.

"I was in the Chemin-des-Dames offensive where 83% and 85% of the bombing planes were lost. We were in a patrol of 300 planes that finally drove the Germans back six miles. I had a feeling of inferiority. Every time volunteers were asked for, I would put up my hand. There were kids there young enough to be my sons."

"At Ham I was ordered to learn if there were any new entrenchments, or railroads running behind the enemy lines. I went in with an escort of four planes that took me over the lines, going on by myself toward Belgium. I suddenly discovered that I was nearly out of gas near Soissons. I didn't have enough gas to get home, so I thought that I had better get across the lines as quick as I could.

"I saw a German balloon and thought I could get a shot at it because I was coming up behind it from Germany. I didn't see four German planes circling the balloon in protection. They gave chase and I ran out of gas just over the lines, at 100 feet. I landed with a dead engine and walked about a mile to the nearest airfield for a can of gas. When I got the plane started I flew back to the field to get filled up. Then a thunderstorm came up and I had to wait for it to go by before flying home.

"I finally got back to the field five hours late, around 6:00 p.m. Everyone was having dinner when I taxied up to a hangar. Walking around a corner of the hangar, I ran into Kenneth Marr. 'Why you old s.o.b.,' he said. 'We thought you had been shot down.' 'Wait a minute,' I said. 'You're wearing my uniform!' Already he had let down the sleeves. He was also wearing my boots. That's the kind of greeting I got."

After thirty hours of flying time at the Front, *Caporal* Hinkle was promoted to *Sergent*. Not long after, he and Harold B. Willis designed the warlike Sioux Indianhead to replace the Squadron's passive Seminole insignia which had been copied from a label off a Savage Arms ammunition box.

"It (the Seminole insignia) looked like an old woman with a drooping bonnet, and it wasn't very warlike. When I was a kid in Minneapolis, my uncle, a soldier, often told me stories about fighting the Sioux Indians. He had given me a head-dress taken in the wars.

"I made a sketch of an Indian giving a war yell, wearing that head-dress. I submitted the design to Captain Thenault, and he and the French Government had it approved. I made a full-size drawing and was having difficulty preparing the stencil for applying the insignia to fuselages when Harold Willis happened along, and showed me the correct way to make the aluminum form. I am happy to know that the insignia is still being used in the French Air Force today."

Hinkle spent a little more than three months with the *Lafayette Escadrille*, flying a number of combat patrols while engaging enemy fighters on several occasions. Hinkle had two unconfirmed victories. He remembered later that:

"One was a two-seater and one was a single-seater that I know were going down. Then there were four others that were doubtful. I followed them down pretty close to the ground. I didn't actually see them hit I don't know whether they crashed. They were behind German lines, so there was no confirmation."

Hinkle saw several of his friends killed while serving with the Squadron, including Edmund Genet, the first American to be killed in action following the U.S. declaration of War. Hinkle eventually became grounded from chronic bronchitis following a severe bout with pneumonia and was forced to leave the Squadron for hospital on 12 June 1917. Upon his release in September he took a train to the Front with James Norman Hall who had also just been released from hospital. But the *Escadrille* had no vacancies so Hinkle returned to the pilot pool while Hall went on to *Spa.112*.

Thus *Sergent* Hinkle's *Lafayette* service came to an end. But his young American comrades continued to remember him with fondness and spoke of him as "a gallant and inspiring gentleman" and "a very courageous fellow." He was later awarded the Montenegran War Cross for his *Lafayette* service.

Sergent Hinkle served as an instructor and a ferry pilot until he left the French Air Service on 18 May 1918. He then tried to join the U.S. Air Service but was rejected because he was twelve years beyond the age limit and had defective eyesight and hearing. Undeterred, he next tried to transfer to the U.S. Naval Aviation but was again refused.

In July of 1918 Hinkle returned to the United States. At Detroit, Michigan he headed up the Department of Experimental and Miscellaneous Engineering in the Bureau of Aircraft Production.

Following the Armistice he lost a fortune trying to market a steam-driven car. A gas-saving carburetor and a self-adjusting auto brake were two inventions he was prevented from marketing by industry opposition.

For a brief time Hinkle lectured at the Army War College. He also served as a reserve air officer in the U.S. 85th Division at Selfridge Field, Michigan from 1925 to 1930 when he resigned his Major's commission to protest the "politicking" he found taking place in the peacetime military.

From 1928 to 1941 Hinkle work as an architect. For the next four years (1941-1945) he was a free lance design engineer at the Ford Motor Company's Willow Run bomber plant, Willow Run, Michigan. While attached there, he was credited with the design of devices for and modifications in aircraft and engine design, including one of the first deicing methods.

In 1949 Hinkle moved to Truth or Consequences, New Mexico for health reasons. He lived in a small house overlooking the Rio Grande.

In July 1963 the French Government flew Hinkle to Pasadena, California where he was made a Knight of the *Legion d'Honneur* at the home of French consul general, Baron Louis de Cabrol. Cabrol pinned the decoration on Hinkle on 14 July, Bastille Day, France's Independence Day.

Hinkle always recalled with fondness his days in service. "Naturally I am proud of my association with the Lafayette Escadrille. My days with the squadron were the finest of my life, and I spent them with some of the greatest men I have ever known."

Hinkle died at age 90 on 20 January 1967 in a Truth or Consequences nursing home of complications brought on by emphysema. He had willed his body to the University of New Mexico medical school at Albuquerque.

THOMAS HITCHCOCK

Thomas Hitchcock, Jr. was born in Aiken, South Carolina on 11 February 1900, the son of Thomas Hitchcock, a land developer and polo player, and Loulie (Eustis) Hitchcock. He attended the Fay School, Southboro, Massachusetts, and the St. Paul School, Concord, New Hampshire.

Hitchcock left the St. Paul School in the spring of 1917 to join the Lafayette Flying Corps. He sailed on the *Touraine* and arrived in France on 10 June 1917. He enlisted in France's *Service Aeronautique* on 25 June. From 29 June to 8 December 1917 he attended the aviation schools at Avord, Pau, and the *G.D.E.*, receiving his *brevet* on the Caudron on 17 October 1917. *Caporal* Hitchcock was assigned to the Front on 12 December 1917 with *Nieuport 87* at Luneville.

On 6 January 1918 Hitchcock shot down a German two-seater for which he was awarded France's *Croix de Guerre*, with Palm. On 19 January he knocked down a second observation plane.

On another patrol with his American comrade, William Wellman, he flew deep inside enemy lines pursuing a two-seater he and Wellman had shot up over Nancy. They killed the plane's observer and dogged the enemy pilot fifteen miles back to his aerodrome. When he landed his crippled aircraft, they continued to attack the plane, and at near-ground level emptied their machine guns into the enemy hangars and barracks.

On 6 March 1918 during a melee involving 15 enemy aircraft, Hitchcock was shot down by an Albatross fighter and crashed near Chateau Salins in German territory. Shot through the right buttock and thigh during the combat, Hitchcock was quickly captured by enemy ground troops.

Hitchcock was first taken to the prison hospitals at St. Avold before being moved on 1 April to the hospital at Saarbruken. From there he was moved to various prison hospitals in Germany. On 15 May he was moved to prison at Landshut, and on 13 July to the prison at Lechfeld. While being moved to the Rastadt Camp in Baden he escaped by jumping from train near Ulm, and on 28 August he regained his freedom in Switzerland.

Sous-Lieutenant Hitchcock returned to Paris and spent the rest of the war seeking re-assignment to the Front. For his war service, he was awarded a total of three Palms to his *Croix de Guerre*.

Following the November Armistice Hitchcock sailed to the United States and returned to his parent's home at Westbury, New York. In the spring of 1919 he enrolled in Harvard and studied chemical engineering. That same year he also began playing polo again, the sport of his youth.

In October 1920 Hitchcock entered Oxford University. He left that institution in the spring of 1921 to join the American polo team which would compete for the Westchester Cup matches. From that point on Hitchcock dedicated his life to polo, becoming one of the top players in the world while dominating the sport for the next two decades.

Hitchcock became involved in the mining and fuel business with George Moore. On 15 December 1928 he married Peggy Mellon Laughlin, widow of Alexander Laughlin, at the Plaza Hotel in New York City. Two daughters and two twin sons were born to this union.

In 1929 Hitchcock became involved with Moore in the Prestcoke Fuel Company in Chicago. He and Moore also formed the United Electric Company. In 1931 he joined the bond department of the Banker's Trust Company. In 1932 he became an executive for Lehman Brothers. He later formed an aviation company.

In January 1942, one month after the U.S. entry into World War Two, Hitchcock was commissioned a Major in U.S. Air Intelligence. After serving briefly with the Air Ferry Command in Rio de Janeiro, he was assigned in April of 1942 to the American Embassy in London as liaison officer between the American Air Force and British Fighter Command. At Duxford, England he became involved in the development program for the P-51 Mustang fighter. From November 1942 to February 1943 Lieutenant-Colonel Hitchcock had charge of seeing that its production order was carried out.

Hitchcock was then assigned to command the 408th Fighter Group forming in Abilene, Texas. He arrived there Christmas Eve 1943. His group was to fly P-51 Mustang fighters but it was disbanded in February 1944, and Hitchcock was assigned to research and development work on the Mustang.

Colonel Hitchcock was ordered back to England to become deputy chief of staff of the 9th Air Support Command, in charge of tactical research and development. He was killed on 12 April 1944 at age 44 while test-piloting a Mustang fighter near Salisbury, Southwest of London. He was posthumously awarded the U.S. Legion of Merit, the Bronze Star, and the Distinguished Flying Cross.

WARREN HOBBS

Warren Tucker Hobbs was born in Worcester, Massachusetts on 3 November 1895, the son of Wilbur W. Hobbs and Marina (Tucker) Hobbs. He was educated at the Worcester Classical High School, and at the Worcester Academy, Class of 1915. While a schoolboy there, he excelled at track and was considered one of the greatest hurdlers and high jumpers in prep school.

Warren Hobbs entered Dartmouth College, Class of 1919, where he captained the track team while still a sophomore. Called the "one man track team", he set a world's indoor hurdling record in competition with the best college runners.

On 5 May 1917 Hobbs left school to join the Dartmouth unit of the American Field Service. He was attached to TMU 526 until 8 June 1917. He enlisted in France's *Service Aeronautique* on 10 June 1917. From 2 August to 8 December 1917 he attended the aviation schools at Avord, Tours, Pau, and the *G.D.E.*, receiving his *brevet* on the Caudron on 29 September 1917.

From 11 December 1917 to 15 January 1918 *Caporal* Hobbs was attached at the Front to Escadrille *N.153*. From 15 January 1918 to 16 March 1918 he flew with Escadrille *N.158*.

Hobbs entered U.S. Aviation and was commissioned a First Lieutenant on 3 March 1918. On 31 May he was assigned to the 103d Aero Squadron, USAS. On 30 January 1918, while in flight, he ran out of fuel and was forced to land in rough terrain. His fighter flipped over, and he was badly injured. Shortly after leaving hospital he was involved in another fall but escaped serious injury.

On 26 June 1918 Lieutenant Hobbs was patrolling southeast of Ypres. A misfiring engine forced him to cross low over German territory. Shortly

thereafter his SPAD fighter was struck by antiaircraft fire, sending it groundward in flames, instantly killing Hobbs. He was 22 years of age at the time of his death. He was buried with full military honors in the British Military Cemetery, Poperinghe, West Flanders, Belgium. In 1928 his remains were moved to the Lafayette Flying Corps Monument, St. Cloud.

ROBERT HOEBER

Robert Bentley Hoeber was born in Nutley, New Jersey on 8 September 1892, the son of Arthur Hoeber, a noted artist, author, and art critic, and Mary Gray (Wildman) Hoeber. Bentley Hoeber attended the Stevens Institute at Hoboken, New Jersey but left there before completing his degree. He was working as an engineer when he left for France on a bogus journalist passport to join the French air service.

Hoeber enlisted in France's *Service Aeronautique* on 10 July 1917. From 10 July to 11 December 1917 he attended the aviation schools at Avord, Tours, Pau, and the *G.D.E.*, receiving his *brevet* on the Caudron on 20 October 1917.

On 19 December 1917 *Caporal* Hoeber was assigned to the Front to *Escadrille Spad 103* Rene Fonk's squadron. He flew with that squadron until the 11 November 1918 Armistice, and was credited with the destruction of one enemy aircraft. *Sergent* Hoeber remained in the service of France until February of 1919 and was awarded that country's *Croix de Guerre*, with Palm.

Following his World War service, Hoeber returned to the United States where he became a representative of a French Aviation Company. During Rene Fonk's 1926 tour of the United States, Hoeber served as his old squadron commander's guide and interpreter. In 1925 Hoeber married Carter Mullikin Lockwood. From this union, a son, Robert, Jr was born. The Hoeber's later divorced. Hoeber eventually left the aviation industry and became involved in the brokerage and investment business with Johnston, Lemon & Co. In 1938 he married Ida Lutz Clagett. From this union a daughter, Gray, and a son, Charles, were born.

During World War Two, Hoeber entered the U.S. Army Air Corps as a Major on 20 May 1942 at Washington, DC. He served with Headquarters, Army Air Force, Counter Intelligence, at Bolling Field. He was honorably discharged on 5 April 1944 at Columbia, South Carolina.

Following World War Two, Hoeber worked as a landscaping engineer. He died of a brain abscess on 14 January 1956 at the VA Medical Center in Washington, DC at age 63. He is buried at the Washington National Cemetery, Arlington, Virginia.

MILTON HOLDEN

Milton Whitely Harrison Holden was born in Camden, New Jersey on 19 February 1890, the son of Albert W. and Martha (Virdon) Holden. He attended Drexel Institute of Technology, Philadelphia, Pennsylvania, and was employed as a bond salesman when he quit his job and sailed for France to enter the European War.

On 2 June 1917 Holden joined the American Field Service. On 25 June 1917 Holden enlisted in France's *Service Aeronautique*. In January 1918 he was released because of injuries received in a flying accident. Holden later became an aid to General Pershing and served as a military aid to the American delegation at the Versailles peace conference.

Following World War One Holden became a stock broker in Philadelphia. On 2 December 1926 Holden married 24-year-old Josephine (Fifi) Widener Leidy in Philadelphia. The recently divorced Leidy, the daughter of Joseph Widener, sportsman and art collector, was the heiress to the Widener fortune. The Holden's divorced in 1932. Holden eventually became a partner in the Wall Street stock brokerage firm of Bull, Holden, and Company. Holden died in West Palm Beach, Florida on 29 May 1974 at age 84.

DABNEY HORTON

Charles Dabney Horton was born in St. Louis, Missouri on 3 June 1891, the son of Charles Dabney Horton and Charlotte Lockwood (Buel) Horton. "Opie" Horton was educated at St. John's School, Ossining, New York; at Blair Academy, Blairstown, NJ; and at Dartmouth College, A.B. 1915. While at Dartmouth, he was a member of the ski team and won several competitive awards.

Following his graduation from Dartmouth, Horton worked for a New York newspaper. He was fired a few months later and sailed for Europe where he became a member of the Norwegian ski team. Desiring to enter the European War, Horton joined the Norton-Harjes Ambulance Service as a driver.

Horton enlisted in France's *Service Aeronautique* in Paris on 16 August 1916. From 16 August 1916 to 10 July 1917 he attended the aviation schools at Buc, Juvisy, Avord, Pau, and the *G.D.E.*, receiving his *brevet* on the Caudron on 17 March 1917.

On 13 July 1917 *Caporal* Horton was assigned to the Front to *Escadrille C. 17*, serving as a bombing and observation pilot with that squadron until 5 January 1918. From 5 January to 18 February 1918 Horton was assigned to *Escadrille Sop. 255*. On 15 September 1918 *Sergent* Horton was assigned to *Escadrille Spad 75*, flying with that squadron until the 11 November Armistice. While a pilot, he took part in engagements in Flanders in 1917; at Fismes, Chemin des Dames, and St. Quentin in 1918.

Sergent Horton was awarded France's *Croix de Guerre*, with Star for his World War service which consisted of flying virtually every kind of combat mission: photography, reconnaissance, artillery *reglage*, trench strafing. He was discharged from the French Army on 20 January 1919 in Paris, France.

Sergent Horton had married Helen Wheelock Hubbard in Paris on 7 January 1917. From this union a son and a daughter were born.

Following World War One Horton wrote sportsman stories for pulp magazines published by Street and Smith publications. He also worked as a college instructor, and an advertizing copy writer. As a freelance writer, he published numerous stories and articles.

At the outbreak of World War Two Horton became a machinist and worked for Republic Aviation Corporation in their air frame plant. Following World War Two he settled in Northport, New York and continued to work as a machinist until his retirement in 1957. Mrs. Horton died in 1964. Dabney Horton died on 21 March 1968 at Huntington Hospital, Long Island, New York at age 77.

Sergent Dabney Horton and Helen Hubbard Horton shortly after their marriage in Paris in January 1917.

Ronald Hoskier

ESCADRILLE LAFAYETTE

Ronald Wood Hoskier, the 19th volunteer of the *Escadrille Lafayette*, was born on 21 March 1896 in South Orange, New Jersey, the son of Herman Charles Hoskier, a London-born financier and author, and Harriet Amelia (Wood) Hoskier of South Orange, New Jersey.

Hoskier attended St. George's School, Newport, Rhode Island, where he was remembered by his classmates as having a "retiring, reserved disposition." His interest were literary, and he frequently contributed prose and poetry to the school paper, eventually becoming its editor-in-chief.

Hoskier vacationed in Europe in the summer of 1914. After climbing a peak in the Swiss Alps, he was preparing for his return to the United States when war broke out in France. Ronald entered Harvard University that fall, a member of the Class of 1918.

In early 1915 Ronald's father, Herman Hoskier, sailed for France with Mrs. Hoskier to work with Richard Norton's Ambulance Service while Ronald had remained in the United States to complete his freshman year at Harvard. At the end of the term Ronald left for England and began studying aviation at Hendon. He returned to Harvard for the fall semester but left college in February of 1916 and sailed for France with the intention of joining French Aviation.

Hoskier enlisted in the Norton-Harjes Ambulance Corps and drove an ambulance until his enlistment in the Foreign Legion on 5 April 1916. He was sworn in with fellow American Edwin Parsons but the two were mistakenly sent to Pau instead of the Buc aviation school which caused a delay in their training. Together they reached the Buc aviation school around 15 May and since both had been pre-war flyers, they had little trouble mastering the Penguin course before progressing onward through the schools at Buc, Avord, Cazeaux, and Pau. Hoskier ranked at the head of his class which included fellow Americans Stephen Bigelow, Edmond Genet, and Edwin Parsons. He was greatly admired by his comrades, and inspired one of his friends to write of him:

Sergent Ronald Hoskier before his Nieuport scout at Ravenel, February 1917.

"He has made a deep impression upon me, and upon the other men as well. His fine, manly face is a clear index of his character, and his eyes are so fearless and honest that one knows with absolute certainty that he is a man to be trusted in any sort of emergency."

Caporal Hoskier received his *brevet militaire* on the Bleriot on 13 August 1916. On 28 November 1917

Sergent Hoskier's parents also served in France while their son was in the Lafayette Escadrille: Herman Hoskier with Richard Norton's Ambulance Service, and Harriet Hoskier in Paris as a Red Cross volunteer.

he left Pau for the G.D.E. pilot pool to await assignment to the Front.

Hoskier joined the *Escadrille Lafayette* at Cachy on the Somme front on 11 December 1916. The winter was bitter and the temperature so cold that at high altitudes the oil froze in the pilots' engines. This happened to Edmond Genet on a 2 February 1917 patrol with Hoskier over Mondidier, shortly after the Squadron had moved to Saint-Juste to support the planned Nivelle offensive.

Ronald, on his next leave to Paris, with funds borrowed from his father, purchased a horizon-blue greatcoat with a huge fur collar to wear over his excellently-tailored uniforms. On his head he wore the French beret, one of only three *Escadrille* pilots to adopt this head covering.

On 15 February Hoskier, Genet and Parsons were flying the morning patrol over Roye when they spotted two German "Valfiches" (Roland Walfisches) at low altitude. They immediately attacked and managed to drive the enemy bi-places groundward before they were driven off by furious enemy antiaircraft fire.

On 17 March 1917 Hoskier was again on patrol with Genet when the two hit severe weather and expended their petrol. They were forced to make an emergency landing in a little town sixteen kilometers north of Beauvais. Hoskier telegraphed for gas, but the message was improperly relayed, and as he and Genet waited for help, both were reported as missing in action. Eventually petrol arrived by truck and the two Americans flew safely back to Saint-Juste.

As *Caporal* Hoskier and the other *Lafayette* pilots continued to fly reconnaissance patrols over the German lines, throughout March they were constantly hindered by thick mist, heavy gales, and snowstorms. Heavy shelling from enemy antiaircraft batteries added to the pilots' misery.

Tragedy struck the Squadron on 19 March. James McConnell, one of its original members, was killed while combatting two German two-seaters near Jussy.

On 21 March Hoskier's parents visited him at the Front. By then his father had command of Section 7 of the Norton-Harjes Ambulance service, and his mother was working as an auxiliary nurse in a French hospital. That night they dined together at the Squadron's mess, and the next morning Hoskier arose early for the first patrol. Ronald left the airstrip into a threatening sky and attempted to fly the mission. But a snowstorm soon forced his fighter down at Mondidier.

On 8 April 1917 the Squadron arrived at Ham in the Somme sector. Shortly after the transfer Ronald Hoskier was promoted to *Sergent*. At that time the Squadron was flying a variety of aircraft over the Front. One that Hoskier chose when a Nieuport or a SPAD was unavailable was a Morane Parasol two-seater which had seen service as a training and gunnery ship prior to its assignment to the Front. Hoskier flew the biplace more than the other pilots, and by 13 April when he again took the Morane to the air, it had become known as "Hoskier's plane."

Hoskier found the Morane an easy plane to fly despite the opinion of other pilots who feared its inher-

Caporal Hoskier at the Buc school of Aviation, 6 June 1916. (L.r.) Hoskier, Rocle, Frey, Dowd, Bigelow, Lt. Behiere-Chef Pilote, Worthington, moniteur, Terres, Chatkoff. (Front row) Genet, Parsons, Boal, Littauer.

ent structural weaknesses and felt it inadequate for use as a scout plane. Whenever Ronald took it over the Front, *Caporal* Jean Dressy, Lieutenant deLaage's orderly, accompanied him as gunner.

Hoskier's friend, Edmond Genet was killed in combat on 16 April north of Montescourt. The grieving Hoskier was granted a short leave to Paris to visit his mother and attended the Harvard dinner while there. He and Genet had constantly been in the air together, and he keenly felt his loss. Paul Ayres Rockwell, brother of Kiffin Rockwell, and the official historian of the Squadron, wrote this of Hoskier at that frantic time:

"From the day of his departure for the Front, every time I have met one of the pilots or have received news from the Escadrille Lafayette, Hoskier has been mentioned as one of the most active members of the unit. Since his arrival, the Squadron has not made a single sortie in which he has not taken part. He had innumerable combats and I have heard so much of these that I am always afraid of receiving news that he has been killed."

On 23 April Hoskier requested permission to fly the Morane two-seater on its last mission over the Front. It was soon to be withdrawn from the Squadron and replaced by a new SPAD S.VII fighter. *Caporal* Dressy also asked to go along on the Morane's last flight. Hoskier had often expressed to the other pilots that he felt a two-seater could do more damage in a dogfight than a single-seater fighter, and he was anxious to prove his theory correct and "really do some fighting." Squadron Commander Thenault reluctantly granted the young American permission.

Sergent Hoskier and *Caporal* Dressy took the Morane to the air that same day above Saint Quentin as part of the Squadron's regular reconnaissance mission. Harold Willis was the patrol leader and was accompanied by Willis Haviland and William Thaw. Hoskier, in the slower Morane, took the tail-end of the formation.

As Ronald entered thick cloud covering his puttering Morane with its 70 m.p.h. maximum limit soon became separated from the faster Nieuports of the patrol. When he exited into clear sky, he evidently spotted what he thought was a lone German fighter below him. He nosed his Morane down. By the time he realized his error, the trap had been sprung. Three German Albatros D-III fighters swooped down on Hoskier and opened fire.

As Hoskier circled, he and Dressy fired their machine guns. They desperately looked for their friends but were unable to spot the other three pilots who by

Paul Pavelka, Ronald Hoskier, and James McConnell at Cachy, December, 1916.

The smashed wreckage of Sergent Ronald Hoskier's Morane Parasol two-seater shortly after having been shot down by three German Albatros D-III fighters. Both Hoskier and his French gunner, Caporal Jean Dressy, lost their lives in the combat which saw their plane hit the ground full-motor just inside allied lines.

now had engaged other enemy fighters. Ronald met his enemy head on. In the ensuing violent battle which lasted fifteen minutes, one of the enemy fighters was struck and spun out of the melee.

Hoskier expended all the ammunition for his forward machine gun while Dressy fired all but thirteen rounds from the drums of his twin Lewis guns. Suddenly, Hoskier was struck in the head by an enemy round, and from an altitude of 8,000 feet the two-seater fell into a spin. Its wings folded up and the Morane dropped straight down, its motor roaring. It struck the ground and crumpled up just inside the forward French lines.

Back at the American aerodrome a French mechanic, upon learning of Hoskier's fate, had rushed outside to announce his death to a half-dozen of Hoskier's comrades standing near the airfield. The Americans quickly broke and ran for Hoskier's locker, remembering that his father had kept him well-supplied with expensive Corona cigars.

All day long the enemy guns bombarded the spot where the Morane had fallen. At midnight Dudley Hill, Edwin Parsons, and "Doc" Rockwell took an ambulance to the death scene and recovered the two smashed bodies. They then transported them to the little mortuary at Ham where they were placed in simple pine coffins.

Two days later Hoskier's parents arrived at the Compiegne railroad station to attend the double funeral for their son and *Caporal* Dressy. Edwin Parsons and Lieutenant DeLaage met them there and accompanied them back to the Squadron at Ham.

Prior to the funeral services, Mr. Hoskier had asked Parsons to direct him to the chapel where his son's body was resting. The two entered the cold room together and Parsons backed away in deference to the grieving father. As Mr. Hoskier silently approached the two trestles supporting his son's coffin, *Sergent* Parsons read the "infinite sadness" in the father's gaze. Then Mr. Hoskier straightened up and backed away, and they turned to leave the chapel.

A barely audible sound caught Parson's ear. He looked back at the coffin and felt a "cold shock." Blood was dripping in a steady stream onto the cement floor from the corner of the coffin that held Hoskier's "cold, stiffened, battered body" now three days dead. Parsons read this as "a sign from the dead to the living loved ones."

The funeral of Sergent Hoskier at Ham, 25 April 1917. Mr. H.C. Hoskier, the pilot's father, stands behind the casket with Capitaine Thenault and Lieutenant Thaw.

While Parsons had been escorting Mr. Hoskier, other American pilots had hovered around the grieving mother. One said to her:

"Mrs. Hoskier, almost all men have two sides, one that is known to their mothers, and one that is known to their men friends. Ronald had but one side, and that was the one known to his mother."

Sergent Ronald Hoskier and *Caporal* Dressy were buried at Ham near Edmond Genet in a full military ceremony. The Squadron had lost one of its best and most highly-regarded pilots. His honesty, fearlessness, and good character had been admired by all who had known him. Paul Ayres Rockwell said this of him:

"Hoskier was an outstanding pilot, a gallant young man, and equally important, a gentleman in the true sense of the word."

Sergent Hoskier was posthumously awarded France's *Croix de Guerre*, with Palm on 1 May 1917 for his 23 April action. Following the War his parents financed the reconstruction of the church at D'Etalon as a memorial to their dead son. The village square was renamed the Ronald Hoskier square in his honor. Today, Ronald's remains rest in the Lafayette Flying Corps Memorial near Paris.

EDWIN HOUGH

Edwin A. Hough was born in Brooklyn, New York, on 2 June 1885, the son of Alonzo D. Hough. Hough attended Long Island schools and was working as a chauffeur and and mechanic when he enlisted in the American Volunteer Motor Ambulance Corps in New York City.

On 23 December 1916 Hough sailed for France aboard the *Touraine*. While serving in France with the Ambulance Corps Hough was awarded the French *Croix de Guerre*.

On 3 June 1917 Hough enlisted in France's *Service Aeronautique*. He was released on 1 August 1917 because of inaptitude.

JOHN HUFFER

John William Francis Mary Huffer was born in Paris, France on 15 June 1895, the son of William Huffer, a tobacco importer who held the monopoly for Virginia tobacco to France and Italy, and Louise Sylvia (Gignoux) Huffer. John Huffer attended private schools and the Gerson school of Paris.

In late 1914 Huffer joined the American Ambulance Field Service in France and served with SSU 2 for 9 months. He enlisted in the French Foreign Legion on 28 September 1915 in order to enter the *Service Aeronautique*. From that date until 1 April 1916 he trained at the aviation schools at Avord, Cazeaux, and Pau, receiving his brevet (no. 2237) on 1 January 1916.

On 1 April 1916 *Caporal* Huffer was assigned to the Front with *Escadrille N. 95* and flew with that squadron until 16 June 1916 when he was assigned to *Escadrille N. 62*. Huffer remained with that squadron until 15 March 1917.

Huffer had an unusual experience when he was shot down early in the war. His plane was brought down while flying over enemy lines. He was captured and taken to a German airfield where he was invited for dinner by the German commanding officer. Next morning he was escorted to the front lines and was turned over to French soldiers. That same day he returned to his squadron's airfield.

From 13 July to 14 September 1917 *Sergent* Huffer flew Farman observation planes with *Escadrille F. 36*. *Sous-Lieutenant* Huffer returned to *SPAD 62* on 4 October and served with that Squadron until 18 February 1918. On 7 November 1917 he was commissioned a Major in the USAS. On 17 May 1918 he took command of the 94th Aero Squadron, USAS. While with that squadron, he was instrumental in designing the famous hat-in-the-ring insignia for the 94th aircraft. He and his pilots were discussing an appropriate insignia when Huffer noted the French employed a circle for their insignia. He suggested the 94th use a red outer circle surrounding Uncle Sam's top hat. This would show Uncle Sam tossing his hat into the war.

After two months of active service at the Front, Major Huffer was transferred to the 1st Air Depot at

Columbey-les-Belles as Assistant Operations Officer. On 25 July 1918 he returned to the Front to command the 93rd Aero Squadron, USAS and remained in that position until the 11 November 1918 Armistice. Huffer was officially credited with the destruction of four enemy aircraft (French records indicate three.) He also had three unconfirmed victories. For his splendid war service he was awarded France's *Medaille Militaire* and *Croix de Guerre*, with three Palms and two Stars.

Following his discharge from the U.S. Air Service, in 1919 Major Huffer sailed to the United States. Shortly thereafter he returned to France where he worked for a coal company in Mulhouse until 1936. That same year he married Renee Pophillat, and from this union a son, William, and daughter, Elizabeth, were born. In 1936 Huffer also founded a radio company which he directed until 1939.

From 1939–1940 he served in the French Air Corps at the rank of Major and was stationed at the Etampes Air Field. Prior to the German occupation of France, he returned to the United States and rejoined the U.S. Army Air Force. From 1943–1946 he served as liaison officer between the French and U.S. Armed Forces. His World War Two decorations include France's *Legion of Honor,* and *Croix de Guerre*, and the U.S. Bronze Star.

From 1947–1949 Lieutenant-Colonel Huffer was assigned to the Pentagon, Washington, D.C. From 1949–1950 he was assigned to the U.S. embassy in London. From 1949–1953 Colonel Huffer was stationed at the U.S. Air Force European Headquarters at Wiesbaden, Germany. In 1954 he joined the U.S. State Department in Paris where he remained until his retirement in 1955.

The Huffers divorced in 1960, and he later married Margaret Castaing, the famous sculptress. They moved to the United States, to Florida. Following her death he returned to Paris. He died there at the American Hospital on 23 April 1986 at age 91. He is buried in the Huffer family tomb twenty miles north of Paris.

DANIEL HUGER

Daniel Elliott Huger was born in New York City, New York on 24 July 1897 the son of Lynch Prioleau Huger, a civil engineer, and Mary (Bailey) Huger. He attended the Kent school in Connecticut through the 11th grade.

Huger joined France's *Service Aeronautique* on 26 January 1917. From 3 February to August 1917 he attended the aviation schools at Avord, Pau, and the *G.D.E.*, receiving his *brevet* on the Caudron on 26 June 1917. While awaiting assignment to the Front, Huger came down with a severe case of bronchitis and was sent to the south of France for a long convalescence.

Upon regaining his strength, Huger secured his release from French aviation and on 13 March 1918 he enrolled in the Naval Reserve Flying Corps as an Ensign in Paris, France. He trained on flying boats at Moutchic and trained further at Avord. On 19 July he became Naval Aviator No. 1007 (HTA) at Lake Bolsena, Italy where he was taking further training on Italian aircraft.

Huger was badly injured in a flying accident at Lake Bolsena and spent three months in the U.S. Naval Hospital at Genoa. Upon his recovery, he was assigned to duty at the U.S. Naval Air Station at Porto Corsini, Italy. Not long after his arrival there, the Armistice was signed, ending hostilities on the Italian front. Following the Armistice Huger attended the Sorbonne in Paris for one year and the University of Paris for one year. He returned to the United States in early 1921. He was discharged from the Navy as an Ensign on 30 September 1921, but was later awarded the rank of Lieutenant-Commander.

When he returned to the U.S. Huger worked as a stock broker in New York City. He married Elizabeth Kress, and from this union a son and daughter were born. This marriage ended in divorce after a few years. Huger later attended Fordham Law School for one year. He married Margaret Crosby in 1945, and this marriage ended in divorce in 1951. In 1953 he married Evelyn Fortune, at which time he retired to Beverly, Massachusetts. Huger died in a Beverly hospital of cancer on 10 October 1967 at age 70.

Earl Hughes

Earl Wayland Hughes was born in Capmish, Michigan on 31 December 1894, the son of William Hughes, a custodian and sports promoter, and Myrtle (Bartlett) Hughes. Hughes attended Detroit public schools. Following graduation from high school, Hughes briefly attended the Michigan State College of Agriculture at East Lansing, Michigan. He left that institution to work as a motor mechanic and was employed in that capacity in Detroit when he decided to enter the European War.

Hughes enlisted in the American Ambulance Service in the United States. He sailed for France and served briefly with Section 11 before enlisting in the *Service Aeronautique* on 13 June 1917. From 5 August to 10 January 1918 Hughes attended the aviation schools at Avord, Tours, and the *G.D.E.* He received his *brevet* on the Caudron on 22 October 1917.

On 14 January 1918 *Caporal* Hughes was assigned at the Front to *Escadrille Br. 66*. He flew numerous bombing runs with that Squadron in the region of Noyon and Montdidier until 15 June 1918, having fought his way home many times through a curtain of Albatros fighters.

Hughes was grounded on 15 June after bursting his eardrums on a high-altitude flight. Not wishing to give up aviation, on 6 October *Sergent* Hughes asked for and was transferred to *Escadrille F. 110*. Thus, he became the only Lafayette pilot to fly the two-motor Farman, and to do night bombing work.

Sergent Hughes flew numerous raids with his Farman squadron until the 11 November 1918 Armistice. For his diligent war service he was awarded France's *Croix de Guerre*, with Star.

Following World War One Hughes worked as a security guard in Detroit, Michigan, for the family of the

Alwrey Bakery Co. He later worked as a guard at the Jackson, Michigan prison, and worked security for the Timkin Axle Company. In his off-time Hughes played catcher on several Michigan amateur baseball teams. Hughes later married Grace Ward. He retired to Baldwin, Michigan. He passed away from an intestinal hemorrhage at the Munson Hospital in Traverse City, Michigan on 29 October 1967 at age 72. He is buried at Pleasant Plains Cemetery in Baldwin, Michigan.

MARK HULL

Mark Leslie Hull was born in Austin, Illinois, on 30 June 1892, the son of Harry S. Hull. He attended schools in Cleveland, Ohio, before entering Princeton University, Princeton, New Jersey. He was working as a manager in New York State when he enlisted in the Lafayette Flying Corps in New York City on 22 May 1917.

Hull sailed for France on the *Rochambeau* on 26 May 1917. On 15 June 1917 Hull enlisted in France's *Service Aeronautique*. He was released on 20 September 1917 because of inaptitude.

SERENO JACOB

Sereno Thorp Jacob was born on 5 August 1895 in Brooklyn, New York, the son of George Jacob, and Alice T. (Allen) Jacob. Jacob attended New York schools. On 25 October 1916 he enlisted in the Norton-Haries Ambulance Service and served in France with Section 5 until April 1917. While still in the Ambulance Service, on 20 March 1917 Jacob enlisted in France's *Service Aeronautique*. From 19 April to 24 December 1917 he attended the aviation schools at Avord, Tours, Pau, and the *G.D.E.* On 21 October 1917 he received his *brevet* on Caudron.

On 26 December 1917 *Caporal* Jacob was assigned to the Front to *Escadrille Nieuport 157*. He flew with that squadron until 8 September 1918, the day he was commissioned a First Lieutenant in the U.S. Air Service. Jacob was awarded France's *Croix de Guerre*, with Palm for his destruction of a German kite balloon. While diligently flying three patrols a day, he was credited with bringing down a total of three enemy aircraft.

Following World War One Jacob owned a Cadillac agency in Belgium. He returned to the United States, began barnstorming, then started an airline service which ran from Connecticut to Springfield, Massachusetts. He next worked for Eastern Air Transport Co. as a test pilot. Several years later he became manager of Curtiss's Bridgeport, Connecticut division then became chief pilot for New England and Western Aviation. He also worked as a carpenter, an assessor, and a real estate and insurance broker. He married Marion C. Wakeman. From this union a son, Merrit, was born. Jacob died in Ridgefield, Connecticut of kidney failure on 18 April 1947 at age 51.

CHARLES JOHNSON

ESCADRILLE LAFAYETTE

Charles Chouteau Johnson, the 10th volunteer of the *Escadrille Americaine*, was born in St. Louis, Missouri on 18 September 1889, the son of U.S. Army Captain David Dick Johnson and Anne Victoire (Chouteau) Johnson, a direct descendent of the founder of St. Louis. Captain Johnson had been a graduate of the U.S. Military Academy at West Point, class of 1868. He had received an "appointment at large" from President Lincoln, a personal friend of his father.

Charles, "Chute" Johnson attended public schools in St. Louis before enrolling in the University of Virginia in 1909. From there he received a bachelor of laws degree in 1913. James McConnell, later to become one of the first four original members of the *Escadrille Americaine*, and Andrew Courtney Campbell, also destined to become an *Escadrille* pilot, were classmates of Johnson at Virginia. Following his graduation, Johnson headed for New York City where he and McConnell together briefly involved themselves in business.

Johnson sailed for France not long after war was declared between France and Germany. He enlisted in the American Ambulance Service early in 1915. He joined the French Foreign Legion that summer, following the example of his uncle, Pierre Chouteau who had served in the Legion in the Franco-Prussian War of 1870-71. Johnson did not intend to serve in the trenches, however, but transferred to French Aviation on 2 September 1915.

Johnson attended the aviation school at Pau from 29 September 1915 until 2 January 1916 when he received his *brevet militaire* on the Bleriot with fellow Americans H. Clyde Balsley and Jean Huffer. But as Balsley noted in his diary entries, Johnson had a difficult time in school.

Dec. 6, 1915. "Johnson has had a run of hard luck. He thought he was a bit better than he really was, rose off the ground too high, slipped on a wing and fell to the ground. The plane dug into the ground and turned over on top of Johnson but he was held in by his belt and not hurt. Then the next day just as he was starting

Sergent Charles Chouteau Johnson (right) with Lieutenant William Thaw at Bar-le-Duc.

to land an eighty h.p. passed over his head and blew him down to the ground and broke the machine very badly. For the next day Johnson almost lost his sense of the feel of the apparatus but now he is going just as well as before."

Dec. 16, 1915. "This morning Johnson had more hard luck. He got in a plane, got his glove caught in the notches of the throttle and could not get enough speed to go up high and could not stop. In working to get his glove loose he made a very sharp turn, landing on a road and went into a ditch while going about 60 kilometers an hour. It is a wonder he wasn't killed but he didn't get a scratch. The machine naturally was smashed to bits."

Following a Paris leave, Johnson and Balsley enrolled in the school at Amberieu on 19 January 1916 and there learned to fly the Voisin. Upon completion of their training on 18 February they volunteered for service with the Air Guard of Paris. On the 26th they were assigned to *Escadrille V.97* and arrived at the squadron's field twelve kilometers from Paris and soon began protecting the city from Zeppelin attacks.

While in the vicinity of Paris "Chute" Johnson and Clyde Balsley heard of the possible formation of an "Escadrille Americaine." They immediately put in their applications for Nieuport training in order to qualify for membership in this proposed fighter squadron. They soon learned to their elation that they were intended to be of the first nine members of the American squadron, according to the 14 March 1916 statement issued by Colonel Henri Regnier, director of French military aeronautics.

Johnson and Balsley left *V.97* on 1 April and arrived at the G.D.E. (*Groupe des Division d'Entrainement*) at Plessis-Belleville on 3 April. There they trained on the Nieuport 23 until 25 May. They arrived at Bar-le-Duc in the Verdun sector on the morning of the 27th and officially joined the roster of *N.124* on 29 May 1916.

Johnson and Balsley sat grounded on the Behonne field for lack of fighters. When the planes did arrive, they had no machine guns, so they spent time applying their personal insignia to their Nieuport 11s. Balsley chose white stars for the sides of his fuselage and wheels to represent his home state of Texas, while Johnson selected a white dice showing a single black dot.

By 8 June their machines had been properly armed. On that date they made their first test flight with Norman Prince as leader across the Meuse River toward Verdun. For approximately the next nine days, whenever the inclement weather eased, the two novice fighter pilots learned "the lay of the lines and points of direction."

Then on 18 June tragedy struck Clyde Balsley. He was seriously wounded 3500 kilometers above the forts of Verdun and was shot down while flying fighter protection for "artillery reglage machines."

Johnson keenly felt the absence of his close friend, Balsley. Between missions he and the other American pilots visited Balsley at the French evacuation hospital ten miles from the Verdun front. On 23 June Victor Chapman had a bag of oranges in his Nieuport which he had planned to deliver to Balsley following an afternoon patrol. Instead he became the Squadron's first fatality when he was shot down while engaging several German fighters above enemy lines at Haumont.

Sergent Johnson's Nieuport 11 fighter armed with LePrieur rockets used to attack German observation balloons. Johnson's personal "white die" insignia is visible on the fuselage and wheel of his scout. Bar-le-Duc, July 1916.

Throughout the summer of 1916 Johnson and the other *N.124* pilots often flew two patrols a day along the Verdun front. During the months of July and August their field at Behonne and the nearby town of Bar-le-Duc were struck by large groups of German bombers.

On 14 September the Squadron was ordered to Luxeuil to fly cover for combined allied bombing raids into Germany. On the 23rd Kiffin Rockwell was killed near Rodern in combat with a German two-seater. On 12 October a massive raid was carried out against the Mauser works at Oberndorf, Germany. Raoul Lufbery, Didier Masson, and Norman Prince were part of the fighter escort. While attempting a landing in the twilight over the Corcieux airfield, Prince had been seriously injured when he had crashed his Nieuport fighter. He slipped into a coma and had died on 15 October. Of the four original members of *N.124* now only James McConnell remained.

On 18 October *Sergent* Johnson flew with the Squadron as it transferred to Cachy, near Ameins, to support the Somme Offensive. By the time it had reached Cachy, the Squadron had scored seventeen confirmed victories while fighting 157 engagements. Once there, it became attached to French *Groupe de Combat* 13 and served as part of that group until the end of its service.

Johnson and the other American pilots found the Cachy barracks barely habitable. Even the most basic necessities were lacking, such as blankets and cooking facilities. Regular patrols continued only into mid-November when fog shut out the sky and winter rain whipped the airfield to knee-deep mud accessible only by duckboards.

On 16 November 1916 the *Escadrille Americaine* became the *Escadrille de Volontaires* On 6 December it officially became the *Escadrille Lafayette*. By then the Cachy field had frozen solid into tire-bursting ruts disturbed only by the bombs from occasional German raiders.

On 26 January 1917 the Squadron left the bitter hell of Cachy for the Ravenel airfield near St. Juste-en-Chausse. The weather there was little improved. The Nevelle offensive had been planned for Spring. But because the pilots had been ordered to stay out of sight, February proved a "dud month."

On 19 March 1917 *Sergent* James McConnell, the last of the "first four" was killed in combat in the air near Jussy. On 7 April the Squadron moved forty miles in the direction of the rapidly advancing allied front to the former enemy airfield at Ham. On 16 April the Squadron's youngest pilot, *Caporal* Edmond Genet, was seriously wounded by antiaircraft fire south of St. Quentin. His fighter plowed into the ground five miles within French lines, instantly killing him.

Ronald Hoskier was the next young American to die, killed in combat with his observer while flying a Morane Parasol above St. Quentin on 23 April. Three days later, on 26 April, *Sergent* "Chute" Johnson gained his first and only confirmed victory east of St. Quentin while on patrol with Lieutenant William Thaw and *Sergent* Willis Haviland.

The early spring sky had been massed with clouds as William Thaw had led a sunset patrol above the lines in the war-torn sector of the Somme. There, the three Americans had run head-on into an Albatros formation.

Both groups had immediately climbed for advantage. The American formation became dispersed in the clouds, and as the pilots broke clear into brief pockets of blue, sudden violent combats ensued as foe surprised foe.

Johnson, in this hide-and-seek fashion, had just roared out of a cloud into a patch of blue when an Albatros fighter had passed right before his propeller. He poured machine gun fire into the enemy bus before it disappeared into a cloud bank, tainting its purity with black oily smoke. Willis Haviland also met and shot down a German bi-place in much the same fashion. By the time the pilots had returned to their aerodrome, their victories had already been unofficially confirmed.

On 15 May 1917 *Sergent* Johnson was awarded France's *Croix de Guerre*, with Palm for his 26 April victory in which he had "attacked and downed an enemy plane." On 23 May the Squadron's second in command, French Lieutenant deLaage de Meux became the next fatality when the engine of his new SPAD S.VII fighter had quit at an altitude of 250 feet and he had been killed in the subsequent fall.

Johnson was granted a short Paris leave as a reward for his 26 April victory. As was his custom, he spent much of it with his aviator comrades at Chatham's, a favorite Paris "watering hole" frequented by the Lafayette Flying Corps pilots.

Johnson's presence at Chatham's was always considered a treat by the *eleves-pilotes* who were lucky enough to be there when he walked through the door. They would quickly huddle around this experienced warrior and intimate friend of such legends as Thaw, Prince, and Lufbery, anxious to draw from this "minor deity" his related wealth of experience with the *Lafayette* Squadron.

On 3 June the Squadron again relocated, to the spacious airfield at Chaudun. By now the Squadron designation had become *SPA.124* since most of the pilots in the *Escadrille* had been issued the new SPAD S.VII fighter.

On 17 July the Squadron was ordered to the airfield at St. Pol-sur-Mer, two miles south of Dunkirk on the edge of the North Sea. But heavy rains washed away any hopes of the Squadron participating in the 31 July Ypres Offensive. Instead, the period there served as an interlude for the pilots as they involved themselves in relaxation.

On 11 August 1917 *SPA.124* flew to the village of Senard on the edge of the Forest of Argonne. There the pilots were to partake in the limited French offensive at Verdun. The weather proved perfect for flying, and for most of a month *Sergent* Johnson and his comrades flew three patrols a day.

On 18 September the Squadron returned to Chaudun to support the 3rd Aisne Offensive planned for 10 October. On 24 September Douglas MacMonagle was killed on a morning patrol near Triaucourt when two German fighters attached themselves to his tail and put a bullet through his head. Andrew Courtney Campbell would be the next pilot to die, on 1 October, shot down and killed north of Soissons.

The Aisne offensive, and the pilots' participation in it continued throughout October. *Adjudant* Chouteau

Escadrille Americaine at Bar-le-Duc, July 1916. Left to right: Lt. deLaage, Charles Johnson, Laurence Rumsey, James McConnell, William Thaw, Raoul Lufbery, Kiffin Rockwell, Didier Masson, Norman Prince, and Bert Hall.

Johnson began to feel the strains of many months of combat and the deaths of so many of his comrades. One colleague wrote of him:

"With the exception of William Thaw, he has probably attended more funeral ceremonies for aviators than any other man in the Lafayette Corps. A very little of that sort of duty is far more than enough, and but few men could have performed it with Chouteau's stoicism."

In the fourteen months since he had joined the Squadron, he had attended the funerals of Victor Chapman, Kiffin Rockwell, Norman Prince, James McConnell, Edmond Genet, Ronald Hoskier, Lieutenant deLaage deMeux, Douglas MacMonagle, and Andrew Courtney Campbell. Several other of his comrades had been wounded or were missing in action. So when the post of flying instructor was offered to him at the Second American Aviation Instruction Center at Tours, Johnson accepted it. He was commissioned a First Lieutenant in the United States Army then was promoted to Captain in early summer of 1918. Shortly thereafter he was sent to the United States for duty as an instructor.

Following his war service, Johnson entered the stock brokerage business in San Francsico. It was there that he married Ursula Hooper, daughter of Mr. & Mrs. J.G. Hooper, on 3 July 1920. Johnson's only child, Barbara Ann, was born on 8 December 1922. Unable to control his drinking, the Johnsons soon divorced.

In March of 1928, Clyde Balsley, who was then living in San Francisco, wrote the following to a Lafayette Colleague in New York:

"Chute lives at the University Club in San Francisco. He is selling bonds, I believe. I haven't seen him in several years but I am continually meeting friends who tell of having been on a party with him, and we exchange greetings in that way."

Eventually Johnson moved to New York and became a Wall Street broker. On 7 November 1932 he married Margaret D.B. Dearborn in New York City.

Johnson pursued the profession of stock broker until he was stricken with throat cancer in the late 1930s. Sometime prior to his illness Margaret Johnson passed away. Johnson fought his illness with great courage and with the same stoicism he had displayed as a combat pilot. He spent much of the last year of his life at Johns Hopkins Hospital in Baltimore, Maryland. Debt-ridden, he died there at age 50 on 10 October 1939.

Shortly after his death, Carl Dolan wrote the following to Clyde Balsley:

"Chouteau was a swell person and a fighter, and he fought this thing for two years before it licked him, and I just want to say that it was the memory of his wife that kept him fighting."

Dolan, Balsley, Dudley Hill, and other former *Escadrille Lafayette* pilots had created a fund to support Johnson in his illness. Upon his death, they saw to his burial in New York and paid his funeral expenses.

Paul Ayres Rockwell, brother of Kiffin Rockwell, and official historian of the *Lafayette* Squadron, later wrote this of his friend, Johnson:

"Charles Chouteau Johnson was a splendid man and devoted to the Allied Cause, to which he rendered real service. All of us who knew him well liked him and respected him."

HARRY JOHNSON

Harry Firmstone Whelen Johnson was born in Embreeville, Tennessee on 24 April 1896, the son of Guy Roche Johnson, a steel blast-furnace design engineer, and Edith (Whelen) Johnson. When his family relocated to Bethlehem, Pennsylvania, he attended local schools there. Harry enrolled in Lehigh University where he studied chemical engineering but left school in February of 1917 to join the American Red Cross. He served for two months in France with a Norton-Harjes ambulance section.

On 25 June 1917 Johnson left the ambulance service to join France's *Service Aeronautique*. From 1 August to 10 December 1917 he attended the aviation schools at Avord, Tours, Pau, and the *G.D.E.* He received his *brevet* on the Caudron on 1 October 1917.

On 12 December 1917 *Caporal* Johnson was assigned at the Front to *Escadrille N. 85*. He flew with that squadron until 9 January 1918. On 9 January he was assigned to *Nieuport 98*. On 20 January 1918, while engaged in combat against heavy odds, Johnson was shot in the stomach but managed to land his machine undamaged near a French hospital. For his coolness and courage, *Caporal* Johnson was awarded France's *Medaille Militaire*, and *Croix de Guerre*, with Palm.

On 12 April 1918, while Johnson was convalescing, he was commissioned a First Lieutenant in the U.S. Air Service. That same day he was re-assigned to the Front with *SPA 168*.

Lieutenant Johnson flew combat patrols with that squadron until 21 May 1918. On that date, while on a morning patrol deep into enemy lines over Suippes, Johnson flew alongside patrol leader, Thomas Cassady, and signaled motor trouble. Johnson banked his *SPAD* and volplaned out of sight of his comrades.

French ground troops later reported that around 10:30 a.m. Johnson approached the forward trenches and began frantically waving for the poilu to clear away from his landing path, and even fired his machine gun to get their attention. His landing gear snagged the barbwire and was torn away. His *SPAD* "capoted" and Johnson was thrown out of his cockpit and landed on his neck. Knocked unconscious, his spine fractured, he died a few moments later. He was buried with full military honors at the Mont Phernet cemetery. He was 21 years of age at the time of his death. In 1928 his remains were re-interred at the Lafayette Flying Corps Memorial near Paris.

ARCHIBALD JOHNSTON

Archibald Burtt Johnston was born in Pittsburgh, Pennsylvania on 29 December 1891, the son of Charles Marshall Johnston and Sara Ann (Murphy) Johnston. Johnston attended Pittsburgh public schools and the Shadyside Academy. He enrolled in Cornell University, Class of 1914. He served as art editor for the *Widow* paper. He and future Lafayette Flying Corps comrade Austen Parker, the literary editors for the *Widow*, assembled what was deemed a pornographic issue of the paper and were expelled from school. Johnston was readmitted following a public apology and received his A.B. degree.

In the spring of 1916 Johnston served in France for three months with Section 3 of the American Ambulance Field Service. He enlisted in France's *Service Aeronautique* on 28 July 1916. From 15 August to 24 April he attended the aviation schools at Buc, Juvisy, Avord, Pau, and the *G.D.E.* On 25 January 1917 Johnston received his *brevet* on the Caudron.

On 27 April 1917 *Caporal* Johnston was assigned at the Front to Escadrille SPAD 83, the only American in the squadron. He flew with that squadron until 12 September 1917. Johnston was awarded France's *Croix de Guerre*, with Palm and Star for his outstanding service with SPAD 83, which included many missions deep inside enemy territory disrupting German troop movements.

Upon leaving SPAD 83, *Sergent* Johnston enlisted in the U.S. Air Service. He was ordered back to the United States where he participated in bond drives. On 23 December 1917 he was commissioned a Captain in the U.S. Air Service. He married Margaret Watson Richard on 9 March 1918.

Captain Johnston served as Chief Instructor in charge of the 1st Aerial Combat and Gunnery School at Gerstner Field, Louisiana. From 1918–1919 he was Chief Instructor and Chief of Control Board at Carlstrom and Dorr Fields. While at Wilbur Wright Field, Dayton, Ohio, he compiled and authored the U.S. Air Service textbook, *Pilot's Gunnery & Combat Training*."

Following World War One Johnston returned to Pittsburgh. In 1919 he started the Johnston Aviation Company there, using war surplus aircraft. His company connected Pittsburgh to New York City. After two years Johnston joined the Aero Import Corporation as that company's sales manager.

In 1922 he became sales manager for Huff Daland Aero Corporation of Kansas City, Missouri. He joined Moth Aircraft in Lowell, Massachusetts in 1928 and sold aircraft for that organization until 1930. The company's plane was patterned after the DeHaviland "Gypsey Moth.", which was a bi-plane with a folding wing.

As sales manager for that firm, Johnston flew to aviation country clubs to demonstrate and to sell the planes. One of his clients would be Charles Lindberg who bought a moth to teach his new bride, Ann Morrow, how to fly.

In 1929 Johnston installed streamlining and a power-boosted Wright Gypsey engine into the Moth. He then flew it to the Cleveland Air Races, and won the speed race around the pylons against Avro Avions and other planes in the class.

When Curtiss-Wright bought the Moth firm, Johnston joined them and sold their Robin aircraft. Johnston left Curtiss-Wright following a short period of employment to try his hand at writing aviation articles for aviation pulp magazines.

Johnston returned to aviation, doing engineering and design work in California for the Douglas and Lockheed companies. He moved to Houston, Texas and retired there. He died in Kerrville, Kerr County, Texas on 3 August 1967 at age 75.

CHARLES JONES

Charles Maury Jones was born in Red Bank, New Jersey on 28 June 1894, the son of William Strother Jones, III, a stockbroker and farmer, and Mary Grace (Russell) Jones. Jones attended St. Paul School, Concord, New Hampshire, Class of 1913. He planned to enroll in Princeton University but instead joined Dr. Wilfred Grenfell on Labrador as a student volunteer in his aid program. In early 1917 Jones sailed for England to join the Royal Flying Corps but was rejected because of his six-foot, three-inch height. He crossed the Channel to France and on 26 March 1917 he joined France's *Service Aeronautique*.

From 1 April to 12 August 1917 Jones attended the aviation schools at Avord, Pau, and the *G.D.E.*, receiving his *brevet* on Bleriot on 16 June 1917.

On 15 August 1917 *Caporal* Jones was assigned at the Front to *Escadrille Spad 73* and flew with that squadron until 21 January 1918. That same month *Sergent* Jones was commissioned a First Lieutenant in the ASSRC.

On 22 January Lieutenant Jones joined the 103rd Aero Squadron, USAS and flew combat patrols with that unit until 8 June 1918. He was awarded France's *Croix de Guerre*, with Star for his excellent work as a patrol leader.

On 12 June 1918 Jones was assigned to the 13th Aero Squadron, USAS, and was a patrol leader with that Squadron until 13 August 1918. That same day Lieutenant Jones was transferred to the 28th Aero Squadron, USAS, and on 2 September 1918 was made its commanding officer. On 3 October 1918 Jones was promoted to Captain. He continued to fly combat patrols until the 11 November 1918 Armistice.

Captain Jones was discharged from the U.S. Air Service on 26 February 1919. Following his World War service he pursued a career as a stockbroker, becoming a partner in the New York Stock exchange brokerage firm of Carlisle and Meelick and successor firms. On 18 March 1920 he married Katharine Hoagland. From this union two daughters and a son were born.

During World War Two Jones again entered military service, as a Colonel in the U.S. Army Air Force. He was Commander of the Air Tactical Command base in Ostros, France; Deputy-Commander for the Air Tactical Command, Mediterranean Area; and Commanding Officer of the British-American Airport for the Potsdam Conference in 1945.

Following World War Two Jones returned to private business. He also pursued his hobbies which included fox hunting, sailing, fishing, and farming. He died in Bedminster, New Jersey on 13 October 1978 at age 84.

HENRY JONES

ESCADRILLE LAFAYETTE

Henry Sweet Jones, the 33rd volunteer of the *Escadrille Lafayette*, was born in Harford, Pennsylvania on 6 June 1892, the son of Pennsylvania politician Edward E. Jones, who later became Register of the U.S. Treasury, and Hattie (Sweet) Jones. Hattie Jones died at age 27 when Henry was just six years of age.

Henry attended local schools in the small community of Harford before enrolling in Lehigh University in pursuit of an engineering degree during the academic year of 1913-1914. Desiring to participate in the European War, Jones withdrew from school, and in early 1916 he sailed on the *Lafayette* from New York to Bordeaux, France to enlist in the American Ambulance Field Service. Upon his arrival in Paris, Jones fell ill with jaundice and spent a month recuperating at the American Hospital at Neuilly. He then served six months with S.S.U.1 as an ambulance driver, participating in the battles of Verdun and Fleury.

Jones sought a more active role in the war. While he was in Paris at the Hill Brothers Outfitters purchasing a new Ambulance Service uniform, he ran into Dudley Hill, then a pilot with the *Escadrille Lafayette*. They became close friends, and this association inspired Jones to move from the Ambulance Corps to French aviation. When his six-month contract with the Ambulance Service expired, Jones joined French aviation via the Foreign Legion on 27 October 1916. He began his training on 28 November and progressed through the schools at Buc, Etampes, Avord, and Pau. He received his *brevet militaire* on the Maurice Farman on 16 March 1917 and was at the G.D.E. on 10 May when he was notified to report to *SPA 124*.

Sergent Henry Jones, his face reflecting the stress of numerous combat patrols over the Verdun sector during August and September 1917

Caporal Jones reached the *Escadrille Lafayette* at Ham on 12 May 1917. Shortly after his arrival, on 3 June the Squadron moved to the newly constructed aerodrome at Chaudun in the Aisne sector which was located midway between Villers-Cotterets and Soissons. The Chaudun field was smooth and spacious. From there, the pilots of *SPA 124* flew protection for the five French reconnaissance squadrons working the area. When not flying escort, the pilots were to fly fighter reconnaissance missions in sectors deemed too dangerous for the lumbering French observation planes. In the forty-three days the Squadron spent in the Aisne sector, its pilots flew 400 sorties and fought 62 individual combats against enemy aircraft. However, the

Willis Haviland, Courtney Campbell, Henry Jones, Edwin Parsons, and Dudley Hill at Chaudun, July 1917.

Squadron was credited with only one confirmed victory, Raoul Lufbery's tenth, on 12 June 1917 when he brought to ground a German fighter protecting a biplace regulating artillery fire.

On 17 July 1917 *SPA 124* was transferred from the Aisne sector to the Flanders sector, to the small village of Saint Pol-sur-Mer two miles from Dunkirk on the Channel coast. There the Squadron became part of the planned Allied offensive set for 31 July which was later to become known as the Third Battle of Ypres. The pilots' particular mission was to bomb and strafe German planes on their airfields. *Caporal* Jones—"Jonesy" as he was then called- and the other pilots trained diligently, but when the great offensive was launched, their own participation was shut down by foul weather.

The pilots flew intermittently when the weather allowed and engaged in desultory combat as the troops below slugged it out in the mud. The offensive soon settled into a deadlock. And by mid-August 1917 the Squadron found itself 250 miles distant, in the village of Senard in the Verdun sector on the edge of the Argonne Forest.

The evening of 16 August saw the Squadron's first patrol, flying reconnaissance and striking ground targets. While the missions proved uneventful, two days later, on 18 August, the air turned hot with machine gun fire as the *Lafayette* pilots flew escort for bombers striking enemy fortifications at Dun-sur-Meuse. Almost at once casualties mounted. Harold Willis was shot down and made a prisoner of war. Stephen Bigelow was seriously wounded. *Sergent* Jones, Douglas MacMonagle, and Kenneth Marr had a fierce combat with four German fighters. On 24 September 1917 MacMonagle was killed in combat against members of the Richthofen squadron near Triacourt.

Jones held up well under the strain of combat. And on occasion he took patrols for other pilots. He was rooming with James Norman Hall when Hall was writing for the *Atlantic Monthly* the series which would later become *High Adventure*. Hall was usually behind on his installments, and had periods of inspiration when he would write for hours at a time. During one such period he was due to fly a dawn patrol and asked Jones if he would take it for him. Jones agreed.

Jones returned from patrol, his plane shot full of holes from an aerial encounter. Hall was aghast when he saw the condition of Jones's fighter. He said to Jones, "My God, Henry, what if you had bought it on a patrol that I should have been flying. It would have been on my conscience the rest of my life. I'll never ask anyone to fly a patrol for me again."

The Verdun sector had proved both tragic and fatiguing to Jones and his *Lafayette* comrades. When they were transferred back to Chaudun in late September, they left Senard having flown 420 missions in six weeks and were credited as a squadron with the destruction of five German planes.

SPA 124 had been moved back to Chaudun to play a role in the forthcoming French assault at Malmaison. While they awaited the preparations, they were assigned to fly protective cover for photo reconnaissance and observation planes. On 1 October Jones had the most harrowing experience of his war career when he and his two comrades flew into a formation of four Hanoveranner reconnaissance buses above Chemin-des-Dames.

The combat began between 15,000 and 18,000 feet. Flight commander Carl Dolan led his two wingmen, Jones and Andrew Courtney Campbell, into a dive to 8,500 feet to where the German two-seaters were circling, apparently conducting artillery reglage. Dolan was forced to quit the fight with engine trouble soon after the engagement began. As Jones and Campbell engaged the two-seaters, Jones closed too quickly on his selected victim's tail and was forced to throttle back to keep from crashing into the enemy two-place. Then out the corner of his eye, not fifty yards distant, he picked up the closing form of another bi-place drawing parallel to his fighter.

The gunner sprayed fire from the rear cockpit. Jones nervously followed the penciled lines of smoke from tracers arcing behind his tail. He hunched into his cockpit in a futile attempt to make himself a smaller target. Realizing the gunner would soon find the correct range, Jones swooped under the enemy machine ahead of him

Sergent Henry Jones beside his SPAD S.VII fighter plane, Senard, August 1917.

to attack its vulnerable underbelly. But at the instant he prepared to fire into his victim's blind spot, a panel slid open in the bottom of the plane and the observer began raking Jones's SPAD with his machine gun.

Three loud whang-g-gs told Jones three bracing wires on his right wing had been shot away. The flac! flac-flac! of bullets puncturing fabric filled his ears. A part of his wing dropped off. Jones, glancing quickly about to learn of Campbell's fate, quit the fight and volplaned his wounded machine above the trenches, eventually managing a safe landing at the Chaudun field.

Dolan, who had been standing on the tarmac awaiting Jones's return, later recalled: "I remember Jonsey running over to me after his landing and yelling, 'Dolan, take a look at this.' I went to him and he had the shattered wing in his hand." The wings and fuselage of Jones's SPAD had been struck in thirty-four places with 7.92mm rounds.

Both pilots stood on the field waiting for Courtney Campbell's return. Darkness arrived first. It would become a long wait. Weeks later there was still no word of their missing comrade. It was into November before his status of "presumed dead" was confirmed when a German airman dropped a map with a message indicating when and where Campbell had been shot down, and that he had been buried near Pargny.

On 23 October the impending Malmaison offensive was launched. But the prevailing fog and drizzle kept *SPA 124* grounded. However, the following day the skies opened up, and the Squadron flew twenty-one missions. On 31 October Henry Jones received an unconfirmed victory when he was credited with forcing an enemy fighter into the ground behind German lines. During his service, Jones, in all probability, shot down at least the five planes needed to qualify as an ace. But because of the *Service Aeronautique's* stringent confirmation requirements, he was not credited with a single victory. But as Carl Dolan later recalled: "They (the French) required that two ground officers had to witness and confirm a plane shot down. I saw Jonesy shoot down at least three or four Germans when I was with him."

By 2 November 1917 the Allied offensive had officially ended with the withdrawal of the German Seventh Army by its commander. The number of patrols the pilots flew diminished at the first sign of early winter. The opportunity arose for some of the Americans to take short leaves to Paris. Jones was one pilot selected.

Considered by his comrades "the liveliest of the crowd," Jones knew how to relax with a bottle of pinard, or barring that, a concocted substitute of "bay rum, olive oil, and vinegar." He had earned the forty-eight-hour Paris pass granted him by squadron commander Thenault. But he began celebrating his good fortune early "in a champagne binge" and when he arrived at the train station, he boarded the wrong car and wound up instead in Verdun with a terrible hangover. By the time he did arrive in Paris, his leave time had nearly expired, so he turned around and headed back to Chaudun following a 600-mile rail trip.

On another occasion when he and Dudley Hill were on leave together in Paris, they stopped at Henry's Bar, a favorite Squadron watering hole, to repay to Henry, who had acted as their banker, several thousand francs they owed him. Henry, a German, was under suspicion by the authorities of being a spy. Around midnight he committed suicide by jumping from Dudley Hill's bedroom balcony through the glass roof of Jones's room, onto the stone floor below. Jones later said to Hill "If we knew he was going to commit suicide, we'd never have paid him off."

On 7 December 1917 the Squadron moved to La Noblette where it flew a large number of patrols but fought few combats. One frustrating engagement involved *Sergents* Jones, David Peterson, James Norman Hall, and Ray Bridgman opposing two German reconnaissance machines inside enemy territory. Peterson and Hall teamed up and dove from 19,000 feet in an attempt to drive the enemy planes to Jones and Bridgman circling at 13,000 feet. The Peterson-Hall team killed the observer of one plane before their Vickers guns stopped up from the cold, a common occurrence above 5,000, due to the thickening of the lubricating oil.

Jones and Bridgman killed the observer in the second plane before their guns also jammed. Meanwhile, the two German pilots beat a safe retreat home, later causing Peterson to complain "there we were, four against two and not a gun in the outfit."

Air activity ground to a halt as the American pilots prepared to transfer to the U.S. Air Service. By mid-December the pilots were advised to seek their release from the *Service Aeronautique* so they could take U.S. commissions. Jones was awarded France's *Croix de Guerre*, with Palm on 14 January 1918 for having been cited on 1 October and 31 October 1917. He was commissioned a First Lieutenant in the U.S. Air Service on 31 January 1918, a rank which failed to adequately reflect his many hours of combat service at the Front.

The *Escadrille Lafayette* officially passed out of existence on 18 February 1918 when it was assimilated into the U.S. 103rd Aero Pursuit Squadron. Jones flew with the 103rd from that date until 8 June 1918. During his nearly thirteen months as a combat pilot he logged more than 400 hours in the air above France. This is how he remembered his last day with the squadron:

"On the day I received my orders to leave to the United States for duty as a combat flying instructor, I had an early 4 a.m. patrol which I was to lead as flight commander. As I reached my SPAD I discovered that someone had removed the spark plugs preventing, of course, my making the flight. Hobey Baker, a pilot in my flight, later confessed, claiming that no pilot with his orders to return to the States should have to fly his last combat patrol. Baker consequently led the flight. William Thaw, C.O. of the 103rd cut the Indian head insignia from my plane as a souvenir and gave it to me."

When Jones was preparing to return to the U.S., he and Lt. Charles Wilcox, who had also been reassigned to the States, boarded the train from La Noblette to Paris. They took seats in the only steel car, the rest of the cars being of wood construction and filled with French soldiers. He and Wilcox were asleep when the train collided with another train coming from the opposite direction. Jones and Wilcox awoke suddenly when two French officers occupying the seat opposite them were thrown into them. Thinking they were being assaulted, they began fighting with the Frenchmen until they realized what had happened.

The train was held up for some time, as many soldiers in the wooden cars were injured or killed. Jones and Wilcox left the scene to eat breakfast, and when they returned they discovered the train had gone on with their baggage. Fortunately, when they arrived in Paris, they found that the two French officers who had been thrown into them had checked their baggage at the station.

Lieutenant Jones sailed for the United States on 4 July 1918 to serve in the capacity of instructor and experimental tester. He served in this dual role at Lake Charles, Louisiana and at Dorr Field, Florida. He received a Star to his *Croix de Guerre* on 17 November 1918 in recognition of his past "victorious combats back of the enemy lines" while with the 103rd. He would later be awarded France's *Legion d'Honneur*. He was discharged from the U.S. Air Service as a First Lieutenant on 27 December 1918.

Following his war service, Jones worked for the Pennsylvania Railroad before joining Transcontinental Air Transport-Maddux Air Lines. He worked for this line and other airlines for several years. He married Helen Adams on 19 August 1923. They had a son, Henry, Jr. in 1924. Jones then went to work for the F.W. Woolworth Co., managing stores in Erie, Pennsylvania, Newark, New Jersey, and Kingston, New York. He retired from Woolworth's in 1960 and settled in Clearwater, Florida. His wife, Helen, passed away in 1970. Jones died in Clearwater on 29 March 1972 at age 79.

David Judd

Edward David Judd was born in Boston, Massa chusetts on 31 August 1894, the son of David H. Judd, and Lillian M. (Hanna) Judd. He graduated from Roxbury Latin School, 1912, and from Harvard College, A.B.(Economics) 1916.

On 6 January 1917 Judd enlisted in the American Ambulance Field Service and served in France with S.S.U. 4 in the Argonne and Champagne sectors until 28 January 1917. On 6 July 1917 Judd joined France's *Service Aeronautique*.

From 30 June to 26 November 1917 Judd attended the aviation schools at Avord, Pau, and the *G.D.E.*, receiving his *brevet* on the Caudron on 1 October 1917.

From 1 December to 18 December 1917 *Caporal* Judd was assigned at the Front to *Escadrille SPAD 73*. From 18 December to 22 January 1918 he flew with *Escadrille SPAD 3* at Dunkirk. While he was with the squadron at Beauzee, just behind Verdun, he was released from the French Service to join U.S. Naval Aviation.

On 19 January 1918 Judd was commissioned an Ensign and was enrolled as Naval Aviator (HTA) #565. He was sent to the Naval Air Station at Dunkirk where he served as a seaplane pilot flying anti-submarine patrols over the North Sea along the Belgian coast.

On 28 May 1918 Judd was transferred back to day-bombardment machines. Following one month of training, he was attached to English Squadron 218 at Calais. While with this squadron, Judd made raids over Zeebrugee, Bruges, and Ostend. He was awarded the U.S. Navy Cross for his "distinguished and heroic service" while a pilot with the Northern Bombing Group for his strikes against enemy emplacements on the Belgian front.

In August of 1918 Judd was transferred to the Paris staff of Captain D.C. Hanrahan, Commander of the Northern Bombing Group. On 6 September 1918 Judd was returned to the United States to instruct in day bombing. He was assigned to the Marine Flying Field, Miami, Florida, and remained there until 28 February 1919, when he was put on inactive status of the U.S.N.R.F. as a Lieutenant, j.g.

Sergent Judd (right) and Adjudant de Curnieu at Avord.

Following World War One, Judd returned to Paris where he worked in the New York Bankers Trust office for five years. He came back to the United States to join the First National Bank of Boston where he remained until 1930. From 1930 until 1959, the year of his retirement, he was an executive involved in wool marketing. Judd married Muriel Powers. He had a son and a daughter. Judd died in Roxbury, Massachusetts on 17 April 1975 at age 80.

HUGO KENYON

Hugo Alden Kenyon was born in Peacedale, Rhode Island on 6 December 1892, the son of Frederick Wilson Kenyon, a synthetic-rubber executive, and Helen Augusta (Dee) Kenyon. As a youth, Kenyon attended Providence schools and was educated in England where his father had gone on business. As a teenager, Kenyon became a proficient violinist and performed on the Providence concert stage.

Kenyon attended Brown University, Class of 1916, but was expelled from that institution and joined the American Ambulance Field Service. In 1915, and again in 1917 he served in France as a driver, his second tour being a nine-month stint with SSU I.

On 5 July 1917 Kenyon enlisted in France's *Service Aeronautique*. From 19 July 1917 to March 1918 he attended the aviation schools at Avord, Juvisy, Pau, and the *G.D.E.* He received his *brevet* on the Caudron on 16 October 1917.

While Kenyon was taking his acrobatic training at Pau, his Nieuport trainer caught fire at 3000 meters. Kenyon cooly and courageously brought the aircraft safely to ground and removed the instruments and cushion from the flaming fuselage.

On 26 April 1918 Kenyon was commissioned a Second Lieutenant in the U.S. Air Service. While awaiting assignment at the Front, he ferried aircraft from the factories to the airfields at Orly and Issoudun. On 1 June 1918 he was assigned to the 103rd Pursuit Squadron, USAS, and flew combat patrols with that unit until the 11 November 1918 Armistice. On 6 November 1918 Kenyon was promoted to First Lieutenant.

Following World War One, Kenyon became a planter in India. He later worked as a Wall Street broker and headed the foreign-loan department for the Shawmut Bank of Boston. He was married three times: to Dotha Merrill Blake; to Germaine Marie Paule Goullet; and to Sylvaine Loezere. Kenyon was the father of five children; three sons, and two daughters.

During World War Two Kenyon re-entered the U S. Army and rose to the rank of Colonel in the Military Intelligence. Following the War, he was associated with a brokerage firm in Philadelphia before being recalled as a civilian intelligence consultant at the Pentagon to the military forces in Western Europe. Kenyon died suddenly at his home at Falls Church, Virginia on 11 November 1953 at age 60.

CHARLES KERWOOD

Charles Wayne Kerwood was born in Chicago, Illinois on 1 August 1896, the son of Mr. & Mrs. M.E. Kerwood. He was educated at the Western Military Academy; the Haverford School; the Grenoble School, France; and at New York University.

In 1916 Kerwood attended the Curtiss Flying School at Newport News, Virginia. He sailed for France in early 1917 and on 18 February enlisted in the *Service Aeronautique*. From 25 February to 18 November 1917 Kerwood attended the aviation schools at Avord and the *G.D.E.* He received his *brevet* on the Caudron on 26 August 1917.

On 21 November 1917 *Caporal* Kerwood was assigned at the Front to *Escadrille Br. 117.* He flew with that squadron until 31 March 1918, the day he was shot down in combat.

On that date *Sergent* Kerwood was on a bombing mission over German lines when he was attacked by four enemy aircraft. Kerwood was wounded and forced down during the melee. He was captured and sent to the prison camp at Landshut, Bavaria. While a prisoner, Kerwood attempted an escape and was shot in the leg by a guard. Following the Armistice he limped into Paris with the wound healed over and the bullet still in his leg. Kerwood was awarded France's *Croix de Guerre*, with Star for his bold and courageous service at the Front.

Following World War One, Kerwood worked as a barnstormer, and as an aero engineer and test pilot for Franco-Rumanian Lines, Europe. He served as a field director for the American Relief Association, Russia and the Balkin countries. Throughout the early 1920's he worked as a pilot for various commercial aircraft firms. In 1925 he was one of the organizers of the Volunteer American Squadron during the French-Riff campaign, flying with the *Escadrille de la Garde Cherifienne* in Morocco against Sultan Abd-el-Krim.

When not flying, Kerwood wrote numerous articles on military and aero subjects for newspapers and magazines. During World War Two Kerwood entered the U.S. Military and flew combat missions with the U.S. Army Air Corps in Europe, in Africa, and in the Pacific. He was later assigned to the Air Corps staff in the capacity of public relations to boost the morale of defense workers. He retired from the military at the rank of Colonel. Kerwood died in Washington, DC on 15 December 1976 at age 80.

Charles Kerwood (center) standing before an Italian Ansaldo S.V.A. 5 reconnaissance plane.

CHARLES KINSOLVING

Charles McIlvaine Kinsolving was born on 19 January 1893 in Rio Grande do Sul, Brazil, the son of Lucien Lee Kinsolving, Episcopal bishop of Brazil and a member of a prominent Virginia family, and Alice (Brown) Kinsolving.

Kinsolving was educated at the Episcopal High School in Alexandria, Virginia. He attended the University of Virginia but left that institution to join the American Ambulance Field Service. In 1916 Kinsolving served in France for five months as a driver with SSU 4 at the Verdun, Champaigne, and Vosges districts. On 13 June 1917 he enlisted in the *Service Aeronautique*. From 19 July to 19 November 1917 he attended the aviation schools at Avord, Tours, and the *G.D.E.* He received his *brevet* on the Caudron on 25 Sept 1917.

On 21 November 1917 *Caporal* Kinsolving was assigned at the Front to *Escadrille Br. 117.* He flew with that squadron as a *Sergent* until 25 February 1918, the day he was commissioned a First Lieutenant in the U.S. Air Service. Kinsolving continued to fly missions with *Br. 117* until 16 June 1918. From 18 June to 28 September 1918 he instructed at the 7th Aviation Instruction Center at Clermont-Ferrand. From 30 September 1918 to the Armistice he was C.O. of the 163rd Day Bombardment Squadron, USAS. Kinsolving was awarded France's *Croix de Guerre*, with Palm and Star for his coolness and courage during bombardment missions along the Rhine and Northern sector of the lines. In 1919 Kinsolving left the U.S. Air Service at Long Island, New York.

Following World War One, Kinsolving became an oil wildcatter in Texas. He established the Sao Paolo Chamber of Commerce in Brazil and was a correspondent for United Press in Rio de Janeiro, Brazil. In 1922 he married Florence Natalie Hogg whom he had met in France during the War where she was working as a Red Cross volunteer. The couple had two sons.

In 1924 Kinsolving was expelled from Brazil for sending coded stories to his news service about the revolution there. He relocated to New York City to work for U.P.I. on its sport desk and as a syndicating manager. Kinsolving left UPI to work briefly on Wall Street. He next founded the public relations department for *Fortune* magazine. He also worked for *This Week* and *Parade*. In 1943 he joined *Newsweek* magazine where he remained for the next 18 years, becoming publisher of its Latin American edition, and its Eastern advertising manager. Kinsolving retired to Carmel, California in 1960 but briefly rejoined *Newsweek* in 1964 as West Coast representative for *Newsweek* International.

Kinsolving was active in his community, serving on the board of the Harrison Memorial Library in Carmel, and with the Monterey Institute of Foreign Studies. He was also historian for the Old Capital Club of Monterey. For eight years he was director of the American Field Service, and for ten years he served as a Trustee of the Lafayette Memorial at St. Cloud, France. Kinsolving died on 24 December 1984 in Carmel, California at age 91.

WILLIAM KIRKWOOD

William Francis Kirkwood was born in Boston, Massachusetts on 30 October 1897, the son of Robert Joseph Kirkwood, a civil engineer, and Selina (Sherry) Kirkwood. He attended the Boston Latin School and was the youngest person to ever graduate from that institution. He then entered as a cadet at the Massachusetts Nautical School, and upon securing his able seaman's certificate in October 1916, he then served as a seaman aboard British and Brazilian merchant vessels. Kirkwood returned to the United States to attended the Massachusetts Institute of Technology for one semester.

Kirkwood left school to do hospital work in France. On 16 February 1917 he secured his passport. He sailed from New York for France on 17 March and arrived in Bordeaux on the 24th. He joined the American Ambulance Field Service and served at the Front for four months with SSU 4.

On 1 August 1917 Kirkwood enlisted in France's *Service Aeronautique*. In October 1917 he was released because of inaptitude. Kirkwood returned to the United States and enlisted as an Ensign in the U.S. Naval Reserve Force. He held the rank of Lieutenant j.g. from 1918–1921 and served diligently at sea with the U.S. Navy.

Following World War One Kirkwood was marine superintendent for the Munson SS lines. From 1924–28 he was managing editor of the *Excelsior* newspaper in Bluefields, Nicaragua. From 1928 to 1930 he served as the Danish Consul for the Government of Nicaragua. While working there, he met and married Hilda Calonje of Spain. From this union two daughters were born.

From 1936–1958 Kirkwood was a Massachusetts parole officer. During World War Two he was Port Representative for the War Shipping Administration in Cristobol, the Canal Zone where he headed up the Recruitment and Manning Organization. He also held the Coast Guard Certificate for over 30 years as a Master of Ocean Vessels. Kirkwood died of pneumonia at the VA Hospital in Brockton, Massachusetts on 2 December 1976 at age 79.

JOHN KOWALL

John Robert Kowall was born in Roxbury, Massachusetts. Kowall attended the University of Pennsylvania. He left school to join the American Ambulance Field Service. On 1 August 1917 Kowall enlisted in France's *Service Aeronautique*.

In November of 1917 he was released because of injuries he received in a flying accident.

THEODORE DE KRUIJFF

Theodore Adrian de Kruijff was born in New York City, New York, the son of Theodore E. and Mary de Kruijff. de Kruijff attended New York City public schools. In anticipation of entering the European War, de Kruijff joined the Curtiss aviation school at Buffalo, New York where he earned a pilot's license. His Curtiss aircraft was involved in a crash while he was flying with another student. de Kruijff broke his leg in the fall. When he had recovered, he continued to fly at Newport News, Virginia until January 1917 when he left for France to join the Lafayette Flying Corps.

On 20 March 1917 de Kruijff enlisted in France's *Service Aeronautique*. From 20 March to 4 December 1917 he attended the aviation schools at Avord, Pau, Cazeaux, and the G.D.E. Despite much trouble from his injured leg, de Kruijff received his military *brevet* on the Nieuport on 6 August 1917.

From 6 December 1917 to 21 May 1918 *Caporal* de Kruijff served at the Front with *Escadrille N. 158*. On 21 May de Kruijff was commissioned a Second Lieutenant in the U.S. Air Service.

From 6 June to 5 July 1918 Lieutenant de Kruijff was assigned as a ferry pilot at the American Acceptance Park, Orly. From 5 July to 1 November 1918 he served at the American Aviation Instruction Center, Romorantin. Lieutenant de Kruijff died of pneumonia at the American Red Cross Hospital #1 in Neuilly on 6 November 1918. He was buried at the A.E.F. Cemetery #34, Paris. In June of 1928 his remains were transferred to the Lafayette Flying Corps monument, St. Cloud.

GEORGE KYLE

George Marion Kyle was born in Madisonville, Ohio on 1 August 1895, the son of George Allen Kyle, a construction engineer who built the railroad between Capetown and Johannesburg, South Africa, and Mildred (Corbin) Kyle. Kyle's mother had been sent back to the States until the birth of her son, at which time she returned with the infant to South Africa.

The Kyle's returned to the States two years later, settling in Portland, Oregon. George attended Portland's Jefferson High School. As a senior he became the State Interscholastic tennis champion. Kyle enrolled in Stanford University to study chemical engineering. In February of 1917 he left his studies to join the Stanford Ambulance Corps No. 1, and served with SSU 14 in France.

Kyle enlisted in France's *Service Aeronautique* on 27 June 1917. From 6 July to 24 December 1917 he attended the aviation schools at Avord and the *G.D.E.* He received his *brevet* on the Caudron on 17 October 1917. On 24 December 1917 *Caporal* Kyle was attached at the Front to *Escadrille Br. 117*, and flew numerous bombing missions from Lorraine over enemy lines. He was given the handle "Lucky Kyle" because he always returned from hair-raising raids. In February 1918 *Sergent* Kyle was commissioned a First Lieutenant, USAS. He continued to fly at the Front with his French Squadron until 1 July 1918 when he was assigned to the U.S. instruction school at Clermont-Ferrand. He remained there until the Armistice. Kyle was decorated with France's *Croix de Guerre*, with Star for his exemplary courage.

Following the War, in 1919 Kyle returned to Portland and worked as a bond salesman before returning to Stanford University to complete his studies. He switched majors to Philosophy and graduated in 1921. He met Dorothy Johnston at Stanford. They married on 3 June 1922 and eventually had two sons.

Following graduation Kyle worked as a reporter for the San Francisco *Examiner* and the San Francisco *Bulletin*. He then became a bond salesman in San Francisco for the Anglo-California Trust Co., and for the Chester Skaags Co. Kyle then moved to Los Angeles to become advertizing manager for the *California Arts and Architecture Magazine*. He left that publication to join the Carl Quill Press as publisher and editor of the *Game and Gossip* sports magazine. Kyle next worked for the Los Angeles Broadcasting Co. selling radio advertizing. At that time he and Mrs. Kyle separated, and he moved alone to San Mateo.

On 22 August 1941 Kyle was attending a house party in San Mateo. When the party ran out of liquor, Kyle and another man volunteered to get more. The man driving their car swerved to miss a car backing out of a driveway and hit a eucalyptus tree head on. Kyle was seriously injured in the accident. He died as a result of his injuries on 7 October 1941 at age 46.

Gorman Larner

Gorman De Freest Larner was born in Washington, D.C. on 5 July 1897, the son of Robert Martin Larner, a prominent newspaperman, and Adelaide (DeFreest) Larner. Larner attended the Force School, Western High School, and the Lawrenceville Prep School before entering Columbia University.

When the United States entered World War One, Larner applied for entry as an officer in the U.S. Signal Corps, but his application was rejected because of his youth. Larner sailed for France, and on 10 July 1917 he joined the *Service Aeronautique*, From 19 July to 1 December 1917 he attended the aviation schools at Avord, Tours, Pau, and the *G.D.E.* He received his *brevet* on the Caudron on 28 September 1917.

From 3 December 1917 to 1 April 1918 *Caporal* Larner served at the Front with *Escadrille SPAD 86*. On 24 April Larner was commissioned a First Lieutenant, USAS, and continued to serve with *Spad 86* until 15 June 1918. While with that Squadron he was credited with the destruction of two enemy aircraft and was awarded France's *Croix de Guerre*, with two Palms.

On 16 June 1918 Lieutenant Larner joined the 103rd Pursuit Squadron, USAS, and flew with that squadron as a flight commander until the November 1918 Armistice. While flying combat patrols he was credited with the destruction of five enemy aircraft, bringing his total to seven. Lieutenant Larner was awarded the U.S. Distinguished Service Cross, with Bronze Oak Leaf which indicated two citations for extraordinary heroism.

On 8 November 1918 Larner was appointed Captain and served as an intelligence officer on the staff of Colonel House at the Peace Commission in Paris. He returned to the U.S. in September 1919. In February of 1920 he joined the Army Air Corps reserve at the rank of Major.

Following World War One, Larner entered Columbia University and received a B.S. degree in 1922. That same year he entered the bond Department of the Guaranty Co. of New York, and for the next fifteen years he worked as a banker. He also served as deputy administrator in the NRA and as a specialist in the field of federal labor legislation and administration.

In World War II Larner entered the U.S. Army Air Corps as a Colonel. He was on active duty for four years, first in London, England as air attache at the American Embassy before being reassigned to the Pentagon. Following the War Larner became General Manager of the National Aeronautical Association and Chairman of Robinson Aviation Inc. of Teterboro, New Jersey. Under his leadership, this company merged with four additional companies to become Robintech Inc. Larner retired from business in 1964. He died in Easton, Maryland on 20 May 1984 at age 86.

He was survived by his wife, Jean (Fleming) Larner and two daughters.

HENRY LEE

Henry S. Lee was born in Cornwall, New York, on 23 May 1889, the son of Henry C. Lee. Upon completing high school in 1909, Lee was employed by the Bethlehem Steel Company and attended the American Pratt Institute, Brooklyn, New York, where he took a course in machine construction. He left that institution in June 1916 in order to enter the European War.

Lee served six months in France as an ambulance driver with the Norton-Harjes Ambulance Service. On 24 April 1917 he enlisted in France's *Service Aeronautique*. On 15 July 1917 he was released because of inaptitude.

On 14 November 1917, in Paris, Lee enlisted as a private in the U.S. Army Signal Enlisted Reserve Corps. He was assigned to the 1st Aviation Training Center. On 24 January 1918 he was assigned to the 94th Aero Squadron, USAS, and served with that unit until 19 November 1918. PFC Lee was then assigned to the 4th Air Park until his assignment to the 1st Replacement Depot, St. Aignan. He was honorably discharged on 28 February 1919.

Schuyler Lee

Schuyler Lee was born in Bloomfield, New Jersey on 29 July 1898, the son of Reverend James Beveridge Lee, a Presbyterian minister, and Mynna (Greenman) Lee. Schuyler attended various schools as a young man, while following his father from parish to parish. He was educated at the German-English Academy, Milwaukee, Wisconsin; the Latin School, Chicago, Illinois; the Haverford School, Pennsylvania; and the Phillips Academy, Andover, Massachusetts, Class of 1918.

On 28 April 1917 Lee, while in his junior year, left the Phillips Academy to join the Andover Unit of American Field Service. He served in France with TMU 526. On 1 June 1917 Lee had enlisted in France's *Service Aeronautique* but continued to serve with his cameon unit until 6 August 1917 when an aviation vacancy opened up. From 16 August to 8 January 1918 he attended the aviation schools at Avord, Tours, Pau, and the *G.D.E.* He received his *brevet* on the Caudron on 21 October 1917.

On 10 January 1918 *Caporal* Lee was assigned at the Front to *Escadrille SPAD 96*. He flew many volunteer combat patrols and on at least one occasion saw his plane cut to pieces by fire from enemy aircraft.

On 12 April 1918, while guarding the rear of a SPAD patrol east of Montdidier, *Caporal* Lee was jumped from behind by a Fokker triplane. His comrades in the air believed that a burst from the enemy fighter instantly killed him, and his plane went spinning earthward. The wrecked aircraft was later found and identified half a mile northeast of Beuvraignes. However, Lee's grave was not located, and his remains were never found. He was 19 years of age at the time of his death. *Caporal* Lee was posthumously awarded France's *Croix de Guerre*, with Palm for his action on the day of his death.

MANDERSON LEHR

Manderson Lehr was born in Petersburg, Nebraska, on 5 July 1895, the son of Henry F. Lehr, a lumber yard operator, and Annie (Hurford) Lehr. As a youth, Lehr attended Albion public schools. He later entered Beloit College, Beloit, Wisconsin, Class of 1918.

Lehr left that institution to join the American Ambulance Field Service. He sailed for France, but upon arrival there sought a more active role in the Great War. On 3 June 1917 Lehr enlisted in France's *Service Aeronautique*. From 10 June to 18 November 1917 he attended the aviation schools at Avord, and the *G.D.E.* He received his *brevet* on the Caudron on 3 September 1917.

On 21 November 1917 *Caporal* Lehr was assigned at the Front to *Escadrille Br. 117*. On 8 March 1918 *Sergent* "Bud" Lehr was commissioned a First Lieutenant in the U.S. Air Service. He continued to fly with his French squadron until 15 July 1918, the day he was killed in combat near Chateau-Thierry.

On that day Lieutenant Lehr was flying his Breguet with two other aircraft on a day-bombing mission along the Marne river between Chateau-Thierry and Dormans. In a torrential rainstorm Lehr dropped his load of 32 bombs on a bridge the enemy was just then crossing in their southward attack. As Lehr turned his bomber toward home, he was attacked from above by ten Albatros fighters.

During the violent combat that ensued, Lehr's engine was struck. Shortly thereafter, Lehr was struck in the head and mortally wounded. Lehr's observer, Lieutenant Carles, was also wounded but managed to take charge of the burning aircraft from his cockpit, flew across the lines, and brought the plane down in a rough wooded field near Coulommiers. The bomber overturned and Carles was thrown clear. Lehr, already dead, burned up in the flaming wreckage of the Breguet. He was 23 years of age at the time of his death. For his heroic actions with *Br. 117*, Lieutenant Lehr was posthumously awarded France's *Legion d'Honneur*, and *Croix de Guerre*, with Palm and Star.

Following the war, in 1923, the wreckage of Lehr's aircraft was discovered. Later, it was learned that his charred remains had been removed to the Oise-Aisne National Cemetery. His body was later removed from that cemetery and re-interred in the Lafayette Flying Corps monument at St. Cloud, near Paris.

David Lewis

David Wilber Lewis was born in Brooklyn, New York on 9 July 1892, the son of David Ward Lewis, a Brooklyn dairyman, and Elisabeth (Corbin) Lewis. "Duffy" Lewis attended New York City schools before enrolling in Harvard College, Class of 1914. After graduation, Lewis rejoined his father's dairy business in New York. Following a failed romantic relationship, he decided to escape to the European War.

On 30 May 1915 Lewis sailed from New York to France. He arrived in Bordeaux on 9 June. He served six months as an ambulance driver with SSU III, the Harvard unit of the American Ambulance Field Service. Upon his return home in December he served six months with the New York National Guard on the Mexican Border. Following that stint Lewis again returned to the family business in New York.

With the New York newspapers filled with news from the European War, Lewis decided to re-enter the conflict. He joined Morgan-Harjes Ambulance Unit No. 21 and sailed for France on 28 April 1917. Lewis arrived in France on 2 May. On 14 June he left his unit for transfer into France's *Service Aeronautique*. On 17 June 1917 Lewis enlisted in the Foreign Legion and on 21 June he transferred to Aviation.

From 22 June 1917 to 19 February 1918 "Duffy" Lewis attended the aviation schools at Avord, Pau, and the *G.D.E.* He received his *brevet* on the Caudron on 13 November 1917.

On 26 February *Caporal* Lewis was assigned to *Escadrille Spad 79* at Maison-Neuve. He flew reconnaissance patrols with his observer along the Front in a biplace *SPAD*, reporting on the movement of enemy troops and equipment. On 29 March 1918 he was commissioned a Second Lieutenant in the USAS but remained with *Spad 79*. In April he was given a *SPAD XIII* fighter to fly. He soon began flying offensive patrols, straffing German troop columns. In late May the squadron was issued Breguet reconnaissance aircraft, and Lieutenant Lewis again began flying photo and reconnaissance patrols.

In early September 1918 Lewis was again issued a *SPAD XIII* fighter and on 7 September he shot down an enemy Siemens-Schuckerts fighter. He was slightly wounded in the melee. On 22 September 1918 Lewis was released from *Spad 79*. On 1 October 1918 he was assigned to the First Air Depot at Colomby-les-Belles to serve as test pilot. On 25 October 1918 Lieutenant Lewis joined the 25th Aero Squadron, USAS, at Toul and remained with that unit until the 11 November 1918 Armistice. For his exemplary war record, Lewis received France's *Croix de Guerre*, with Palm (Ambulance), and *Croix de Guerre*, with Star (Aviation).

On 12 May 1919 Lewis was promoted to First Lieutenant in the USAS. He returned to the United States on 19 June 1919. He remained in the service for several years and was promoted to Major on 31 July 1923.

Upon the death of his father in 1925, Lewis left the military to take over the family diary products business in New York State. Lewis sold the business in 1936.

Lewis tried to re-enter the Army Air Force during World War Two but was rejected for medical reasons. Following the War he sold automobiles in New York. He eventually moved to Englewood, Florida and worked in the post office there for a number of years while writing a feature column for the Venice *Gondolier*. He married Doris Elizabeth Petitt in Englewood on 22 October 1961. Lewis died on 6 June 1982 at his home in Englewood, Florida at age 89.

KENNETH LITTAUER

Kenneth Proctor Littauer was born in Newark, New Jersey on 13 March 1894, the son of Alfred Littauer and Evelyn (Proctor) Littauer. He was educated at Central High School, Washington, DC. At the outbreak of World War One, Littauer had been a student in France. He left school to work as a journalist, magazine writer, and war correspondent.

On 29 March 1916 Littauer enlisted in France's *Service Aeronautique*. From 1 April to 14 October 1916 he attended the aviation schools at Pau, Buc, Chateauroux, and the *G.D.E.* On 24 July 1916 he received his *brevet* #4023 on the Caudron.

On 16 October 1916 *Caporal* "Kepi" Littauer was assigned to the Front south of Dunkirk to *Escadrille C. 74*. There he piloted Caudron G.4, and G.6 aircraft on numerous photo-reconnaissance and artillery-reglage missions. On 8 February 1917 *Sergent* Littauer's bus was attacked by four Albatros D.II fighters. During the melee a bullet struck the center of his wind-screen, missing his head "by the thickness of a cigarette paper."

On 2 January 1918 *Sergent* Littauer left *C.74*. He was commissioned Captain in U.S. Aviation on 19 January 1918. On 14 February he was made Executive Officer for the 88th Aero Observation Squadron, USAS, at Amanty. He served in that capacity and as a flight commander until 4 July 1918. From 4 July to 20 September he commanded the 88th Aero Squadron. From 20 August to 20 September he also served as Acting Chief of Air Service, 3d Army Corps. From 17 September to 24 October 1918 Captain Littauer was commanding officer of the 3d Corps Observation Group. From 24 October to the 11 November Armistice Littauer was Chief of Air Service, 3d Army Corps. He was promoted to Major in November 1918. In spite of his rank and position, Littauer continued to participate in all dangerous missions, unable to order his pilots to do so unless he also faced the risks.

Major Littauer received several Allied decorations in recognition of his distinguished and heroic service. These included France's *Croix de Guerre*, with Palm and Star; Belgium's *Croix de Guerre* while being made a *Chevalier de l'Ordre de Leopold*. He was also awarded the U.S. Distinguished Service Cross for "repeated acts of heroism in action near Conflans, France, September 14, 1918, and near Doulcon, France, October 30, 1918.

Following World War One Littauer returned to his previous occupation as magazine writer, working for Street and Smith magazine publishers. In June 1919 he married Helen Averill. From 1928-1948 he was fiction editor for *Colliers* Magazine.

During World War Two Littauer reentered military service as a Colonel. He served as senior intelligence officer with the 8th Air Force in England and helped to plan the D Day bombings.

Littauer left *Colliers* to found the New York literary agency of Littauer and Wilkinson. He worked actively in this firm until his death on 14 July 1968 in Norwalk, Connecticut at age 74.

RALPH LOOMIS

Ralph Lane Loomis was born in Bedford, Massachusetts on 13 April 1887, the son of Judge Elihu G. Loomis, and Marion H. (Fitch) Loomis. Loomis attended Bedford public schools, the Concord High School, Boston Latin School, the Phillips Academy, Andover, and Amherst College, Massachusetts, BA 1908. In 1911 Loomis graduated from the Harvard Law School, and that same year was admitted to the Suffolk Bar, Boston.

Loomis practiced law for one year before engaging in the Boston paper packaging business of Loomis & Co. On 25 June 1917 he sailed for France with the American Field Service. Upon reaching France, Loomis enlisted in the *Service Aeronautique* on 21 July 1917.

From 26 July 1917 to 1 January 1918 Loomis attended the aviation schools at Avord, Tours, Cazeaux, and the *G.D.E.* He received his *brevet* on the Caudron on 21 October 1917.

On 9 January 1918 in Paris Loomis transferred to U.S. Naval Aviation with the rank of Ensign. On 1 February 1918 he was made Naval Aviator (HTA) No. 578. On 14 February 1918 Ensign Loomis began service with the Northern Bombing Group, attached to the Royal Naval Air Service at Dunkirk. He made numerous flights out of that station to attack enemy fortifications at Zeebrugge and Ostend. Loomis's combat service ended on 22 August when his unit was withdrawn to allow U.S. Marine Corps pilots to take its place. Ensign Loomis was awarded the U.S. Navy Cross for his "Distinguished and Heroic service."

Following World War One Loomis returned to his Boston-based Loomis & Co. He also founded and became director of Loomis Wadding Co, Inc. He met and married Emma Hawkridge of Newington, New Hampshire. Ralph Loomis died of renal failure on 3 November 1968 in Newburyport, Massachusetts at age 81.

WILLIAM LOOMIS

William Fitch Loomis was born in Bedford, Massachusetts on 27 April 1894, the son of Judge Elihu G. Loomis, and Marion H. (Fitch) Loomis. Loomis was educated at Bedford Schools, at Concord High School, the Lawrence Academy, Groton, and Phillips Academy, Andover. He graduated from Amherst College, Massachusetts, B.A. 1917.

On 19 May 1917 Loomis sailed for France to serve in the American Field Service. On 9 June 1917 he enlisted in France's *Service Aeronautique*. From 16 June to 20 November 1917 he attended the aviation schools at Avord, Pau, and the *G.D.E.* He received his *brevet* on the Caudron on 28 September 1917.

On 24 November 1917 *Caporal* Loomis was assigned at the Front to *Escadrille N. 153*. He flew with that squadron until February of 1918.

On 21 February 1918 Loomis was commissioned a First Lieutenant in U.S. Aviation. He served as an aviation instructor at Issoudun until 1 March 1918 when he joined the 94th Aero Pursuit Squadron, USAS. On 11 August 1918 Lieutenant Loomis left the 94th for the 213th Aero Pursuit Squadron, USAS. He remained with that Squadron until 19 November 1918. During his service at the Front Lieutenant Loomis saw action in the air as a flight leader over Chateau-Thierry, St. Mihiel, and Verdun. He received France's *Croix de Guerre*, with Star his exemplary war service.

Following the Armistice Lieutenant Loomis served as an instructor at St. Jean De Monts, France until March 1919. On 10 April 1919 he was honorably discharged from the military.

Following World War One Loomis worked briefly as a bond salesman before becoming involved in commercial aviation. He became a sales engineer for United Aircraft Corporation, working in France, and in the Netherlands. While in France Loomis married Danielle Giraud-Côndé of Neuilly-Sur-Seine on 18 November 1922. Eventually two sons were born of this union.

Loomis returned to the United States, to East Hartford, Connecticut, to became sales manager for the Pratt & Whitney Aircraft Division of the U.S. Aircraft Corporation. He remained with that firm for 25 years until his retirement on 1 June 1960. William Fitch Loomis died in Paris, France, on 19 December 1964 at age 70.

EDWARD LOUGHRAN

Edward J. Loughran was born in the Harlem borough of New York City in 1893, the son of James Patrick Loughran, a saloon keeper. As a young man from a large family, Loughran had little time for education. He and his brothers had to start work at an early age. At the outbreak of World War One he was working as a railroad section hand with his brother, Edward, in Desoto, Kansas.

Loughran left Kansas to participate in the Great War. He served in France with the American Red Cross for six months prior to enlisting in France's *Service Aeronautique* on 20 March 1917. From 27 March to 26 October 1917 he attended the aviation schools at Avord, Pau, and the *G.D.E.* He received his *brevet* on the Caudron on 31 August 1917.

On 29 October 1917 *Caporal* Loughran was assigned at the Front to *Escadrille Spad 84*. He flew diligently with that squadron, even passing by accumulated leave, until 18 February 1918, the day he was killed in combat southeast of Minaucourt.

On that day, while returning from a morning reconnaissance patrol, *Sergent* Loughran's *SPAD* was attacked by three enemy fighters. During the ensuing combat, Loughran was evidently wounded but still managed to reach Allied territory where he began a normal descent. However, 500 meters from the ground, Loughran's machine suddenly nose-dived and crashed with great force just behind the French defenses south of Minaucourt.

Sergent Loughran's body was recovered from his fighter and he was buried in the war cemetery at Mont Frenet. He was 25 years of age at the time of his death. In 1928 his remains were removed to the Lafayette Flying Corps monument just outside Paris.

WALTER LOVELL

ESCADRILLE LAFAYETTE

Walter Lovell, the 23rd volunteer of the *Escadrille Lafayette*, was born on 9 September 1884 in Newton, Massachusetts, the son of Wallace D. Lovell and Josephine (Hastings) Lovell. He attended Milton Academy, Newton High School, and the Stone School, Boston, before enrolling in Harvard College where he received an A.B. in 1907. While attending Harvard, he served as a member of the 1st Corps of Cadets, Massachusetts Volunteer Militia.

Following graduation, Lovell entered the brokerage business. Then in late 1914 he sailed for France to join the American Ambulance Field Service where he became an ambulance driver and later assistant-director of Section 2, serving from February of 1915 to June of 1916 on the Bois le Pretre and Verdun Fronts. Four other future *Lafayette Escadrille* pilots served with Lovell in Section 2: Willis Haviland, James McConnell, Kenneth Marr, and Harold Willis. Lovell was cited and was subsequently awarded the *Croix de Guerre*, with Star while with the Sanitary Service of the 73rd Division. His citation reads:

"Lovell, Walter, second in command of the American Automobile Sanitary Section No. 2: Has always given proof of a noteworthy spirit; has constantly set the example of courage to the other drivers, and has been an invaluable assistant to the commander of his Section."

Lovell's desire to serve more actively in the cause for France brought about his resignation from the Ambulance Service. He enlisted in the French Foreign Legion in Paris, and on 22 May 1916 he transferred to French aviation and was detailed to the aviation school at Buc on 29 June 1916. The following day he was joined by his Ambulance Corps comrade Harold B. Willis.

Lovell received his *brevet militaire* on the Bleriot on 7 October. He then attended the aviation school at Avord, followed by gunnery school at Cazaux, and combat and acrobatic school at Pau. He was then sent to the G.D.E. pilot pool at Plessis-Belleville and remained there until 26 February 1917 when he joined the *Escadrille Lafayette* at Ravenal.

Not long after Lovell joined *N.124* the specter of death seemed to hang over the Squadron. Its first victim was his good friend, James McConnell, killed in combat near Jussy on 19 March. On 7 April the Squadron moved to Ham to support the Nivelle Offensive. Death followed the pilots there.

Caporal Edmond Genet was Lovell's next friend to die, killed on 16 April when his fighter plunged into the ground south of Saint Quentin. Seven days later

Walter Lovell at the Buc School of Aviation. On his tunic he wears the French Croix de Guerre, which he won while serving in the American Ambulance Service.

Sous-Chef Walter Lovell of the American Automobile Sanitary Section No. 2 beside his Ford ambulance, Bois le Pretre Front, spring of 1916. Right: Walter Lovell in his American Ambulance uniform.

Sergent Ronald Hoskier and his gunner, *Caporal* Jean Dressy, were killed on 23 April near Saint Quentin. French Lieutenant Alfred deLaage deMeux, the Squadron's ever popular second-in-command, was the next pilot to be killed, crashing his SPAD into the Ham airfield on 23 May.

Attrition had thrust *Sergent* Lovell into the role of flight leader, a position for which he was well suited. A natural leader, since joining the Squadron he had demonstrated a willingness to take to the air, and he and colleague *Sergent* Harold Willis continued to fly more than their share of patrols.

Raoul Lufbery, the Squadron's top ace, would assemble into a special formation Lovell, Willis, and Ted Parsons. Weather permitting, with Lufbery as flight leader, the foursome would cross into German territory and pass over the enemy airfields, challenging their pilots to rise and meet them in combat.

When Lovell was not flying formation, he was patrolling solo, or was on a sortie with Harold Willis deep inside German territory where they engaged and destroyed enemy aircraft for which they would never receive official credit, since they were too far from French observation posts for confirmation.

Often times the pilots and mechanics of the *Lafayette* Squadron would stand on the airfield facing the lines in the closing twilight anxiously straining eyes and ears for some sign of Lovell and Willis. Each tick of the Squadron clock weighed ominously on the men's minds as they wondered if that day their comrades would not return, victims of enemy fighters, antiaircraft fire, or emptied fuel tanks. Then almost miraculously they would appear. And as a fellow pilot remembered:

"Some one would shout joyfully. 'There they are!' pointing out two minute specks, scarcely visible in the gathering twilight, and in a much happier frame of mind, we would watch them planing down from an immense height, until we lost sight of them in the shadows closer to the earth. It was their practice to stay out until the last possible moment. Sometimes they had to come home on nourrice, the little emergency gas-tank. Occasionally, with both tanks empty, they reached the field only by planing flatly, and landed with 'dead sticks.'"

But one day the team would be split up; the result of an air battle against superior numbers of the enemy.

On 11 August 1917 the Squadron moved to the Senard airfield. The French had planned an offensive for Verdun, and *Spa.124* had been ordered to fly support. While at Verdun, on 18 August, *Spa.124* was picked to escort a daylight bombing raid over Dun-sur-Meuse, forty kilometers inside German territory. *Sergent* Lovell, Harold Willis, Edwin Parsons, Didier Masson, and Stephen Bigelow were among the pilots chosen to fly protection for the "slow-flying Sopwith bombers."

The *Lafayette* pilots had been assigned rigid flight positions on either side of the bombers. The formation soon ran into enemy flack which drew German fighters to them. Fokker and Albatros scouts swarmed around the flight and a melee ensued which saw Stephen Bigelow seriously wounded and Harold Willis shot down and taken prisoner after knocking down an Albatros. Walter Lovell brought an Albatros down in flames and drove off five others while coming to the aid of a Sopwith bomber flying alone and under attack.

Back at the aerodrome Lovell grieved at the loss of his close friend, Harold Willis, then was greatly relieved when he later learned that Willis had been captured uninjured.

Lovell and the other *Lafayette* pilots flew as many as three patrols a day as the French offensive continued in earnest. Then, after twenty-four days of successive flying, the skies blackened, and for eight straight days rain came in torrents.

On 3 September 1917 the weather changed. Once again Lovell and his comrades left the Senard airfield into a hot, cloudless sky. As Lovell later wrote of it:

"The Boche came out in full force. At different points of our sector Hun observation machines kept crossing our lines at very high altitudes, and we were on the jump all day driving them back. Lufbery and I were out on the same patrol and no sooner had we reached the lines that we found two big fat Hun biplace machines waiting for us. They were using incendiary bullets, and for a while it looked like a Fourth of July celebration. The Boche finally beat it off. In this mix-up, Lufbery received three bullets in his machine—an incendiary was still smoking in his landing gear when we got back.

"Shortly afterwards, we discovered a solitary Hun bi-place trying to sneak into our lines. We pounced on him and gave him a warm reception until four of his comrades came to his rescue. One machine I attacked was apparently badly hit. He went volplaning back into his own lines. If this good weather keeps up, I see some very active times ahead."

On 10 September 1917 *Sergent* Lovell was awarded France's *Croix de Guerre*, with Palm for his 18 August victory over the enemy Albatros. His citation called him "an excellent fighting pilot remarkable for courage and sangfroid" (who) "while protecting a bombardment expedition August 18, 1917, shot down an enemy airplane, which fell in flames."

On 18 September the Squadron moved to Chaudun in support of the forthcoming 3rd Aisne Offensive. *Sergent* Douglas MacMonagle was the next *Lafayette* pilot to die, killed in combat on 24 September in a battle with a bi-place north of Soissons.

Top Photo: Eleve-Pilote Walter Lovell (6.1) at the Buc School of Aviation, September 1916. Edward Hinkle stands second on left. Above: Sergent Walter Lovell beside his SPAD S.VII scout at the Senard airfield, August 1917.

Adjudant Lovell left the *Lafayette* Squadron against his wishes on 24 October 1917. After ten months of exemplary service which saw him "nearly always at the head of at least one of the daily squadron formations," he was attached to the American G.H Q. at

Walter Lovell and Edmond Genet, just having returned from a patrol, discuss an aerial engagement with Raoul Lufbery and James McConnell as McConnell studies the sector map on Lovell's cockpit map holder, Ravenal.

Chaumont. It was a great loss to the pilots of *Spa.124* who had learned to trust his leadership and dependability. Harold Willis later expressed the sentiment of all the pilots of the Squadron who had flown with Walter Lovell when he had said of him:

"In a fight he would never leave you. If his guns jammed he still wouldn't leave you. You could always depend on him."

Adjudant Lovell, in this new capacity at G.H.Q., flew to the various allied squadrons on the Western Front, studying them and reporting on how the newly formed U.S. Air Service could best interact with its Allies. On 1 January 1918 he was commissioned a Captain of the Aviation Section, Signal Corps and continued his duties at General Headquarters A.E.F.

This was a great disappointment to Captain Lovell who had requested reassignment to flight status with a combat squadron. While making the transfer to the American Air Service, his medical tests had indicated color-blindness and a partial hearing loss which had required a personal waiver from General Pershing before he was granted his officer's commission. These defects were used to keep him behind a desk instead of in the cockpit of a SPAD fighter, or as commanding officer of one of the newly-formed U.S. fighter squadrons. Instead, Captain Lovell was assigned to Colombey-les-Belles where he took charge of the assignment of personnel to the A.E.F. Air Service squadrons.

On 15 April 1918 Captain Lovell married Helene du Bouchet in Luxeuil, France. Helene, who had been born in Switzerland, had earned a law degree from the Sorbonne, and was serving as a volunteer nurse under her father, Doctor Charles du Bouchet of Philadelphia, then surgeon-in-chief of the American Hospital in Paris.

Helene had been introduced to Captain Lovell by her brother, Charles Vivian, who had served with Lovell in the American Ambulance Corps. There, the two had become good friends. Charles Vivian would later join the U.S. First Infantry Division and would die in Paris on 16 May 1918 of wounds received in action, just one month after his sister, Helene, and Captain Lovell had married.

In August of 1918 Lovell was promoted to Major. On 27 October 1918 a daughter Vivian was born. Major Lovell returned to the United States and was detailed as a chief aviation instructor and served in that capacity until he resigned his commission in January of 1919 at Washington, D.C.

Following his resignation from the U.S. Air Service, Lovell returned to Paris in the March of 1919 to work as an exporter. There, a son, Charles Vivian was born, on 8 October 1920. Lovell returned to the United States in 1924 where he became involved in a variety of businesses in Boston and in New York before purchasing the Suffolk Blue Coal and Ice agency at Bay Shore, Long Island. On 28 January 1925 a second daughter, Ghislaine, was born. Sometime following the birth of their third child, the Lovell's separated amicably. Neither remarried in their lifetime. Walter Lovell died on his 53rd birthday on 9 September 1937 after a three-month illness from a brain abscess at the Southside Hospital, Bayshore, Long Island, New York.

WILLIAM LUDLAM

William Leslie Ludlam, Jr. was born in Montclair, New Jersey, on 27 September 1878, the son of William Ludlam. Ludlam attended Montclair High School and Montclair Academy.

He was working as a chauffeur and mechanic in Montclair when he decided to take a position as chauffeur for the Jansen Company in Paris.

Ludlam sailed for France aboard the *Touraine* on 18 November 1916. He enlisted in France's *Service Aeronautique* on 20 December 1916. He was released on 16 August 1917 because of ill health.

RAOUL LUFBERY

ESCADRILLE LAFAYETTE

"I only know one certain thing about him. Raoul Lufbery flew, fought and died for revenge."
-Edwin Parsons

Gervais Raoul Lufbery, the eighth volunteer of the *Escadrille Americaine*, was born on 21 March, 1885 at Clermont-Ferrand, France, the third son of New York chemist Edward Lufbery and *Madame* Annette (Vesieres) Lufbery. The senior Lufbery had been educated in Austria, and at the time he had met Annette, he had been working as a chemist and researcher in a Paris chocolate factory. The Lufbery's first two sons were Julian and Charles. *Madame* Lufbery died shortly before the infant, Raoul, reached his first birthday.

Five years after the death of Raoul's mother, Edward Lufbery married *Mademoiselle* Marie Bassddore and took a position in the United States. He placed Raoul and his brothers in the care of relatives in France and became employed as a chemist at the Rubber Works in New Jersey and later at the New York Insulating Company in Wallingford, Connecticut.

While a young man, Raoul worked in chocolate factories in Blois and in Clermont-Ferrand until approximately 1904 when he left France to pursue a series of odd jobs in North Africa, Turkey, the Balkans, and Germany. In 1906 he and his older brother, Charles, decided to visit their father in Wallingford. Both brothers had been assisting Edward financially since the death of his wife, Marie in 1901. Edward had been faced with the care of five small children—a son and four daughters—born of this second union, and he had appreciated his sons' consideration.

Edward Lufbery followed his chemist trade on both sides of the Atlantic. He also supplemented his livelihood by dealing in stamps. In pursuit of specimens either for resale, or to add to his own collection, he would cross the Atlantic Ocean forty-six times in his lifetime. As fate would have it, about the time the brothers were departing their ship at New York harbor, Edward Lufbery, unaware of his sons' arrival, was boarding a ship for France. Raoul Lufbery would never see his father again. Edward Lufbery died in France in November of 1929.

Raoul and Charles worked in Wallingford for two years, making casket handles at the silver factory of Simpson, Hall, and Miller. Then around 1908 Raoul left Wallingford for a life of adventure. His first stop was Cuba. He did not stay there long before going to New Orleans where he found work as a baker. He left New Orleans for San Francisco where he took a position as a hotel waiter. Tiring of that occupation, he enlisted in the U.S. Army and served a stint in the Philippines and in the process became a naturalized American citizen. While in the Army, he won a number of shooting prizes and became the best marksman in his regiment.

After his discharge, Raoul travelled first to Japan then to China and India where he became a ticket collector in the Bombay Railroad station. While in India, at Calcutta, in 1912 he met the French exhibition flyer, Marc Pourpe, and eventually signed on as his mechanic. Together they toured China and Egypt where Pourpe made his famous flight from Cairo to Khartoum and back.

In the summer of 1914 Raoul accompanied Pourpe to France to pick up a Morane Parasol aircraft. They had intended another tour of the Orient. But when war was declared in August, Pourpe enlisted in the French Air Service. Raoul enlisted in the French Foreign Legion as an infantryman on 24 August 1914 in order to be "detached" to the *Service Aeronautique*. He made the transfer to aviation on 31 August to serve at the Front as Pourpe's mechanic. Meanwhile his brother, Charles, had left Wallingford for France and soon enlisted in the French Army, fighting in the infantry until he was gassed and wounded by a grenade.

On 2 December 1914 Marc Pourpe was killed in an airplane accident as he attempted a night landing in fog. Pourpe was probably the closest friend Lufbery had ever had. One observer who was close to them both called their friendship "a veritable cult." For some unexplained reason Raoul blamed the Germans for his comrade's accidental death. Vowing revenge, Lufbery transferred to the aviation pilot school at Chartres and received his *brevet militaire* on the Maurice Farman on 29 July 1915. He next attended the schools at Amberieu and at the R.G.A.*(Reserve General Aeronautique)*, completing his training on the Voisin on 1 October 1915.

On 7 October *Caporal* Lufbery was assigned to the Front as pilot with *Escadrille V.B. 106*. He flew with that squadron until 22 May 1916 when at his own request he returned to the Nieuport Division of the G.D.E. (*Group de Division d'Entrainement*) for retraining as a combat pilot. While there, like other great future combat pilots, he first showed little indication of the consummate fighter he would become. Rather, he demonstrated a clumsiness in a chasse machine which nearly forced his instructors to wash him out and return him to the Front as a bombardment pilot. However, through diligent work, he ultimately overcame his weaknesses and finished Nieuport school on 22 May, 1916, having mastered aerobatics and combat flying.

On 24 May Lufbery was assigned to the *Escadrille Americaine* which had just moved from Luxeuil to Bar-le-Duc in the Verdun sector. The American pilots' first impressions of their new comrade was of a man short in stature—five foot-seven inches tall—who was broad shouldered and powerfully muscled and displayed a charming smile. But they soon learned how deceptive that smile could be. It concealed a modest, almost sullen man who knew no fear.

The mechanical skills Raoul had learned while in the company of Marc Pourpe soon paid dividends as he quickly built a reputation as one of France's premier combat pilots. The fact that he kept his bus in perfect running order considerably helped him to build his large list of victories. *Lafayette* pilot Edward Hinkle said this of Raoul:

"Lufbery was a wonderful mechanic and his plane was always the best in the Escadrille. Anyone would rather have a second-hand Lufbery machine than a new one, anytime."

Another secret to his great success was the painstaking care he gave to his machine gun and its ammunition. He practiced endlessly with his Lewis gun and personally examined each round for defects before sliding it into the drum magazine.

On 30 July 1916, following a two-month dry spell and sixteen unsuccessful combats, *Sergent* Lufbery received his first confirmed victory when he shot down a German bi-plane east of Etain, in the Verdun sector. The following day he attacked at close range a group of four two-seaters near Fort Vaux and sent one of them crashing to ground. His third victory came on 4 August when he again downed an observation craft above Abancourt, near Verdun.

On 8 August he came upon a loan Aviatik near Fort Douaumont. Before pressing the attack, he carefully scrutinized the sky for enemy fighters, not sure if the machine was a decoy. He always cautioned himself before attacking enemy observation machines: "Always remember it may be a trap!"

Convinced that the sky held no further threat, Lufbury brought his Nieuport in just below and behind the Aviatik's tail and emptied a full drum from his Lewis gun into the enemy's vitals. The bi-plane came apart and zigzagged groundward trailing greasy smoke as flame gushed up its fuselage. It slammed into the Ravine of the Viper and was soon consumed. For this and preceding actions, *Sergent* Lufbery was awarded France's *Medaille Militaire* and *Croix de Guerre*, with Palm.

Capitaine Thenault was quick to recognize Lufbery's great flying skills and would always place him above the Squadron's patrols to "dominate the situation." To Thenault, Lufbery was a "superman" who had to be ordered to rest.

Raoul Lufbery at Cachy in October 1916 preparing to patrol in his Nieuport scout. Didier Masson holds the lion cub, Whiskey, just left of the Squadron's Seminole Indian head. Robert Soubiran stands behind Masson.

"To fly high is very fatiguing," Thenault later noted, "as the sudden changes of altitude quickly tire the heart. But never have I met a pilot with more endurance than Lufbery. When the sky was clear he would go up three or four times a day to eighteen thousand feet just for his own pleasure, in a dilettante fashion. Never was he at all ill from it."

Lufbery's closest brush with death came north of Mulhouse on 9 October 1916 shortly after he had broken off an unresolved combat with the skilled pilot of a Fokker single-seater, later thought to have been the great Oswald Bolke. Upon the Fokker's departure, Lufbery's attention had been seized by "a big white two-seater of very substantial appearance" which was heading for home across French soil.

Lufbery rushed to attack, anxious to down the enemy bus within Allied lines. Disregarding caution, he hastily attacked the enemy's tail, firing his machine gun until his Nieuport nearly collided with the bi-plane. He yanked back on his stick and zoomed over the top of the German bus, preparing for a second pass, when suddenly his engine quit, silenced by three enemy bullets.

Lufbery landed on the nearest flying field where he was swarmed by "pilots, observers, mechanics" who had been following the combat. Lufbery ignored their questions as he climbed down and examined his Nieuport. His fighter had been sprayed with enemy fire: his hood was a sieve of holes; his motor destroyed; his gas tank perforated. A strut had been shot to kindling, and his left aileron had been blown away. Other slugs had torn open his left flying boot and had ripped through his flight suit, grazing his chest. But miraculously he himself had escaped without a scratch.

Lufbery officially became an ace, the first American pilot to do so, on 12 October 1916, the same day Norman Prince was mortally injured. He scored his fifth confirmed victory against a Roland C.II three-seater while fighting to protect the allied bombers returning from the great raid against the Mauser Works at Oberndorf, Germany. Lufbery was cited in the Official Communique of the French Army for his action that day and was promoted to *Adjudant* and awarded another Palm to his *Croix de Guerre*.

Following this great raid, on 18 October 1916, the Squadron was moved to Cachy to support the Somme Offensive. On 9 and 10 November Lufbery scored his first two "unofficial victories" with the planes falling too far behind enemy lines to be seen and confirmed by the three required impartial witnesses, either balloon or ground observers.

Raoul Lufbery and Didier Masson taking a break between patrols.

Left: Raoul Lufbery holding two of the Lafayette Squadron's many mascots. Above: Adjudant Raoul Lufbery on leave in Paris seated inside the Hispano Suiza roadster, a courtesy of the company which made the engines for his SPAD fighter planes.

His "unofficial victory" list would grow to enormous proportions. If those many victories could have been confirmed Lufbery undoubtedly would have been ranked near the top of the list with the great Allied aces. Edward Hinkle, who flew with Lufbery, believed his total to have been close to seventy enemy aircraft destroyed. He recalled one day during his own tenure with the Squadron when Lufbery had knocked down seven enemy aircraft and had failed to receive credit for a single plane.

Another *Lafayette* pilot, James Norman Hall, said that the seventeen official victories Lufbery was ultimately credited with was "no more than half the number of planes he actually destroyed."

Carl Dolan also recalled a patrol he had flown with Lufbery in which he had seen Lufbery down three of the five German planes he attacked. And yet, only one of Lufbery's victories was confirmed. When Dolan had asked Lufbery if it had not bothered him that so many of his victories had gone unconfirmed, Lufbery had replied:

"What the hell do I care. I know I got them."

Unquestionably, Raoul Lufbery was the *Lafayette Escadrille's* greatest fighter pilot, and perhaps the greatest American fighter pilot of World War One. On a number of occasions, while out on patrol, he saved the lives of several of his young comrades by being nearby at the right moment. And while his American comrades were proud to call "Luf" one of their own and truly admired his courage and skill, they found him an enigma. Hinkle said this of Lufbery:

"He was a strange man. A solemn man. He was a very quiet man, not easy to know. He never talked very much. I remember taking long walks in the woods with him to gather mushrooms for the cooks. He might not say a word in two hours."

Carl Dolan remembered how Lufbery had been as much a loner on the ground as he had been in the air. When the other pilots would head for leave in Paris, Lufbery would often go home to Clermont-Ferrand where he had spent his youth, to visit family and friends.

"In that part of his life he didn't mix with our crowd one bit," Dolan said. "The French felt he was an American, but Luf felt he was a Frenchman."

Henry Jones remembered Lufbery in a 1969 interview. "Lufbery spoke with a strong French accent, a rough, tough person, but he would do anything for you. He was very well liked, and when he did go to Paris, he raised hell. He drove a Hispano-Suiza roadster which was loaned to him by the company every time he came to Paris with whores hanging all over it The car was always waiting for Lufbery at Henry's hotel when he came on leave as a courtesy from the company."

Edwin Parsons, another *Lafayette* pilot who flew many missions with Lufbery, summed him up this way. "To me, Luf was one of the greatest mysteries of the war. No man alive can truthfully say that he knew him. I ate, slept, drank and fought beside him for months on end. I discussed combat tactics and played bridge and went on binges with him. He saved my hide once when I was a very green young pilot . . .

"I was in daily contact with a figure of flesh and blood, but know him? Not a chance. In contrast to him, the Sphinx was a child's primer. He kept his real self shut up like a clam in a shell. He was a man seemingly devoid of fear or, in fact, emotion of any kind. But what a man he was in the air! He had forgotten more about combat flying than most men ever knew."

Lufbery lived to fight. With no concept of fear, the prospect of death could not dissuade him from his missions. And during his relentless hunts he often took as much as he gave, many times returning to the Squadron's airfield, his riddled plane a testimony to his fierce combats. Only one of the frequent attacks of rheumatism could keep him out of the sky, attacks which at one time became so severe that Lufbery required hospitalization. Often he flew in great pain, his joints and muscles half crippled.

Raoul Lufbery and William Thaw were the only two American pilots of *N.124* to be given officer's commissions by the French Army. And while Thaw's commissioned had come early in his career, it took Lufbery nearly three years to receive the distinction in spite of his many victories, medals, and citations.

One reason for the delay had stemmed from an incident at the train station at Chartres when Lufbery had accompanied a friend out onto the *quai d'arrivee* without the required ticket. The station attendant had tried to stop Lufbery then had physically attempted to eject him. Lufbery had considered the man's actions an insult to his Military Medal and with one powerful blow had knocked the attendant out and broke six of his teeth. Lufbery was arrested and taken before the station's military commissioner who sentenced him to thirty days in the stockade. Only *Capitaine* Thenault's direct intervention secured his release.

On 24 October 1917 the now *Sous Lieutenant* Lufbery had another of his incredible days which Edward Hinkle had referred to. He had been called upon with the other *Lafayette* pilots to assist in stopping a German counterattack which had been launched in response to a French advance 2 1/2 miles into their territory.

The first patrol had begun at 6:40 a.m. and had lasted approximately 1 1/2 hours. During that time Lufbery fought two combats with German fighters, proving victorious in both engagements. He returned to the Chaudun airfield to refuel and rearm his SPAD and again rose to combat, this time fighting it out with three German fighters, tumbling two of them to ground. Just before noon Lufbery left Chaudun for his third sortie and soon faced and fought two enemy fighters, knocking both of them down. And while it was a virtual certainty that Lufbery had shot down six enemy fighters in one morning's work, he only received confirmation for one official victory.

Lufbery, the great marksman, had again made his fighter an extension of himself, turning his entire machine into a flying gun. Blessed with incredible vision, superb reflexes, and a timeless patience while stalking the enemy, these were the components which made up "Luf's private formula." Like all the great air victors who shared these traits, Lufbery never substituted opportunity and favorable position for reckless courage. Those who did so bought an early grave.

On 1 March 1917 while the Squadron was at Saint-Juste, Lufbery was made a *Chevalier* of the French *Legion d' Honneur*. His award document, dated 29 January, called him a fighter pilot of remarkable daring. This, he continued to prove by knocking down many more enemy planes before he left the Squadron to join American Aviation. Edward Hinkle best summed up the career of Raoul Lufbery while he was with the *Escadrille* when he later said of this great ace:

"He preferred to patrol alone. Ninety percent of the time he was all by himself. Once Luf spotted an enemy plane, he took his time maneuvering into precisely the position he wanted. He attacked with the sun at his back, and many an enemy pilot never knew what hit him."

Major Lufbery conversing with Major Jean Huffer, C.O. of the U.S. 94th Aero Squadron. The 94th Hat-in-the-Ring insignia is visible between them.

Sous-Lieutenant Lufbery was commissioned a Major in the U.S. Air Service on 10 January 1918. On 22 January he was sent to the Headquarters of the 1st Pursuit Organization Center at Villeneuve-les-Vertus. He left there on 24 January for the American Aviation Instruction Center at Issoudun where his great talents as a combat patrol leader were squandered by his superiors who gave him a writing pad and pencil and a rolltop desk, and told him to write a pamphlet on "how to kill Germans."

But on 18 February Major Lufbery had escaped his office confines through the intervention of Major Thaw and was assigned to the newly formed U.S. 95th Aero Pursuit Squadron which had just arrived at Villeneuve, the first American-trained pursuit squadron to enter the Zone of Advance. On 5 March the U.S. 94th Aero Pursuit Squadron also arrived from Issoudun and Major Lufbery served as combat instructor as he helped to prepare the novice pilots in the intricacies of patrolling.

On 28 March 1918 Major Lufbery led Lieutenants Douglas Campbell and Edward Rickenbacker on the 94th's first combat patrol. Rickenbacker, soon to become America's top ace in World War One, later would write that "everything I learned, I learned from Lufbery." On the 29th Lufbery also led the 94th's second patrol accompanied by Lieutenants Thorne Taylor and John Wentworth. He continued to lead patrols daily. When not doing so, he was in the air hunting solo.

On 7 April the 94th, now under the command of Major John Huffer, was placed under the VIII French Army as an independent unit and was moved to the Gengoult airdrome, four kilometers northeast of Toul. The squadron's first combat occurred on 12 April near Xiveray when Major Lufbery single-handedly attacked three Albatros fighters and almost assuredly shot one down, adding yet another German plane to his list of "unconfirmed." He had successive combats on 23 April near St. Baussant, and again on 18 May when he fought it out with six German fighters over Norrey until his guns jammed and he was forced to withdraw.

The cottage garden where Lufbery fell after jumping from his burning Nieuport scout. He landed on the picket fence behind the peasant woman. She holds the end of a picket wrapped in paper which she pulled from Lufbery's throat, hoping to save his life.

On 19 May 1918, at approximately 10 a.m., the antiaircraft batteries on top of Mont St. Michel began firing, their white puffs breaking at high altitude. Almost immediately Major Huffer received notice that a German photo-reconnaissance plane was heading toward the 94th's field. Seconds later it appeared.

First Lieutenant Oscar J. Gude, the only pilot on the squadron's airfield who was ready for flight, climbed into his fighter and immediately rose to meet the enemy bus. Gude had yet to fight a combat with the enemy. As he gained altitude, the German bi-plane appeared to be struck by shrapnel. It fell into a spin and headed groundward but at 200 feet it recovered, and its pilot turned back toward his own lines.

Gude followed the German bus. Through inexperience, he began firing with the enemy still far out of range, emptying his machine guns in a long continual burst while causing no damage to the observing plane. Major Lufbery, who had been following the engagement from his barracks, jumped onto a motorcycle and sped to the hangars as the enemy bi-plane again fell under a violent artillery barrage but still managed to maintain a steadily retreating course in the direction of Nancy.

As Major Lufbery slipped into his flight suit, he soon discovered that his own Nieuport fighter with its well-maintained machine gun was out of commission. He climbed into Lieutenant Philip Davis's Nieuport, and with no time to check the machine gun, sped off and rapidly ascended to the attack.

Major Lufbery engaged the enemy bus, a Rumpler, at two thousand feet. He fired several short *rafales* before swerving away, evidently in an attempt to clear a jammed round from his machine gun. Circling, he managed to do so then again attacked the enemy bus from the rear.

From this point in the action on, a number of conflicting reports have been written as to what happened to Major Lufbery after the engagement began. Some eyewitness accounts record that Lufbery jumped from his flaming plane, while other witnesses say his plane

Pilots of the U.S. 94th Aero Squadron with French officials awaiting the arrival of Major Lufbery's casket, 20 May 1918.

was not burning, and that he fell out of the cockpit when his fighter flipped over.

The official report states that his Nieuport shuddered when struck by enemy fire and burst into flame. A round from the enemy observer's guns had pierced his fuel tank then had entered the cockpit and shot away Lufbery's right thumb as he clutched the joystick. Lufbery attempted to put the fighter into a slip to force the growing inferno away from himself. But the wind from the rapidly plummeting plane only fanned the fire.

Lufbery struggled out of the cockpit as flame swept down the fuselage and ignited his flight suit. Lufbery apparently jumped toward a stream beneath him that promised a remote chance for survival from a height of more than 2,000 feet. He missed the stream by a hundred yards, landing on a fence surrounding the garden of a house in the little town of Maron, just north of Nancy.

Some thirty minutes later his comrades arrived at the scene where he had fallen. Already Lufbery's scorched body had been removed by the villagers to the town hall where they had covered it with flowers from their gardens. They gently transported it to the American Hospital near the squadron aerodrome. The next day, following a full military funeral, Major Lufbery was buried in the American Cemetery, Sebastopol Barracks. He was thirty-three years of age at the time of his death.

Gone was the great Lufbery of whom Edwin Parsons had said few could equal and none could surpass. "His cool head, steady nerve and unerring aim were worth a whole squadron. He flew as the bird flies," Parsons said, "without any thought of how it was done."

For his war service, Major Lufbery was awarded France's *Croix de Guerre*, with ten Palms, the *Medaille Militaire*, the *Legion d'Honneur*, and the British Military Medal. French *Capitaine* Georges Thenault, his *Lafayette Escadrille* commander, would later write of him: "His Spad was always the highest and every day he won new victories. He seemed to hardly care about having them confirmed, Calmly he reigned as a sovereign lord in his chosen element and beat down his foes to accomplish his duty and not for the sake of glory." On 4 July, 1928 Lufbery's body was reinterred in the *Lafayette Escadrille* Memorial at Villeneuve, near Paris, there to rest in eternity with his comrades.

GEORGE McCALL

George Archibald McCall was born in Jenkinstown, Pennsylvania on 24 August 1886, the son of George McCall, President of the Catawissa Railroad, a Philadelphia and Reading Railroad subsidiary, and Mildred (Carter) McCall. A colonial descendant of military ancestors, McCall's grandfather, General George A. McCall, commanded the Pennsylvania Reserves in the Civil War. An earlier ancestor, General John Cadwallader, commanded Pennsylvania's troops at the Battle of Princeton in the Revolutionary War.

McCall was educated at the Episcopal Academy, Philadelphia, the Wharton School, and at the University of Pennsylvania, Class of 1909. He left that institution without taking a degree and took up ranching in Mexico.

Desirous of entering the European war, George McCall sailed for France in the fall of 1915 and served for 9 months with Section 4 of the American Ambulance Service. He was awarded France's *Croix de Guerre* while with the Ambulance Service.

On 1 September 1916 McCall enlisted in France's *Service Aeronautique*. From 14 September to 28 May 1917 he attended the aviation schools at Buc, Avord Pau and the *G.D.E.* He received his *brevet* on the Bleriot on 16 March 1917.

From 30 May to 9 September 1917 *Caporal* McCall served at the Front with *Escadrille SPAD 23*. From 15 September to 22 October 1917 *Sergent* McCall flew combat patrols with *SPAD 86*. From 6 November 1917 to 23 April 1918 he was with *SPAD 48*. From 29 May to 30 September 1918 *Sergent* McCall flew with *SAL. 30*. And from 24 October 1918 to the 11 November 1918 Armistice McCall served with *SPAD 103*. McCall served faithfully and with quiet heroism both in the Ambulance Service and in the French Air Service.

Following World War One McCall lived in Paris for two years. He then returned to the United States and was supported by a family remittance. McCall died in Philadelphia on 16 April 1956 at age 69.

JAMES MCCONNELL

ESCADRILLE LAFAYETTE

James Rogers McConnell, a founding member of the *Escadrille Americaine*, was born in Chicago, Illinois, on 14 March, 1887, the son of Judge Samuel Parsons McConnell and Sarah (Rogers) McConnell. McConnell's paternal grandfather was Union General John McConnell. His maternal grandfather was Judge John G. Rogers of Chicago.

Judge Samuel P. McConnell served on the Circuit Court of Cook County, Illinois from 1889 to 1894 when he left the bench for private practice. In 1899 he moved his family, then to include two sons and two daughters, to New York City to become President of the George A. Fuller Company, a large building contracting corporation. He also served as legal representative for the Vanderbilt interests.

While the McConnells were in New York their son, Rogers, died of a fever. Following his untimely death Sarah McConnell became a "nervous invalid," who relegated the responsibilities of her youngest children to her daughter, Julia. Eventually the McConnell's were divorced, and in 1902 Judge McConnell married Mayo Methot, a domestic servant. From that union another son and two daughters were born.

While still in his pre-teens, young James McConnell spent a portion of his childhood in France with his mother and two small sisters. And it was there that he had learned to speak French fluently. In his youth he had suffered from muscular rheumatism which again surfaced with savage intensity in his adult life.

In 1905 Judge McConnell left New York City for Carthage, North Carolina to become president of the Randolph & Cumberland Railway Company. He also acted as general manager of the Deep River Development Company.

When his family had been living in New York, James attended the Morristown School, New Jersey, and the Haverford School, Pennsylvania. Following a summer vacation in Chicago, he and a friend had driven the first automobile from Chicago to New York City. This feat had attracted much attention in the towns and cities along his route as McConnell was often greeted by large crowds whose interest had been stirred by the local press.

James had chosen to enroll in the University of Virginia upon Judge McConnell's relocation to the South. While pursuing a law degree as a member of the class of 1910, he soon became one of the best-known students in Charlottesville by outfitting himself in a Scottish kilt and full regalia in the tartan of his clan. Having taught himself to play the bagpipes, McConnell

Sergent James McConnell standing before his Nieuport fighter dressed in his fur-lined flying "combination."

then paraded about the campus and the streets of Charlottesville piping music which few appreciated and was an irritation to many.

Considered a brilliant student, McConnell easily mastered his studies. However, he developed a prankster's reputation and was nearly expelled from the University when he placed a chamber pot on the head of a statue of Thomas Jefferson prior to its unveiling on the University campus before President Taft and other assembled dignitaries.

While he was at the University of Virginia, McConnell was made "King of the Hot Foot Society" and President of the Aero Club. But he abruptly left that institution in 1910 without his law degree. There are rumors he was expelled.

McConnell returned to New York City and involved himself briefly in business with future *Lafayette Escadrille* pilot, Charles Chouteau Johnson. Eventually he returned to Carthage to work as land and industrial agent for the Randolph & Cumberland Railway Company. James also served as a Second Lieutenant in the State Militia. But in January 1915 the twenty-seven year old McConnell left his position with the railroad in Carthage, and in a spirit of adventure sailed for France to observe the war.

Not wanting to be a mere "sightseer looking on" McConnell joined the American Ambulance Field Service as a driver and served with Section No. 2 from February to December of 1915 through the heavy fighting near Pont-a-Mousson and the Bois-le-Pretre. On 5 October 1915 he was awarded France's *Croix de Guerre*, with Star for "courage and fearlessness worthy of the highest praise."

But while McConnell was serving in the Ambulance Corps with future *Lafayette Escadrille* pilots Harold Willis and Walter Lovell, he felt the need to participate more actively in the war. "The more I saw the splendor of the fight the French were fighting the more I began to feel like an embusque (shirker)" he wrote. "So I made up my mind to go into aviation."

McConnell joined the *Service Aeronautique* on 1 October 1915 Upon receiving a new uniform and equipment at Dijon, he immediately began his aviation schooling at Pau. On 6 February 1916 he received his *brevet militaire* on the Bleriot. Following Nieuport instruction, on 16 April he joined Victor Chapman and Kiffin Rockwell at the R.G.A.(*Reserve General Aeronautique*) near Paris and was immediately assigned to the *Escadrille Americaine.*

McConnell reached the Front on 20 April 1916 at Luxeuil, in the Vosges, one of the first seven American

Future Lafayette Escadrille pilots James McConnell and Willis Haviland while serving in the American Ambulance Service in the Fall of 1915.

pilots to be assigned to the newly-formed *N.124*. Soon after he was given his Nieuport fighter, he had his personal insignia painted on its sides: his nickname, "MAC", and a large white footprint to represent the "Hot Foot Society" of his alma mater.

On the morning of 13 May he flew with *Capitaine* Thenault, Victor Chapman, Kiffin Rockwell, and William Thaw southwest of Mulhouse as part of the Squadron's maiden flight over the lines. The only resistance the patrol met came from antiaircraft fire.

On 20 May the Squadron completed its relocation to Behonne in the Verdun sector where the American pilots were ordered to protect allied observation and artillery reglage machines from enemy fighters. Air activity greatly intensified, and on 23 June McConnell's close friend, Victor Chapman was killed in combat in the air northeast of Douaumont in a battle with five enemy fighters.

McConnell chafed to avenge Chapman's death. On 30 June, while on a morning patrol over Verdun, he and Raoul Lufbery attacked a German machine and sent it groundward out of control. However, it was not officially confirmed by balloon observers or by ground troops.

In August, in addition to his regular patrol duty, McConnell often hunted German *drachen* with Norman Prince. He hovered above Prince, flying cover, while his comrade attacked the balloons with *Le Prieur* rockets rigged to the V struts of his fighter. After several unsuccessful attempts, their teamwork met with success. As McConnell flew cover overhead, Prince dove on a black observation balloon and fired a salvo into it, successfully incinerating it.

In the third week of August with the Squadron still at Bar-le-Duc, McConnell left on a late afternoon patrol with Kiffin Rockwell and Norman Prince. In conjunction with an allied attack near Fleury and Thiaumont, their objective was to drive off enemy observation machines.

McConnell and Prince had remained in the air past dusk to be sure of no further German activity before heading back to their aerodrome. The stars shown brightly above them as they nosed their ships about. Below they watched the spurt of infantry signal flares.

While still some distance from his aerodrome, McConnell's engine sputtered and he immediately suffered a *panne de moteur*. He spotted the flares of another aerodrome below and nosed his Nieuport downward. He approached the airfield at a terrific speed, nearly one hundred miles an hour, misjudged his distance in the darkness, and overshot the landing strip. As his plane sped toward a stand of trees, he glimpsed soldiers running across the field toward his inevitable crash site.

McConnell's luck held. Instead of hitting the trees, he plowed between them and tore off both wings before jumping a road and splintering his propeller against a steep bank. His motor absorbed most of the shock, and his safety belt held as the tail of his Nieuport catapulted skyward and was sliced in two by low-strung phone wires.

McConnell had a first thought himself uninjured and had dined as a guest at the Squadron's mess. But on his return by car to Behonne he found himself in a state of ever-increasing pain and realized he had severely wrenched his back.

Although he continued to fly for several days after his accident, soon he could not dress himself and could only walk with a cane. He took a brief Paris leave, hoping the condition would abate. But by the time he returned to Bar-le-Duc he could not walk.

On 26 August *Capitaine* Thenault ordered him to hospital. McConnell vigorously protested, saying: "Not being able to walk does not prevent my flying." Thenault merely replied, "Jimmy, go to bed." and ordered McConnell to the auxiliary hospital at Vitrey-le-Francois.

While in hospital McConnell wrote a 19 September 1916 letter to a friend in Paris. "I am getting discouraged about my back," he had said. "It seems to be getting well and then flop, she drops back to first condition. If it doesn't build up soon, I am going to tell them I am well."

McConnell became ever more anxious to rejoin the Squadron upon hearing of the death on 23 September of Kiffin Rockwell, his closest friend. But he knew that physically, he could not. He requested his doctor's permission to journey to Paris to continue his recuperation at the residence of Mrs. Alice Weeks.

McConnell arrived in Paris on the 27th and recuperated for 45 days at Mrs. Weeks's apartment at 80 Rue Boissier. He used the time to work on the book he had begun in the hospital concerning his experiences with the *Escadrille Americaine*. On 11 November, although not yet well, he left Paris to rejoin the Squadron at Cachy in the Somme sector.

Inclement weather had kept the Squadron grounded through much of November and early December. Mrs. Weeks had booked passage to the United States, and McConnell hurried to complete his manuscript, hoping she could deliver it to Doubleday Page & Company, his New York publisher. Unable to make the deadline for her departure, he sent on an article with her for "World's Work" magazine while making other arrangements for the manuscript. He wrote this explanation to her in a 11 December letter:

"Between a bad cold, the general cold and the fact that Whiskey chewed up my fingers, I have been unable to write and will have to send over the rest of copy for my book by Dudley." (*Escadrille* pilot Dudley Hill from Poughkeepsie, New York who was soon to return home on leave.)

Throughout the month of January 1917 McConnell fought two undecisive combats. On the 26th the Squadron was transferred from Cachy to Saint-Juste. Upon returning from a Paris leave, on 7 February *Sergent* McConnell wrote a letter to Mrs. Weeks, now in the United States, detailing life at the new aerodrome.

"We have much better barracks but just now it is fearfully cold. The machines freeze up while one is flying. . . . I have never seen such a spell in France before. Worst winter since 1870. Whiskey ate up my fatigue cap and last night my kepi. Yes, Page (Frank C.

Page, editor of Doubleday Page & Company) got the manuscript. Book will be out the 20th of this month."

The aching cold that had nestled in around the aerodrome aggravated the rheumatism in McConnell's back, and by the second week of February Thenault had ordered him back to the hospital. While he was there his book, titled *Flying For France,* was released by Doubleday. It was favorably received in the United States and helped to elevate the patriotic sentiment in a nation about to enter the War. It is a marvelously well-written book and clearly indicated that McConnell had genuine ability as a writer.

Compelled by his sense of duty, McConnell left the hospital a virtual cripple against his doctor's orders. By 21 February he had returned to the Squadron to fly volunteer missions with Lieutenant deLaage against enemy observation balloons. While there, he required help to dress and had to have his mechanics lift him into his fighter. The rheumatism in his back had dangerously restricted the head movement required for total vision, an essential for survival in the air.

The day following his return to the Squadron *Sergent* McConnell wrote a letter dated 23 February to Frank Page, his editor at Doubleday, thanking him for his first royalty check of 861 francs. He went on to say:

"Had rotten luck the other day. Caught a fat Boche five kilometers in our lines and my machine gun wouldn't work. I stayed near him trying to fix it until he regained his lines. I could have cried." McConnell concluded by saying he hoped to soon transfer to the American Aviation Service with anticipation of a high rank, since he and William Thaw were the senior men in the *Lafayette* Squadron.

The eve before McConnell's 30th birthday, 13 March 1917, found him in a pensive mood. The last sentence in his diary that day suggested he had a prescient sense of his fate: "This war may kill me, but I have it to thank for much."

Edwin Parsons later recalled how the morning of 19 March 1917 had been a "lowering, gray day with storm-driven scud." He, *Sergent* James McConnell, and *Caporal* Edmond Genet had been scheduled for the 10 a.m. patrol over the Somme battlefield. McConnell and Genet left the airstrip first, followed by Parsons. Parson's oil line clogged shortly after he had cleared the field, immediately burning up his motor. He brought his fighter down two kilometers from his aerodrome as his two friends continued on toward German territory.

Deep inside enemy lines McConnell and Genet ran upon two German bi-places north of Ham. Genet later related the events in a letter to his mother.

The last sight Genet had of McConnell as Edmond attacked the nearest German bus was of his friend pursuing the second enemy plane. Genet fought a fierce combat between clouds and mist which saw his fighter heavily damaged and his left cheek cut open by a bullet fragment. Stunned, Genet still remained in the zone of combat for fifteen minutes, searching for McConnell. Bleeding, in pain, his fighter in danger of losing its upper wing, Genet only reluctantly returned to the aerodrome, hoping that McConnell had already returned to the field.

When McConnell failed to appear from patrol, his comrades began the long wait on the tarmac. *Capitaine* Thenault hovered near his phone anxiously awaiting word of McConnell's fate. At day's end their only hope was that their friend had been shot down and taken prisoner.

On the night of 23 March all hope was dashed when Commandant Fequant, head of *Groupe de Combat 13* received a telephone call from the Front that a French cavalry patrol had discovered a crumpled Nieuport bearing McConnell's number 2055 on its rudder. The wreckage was in a small apple orchard at the southern edge of the little village of Detroit Bleu. Nearby was a dead pilot, his body stripped by the Germans of his outer clothing, boots, I.d., and watch.

Thenault, Lieutenant deLaage, and several American pilots drove to the crash site to take charge of McConnell's body. En route, deLaage was grieving openly and recalled the previous Christmas at Cachy when the pilots had dined together in their barracks. There had been thirteen men at the table, and McConnell had made reference to their superstitious number. When deLaage had asked his friend what that superstition was, McConnell had explained how "with thirteen at table it was supposed that one of the assembly would die within the year." deLaage had remembered how he had laughed that Christmas, while this day his words returned to haunt him.

"Well," he had told his friend, McConnell "it will hold good with this crowd. But unfortunately there will probably be more than one." deLaage's words proved prophetic. He himself would also be killed, on 23 May 1917, just two months from that very day they had gone to recover McConnell's body.

As McConnell's friends gently picked up his remains for burial, they witnessed several bullet holes in his body, any one of which would have proved fatal. A peasant woman came forward who had seen McConnell's last combat. She related how she had watched him engage a German airplane when a second enemy bus had dived on him from behind and had shot him down.

Capitaine Thenault, upon first learning of McConnell's death, had opened a note the pilot had left to be read in the event of his death. As a last gesture of his generous nature and concern for others, McConnell had willed his sleeping bag to Edwin Parsons, knowing his friend hadn't the francs to purchase one and had suffered from the bitter winter's chill. The note also contained McConnell's last simple request:

"My burial is of no import. Make it as easy as possible for yourselves. I have no religion and do not care for any service. If the omission would embarrass you, I presume I could stand the performance. Good luck to the rest of you. God damn Germany and *vive la France*."

In conjunction with McConnell's wishes, *Capitaine* Thenault assembled the pilots where their comrade had fallen, and together they gently and quietly placed him in the devastated earth at Petit-Detroit, southeast of Ham. The French 165th Infantry used stones from a nearby ruined village to erect a monument to McConnell. Eventually, the citizens of Ham raised the funds to provide a cement tomb and a memorial stone for McConnell.

Sergent McConnell was posthumously awarded France's *Croix de Guerre*, with Palm. The accompanying citation spoke of a "pilot as modest as he is courageous." Paul Rockwell, Kiffin's brother, came to the Squadron to see to Jim's effects. He was heart struck by McConnell's death and said of him "I never had a better friend except Kiffin."

On 2 April 1917 morning services were held in McConnell's honor at the American Church in Paris. Many dignitaries were present including American Ambassador, William G. Sharp. Also represented were members of the Lafayette Flying Corps and the American Ambulance Service. Three tearful *mesdemoiselles* also attended, each one believing she was McConnell's fiance. One would later marry another *Lafayette Escadrille* pilot.

The speech delivered by Bishop C.H. Brent spoke eloquently of McConnell's devotion. But perhaps the most fitting eulogy was delivered several weeks later by Emil Marshall, McConnell's comrade, in a 1 May 1917 letter to a mutual friend:

"I often think of you and poor Jim...He had done so much for me. One of my greatest pleasures would have been to show him some day how much I appreciated it. I had so looked forward to that. Now it is impossible. All I can do is to always remember him as the best and bravest comrade one could have wished."

James McConnell was the first University of Virginia alumnus to die in World War One. A statue in his honor titled *Aviator* rests on the University grounds. It was sculpted by Gutzon Borglum, who later planned the South Dakota Mount Rushmore Memorial. In 1928 McConnell's body was placed in the crypt at the *Lafayette Escadrille* memorial near Paris, there to rest in eternity with his comrades.

James McCreary

James B. McCreary, Jr. was born in Buffalo, New York, the son of James B. McCreary. He graduated from Lafayette High School. He served in France for 18 months with sections 5 and 21 of the American Red Cross. On 23 July 1917 he enlisted in France's *Service Aeronautique*. He was released in April of 1918 because of ill health.

WILLIAM MCGINN

William McGinn was born in Cincinnati, Ohio, on 6 July 1886, the son of John Joseph and Mary McGinn. He attended Cincinnati schools, and Ohio Mechanical Institute. He was working as a journalist for the Cincinnati *Commercial Tribune* when he decided to enter the European War.

McGinn enlisted in the Lafayette Flying Corps in New York City on 15 June 1917. He sailed for France aboard the *Rochambeau* on 23 June.

On 20 July 1917 McGinn enlisted in France's *Service Aeronautique*. He received his *brevet* on 2 October. On 2 November 1917 he was released from French aviation upon his own request.

Left to right: William Bullen, William Gibson, John Endicott and McGinn.

HERSHEL MCKEE

Herschel Jessup McKee was born Indianapolis, Indiana on 19 November 1897, the son of James McKee, and Laura (Rariden) McKee. McKee attended Indianapolis public schools and the Hawthorne High School, graduating in 1913.

For a time McKee worked as an automobile mechanic in Indianapolis. In the spring of 1917 he sailed for France to enter the European war, and on 12 April he enlisted in France's *Service Aeronautique*. From 15 April to 12 October 1917 McKee attended the aviation schools at Avord, Pau, and the *G.D.E.* He received his *brevet* on the Caudron on 30 August 1917.

From 15 October 1917 to 8 February 1918 *Caporal* McKee served at the Front with *Escadrille N. 312*. On that date he was shot down by German antiaircraft fire.

Sergent McKee landed unhurt and was taken prisoner by German troops. He was placed in a prison camp but managed to escape in September 1918. He was recaptured three days later and was held in Germany until after the 11 November Armistice.

Following World War One, McKee became involved in auto and motorcycle racing at Indianapolis. He served as a riding mechanic for such drivers as Andre Boillot, Riley Brett, Frank Elliott and Chet Gardner. McKee was also involved in commercial aviation, flying mail runs to and from Chicago until 1935. He married Viola May Cain. From this union a daughter was born.

During World War II McKee joined the Army Air Corps as a Lieutenant-Colonel. While serving at the Wright-Patterson Air Force Base at Dayton, Ohio, he was credited with helping to perfect drag chutes now used as brakes for jet planes. He also served in England and flew in a raid on the Ploesti oil fields in Romania.

Colonel McKee remained in the Air Force following World War II and retired in November 1958. He died in Indianapolis, Indiana on 13 April 1965 at age 67. He is buried at Arlington National Cemetery.

WILLIAM MCKERNESS

William John McKerness was born in New York City, New York on 5 March 1895, the son of Mr. & Mrs. William H. McKerness. McKerness attended Wallingford, Connecticut public schools. He attended Harvard University for a short time. In 1916 he served on the Mexican Border with Co. F, Connecticut National Guard. He was employed as an office clerk when he decided to enter the European War.

McKerness sailed for France on 26 May 1917, intending to serve with the Harvard Ambulance Corps. Instead, on 13 June 1917 he enlisted in the *Service Aeronautique*. From 20 June 1917 to 10 May 1918 he attended the aviation schools at Avord, and Cazeaux. McKerness was unable to master aviation and was involved in a series of smashups during training. He was offered his release from French service but instead applied to become an aviation machine-gunner. McKerness's request was granted and he was sent to aerial Gunnery School at Cazeaux.

On 12 May *Caporal* McKerness was assigned to the Front as observer and gunner with *Escadrille C. 46*. He was one of only three Americans to serve in this capacity, with Marius Rocle, and Fred Zinn. McKerness flew numerous missions as forward gunner in the three-seater Caudron. Often his pilot was fellow American Glen Sitterly, as together with French gunner, Lacassagne, they flew missions of protection deep inside enemy lines, warding off enemy fighters while the Breguets they were guarding completed their observation runs.

On 15 June 1917 *Sergent* McKerness was wounded on one such mission as the plane Sitterly was piloting was attacked by fifteen German fighters and was shot to pieces. Sitterly managed to bring the Caudron safely home in spite of a jammed rudder and a flaming motor. For his heroic action that day in protecting the aircraft in spite of his wounds, McKerness was awarded France's *Croix de Guerre*, with Star.

On 15 August 1918 McKerness left the aerodrome with a French pilot at the controls. Northeast of

Ribecourt their bus was attacked by eight Fokkers. The enemy's first burst killed the pilot, and the Caudron fell out of control between Ribecourt and Saint-Leger, between allied lines. French troops rushed to the fallen plane, finding the French pilot, Sergent McKerness, and the other gunner dead in their seats. McKerness was 22 years of age at the time of his death.

Sergent McKerness was buried in an allied cemetery. His remains were later exhumed and removed to the United States for reburial.

JAMES MCMILLEN

James Hiatt McMillen was born in Cleveland, Ohio on 1 March 1891, the son of William Ferris McMillen, a Chicago Congregational minister, and Almeada Catherine (Regal) McMillen. McMillen attended Chicago public schools and Chicago's Lewis Institute before enrolling in Oberlin College, Oberlin, Ohio, Class of 1913.

McMillen left Oberlin College at the end of his junior year and began working in Wall Street selling stocks and bonds. At age 25 he decided to join the Lafayette Flying Corps. He sailed for France in February 1917 aboard the *George Washington*.

On 25 June 1917 McMillen joined France's *Service Aeronautique*. From 29 June to 15 February 1918 he attended the aviation schools at Avord, Pau, and the *G.D.E.* On 3 December 1917 McMillen received his *brevet* (#10,260) on the Caudron.

From 15 February to 12 March 1918 *Caporal* McMillen ferried aircraft from the factories to the Front. On 12 March 1918 he was assigned to the Front with *Escadrille SPAD 38*.

Sergent McMillen flew combat patrols with that squadron until 10 July 1918. On 11 July McMillen was commissioned a First Lieutenant in the U.S. Air Service. At the request of his French Commander, *Capitaine* Georges Madon, McMillen continued to fly combat patrols with *SPAD 38* until 27 September 1918 when he was ordered to Issoudun, Field Nine, to serve as an instructor in U.S. Aviation.

In December 1918 Lieutenant McMillen returned to the United States, to New York City, aboard the *George Washington*. Following his discharge at Garden City, Long Island, he resumed his position as a stock broker in a Wall Street firm.

Shortly after returning to New York City, on 15 January 1920 McMillen married Alberta H. Weber. From this union a son and daughter were born. McMillen became involved in stock investment business with the Evans-Stillman Company. This firm later became the Stillman-Maynard Company. McMillen worked for and was a partner in that firm for more than forty years. Following his retirement in 1966, he actively pursued his golfing hobby. He died in New York City on 12 December 1979 at age 88.

DOUGLAS MACMONAGLE

ESCADRILLE LAFAYETTE

"Somehow one had to try and get rid of the thought that a mother wept for every man shot down."
—Oberleutnant Ernst Udet

Douglas MacMonagle, the 35th volunteer of the *Escadrille Lafayette*, was born in San Francisco, California on 19 February 1892, the only son of Dr. Beverley MacMonagle, and Minnie (Corbitt) MacMonagle. His father died while Douglas was still a youth.

MacMonagle attended the Hackley School, Tarrytown, New York in preparation for college. He then attended universities in Switzerland and in Germany before enrolling in the University of California at Berkeley, Class 1917, to study mining. MacMonagle was expelled from that institution after 1 1/2 years during the 1915 fall semester. He then left for New York City and joined the American Ambulance Service.

MacMonagle crossed the Atlantic in the late fall and landed in France in October of 1915. He served with distinction with Ambulance Sections 3 and 8 from 30 December 1915 to 20 September 1916, the date on which he was dismissed from service for "drunkenness." He was awarded France's *Croix de Guerre*, with Star while serving at Verdun with Section No. 8.

On 22 June 1916 MacMonagle and three other Americans had been evacuating wounded at the entrance tunnel of Fort Travannes, two kilometers from Fort Vaux, when a German "380" high-explosive round landed in their midst. The concussion knocked the Americans to the ground, but miraculously none of them was wounded. MacMonagle got up, dusted himself off, and while still under bombardment, continued with his evacuation work. His *Croix de Guerre* citation captures the action.

"Volunteer automobilist Douglas MacMonagle of American A.S.S. No.8: A shell having fallen in the middle of a *poste de secours*, he retained his self-possession, and with the utmost devotion, under bombardment, assisted in loading three wounded men whose removal was imperative."

On 3 October 1916 MacMonagle enlisted in French Aviation via the Foreign Legion. On 17 October he was sent to the Aviation School at Buc. He received his *brevet militaire* on 13 April 1917 on the Caudron. Following further training at Avord and at the G.D.E. he joined the *Escadrille Lafayette* on 16 June 1917 at Chaudun, in the Aisne sector, the same day as David Peterson and James Norman Hall.

On 3 June the *Escadrille* had moved to Chaudun from Ham as part of *Groupe de Combat 13*. Now officially designated *Spa.124*, its assignment was to fly fighter protection for the five French reconnaissance squadrons observing the sector. And while no major battle raged over the ground beneath them, the *Lafayette* pilots found the air full of activity.

The Americans, in their forty-three days in the sector, would fly 400 sorties and fight 62 combats with the enemy. And *Sergent* MacMonagle, in his readiness to accept his share of the Squadron's patrols, immediately won the respect and confidence of his fellow pilots.

While appreciating his skills as a combat pilot, the Squadron's veterans questioned MacMonagle's reckless courage and immediate willingness to engage the enemy on any occasion.

At Chaudun MacMonagle and the other pilots flew missions from 3:30 a.m. to 9 p.m. when the last patrol returned to the aerodrome.

On 17 July MacMonagle moved with *Spa.124* to St. Pol-sur-Mer. The airfield sat near Dunkirk, and the Squadron's mission was to support the 3rd Ypres Offensive. However storms kept the pilots grounded much of the time, and except for occasional shelling from long-range German gun batteries, the Americans were able to spend their time relaxing in tents and bathing in the sea.

But on 11 August the grimness of war returned when the Squadron was sent to Senard near Verdun to fly support for a planned limited offensive. The good flying weather there made patrols a two and three-a-day occurrence.

MacMonagle and the *Lafayette Escadrille* left Senard for Chaudun on 18 September to fly support for the 3rd Aisne Offensive. The fall air had already turned bitter at high altitudes and frequent storms and autumn mists made flying more difficult.

On the morning of 22 September MacMonagle, Henry S. Jones, and Kenneth Marr were jumped by four Albatros fighters while flying patrol over the Verdun-Argonne sector. MacMonagle and Jones each engaged a fighter as two pounced on Marr. The Americans used all their skills to escape with their lives, particularly Marr who had his elevator control cables severed by enemy machine gun fire.

Following the patrol, MacMonagle and Carl Dolan were granted a forty-eight hour Paris leave. MacMonagle had not had an extended leave since joining the American Ambulance Service in 1915. And lately, he had been taking too many chances in the air. Dolan noticed his friend was displaying the symptoms of chronic combat fatigue. So the Paris leave, while brief, was a welcome respite.

Dolan soon found MacMonagle as combative on repos as he had been flying patrol, particularly toward that despised breed, the "barracks-flyer". Impressive in stature, quick-witted, and with a rapier tongue, he enjoyed slicing up those *aviateur des boulevards* "who rolled up flying time while on leave in Paris."

MacMonagle's mother, Minnie, who owned an apartment in Paris, had been working in a Red Cross hospital in Paris as a volunteer nurse. Douglas had spent his morning hours with her before hitting the favorite Paris rest spots frequented by unwinding aviators. Dolan proved a fortuitous companion for the big Scotsman, MacMonagle. Almost as soon as MacMonagle had left his mother's company, he had begun drinking heavily. And by evening Dolan, a teetotaler, had found himself steering his inebriated friend to a bed in a decent hotel.

On the morning of the 23rd MacMonagle arose early. He paid a quick visit to his mother, informing her that he had secured a permission for her to visit him next day at the Squadron. They then set the time when he would pick her up at the Senard railroad station. Then he and Dolan again headed for the favorite watering holes.

It was at one of these stops that MacMonagle had first confided to Dolan that he felt he would soon be killed. However, he said he would go out fighting and doing his best, the way his mother would want it. Late that same afternoon Dolan found himself in the same role as the day before, but this time guiding MacMonagle to the train station for the coach that would return them to Senard on time. As Dolan later remembered:

"He kept running away from me whenever we'd get down to the railroad station.

"'Mac!' I said, 'We'll wind up in jail if you don't come on.'

"And he said, 'Well, look, Doley, I'm going to get killed!'

"I said, 'Well, they'll shoot you at sunrise if you don't come, so what the hell's the difference?' And he said:

"'Well, I'm not going.'

"He'd jump out of the cab and off he'd go and it would take me half a dozen bars to find him."

Dolan finally managed to get his friend onto the train. Once they had found their seats, Dolan had attempted to persuade MacMonagle to catch some sleep. MacMonagle laughed the suggestion off and instead used the opportunity of each station stop to purchase bottles of champagne which he liberally poured out for himself and his comrades-in-arms returning with him to the Front.

It was 2 a.m. by the time the two young Americans had caught a ride on a truck leaving the train station for the Senard airfield. MacMonagle slept fitfully on the final leg of the journey but came to life when the camion reached its destination.

Made bold by champagne, MacMonagle decided to make a pre-dawn call on *Capitaine* Thenault's quarters to announce his return from leave. Dolan physically restrained his friend from his reckless course but finally released MacMonagle and moved back into the shadows as the big American stumbled toward the C.O.'s quarters and slapped open the door.

When MacMonagle had entered the room where Thenault was sleeping, he had groped in the darkness until he gripped the end of the Captain's cot. He then jerked it to knee-height and dropped it to the floor and tore the covers from Thenault, commanding: "Come on, you old sonofabitch, get up!"

Enraged, Thenault swung to the floor, and heatedly ordered MacMonagle into his flying gear for the first morning patrol set to begin in one hour. MacMonagle laughed heavily at Thenault's reaction then stumbled out of his quarters toward his own billet for his flight suit.

In his anger, *Capitaine* Thenault had violated the Squadron's unwritten law that a pilot returning from leave was to be put on *repos* until he had time to sober up. Dolan, realizing that MacMonagle was in no shape to fly, offered to take the patrol for him. But MacMonagle would have none of it. Dolan poured coffee into his friend and lent him his flying cap when MacMonagle could not find his own.

The barracks came to life as the hour for the patrol approached. Hill, Lufbery, Parsons, and "Doc" Rockwell climbed out of their cots and into their flying gear. Dolan helped MacMonagle slip into his sidcot. After asking Lufbery to keep an eye on MacMonagle, Dolan went to bed.

The engines in Hill's and Parson's SPADs refused to start. Lufbery left the field first, followed by Rockwell and MacMonagle. Once the patrol had reached an altitude of 15,000 feet and was in a V formation, Rockwell looked back for his comrade and saw that MacMonagle's fighter was drifting aimlessly about the sky.

The eagle-eyed Lufbery was the first to spot the formation of eight German Albatros fighters flying 1500 feet above their patrol. Their colors told him they were part of the Richthofen "Flying Circus." He gestured for the other pilots to follow him and climbed steeply to get above and behind the enemy buses. Rockwell followed, but MacMonagle turned around directly into the sun to face an enemy fighter that was coming toward him. Blinded by sunlight reflecting along his Vickers barrel, he flew head on into the German formation.

Rockwell immediately broke away from Lufbery's tail and made a desperate running attempt to head MacMonagle off. But before he could reach him, he was in the midst of the German patrol. Two of the enemy got onto MacMonagle's tail and fired. His plane shuddered then corkscrewed in a three-mile power dive, and the Americans followed its descent in horror until it crashed into the woods near Triaucourt.

Parsons having finally started his fighter, had caught up with Lufbery's formation. The three American pilots chased the Albatros pack back toward their aerodrome.

At eight a.m. that same morning *Capitaine* Thenault stuck his head inside Dolan's tent and brutally awakened him by demanding:

"Dolan, go get MacMonagle's body!"

Half awake, Dolan struggled into his clothes and staggered outside toward an idling cameon, uncertain what the Captain had meant. Dolan climbed aboard the Squadron's truck which already held several other *Lafayette* pilots, and they headed off in the direction of the Front.

The Americans found MacMonagle's fighter in the woods between the first and second line trenches. It was remarkably intact after its long spiraling descent. MacMonagle's body was still strapped inside his SPAD, his head still covered with Dolan's flight cap, the only visible sign of trauma, the holes in his temples where a German bullet had entered and exited.

Dolan retrieved the body to the Squadron aerodrome, and with the help of his friends they laid him out in a shed in preparation for burial. Dolan had a second sad task ahead of him. Meeting Douglas's mother, Minnie MacMonagle, at the Senard train station where she was to arrive to pay a visit to her son. The burden fell on him to tell her that he had just been killed.

The next day Mrs. MacMonagle stood stoically near her only son's casket as he was buried in the military cemetery at Triaucourt with elements of the 1st American Infantry Division furnishing the firing squad. Her courageous demeanor was recognized and silently saluted by all the pilots in attendance.

Following the funeral, Mrs. MacMonagle remained with the Squadron long enough to gather her son's personal effects. As she departed her son's comrades and prepared to board the Paris train, she was confronted by the proprietor of the local Squadron hangout who presented her with a 10,000 Franc liquor and bordello

Lafayette Escadrille pilots at Chaudun, 10 July 1917. Standing l.r. Soubiran, Campbell, Parsons, Bridgman, Dugan, MacMonagle, Lovell, Willis, Jones, Peterson, Maison-Rouge. Seated. Hill, Masson and "Soda," Thaw and "Whiskey," Thenault, Lufbery, Johnson, Bigelow, Rockwell. Below: Indian head insignia removed from Sergent MacMonagle's downed SPAD S.VII fighter.

bill. Mrs. MacMonagle quickly settled her son's debts and departed Senard.

The loss of Douglas MacMonagle had a profound effect on his fellow pilots. As James Norman Hall wrote upon hearing of his death: "It is a terrible loss for me, for all of us. He was one of the truest friends and comrades a man could have . . . Mac was a man's man, if there ever was one."

Hall felt he could not rest until he had avenged his friend's death. This he did on New Years day 1918 when he downed a lone Albatros fighter in the freezing air east of Brimont. Following the victory Hall said of it:

". . . I felt no elation whatsoever, only a grim kind of satisfaction that, at long last, I had avenged Doug MacMonagle's death."

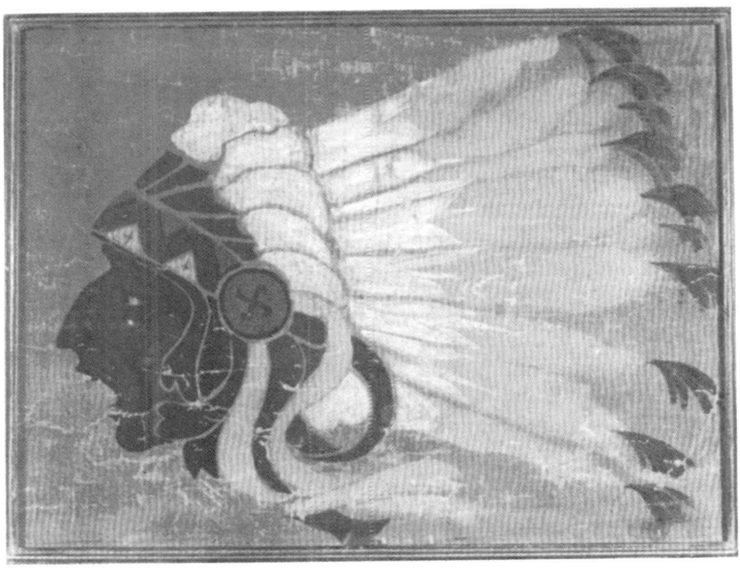

Sergent Douglas MacMonagle was posthumously awarded France's *Croix de Guerre* with Palm on 20 October 1917 for his 24 September action. His body was later transferred to the American Cemetery, Romagne-sous-Montfaucon Meuse. In 1928 it was moved to the *Lafayette Escadrille* Memorial outside of Paris, there to rest in the company of his comrades.

GORDON MACKE

Gordon Bernard Macke was born in Cincinnati, Ohio, on 4 March 1890, the son of B.H. Macke, a banker. Macke attended New York schools. He spent some time in Manila, the Philippines, before enrolling in Valparaiso University, Valparaiso, Indiana, class of 1910. Macke left that institution after one year and returned to Manila for business. He also maintained an interest in civilian aviation and became a member of the Aero Club of America on 6 June 1917.

That same month Macke decided to enter the European war. He joined the Lafayette Flying Corps in New York City and sailed for France aboard the *Touraine*. On 29 June 1917 Macke enlisted in France's *Service Aeronautique*. On 26 September 1917 he received his *brevet*. He was released at the *G.D.E.* at his own request because of ill health.

Left to right: Miller, Bullen, Sittlery, Rogers, unidentified, Winslow, Macke. Avord, summer of 1917.

GUY MAGLEY

Guy Bertram Magley was born in Joplin, Missouri on 17 February, 1892, the son of Emmette Magley, a coal miner and inventer, and Maude (Anderson) Magley. Guy Magley graduated from Joplin High School, and was involved in the theater and in vaudeville when he decided to enter the European War. On 19 July 1916 Magley enlisted in France's *Service Aeronautique*. He was released on 14 October 1916 because of inaptitude.

Following World War One Magley returned to vaudeville as a ballroom dancer, and put on stage productions in England, Australia, and in the United States. He also became a union organizer for the American Federation of Actors which was founded by Sophie Tucker, and which was later taken over by the Movie Union. While working in this capacity, he met eighteen-year-old Mary Johansen in Chicago. They were married in Ohio in 1940.

In 1941 the Magleys relocated to Glendale, California. Guy became a real estate broker and founded the Guy Magley Realty Company. Mary Magley also secured her broker's license and worked with her husband in his office. Their marriage lasted for thirty years until Guy Magley's death. He passed away from heart failure at the Mid-Valley Hospital, in Van Nuys, California on 18 February, 1971, one day after his 79th birthday.

Charles Malone

Charles Thomas Malone was born in Maryland on 19 July 1896, the son of Mr. & Mrs. John E. Malone of Yonkers, New York. Malone attended New York City schools. From 30 September 1916 to 20 March 1917 Malone served in the American Ambulance Field Service as a *conductor*. On 21 March 1917 he enlisted in France's *Service Aeronautique*. From 3 April to 15 September 1917 Malone attended the aviation schools at Avord and Pau. On that date, he was seriously injured in an flight-training accident at Pau.

Malone was 5000 meters above the aerodrome in a 120 H.P. Nieuport when his machine suddenly ran out of gas. During the rapid descent of his plane, Malone blacked out and his machine went into a *vrille*. By the time Malone regained consciousness, his plane had nearly reached the ground. At the last instant he managed to correct his aircraft. Still, the Nieuport struck the chimney of a house, crashed into a ditch and overturned, seriously injuring Malone.

Malone was confined to the hospital at Pau until February 1918. He took a convalescent leave in the United States then sailed back to France in late June 1918. He returned to Headquarters at Pau on 5 July and was informed that he would be discharged because of his disabilities.

Malone was discharged from the French Army at Lyons on 29 October 1918. On 6 November he joined the American Red Cross as interpreter at its Paris headquarters. He was later made information officer. Following the Armistice Malone served as Red Cross transportation officer at Prague, Czechoslovakia until August 1919 when he returned to the United States.

Malone worked as an accountant and auditor in the Washington, D.C. area. He married Ludmila Urban. From this union a son, John, was born.

During World War Two, on 5 May 1942, Malone enlisted in the U.S. Army. On 10 February 1946 he was discharged at the rank of Lieutenant Colonel. Malone died on 24 November 1947 at age 51. He is buried at Arlington National Cemetery.

HAROLD MANIERRE

Harold Lee Manierre was born in Chicago, Illinois on 22 January 1897, the son of William Reid Manierre, a wealthy resident with extensive property on Lakeshore Drive, and Julia (Edson) Manniere. Manierre attended Chicago schools and the Hackley School, Tarrytown, New York. On 15 July 1917 Manierre enlisted in France's *Service Aeronautique*. On 16 August 1917 he was released because of inaptitude.

Following World War One Manierre was given a large trust fund from his father's estate. This allowed him to pursue a career as an artist. He married Ruth Ingalls. In 1935 he moved to Woodstock, New York where he pursued a career in music and sculpture and invested in the stock market. Later in life he married Ruth Ingalls. He died of emphysema at Benedictine hospital in Kingston, New York on 11 June 1976 at age 79. He is buried at the artists cemetery in Woodstock.

KENNETH MARR

ESCADRILLE LAFAYETTE

Kenneth Archibald Marr, the 26th volunteer of the *Escadrille Lafayette*, was born in Oakland, California on 10 June 1885, the son of Archibald Marr and Alberta (Witherell) Marr, both naturalized citizens from Canada. At the time of Kenneth's birth his father had been working in Oakland as a hull superintendent for Southern Pacific Ship Builders, a firm that made ferry boats.

Ken graduated from high school and spent his early adulthood in California. He later went to Nome, Alaska and there met and formed an association with one Scotty Allen who put on trick performances with his malamute dog team.

At the outbreak of World War One Allen contracted to deliver 300 malamutes to the French government for the transportation of supplies and the evacuation of wounded *Chassuers Alpins* out of the snow-covered Vosges Mountains. When Allen turned the dogs over to the French government, Marr enlisted in the American Ambulance Field Service. He served with S.S.U.2 from December of 1915 into July of 1916. He was gassed at the Battle of Verdun while picking up wounded. Then "to get a clean shirt and the blood off my clothes" Marr joined the Foreign Legion and transferred to French Aviation on 20 July 1916.

Marr began his aviation training on 8 August 1916 at the field at Buc. He received his *brevet militaire* on the Bleriot on 7 January 1917. He completed his training at the schools at Avord, Cazeaux, and Pau, and arrived at the G.D.E. on 26 March 1917.

Marr joined the roster of *N.124* on 30 March with William Dugan and Thomas Hewitt while the Squadron was at Ravenal. He soon proved a brave, capable pilot. And through his knowledge of French and his friendly nature, he quickly filled the role of ambassador of good will between the *Lafayette* Squadron and the personnel of the French Air Service and the peasants in the nearby villages.

Caporal Marr was given the nickname "Siwash" by his compatriots shortly after joining *N.124*. They had heard how he had come to France from Alaska in the company of two Alaskan Indian dog handlers which they had assumed were members of the Si-Wash tribe. Thus the moniker "Si" or "Siwash" was attached to Marr, and it stuck with him until his death.

Shortly after "Siwash" joined *N.124* the Squadron moved thirty-two miles northeast to an aerodrome near Ham in the Somme sector to assume the role of aerial reconnaissance. The field, only recently vacated enemy territory, seemed jinxed for the American pilots.

On the morning of 8 April as the Squadron sought to depart on patrol, Hinkle, Willis, Lovell, and Genet all failed to become airborne because of some difficulty with their fighters. Marr, the fifth man on the line, successfully took off and circled the field. However, his engine began skipping. And while attempting a landing, his wheels snagged twisted railroad tracks at the end of the strip, and his Nieuport back flopped and skidded down the embankment, coming to rest as a pile of rubble. Marr escaped the wreckage with bruises, while *Capitaine* Thenault counted his blessings and cancelled further flights for that morning.

In mid-September 1917 the Squadron was flying combat patrols near Montfaucon. On the 19th "Siwash" was cited in orders for his role in a battle with six enemy aircraft near there. In the subsequent engagement he shot down and was given official confirmation for one of the Albatros scouts. Marr's citation for which he was later awarded France's *Croix de Guerre*, with Palm also mentioned his action on September 22nd in what became his closest call while flying patrol over

Sergent Kenneth Marr preparing to leave on patrol, Ravenal, April 1917.

the Verdun-Argonne sector with Henry S. Jones and Douglas MacMonagle.

The three pilots were jumped by four Albatros fighters, with two of them singling Marr out as their victim. He managed to force one Albatros out of the fight, but the other clung tenaciously to his tail and fired until he had severed Marr's elevator control cables.

Jones drove the German off Marr's floundering SPAD while Marr maintained control of his fighter through the use of his throttle to elevate and lower his plane's nose. As he revved the engine the prop wash blasted his detached elevator up and simultaneously raised the nose of his plane. As he cut the r.p.m's, the elevator would drop as would the SPAD's nose.

In this manner, by alternately gunning then cutting his engine, he brought his fighter down in a controlled descent from 6,000 feet to a safe landing in the Forest de Hesse. He then commandeered his rescuer's staff car and sped off for his field to pick up his mechanics and some spare cable. Returning to his SPAD, he made the needed repairs then flew out of the shell-scared forest to his Squadron's field where he refilled his tanks with petrol and headed out on another mission.

On the afternoon of 4 December 1917 "Siwash" Marr scored a probable victory while flying protection for a French photographic plane with David Peterson and Christopher Ford. The pilots' limbs were like lumps of ice from the frigid air, their reflexes slowed as seven Albatros scouts jumped the French two-place right after the flight had entered enemy airspace. The French plane knocked down one of the Germans as the Americans climbed above it to protect it.

The four allied planes made a quick turn toward their lines. The fighting grew fierce as the Americans sought to protect their charge against the assaults of the six remaining enemy fighters. Both Ford's and Peterson's engines were hit by enemy slugs which forced them out of the fray. Marr hung on alone and kept the six Albatrosses occupied until the photo plane reached the safety of Allied lines.

Marr, undaunted, seeing his charge safely home, again returned to enemy airspace. At 6,000 feet he found a German two-seater and quickly placed himself between the sun and the tail of his intended victim. At fifty yards he fired twenty rounds into the ship's vitals, killing the pilot. As the enemy plane nosed down, the observer stood behind his guns and continued to fire at Marr's SPAD. "Siwash" later recalled the victory which also went unconfirmed:

"I certainly felt sorry for the helpless beggar. There he was dashing to a certain death but to the last gamely

Captain Kenneth Marr, USAS, while serving as a flight commander with the famous 94th Hat-in-the-Ring Squadron, Gengoult aerodrome, 17 May 1918. On 9 June Marr would succeed Major Jean Huffer as the 94th's commanding officer, a position he would hold until 25 September 1918, when he would be succeeded by Captain E.V. Rickenbacker. Marr is standing beside a Nieuport 28 scout which had the dangerous weakness of shedding the fabric on its upper wing plane.

trying to give me a fight." The plane "dived steeply, turning sideways, and banged into the ground about three miles inside Germany."

Marr concluded his service with the French Army at the rank of *Sergent*. He was commissioned a Captain in the U.S. Air Service on 26 January 1918. As the old *Lafayette* Squadron was assimilated into the U.S. 103rd Aero Pursuit Squadron, Marr served with that unit from its inception on 18 February until 31 March 1918. On 8 April he was officially assigned to the U.S. 94th Aero Squadron as a flight commander. On 9 June he succeeded Major Jean Huffer as the squadron's commanding officer.

Marr flew a number of combat patrols with the 94th "Hat-in-the-Ring" Squadron. But his closest call occurred on 10 May 1918 as he came in for a landing at the Toul airfield following an afternoon patrol. Lieutenant Rickenbacker, who was over the airfield at the time witnessed Captain Marr and Lieutenant Thorne Taylor simultaneously land their Nieuport 28 fighters at opposite ends of the air strip. Inevitably the two fighters collided in the middle of the field, locking their wings together, and spinning "around and around . . . like a top."

The collision had activated the firing mechanism on Captain Marr's machine gun which then began spewing deadly fire in all directions at the rate of 650 rounds per minute, shooting up everything in sight and sending pilots and ground crew scurrying for cover. Eventually Marr's machine gun either jammed or expended its ammunition and the spinning Nieuports ceased their bazaar waltz. Although Marr's fighter was demolished, neither he nor Taylor nor any other squadron member was hurt by what Rickenbacker called "a gigantic pin wheel shooting out living sparks in every direction."

On 17 September Captain Marr was promoted to Major. But because of injuries he had sustained while flying, and from the debilitating effects of having been gassed while in the Ambulance Service, on 25 September he was succeeded as commanding officer of the 94th by Captain E.V. Rickenbacker. He was invalided home in late October to take charge of a "Liberty Bond Circus" which was scheduled to barnstorm across the United States. His transport ship was the first troop vessel to reach New York harbor after the signing of the Armistice.

On 29 November the French government awarded Major Marr a *Croix de Guerre*, with Star for his service as commander of the 94th Aero Squadron. Following the Liberty Bond tour Marr was made second in command of Rockwell Aviation Field near San Diego, California. In 1919 he was placed in command of the aerial gunnery school at Ream Field, also near San Diego, where he supervised all combat and aerial gunnery work.

Upon completion of his military service, Ken Marr began work at Paramount Pictures as an assistant to director John Ford. While there, he met the movie actress Alice Ward. They were married in 1928. A short time later Ken left Paramount and formed the Marr & Janeway Company, which redrilled dead oil leases. The business was a financial success and Marr transferred some of his assets into lumber and real estate, buying a 5,000 acre sheep ranch in Garberville, Humbolt County, California. He retired to Phoenix, Arizona. He died in the Veterans Administration Hospital in Palo Alto, California of arteriosclerosis on 28 December 1963 at age 78.

DIDIER MASSON

ESCADRILLE LAFAYETTE

Pierre Didier Masson, the 13th volunteer of the *Escadrille Americaine*, was born the son of Jules and Sarah (Chontrier) Masson in Asnieres, Seine, France on 23 February 1886. He attended schools in France until age twelve at which time he began studying as a jeweler's apprentice. He pursued this occupation until 1904 when he enlisted in the 129th Infantry Regiment at LeHavre.

In 1906 Masson left the infantry and began working for a magneto manufacturing firm in Paris and in London. In 1909 he became acquainted with French aviator Louis Paulhan and shortly became his mechanic.

Masson first soloed in a Farman biplane at Chalons (Marne) France in 1909. He then participated with Paulhan in aviation meets in London before coming to the United States in early 1910 for the first Los Angeles meet held at Dominguez Field. There, in January, Paulhan set an altitude record and won $19,000 in prize money. Paulhan remained in the U.S. for six months before returning to Europe to enter and win the London-Manchester race. Masson stayed on in America, flying exhibitions and testing aircraft for various manufacturers.

That fall of 1910 he flew as an instructor in Pineus Brauner's Curtiss-type aircraft at Mineola, Long Island, New York. In November Masson began flying the Curtiss aircraft of N.C. Addosides. After Addosides shipped the machine by rail to the west coast, Masson flew it on 7 January 1911 from Los Angeles to San Bernardino, a distance of sixty air miles in five hours and forty minutes, carrying a load of Los Angeles *Times* as cargo.

One feat Masson performed while flying exhibition was to bring his plane down in the San Francisco Square which was surrounded by trees, using Market Street as his approach. That same year, 1911, he crashed while testing a new machine, fracturing several ribs and barely escaping with his life.

Following a period of convalescence with friends in California, Didier went into the business of selling animal skins. However, he maintained his aviation interest, and in January of 1913 he received his Aero Club of America Certificate #202. He then joined Glenn L. Martin—later the founder of the Martin Marietta Corporation—at his school at Balboa as an aviation instructor.

It was there that Masson was introduced to Colonels Camberos and Alcalde, representatives of Mexican revolutionary General Obregon, by California Aero Club president C.C. Griffith. They offered Masson a captaincy and head of their revolutionary air force at a monthly salary of $300 plus flight bonuses of $50 for each reconnaissance mission and $250 for each bomb-

ing mission he would fly against General Huerta's Federalists.

When Masson accepted their terms, for $5000 the Colonels bought an American-built Martin biplane powered by a 75 h.p. Curtiss engine and placed it and Masson under the orders of Gral Pasquiera, Governor of Sonora.

Before entering the Carranza revolution, Masson hired one Thomas Jimmy Dean as his mechanic. Dean trucked the Martin pusher, now named the *Sonora*, from Tuscon to a point between Tuscon and Nogales where the sheriff of Tuscon intercepted and stopped his truck. The sheriff left his deputy, a one-legged man named Hopkins, to guard over the truck and plane. Dean won Hopkins over to the revolutionary cause with a bribe of money and the offer of the rebel rank of major. Hopkins then accompanied Dean and the plane on into Hermosillo, Mexico.

The rebels had chosen the Federal base of Guaymas on the Gulf of California as Masson's first target. Their airfield was forty miles away at Moreno. It was there that they converted the Martin to a bomber with the addition of a bombardier-observer's seat and a bomb rack to contain up to 150 lbs of 18-inch-long pipe bombs. They then selected Masson's targets, three Federal gunboats (the *Guerrero*, the *Morelos*, and the *Tampico*) guarding the Guaymas port.

Masson left the Moreno airfield with Captain Joaquin Bauche Alcalde as his bombardier on 10 May 1913 to begin what would be the world's first aerial attack against an armed vessel. They flew 2500 feet above the *Guerrero* and dropped their load under a hail of fire from the gunboat's crew. But the dynamite-filled pipe-bombs exploded harmlessly in the water.

Masson and Alcalde repeated their performance against the *Guerrero* the next day and in the weeks that followed. They made their final run on 4 August, the day the *Sonora's* engine quit under enemy fire, forcing Jimmy Dean as bombardier to jettison his bomb load as Masson glided the Martin to safety.

Didier quit the cause the following day, disgruntled that he had not been paid for his previous month's service and refusing to meet Obregon's demands that he start bombing cities.

Masson had taught one Gustavo Salinas the handling of the *Sonora* and he turned the plane over to him and Alcalde. Salinas flew with Alcalde until Alcalde was later executed by Obregon when he changed camps to fight for Pancho Villa. Dean then became Alcalde's replacement and flew bombing missions with Salinas against the gunboats at Mazatlan until Salinas wrecked the aircraft beyond Dean's ability to repair it.

Didier Masson (left) standing beside the Sonora, a Curtiss-powered American-built Martin bi-plane. The location is probably Moreno, Mexico. Masson converted the Martin to an aerial bomber. On 10 May 1913 following a forty-mile flight from Moreno to Guaymas on the Gulf of California, Masson and his Mexican bombardier conducted the world's first aerial bombing raid by dropping pipe bombs on the Federal gunboat Guerrero. Photo courtesy Greg Whipple.

About the same time that Masson was flying for Obregon, American Edwin Parsons, another future *Lafayette Escadrille* pilot, was serving in Mexico as a captain of aviation in Pancho Villa's opposing Armee del Norte, flying reconnaissance missions and training Villista officers how to pilot a Curtiss two-place. Fortunately for the *Lafayette Escadrille* the two future *N.124* pilots never met in mortal combat.

Didier Masson remained in Mexico until early September 1914, traveling around rebel-controlled territory conducting business. Then on 8 September 1914 he left Mexico for France to rejoin his infantry unit as it entered the war against Germany.

He served in the 129th and the 36th Infantry Regiments into October of 1914. He then enlisted in French aviation and trained at Pau and the R.G.A., receiving his *brevet militaire* on the Caudron on 10 May 1915, the second anniversary of his first Mexican bombing raid.

Caporal Masson's first assignment at the Front was as pilot with *Escadrille C.18* from March to September of 1915. He then attended pursuit school and joined the roster of *N.68*, flying with that squadron into April of 1916. He served as *moniteur* at Cazeaux from 16

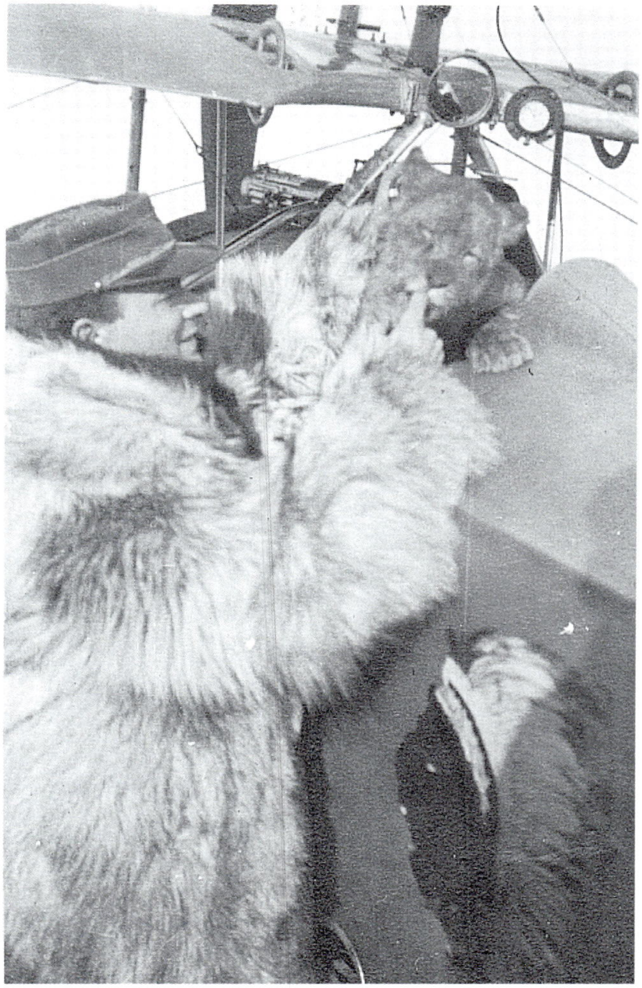

Adjudant Didier Masson with the lion cub Whiskey in the cockpit of his Nieuport 17 C.1, Cachy, Nov. 1916. The Escadrille Americaine Indian head was visible on the fuselage of his fighter.

April to 19 June 1916 when he joined the *Escadrille Americaine N.124* at Bar-le-Duc in the Verdun sector the day after Clyde Balsley had been seriously wounded and shot down over Verdun, becoming the Squadron's first casualty.

Four days later on 23 June the Squadron suffered its first fatality when Victor Chapman was shot down and killed at Haumont above enemy lines.

On 15 September the Squadron returned to Luxeuil and on 12 October it participated in a joint allied bombing raid against the Mauser arms works at Oberndorf, Germany. Masson was chosen to fly fighter escort for the bombers along with Lieutenant de Laage, Raoul Lufbery, and Norman Prince.

As the Breguet and Farman bombers returned home from their mission they were struck by a Fokker formation. The four *Escadrille* pilots immediately engaged the enemy, and although outnumbered three-to-one, managed to drive off and destroy a number of the attacking force. It was on this Oberndorf run that Didier Masson demonstrated what a truly excellent pilot he was and affirmed his nickname "Rubber Man" which had been bestowed upon him by his colleagues for his marvelous escapes from crippled aircraft.

In a fast, furious dogfight in an effort to protect the bombers, Masson had locked a Fokker into his sights and was about to fire his Lewis machine gun when his engine stuttered and abruptly quit for lack of fuel. Evidently his gas tank had been nicked. Presently above enemy lines, Masson quickly calculated that he had just enough altitude to glide back to allied territory.

His intended German victim instantly sensed Masson's predicament and switched roles from hunted to hunter. He hooked onto Masson's tail and fired a burst from his Spandau, riddling Masson's upper wing and fuselage and chopping away his windscreen and instrument panel. Miraculously Masson escaped the fire storm untouched. Still, he was at the German's mercy and was forced to maintain his gentle glide since any evasive maneuver would drop him into enemy lines.

The Hun pilot in his haste for victory, dove under and overshot Masson's Nieuport, surfacing before his nose. Masson instantly saw his opportunity and banked with the enemy, emptying the drum of his Lewis gun into the Fokker's vitals. The Fokker shuddered. Its pilot glanced back in shocked surprise then slumped forward over his controls. The enemy plane rolled away out of control. Minutes later it slammed into the ground and burst to flames.

Masson, continuing his motorless glide, passed so low over enemy lines that German riflemen nearly shot him down. Machine gun rounds pierced his wings, but again by some miracle he slipped unscathed through the maelstrom. He flew so low over the French front line that his wheels nearly snagged on the barbed wire defenses. He plopped his riddled Nieuport down in the midst of shell craters and hastily threw off his safety harness. He scrambled out and away from his Nieuport barely in time to escape the enemy artillery barrage which soon pulverized his plane.

Edwin Parsons, fellow *Escadrille* pilot and an eight-victory ace, later wrote this of Masson's remarkable feat: "Masson had accomplished the impossible. In a motorless glide, helpless to maneuver, he had not only escaped the almost certain death for which he had been fated, but in addition he had gotten his man. It was one of the greatest feats performed during the war, for it was against an enemy who held all the advantage."

But not all the returning pilots had been so lucky. Norman Prince had snagged his Nieuport in wires

Adjudant Didier Masson (rt.), Sergent Fred Prince and Capitaine Thenault with a French infantry officer in the trenches on the Somme attempting to confirm a pilot victory, December, 1916.

above the Corcieux airfield and had died of injuries the following day from the subsequent crash.

Adjudant Masson was cited on 8 November for his 12 October action. He was awarded France's highest combat decoration, the *Medaille Militaire*, and in addition the *Croix de Guerre*, with Palm. His accompanying citation reads:

"Very old pilot; after having taken part in numerous flights of artillery regulation and reconnaissance, has participated valiantly in the pursuit operations of the Verdun group. The 12th of October, 1916, during the protection of a bombardment, brought down an enemy plane. Accomplished his mission to the finish, despite fuel exhaustion which befell him over German lines and forced him to return by volplaning. The above citation carries with it the Croix de Guerre with palm."

On 18 October the Squadron moved to Cachy in the valley of the Somme to spend a winter of discontent. The pilots' barracks stood in appalling condition. The men even lacked blankets for warmth and basic cooking facilities. *Adjudant* Masson was designated *chef de popote* (mess officer) and he and William Thaw journeyed to Paris, and with a donation from William K. Vanderbilt, purchased the necessary items with which to set up a proper kitchen. Under Masson's management, and with the assistance of the French chef, Sampson, who had cooked both at the British embassy and at the Hotel Ritz in New York City, the Squadron's mess grew famous all along the Front and the offer to dine at one of Masson's excellently prepared tables became a keenly sought-after invitation.

On 26 January 1917 the *Escadrille Lafayette* moved to Ravenal, near Saint-Juste-en-Chausse to prepare for a planned spring offensive. On 15 February *Adjudant* Masson left the Squadron with the flu and upon recuperating was sent to Issoudun as an instructor. He remained there until 15 June when he was reassigned to *N.124*, then at Chaudun. In his absence James McConnell, Edmond Genet, Ronald Hoskier, and Lieutenant deLaage had all been killed. He also discovered that the Nieuport he had flown had been replaced by a new SPAD S.VII fighter along with the Squadron's designation from *N.124* to *Spa.124*. He had a red clockwise swastika, his personal insignia—then a symbol of good luck—painted on the fuselage of his SPAD. He took the new fighter out on patrol and later recorded the day's action in his laconic, spare style.

"Saw 2 enemy aircraft vicinity of Anizy at 23 hours 10. Accompanied by Sergeant Parsons and Corporal Dugan attacked at once. Combat endured for five minutes. Fired twenty rounds at close quarters with Rumpler. Observer was killed by first burst. Motor was hit and plane went into spin. Final result—one enemy aircraft observed to crash and burn in vicinity of Longport at 23 hours 20, the other disappeared at low altitude in his own lines. Confirmation requested."

On 17 July 1917 the Squadron was ordered to St. Pol-sur-Mer on the North Sea. A period of relative inactivity and deserved rest followed as the American pilots unwound from the rigors of combat.

On 11 August it was back to Senard at the edge of the Argonne. The *Lafayette* pilots flew many patrols and frequently encountered their German adversaries flying Albatros D.III and D.V fighters. During this violent time Harold Willis was shot down and made prisoner of war and Douglas MacMonagle and Andrew Courtney Campbell were killed in engagements with the enemy.

The end of September found the Squadron back at Chaudun. *Adjudant* Masson, suffering from flying fatigue and a chronic cold, left the Squadron for the final time on 8 October 1917, but not before securing a second Palm to his *Croix de Guerre*.

Didier Masson's "soft-spoken, reticent, and unassuming" nature had made him one of the favorites of the Squadron. Edwin Parsons, who had flown a number of patrols with him, described him in the following way:

"There was nothing spectacular about his war flying: no fancy work, frills or brilliant acrobatics. But you felt absolutely certain that when you started out on a patrol with Masson, you'd never be deserted. No matter what happened, Didier was going to be there til the last dog was hung.

"He was five feet six of human dynamite. Characterized by an iron nerve and an innate courage that laughed at danger. His calm, even disposition, unfailing good humor and keen judgment in a crisis made him a favorite with everyone."

Upon leaving the Squadron *Adjudant* Masson was assigned to Paris-defense *Squadron N.471* from 10 October to 28 October 1917. That day, 28 October, he became an instructor at the American Aviation Instruction Center at Issoudun and served in that capacity until 1 November 1918. While instructing there, he was able to pass on his valuable experience learned in years of flying combat which dated back to the 1913 Mexican Revolution. Following his Issoudun service he ferried planes from the French factories to the Front until the 11 November 1918 Armistice.

In January of 1919 Masson left the *Service Aeronautique* and returned to the United States. He became involved in various endeavors including the exportation of chewing gum. Some time in the early 1920s he met and married Modesta Escalante. Around 1928 he became the field manager for Pan American Airways at Belize, British Honduras. It was there that his son Didier was born on 12 July 1929.

Around 1935 Didier was appointed the Consul of France for British Honduras. In 1936 he formed the Belize Canning Company to market crayfish but was unable to secure the required financial backing to make it a success.

When Pan American closed its operations in Belize, Masson acquired various agencies in Europe which he subsequently lost as the countries became occupied by the Nazis. Then in June of 1940 when the Governor of Vichy signed an armistice with Germany, Masson resigned his post as Consul of France.

In 1942 Didier left British Honduras for Chetumel, Mexico where he worked as the manager for the Hotel Iris. He spent the last eight years of his life there before dying at age 64 on 2 June 1950 in Merida, Yucatan, Mexico.

WILLIAM MEEKER

William Henry Meeker was born in New York City, New York on 5 January 1894, the son of Henry Eugene Meeker and Jenny (Royce) Meeker. Meeker prepared for college at the Pomfret School, Pomfret Connecticut, from which he graduated in 1913. Meeker then entered Harvard College, Class of 1917. While attending that institution, Meeker served as a captain in the Harvard Reserve Officers' Training Corps. He spent the summer of 1916 as a student at the Thomas-Morse Aeroplane Company's flying school at Ithaca, New York where he earned 300 minutes flying experience.

On 23 April 1917, while still a student at Harvard, Meeker enlisted in the U.S. Signal Officers' Reserve Corps, Aviation Section, at Cambridge, Massachusetts. He left College in May, having passed the exams for his bachelor's degree, and sailed for France aboard the *Chicago* on the 19th of that month.

On 3 June 1917 Meeker enlisted in France's *Service Aeronautique*. From 13 June to 11 September 1917 he attended the aviation schools at Avord and Pau. He received his *brevet* on the Caudron in nearly record time on 26 July 1917.

On 11 September 1917 *Caporal* Meeker was at Pau training on an 18-meter Nieuport. His motor stalled as he was attempting a vertical spiral, and he fell into a wing slip. Unable to pull out in time, he crashed into the ground and was instantly killed. He was 23 years of age at the time of his death.

ALVIN MILLER

Alvin Ford Miller was born in Richmond, Indiana, on 14 December 1888. He was working in New York City, New York, as a travel stamps manager when he sailed for France on business aboard the *Philadelphia* on 20 May 1916.

In September 1916 he enlisted in France's *Service Aeronautique*. In May 1917 he was released because of defective eyesight.

GEORGE MILLER

George Miller was born in Greeley, Kansas, on 6 April 1884. He was the son of Thomas Miller. As a young man, Miller served in the U.S. Engineer Corps and was honorably discharged in May of 1914. He then worked as a seaman.

Miller was residing in New York City, New York, when he sailed for France aboard the *St. Louis* on 9 December 1916.

On 29 December 1916 Miller enlisted in France's *Service Aeronautique*. On 10 June 1917 he was released because of inaptitude.

George Miller with other Americans eleve pilotes at Avord, March 1917: Standing l.r. McCall, Wilcox, Adams, Scanlon, Molter, Chadwick, Wilson, Carrere, Hall, Kerwood, Donze, Pollock, Rounds, Huger, Turnure, Aten, Dolan. Kneeling: Baer, Pelton, Stehlin, Rheno, George Miller, Wells.

WALTER MILLER

Walter Bernard Miller was born in New York City, New York on 9 November 1893, the son of Bernard and Valeska (Hager) Miller. Of German-Irish parentage, but a citizen of the United States, Miller was left an orphan at an early age with the death of his parents in the Slocum disaster.

Miller was educated in New York schools. In 1911, at age eighteen, he enlisted in the United States Navy, and was attached to the *U.S.S. Des Moines* and the *U.S.S. Leonida* where he served in the West Indies, in Central and South America, and in Tampico during America's conflict with Mexico. Miller was honorably discharged from the Navy in 1914.

From May to August 1916 Miller served as a cadet officer with the International Mercantile Marines Lines aboard the *S.S. Siberia* and *Philadelphia*. On 2 December 1916 he sailed for France aboard the *S.S. Rochambeau* to join the American Field Service and was attached to the Vosges Detachment as an ambulance driver until 2 June 1917. While serving in France Miller married. From this union, a son, Walter Bernard Miller, Jr., was born.

On 10 June 1917 Miller enlisted in France's *Service Aeronautique*. From 16 June 1917 to March 1918 he attended the aviation schools at Avord, Juvisy, and the *G.D.E.* He received his *brevet* on the Caudron on 10 October 1917.

On 1 April 1918 *Caporal* Miller transferred to U.S. Aviation and was commissioned a Second Lieutenant. That same day he was attached to the First Aero Squadron, First Observation Group.

On 3 August 1918 Lt. Miller and Lt. James Sykes, his observer, went out on a mission to photograph the region north of *Fere-en-Tradnois*. They were escorted by a formation of seven chasse planes of the First Pursuit Group. Miller and his comrades met thirty enemy machines between Soissons and Fismes. During the aerial melee which followed Lieutenant Miller and his observer were shot down and killed in combat north of Chateau-Thierry. Five of the chasse protection planes were also shot down.

Miller's bus crashed into a railroad cut. He and Lt. Sykes were buried nearby. He was 24 years old at the time of his death. Later Miller's body was exhumed and reburied in the Suresnes American Cemetery. In 1928 he was disinterred and on 4 June he was placed in the Lafayette Flying Corps Monument, St. Cloud.

GORDON MILLS

Gordon R. Mills was born in Chicago, Illinois, the son of H.I. Mills. He attended Chicago schools and Yale university. At the time he decided to enter the European War he was employed as a stock broker. On 6 July 1917 Mills enlisted in France's *Service Aeronautique*. In October 1917 he was released because of ill health.

Gordon Mills (right) with George Dock at the Bleriot course, Avord, September 1917.

BENNETT MOLTER

Bennett Arthur Molter was born in Chicago, Illinois on 23 May 1890, the son of Nicholas Molter and Bertha (Hett) Molter. While a youth, Bennet's parents moved to Wausau, Wisconsin where Bennett attended public school. As a young man Molter left Wisconsin for California where he worked in the motion picture industry. Desirous of entering the European War, Molter sailed for France and enlisted in the *Service Aeronautique* on 2 November 1916. From 8 November to 16 July 1917 he attended the aviation schools at Buc, Avord, Pau, and the *G.D.E.*, receiving his *brevet* on the Bleriot on 12 May 1917.

On 20 July 1917 *Caporal* Molter was assigned to the Front with *Escadrille SPAD 102*. He flew with that squadron near Dixmude until 1 August, the day he was injured in a flying accident at the Squadron's aerodrome. That same month he was granted permission to return to the United States and applied for transfer to the American Air Service.

While awaiting his transfer, Molter completed his manuscript titled "Knights of The Air" which dealt with the training and operations of the pilots in the French Air Service. Molter's manuscript was published by D. Appleton & Company, in 1918.

On 2 November 1917 Molter was commissioned a First Lieutenant in the U.S. Aviation. He served as aviation instructor and gunnery officer at a number of Stateside Airfields. On 9 October 1918 he was promoted to Captain and was honorably discharged on 12 December 1918.

On 2 April 1919 Molter married Mona Pujo, daughter of a Louisiana Congressman. From this union a daughter and son were born. From 1919–1920 Molter served as Director of Latin American Sales for the Paige-Detroit Motor Car Company. In 1924 he left the automobile export business and became an investment banker in New Orleans, Louisiana. Molter became a prominent business and civic leader in that city and later served as vice-president of the International Lubricant Corporation.

Molter remained in the Air Reserve, was promoted to Major, and became the National Secretary of the Reserve Officers Association. From 1940–1941 he was chairman of the New Orleans Airport Commission.

On 3 March 1941 Molter re-entered active military service. During the War he served in Trinidad and in England before returning to the United States where he helped to draft the first U.S. program to train American pilots in gunnery and aerial combat. He also served as Air Base Commander and rose to the rank of Colonel. He was released from active service on 10 June 1945.

Following the war he returned to New Orleans and in 1948 was an unsuccessful candidate for that city's Commission Council. Bennet Molter died of heart disease in New Orleans, Louisiana on 25 April 1950 at age 57.

ROBERT MOORE

Robert Louis Moore was born in Paris, Texas, on 12 October 1895, the son of Mr. & Mrs. Charles H. Moore. Moore attended Paris public schools. Following high school, Moore engaged in competitive bicycle racing, worked as a jockey, an amateur boxer, and as an automobile mechanic and racer, competing on both American and European tracks.

In the fall of 1916 Moore was working as a motor inspector for the Goodrich company when he left that firm to join the Norton-Harjes Ambulance Service. Moore served for six months in France as an ambulance driver with Section 11. Seeking a more active role in the War, on 24 May 1917 Moore enlisted in France's *Service Aeronautique*. From 9 June to November 1917 he attended the aviation schools at Avord and the *G.D.E.* He received his *brevet* on the Caudron on 26 September 1917.

From November 1917 to 1 January 1918 *Caporal* Moore served at the Front with *Escadrille C. 305*. From 6 January to 1 May 1918 *Sergent* Moore served with *Escadrille SPA 96*. On that date he was released from French service on account of ill health.

Other accounts of Moore's service indicate he was shot down in July 1918 and for a time he was officially reported "killed in action." Further reports state that he was badly burned and was awarded France's *Croix de Guerre*, with Palm while recovering in a French hospital from July to December 1918. Upon his release it was reported that he was given a furlough to the United States in January 1919 and that he was sent to Dr. Kellogg's Sanitarium in Battle Creek, Michigan by his former employer, the Goodrich Rubber Company. Following a two-month recuperation, he was officially discharged from the French Army in April 1919.

Following the War Moore worked as a commercial pilot making flying pictures for the Famous Players-Lasky Corporation. He then joined the Miami-based Rogers Air Lines, flying a route along the Florida coast, and establishing the speed record between Miami and New York. In March 1932 Moore was flying five passengers from Miami to the Bahamas when his plane was forced down in the Gulf stream. He and the others clung to the plane's wing for many hours until each one eventually succumbed to exhaustion and slipped into the sea. Moore, the sole survivor, was picked up 56 hours after the forced landing, desperately ill, clinging to the bobbing wing of the plane. From 1932–1934 Moore was in charge of all the Curtiss-Wright Corporation's marine operations in Miami.

In February 1934 Moore fell ill with a stomach ailment. Following a series of operations, he died in the Victoria Hospital, Miami, Florida, on 12 April 1934 at age 38. He was survived by his wife, Christine (Bartlett) Moore.

GEORGE MOSELEY

George Clark Moseley was born in Chicago, Illinois on 5 April 1894, the son of Carleton Moseley, President of the Chicago office of Chase and Sandborn Coffee, and Marion (Wilder) Moseley. George attended the Hill School before entering Yale University, Class of 1917. While there he was an All-American left end on the 1916 Yale football team. Moseley weighed only 170 lbs, but he had "iron hands." He received his PhB from Yale in 1917.

On 6 April 1917 George "Mose" Moseley enrolled as a Carpenter's Mate 3d Class in the Aviation Section, First Battalion, New York Naval Militia. He took aviation training at Bay Shore, Long Island, New York. He was discharged in June of 1917 and sailed for France to join the Lafayette Flying Corps.

Moseley enlisted in French Aviation on 10 July 1917. From 18 July to 25 December 1917 he attended the aviation schools at Avord, Tours, Pau, and the *G.D.E.* He received his *brevet* on the Caudron on 20 October 1917. From 27 December 1917 to 4 February 1918 he was assigned for duty at the Front with *Escadrille SPA 150.*

On 4 February 1918 *Caporal* Moseley was released from the Lafayette Flying Corps to enroll as an Ensign and Student Naval Aviator, Naval Reserve Flying Corps, at Paris. On 11 March 1918 he was assigned to the Naval Air Station at Moutchic for "boat" training, the same day he became Naval Aviator 566(HTA). From 15 March to 15 June he served at the Naval Air Station at Dunkirk, flying *chasse* hydro-aeroplanes over the North Sea. On 15 June he was sent to the U.S. Army Bombing School at Clermont.

In July 1918 Ensign Moseley was assigned to the Northern Bombing Group, attached to 218 Squadron, Royal Flying Corps. In September he was attached to French *Escadrille de Saint-Pol* and remained there until 5 November 1918. For his war service, Moseley was awarded the U.S. Navy Cross.

His citation describes his service:

"For distinguished and heroic service as an aviator in an aeroplane in active operation cooperating with the Allied Armies on the Belgian Front during September, October and November, 1918, bombing enemy bases, aerodromes, submarine bases, ammunition dumps, railroad junctions, etc. Attached to Northern Bombing Group."

In December of 1918 Ensign Moseley was returned to the United States, to Great Lakes for release from active duty. On 3 February 1922 he was honorably discharged from the Navy as a LTJG Naval Reserve Flying Corps.

Following his World War Service, on 18 December 1918 Moseley married Ethel Maude Spencer, the sister of his close friend, Dumaresq Spencer. From this union three sons were born.

From 1920–1928 Moseley worked as assistant factory superintendent for Chase & Sandborn in Chicago, Illinois. From 1929–35 he was manager of the Chicago brokerage office of Smith, Graham & Rockwell. From 1936–38 Moseley was a partner in the brokerage firm of Messick & Co. In 1938 he joined the Moseley & Co. coffee business. Mrs. Ethel Moseley died that same year.

On 15 March 1939 Moseley married Elaine Rosenthal Reinhardt in Chicago. Prior to their marriage Mrs. Moseley had been the Women's Western golf champion and the North and South amateur champion. During World War One she toured the U.S. as one of the "Dixie Kids," a group of young golfers who raised $150,000 to benefit the American Red Cross.

In 1941 George Moseley became assistant vice president and sales manager for the General American Transportation Corp. He remained with that firm until his retirement in 1959. Moseley continued to work part-time with the Central Realty and Investment Company from 1960 until his death in Chicago, Illinois on 28 January 1976 at age 82.

OSCAR MOUVET

Oscar Louis Mouvet was born in New York City, New York on 20 November 1887, the son of Oscar Mouvet and Emma (Fald) Mouvet. He attended New York schools and Manchester College, North Manchester, Indiana. In 1914 when war broke out between France and Germany he left the United States, where he had been working as a dancer, and sailed for Paris. On 15 August 1915 Mouvet enlisted in the French Foreign Legion (Infantry). He was wounded on 4 July 1916 during the Battle of the Somme near the village of Belloy-en-Santerre.

Mouvet was sent to a Paris hospital. Upon his recovery from his wounds, on 19 October 1916 Mouvet enlisted in France's *Service Aeronautique*. In April 1917 he was released because of inaptitude. He returned to the Foreign Legion and in July 1917 he participated in the Battle of Verdun. He was cited in Regimental Orders and was promoted to corporal. On 8 December 1917 Mouvet was wounded in the face by a shell splinter. He returned to the United States to recuperate.

Following World War One Mouvet managed night clubs in Paris. He died of tuberculosis in the Pyrenees in January 1930 at age 42.

American Volunteers July 4, 1917

Left to right standing: Fred Zinn, Adjutant Albanel(Zinn's French pilot), Eugene Jacob, Andrew Walbron, Christopher Charles, Oscar Mouvet, Jack Moyet, William Paringfield. Seated: Robert Mulhauser, Guy Agostini, Raoul Lufbery, Paul Rockwell, Willis Haviland, Algernon Sartoris.

Curtis Munson

Curtis B. Munson was born in Washington, D.C. on 9 February 1892, the son of Robert F. Munson, a bond broker, and Olivia (McKee) Munson. As a youth he lived in Philadelphia, in New York, and in Michigan. He attended Yale University, Class of 1916.

Munson left college in 1915 to serve in European war with the Morgan-Harjes Ambulance Service. After four months service, he returned to the United States and finished school then began work as a newspaperman.

Munson returned to France and enlisted in the *Service Aeronautique* on 23 July 1917. In October 1917 he was released at his own request.

Prior to World War One, Munson had been romantically involved with Frances Grahame Hunter. They met again at the Front when Munson was in the Ambulance Service and Miss Hunter was working as a nurse. In October 1917 Munson and Miss Hunter were married in a chapel in Paris. Eventually three daughters were born of this union.

In November 1917 Munson was commissioned a Second Lieutenant in the Field Artillery. He attended the Field Artillery School at Saumer, and the Tractor School at St. Maur. He was then assigned to Battery C, 55th Artillery, Coast Artillery Corps. Lieutenant Munson participated in the 2nd Battle of the Marne, operations on the Vesle, and in the Meuse-Argonne Offensive.

Following World War One Munson returned to the United States, then to Edmundton, Alberta, Canada where he purchased a newspaper. He also took over his father's coal mine, and later acquired timber interests, and a grain farm.

In 1927 the Munsons divorced. In 1933 Curtis Munson married Edith Cummings. She was a champion golfer from Chicago, and was the first woman to grace the cover of *Time* Magazine.

At the outbreak of World War Two Munson entered service as a Lieutenant-Commander in the U. S. Naval Reserve. As an Assistant Naval Attache at London he advanced to Commander, Combat Intelligence, and was involved in the American landings at Sicily, and at Salerno, Italy. In November of 1942 he was given a special Naval commendation for his work as part of the Naval advance party during the assault of Arzew, Algiers. With the 11th Amphibious Force he participated in the assault at Omaha Beach, at Normandy, and was awarded the U.S. Bronze Star and France's *Croix de Guerre*.

Wedding photo of Curtis Munson and Frances Grahame Hunter, Paris, October 1917.

Following the War Munson continued operating his coal mining interests in Canada, spending his time there, and in Washington, D.C. where he was involved in stocks and bonds, and at his home in Hobe Sound, Florida. Curtis Munson died in Washington, D.C. on 13 October 1980 at age 88.

ALAN NICHOLS

Alan Hammond Nichols was born in Ann Arbor, Michigan on 7 January 1897, the son of Walter H. Nichols, a California superintendent of schools, and Esther (Connors) Nichols. Nichols attended Monrovia and Pasadena schools before enrolling in Leland Stanford University, Class of 1919.

In early February 1917 Nichols joined the Stanford contingent of the American Ambulance Field Service. Following a train trip from Palo Alto to New York City, on 14 February Nichols and his 20 comrades sailed for Europe aboard the *Espagne*. The group arrived in Bordeaux, France on 26 February.

Once in Paris Nichols's group became Section 14 of the American Ambulance Field Service. Nichols served five months with Section 14 in Paris and in the field.

On 1 July 1917 he enlisted in France's *Service Aeronautique*.

From 26 July to 17 December 1917 Nichols attended the aviation schools at Avord, Tours, Pau, and the *G.D.E.* He received his *brevet* on the Caudron on 18 October 1917.

On 19 December 1917 *Caporal* Nichols was assigned to the Front to *Escadrille SPA 85*. Shortly after joining the squadron, *Sergent* Nichols was credited with the destruction of a German aircraft.

On 2 June 1918 Nichols and his comrades flew a dawn patrol over enemy lines beyond Soissons. Near Montdidier, Nichols spotted two enemy chasses above his formation and rose to attack. As he opened fire, a third unseen enemy fighter opened up on him from behind. Shot though the stomach, *Sergent* Nichols disengaged himself and managed to land within French lines.

At 2:30 that afternoon Nichols was transported by ambulance to the Royallieu Hospital, near Compiegne. Mortally wounded, *Sergent* Nichols died just before midnight that same day. He was 21 years of age at the time of his death. For his heroic war service, *Sergent* Nichols was awarded France's *Croix de Guerre*, with two palms.

Nichols was buried at Royalieu with fully military honors. His body was transferred to the American Cemetery near Ploisy, Aisne. In 1928 his remains were re-interred at the Lafayette Flying Corps Memorial near St. Cloud, France.

Charles Nordhoff

Charles Bernard Nordhoff was born in London, England on 1 February 1887, the son of Walter Nordhoff, a California cattle rancher, and Sarah Cope (Whitall) Nordhoff. Charles worked on his father's ranch while attending California schools. He enrolled in Harvard University, Class of 1909. Following graduation, Nordhoff returned to work for his father in the cattle ranching business.

In December 1916 Nordhoff sailed for France and served six months as an ambulance driver in the American Ambulance Field Service in the Vosges Detachment. On 3 June 1917 he enlisted in France's *Service Aeronautique*.

From 14 June to 12 January 1918 Nordhoff attended the aviation schools at Avord, Pau, and the *G.D.E.* He received his *brevet* on the Caudron on 30 October 1917.

American Eleve Pilots at Avord Cafe Summer 1917
l.r. Charles Nordhoff, unidentified, John Cotton, waitress, David Lewis, Alan Ash, Fearchar Ferguson, Cyrus Chamberlain.

On 15 January 1918 *Caporal* Nordhoff was assigned at the Front to *Escadrille N. 99*. He flew with that squadron as an enlisted pilot until 18 February 1918. On 19 February he was commissioned a Second Lieutenant in the Aviation Section, U.S. Signal Corps and continued to fly with his old French squadron. He was awarded France's *Croix de Guerre*, with Star for his 29 May 1918 destruction of an enemy aircraft.

On 11 July 1918 Lieutenant Nordhoff was ordered to serve on the executive staff of the U.S. Air Service. He remained there until the 11 November Armistice. Following the war, Nordhoff and James Norman Hall were ordered by Colonel Edmund Gros to write the history of the Lafayette Flying Corps and its pilots. On 20 February 1919 Nordhoff was promoted to First Lieutenant. That same month he returned to the United States aboard the *S.S. Mauretania* and on 10 March 1918 he was discharged from the U.S. Air Service at Garden City, Long Island, New York. Nordhoff and Hall completed the two-volume *Lafayette Flying Corps* history at Martha's Vineyard, Massachusetts and sent the manuscript to the publisher in the fall of 1919.

This collaboration between the two writers was the first effort of a partnership that would last through most of their professional careers. In late January 1920 Nordhoff sailed to Tahiti. Hall joined him there, and a number of jointly written books followed including *Falcons of France*, *Mutiny on the Bounty*, *Men Against the Sea*, *Pitcairn's Island*, and the *Hurricane*.

Shortly after his arrival in Tahiti, Nordhoff married a native girl, Vahine Tua Tearae Smidt. From this union seven children were born. In 1940 Nordhoff broke off his partnership with Hall, abandoned his family, and relocated to Santa Barbara, California. He divorced Vahine there and married Laura Grainger Whiley on 12 June 1941.

While in California, Nordhoff continued to work as a writer. He died of congestive heart failure at his home in Santa Barbara on 10 April 1947 at age 60.

NATHAN OAKES

Nathan Prince Oakes, Jr. was born in Pomona, California on 22 February 1889, the son of Nathan Prince Oakes, a descendant of a family of ship builders, and Ida (Herrick) Oakes. Nathan's parents had moved to California in 1880. He attended Pomona public schools then worked for several years as a scenario writer for California film companies. In 1916 when Nathan's parents returned to Providence, Rhode Island, the place of their origin, he returned with them and attended the Providence College's School of Design.

In March of 1917 Oakes sailed from New York to Bordeaux, and on the 27th he joined the American Ambulance Field Service in Paris. His duty was to act as official photographer for that organization. When this did not materialize, he joined the French Foreign Legion on 31 May with the intention of transferring to aviation.

On 19 June 1917 Oakes enlisted in France's *Service Aeronautique*. Oakes received his *brevet* on 15 October 1917. On 12 December 1917 he was released upon his own request after developing ulcers.

Oakes returned to the United States and was commissioned a Second Lieutenant in the U.S. Air Service. He served as an aviation instructor at Lake Charles, Louisiana. Following the November 1918 Armistice, he joined the Victory bond drive and did aviation recruiting work by stunt flying and dropping aviation recruiting circulares from his Curtiss bi-plane along the eastern seaboard.

Lieutenant Oakes left the U.S. Air Service in the fall of 1919. He married, and continued his studies at Providence, Rhode Island. The Oakes's had a son and a daughter. He and his wife divorced, and he married Marjorie King. Oakes became an executive with DuPont's paint division. He also worked for General Motors. In 1941 he moved to Harlingen, Texas where he became a citrus farmer and raised grapefruit. Oakes died in Harlingen on 24 March 1957 at age 68. He was interred at the Restlawn Cemetery, La Feria, Texas.

Carter Ovington

Carter Landram Ovington was born in Baden-Baden, Germany on 1 May 1897, the son of Edward J. Ovington, the overseas purchasing agent for Ovington Brothers Store of New York City, and Georgia (Maize) Ovington. Carter Ovington attended schools in Paris and was a student at the *Ecole Superieur de l'Electricite* when he decided to enter military service.

On 20 May 1917 Ovington joined France's *Service Aeronautique*. From 8 May to 10 December 1917 he attended the aviation schools at Avord, Pau, Cazeaux, and the *G.D.E.* Ovington received his *brevet* on the Caudron on 31 August 1917.

On 12 December 1917 *Caporal* Ovington was assigned at the Front to *Escadrille Spa 85*. He flew with that squadron until 9 January 1918. On 9 January *Sergent* Ovington was assigned to *Spa 98*. He was commissioned a First Lieutenant in the U.S. Air Service on 1 April 1918 and was ordered to serve as an instructor at the American Acceptance Park, Orly.

Lieutenant Ovington petitioned to be returned to the Front. With the assistance of his former French commander, he was shortly returned to his old French squadron where he remained until 29 May 1918.

Lieutenant Ovington participated in many patrols with *Spa 98*. On 29 May the Germans launched an offensive in front of Chateau-Thierry. Caught by surprise, the French forces had few squadrons in the sector, and Ovington's squadron was called upon to fly numerous patrols.

Lieutenant Ovington and his comrades confronted large groups of enemy machines. On one sortie he was involved in four separate combats. Around noon of that day Ovington returned to his aerodrome to refuel and rearm. He was immediately given command of a patrol which had been ordered to attack enemy observation balloons deep inside German lines.

Ovington returned from the mission enveloped in low-flying clouds. Blinded, he collided head on with another French machine piloted by *Sergent* Hoor. The wreckage of both machines fell on enemy occupied territory in the region of Lagery, north of Chateau-Thierry. He was 21 years of age at the time of his death.

When Allied troops seized the territory, a search was conducted for Lieutenant Ovington's body and the remains of his aircraft. Neither was ever found. Lieutenant Ovington was awarded France's *Croix de Guerre*, with Palm for his heroic service the day of his death. In 1928 he was posthumously awarded France's *Legion d'Honneur*.

Georgia Ovington, the mother of Lieutenant Ovington, became the secretary of the Lafayette Flying Corps. She gave much of her time and energy in assisting her son's American comrades, and helped to perpetuate the history of the Corps until her death.

DAVID PADEN

David Sheldon Paden was born on 4 December 1895 in Evanston, Illinois, the son of Joseph E. Paden, politician and one-time Mayor of Evanston (1907–1913), and Charlotte (Colt) Paden. David Paden attended Evanston schools before entering the University of Michigan in the fall of 1915, in pursuit of a degree in sales engineering.

While a student at the University of Michigan, Class of 1917, David Paden left his engineering studies to join the American Ambulance Field Service. He served with SSU 8 for four months before enlisting in France's *Service Aeronautique* on 30 July 1917. From 1 August 1917 to 4 September 1918 Paden attended the aviation schools at Avord, Tours, Chateauroux, Pau, and the *G.D.E.* He received his *brevet* on the Caudron on 26 May 1918.

On 6 September 1918 *Caporal* Paden was assigned at the Front to *Escadrille Spa 163*. He flew with that squadron until the 11 November Armistice. While a member of this squadron, *Sergent* Paden was shot down twice and was awarded France's *Croix de Guerre*, with two Palms, having been cited in orders in September and November of 1918.

While in service with *Spa 163*, *Sergent* Paden flew the entire western front from Chateau Thierry to Verdun. He was discharged in Paris on 8 February 1919.

Following World War One Paden worked as a sales engineer. He married Margaret Desmond. Two children were born from this union: a daughter, Jeanne Reed, and a son, Desmond. Upon his retirement, Paden moved from Illinois to Florida. He died in South Pasadena, Florida on 14 August 1976 at age 80.

HENRY PALMER

Henry Brewster Palmer was born in Rochester, New York, on 25 December 1887, the son of Charles H. Palmer and Mary (Allis) Palmer. Palmer was educated in Rochester's Lewis School and Bradstreet's Preparatory School. In 1903 he entered St. George's School, Newport, Rhode Island, before enrolling at Harvard University, Class of 1910. While at Harvard he participated in all sports. Following his graduation, he continued his athletic career at golf, riding, and mountain climbing while working in the New York bond department of the White, Weld & Co. banking house.

In 1912 Palmer moved to San Francisco and was associated with the bond business of the William R. Staats Company. He returned to New York two years later, and on 24 June 1916 Palmer joined the American Ambulance Field Service as an ambulance driver.

Palmer served with Section Three at Pont-a-Mousson, France and in the Balkins to 11 May 1917. He was awarded France's *Croix de Guerre*, with Star for his "courageous action in removing wounded in the region of Monastir between October and December, 1916."

On 25 May 1917 Palmer enlisted in France's *Service Aero-nautique*. From 9 June to 12 November 1917 he attended the aviation schools at Avord and Pau. He received his *brevet* on the Caudron on 30 September 1917.

Caporal Palmer arrived at the Pau acrobatic school ill with pneumonia. Weakened from his service in the Balkins, he was unable to rally against the illness. On 5 November he entered the Pau hospital. On 12 November 1917 *Caporal* Palmer died of pneumonia at the hospital and was buried in the cemetery there. He was 29 years of age at the time of his death. His remains were later removed for burial to the Lafayette Flying Corps Memorial near St. Cloud, Paris.

AUSTEN PARKER

Austen Gilette Parker was born in Great Falls, Montana, on 11 September 1892, the son of Willis F. Parker, an attorney and gold speculator, and Florence L. (Austen) Parker. Parker attended Great Falls and Helena public schools. For a short time he worked as a cub reporter for the Helena Daily *Independent* before attending Cornell University, Class of 1914. In his junior year, Parker served as the literary editor for the Cornell *Widow*. He and Arch Johnston put out an issue that the University considered pornographic, and both men were expelled. Parker worked for a time as a reporter for the New York *Times* and *World*.

Parker left the newspaper business to manage a Washington Square theater with his friend, Dudley Tucker. The two men decided to take a trip to the Orient to study theater there. When their steamer reached Panama, they left the ship to visit the ancient mines and ruins at Darien. As they wondered through the jungles, they became lost and emerged on the plantation of a German who held them prisoner for several weeks. They made an escape and took a dugout canoe on a 150-mile voyage to Panama. There, they cancelled their trip to the Orient and sailed for Bordeaux, France to join the Lafayette Flying Corps.

On 2 May 1917 Parker enlisted in France's *Service Aeronautique*. From 8 May to 15 December 1917 he attended the aviation schools at Avord, Pau, and the *G.D.E.* He received his *brevet* on the Caudron on 24 September 1917.

From 19 December 1917 to 9 January 1918 *Caporal* Parker flew at the Front with *Escadrille Spa 85*. From 9 January to 13 April 1918 *Sergent* Parker was assigned to *Spa 98*.

On 24 May 1918 Parker was commissioned an Ensign in U.S. Naval Aviation. On 24 May 1918 he became Naval aviator HTA (#1030) in Paris. He served at the U.S. Naval Air Station at Lake Bolsena, Italy, and at the Naval Air Station at Porto Corsini, Italy, attached to the 241st (Italian) Combat Squadron. For his outstanding War Service, Parker was awarded the U.S. Navy Cross and Italy's War Cross.

Ensign Parker was discharged in the United States in February 1919. That same year he married Phyllis Duganne, the well-known writer. From this union a daughter, Jane, was born. Parker worked as a fiction writer of aviation stories for the *Saturday Evening Post*, *Colliers*, and other periodicals. In 1925 he rejoined the French service as a lieutenant-colonel to fight in Morocco in the Riff War against Sultan Abd-el-Krim.

Following his return from Morocco, Parker settled in New York to continue his fiction-writing career. He also worked as a playwright and met the actress Miriam Hopkins in this capacity. Parker and Phyllis Duganne were divorced, and he married Miriam Hopkins in 1929. He followed her to Hollywood, California, where he became a script and scenario writer. Parker died in Los Angeles of a cerebral hemorrhage on 20 March 1938 at age 45.

EDWIN PARSONS

ESCADRILLE LAFAYETTE

"I have no hesitation in confessing that I had gotten into this scrap through a pure desire for adventure and without any clear idea of what I was letting myself in for."

Edwin Charles Parsons, the 21st volunteer of the *Escadrille Lafayette*, was born in Holyoke, Massachusetts on 24 September 1892. Parsons was the son of Springfield, Massachusetts insurance broker Frederick Dwight Parsons and Grace M. (Steele) Parsons. He was a direct descendant of Coronet Parsons, a founder, in 1636, of Springfield.

As a youth Parsons attended Springfield public schools, and the Phillips Exeter Academy from 1908–1910. In the fall of 1910 he briefly attended the University of Pennsylvania at his father's request to train for a future partnership in his insurance business.

Parsons soon became bored with school and abruptly dropped out and headed for Southern California. Once there, while in his late teens, he married briefly and found work as a ranch hand and gold miner.

In 1912 young Parsons met pioneer aviator Glenn H. Curtiss near Los Angeles. As a result of this meeting, Parsons developed a keen interest in aviation. He made his first flight on 12 July 1912 at Los Angeles's Dominguez Flying Field and soloed shortly thereafter.

By 1913 Parsons had mastered Curtiss's aircraft. Late that same year in Los Angeles he met agents of Mexican revolutionary Francisco "Pancho" Villa who offered gold to Parsons in exchange for his coming to Mexico to teach Villa's soldiers how to fly.

The end of the year found Parsons at the Sheldon Hotel in El Paso cutting a deal over whiskey with another American, one Jeff de Villa, who presented himself as Pancho Villa's financial representative. As a result of their meeting, Parsons was given the rank of Captain in the Aviation Corps *Armee del Norte*, Mexico. He was also given a first month's salary advance of $200 in gold plus enough gold to buy an airplane. When the men had finished with their meeting, de Villa remained in the States with Parsons to see to the delivery of the plane.

Parsons returned to Los Angeles where he bought and shipped a dismantled Curtiss bi-place on a flatcar to Ciudad Juarez, Mexico. Following its inspection, it was pulled to a siding in Chihuahua City where Parsons directed its assembly. Then for the next eleven months he coaxed reluctant Villista officers into the plane's cockpit, instructing them on the controls, but finding few with the courage to rise with him into the air.

Parsons flew a number of reconnaissance missions over contested territory. Ironically, at about the same time, another future *Lafayette Escadrille* pilot, Didier Masson, was also flying reconnaissance and bombing missions in Mexico, but for an opposing group of Mexican insurgents. Fortunately, the two never met in combat.

Following a crash-landing in Zacatecas, Mexico, with Jeff de Villa in the rear cockpit, Parsons caught the train to Juarez to hunt down Curtiss aircraft parts for those damaged in the landing. During this mission he met a German agent posing as a cigar store owner in Chihuahua who had knowledge of Villa's future intentions. He warned young Parsons that Villa would soon clash with the U.S. government, and that Parsons, as an American mercenary in Mexico, could find himself in serious jeopardy back home. Weighing the consequences of his present course, Parsons decided to ship the Curtiss parts on to de Villa while he hopped the streetcar across the Rio Grande to El Paso.

Parsons, the prodigal son, returned to his forgiving family in Springfield in time to re-enroll in a business curriculum for the 1915 spring semester at the University of Pennsylvania. But his thoughts of school dissipated when Parsons learned in Philadelphia from a lady reporter that a squadron of American aviators had been formed in France.

However, Parsons soon learned the lady reporter "was a little bit ahead of her time." While efforts were then being made by Americans William Thaw, Norman Prince, and others to form an all-American volunteer squadron, it would not become a reality until the following year. But Parsons would not learn of this until sometime later when it was too late to turn back. His conversation with the reporter had convinced him that he must shed the confining business career his father had mapped for him and cast his fortunes toward France.

Parsons's decision to serve in France mirrored those decisions of several of his young colleagues who had also chosen to serve the Republic. First and foremost was a desire for adventure, coupled with the desire to punish "the brutal hun" who had been receiving adverse publicity in the American press. Frequently, the desire to rebel against parental authority had also played a part in their decision.

When Parsons had told his father of his plans to sail for France then had asked him for passage money, he remembered how he had "flatly stated that it was not our war and he wanted no part of my wild scheme." Thereupon, the only consideration entering Parson's mind was how he was going to secure his transportation to Europe.

Caporal Ted Parsons, Lt. Rene Simon, Caporal Stephen Bigelow, Caporal Edmond Genet at the Simon villa at Pau, France, 3 December 1916. Simon was chief acrobatic instructor at Pau who had married an American woman.

Parsons later said, "I don't know why I particularly wanted to go and get into another war. Perhaps it was because of the fact that I'd always liked French and the study of it came easily to me, whereas I disliked German. But in any event I decided to get into the scrap."

For whatever the cause, Parsons left Springfield for New York City in December of 1915. Once there, he received from the French Consulate the names of shipping firms which might provide passage across the Atlantic. Parsons chose the *Carpathia* which was about to leave New York for St. Nazaire with a cargo of 2,000 horses. Parsons soon conned his way aboard as an "assistant veterinarian." It proved an unpleasant voyage, with much of his passage, including Christmas and New Years, spent in a hold full of horse manure.

Upon his arrival in St. Nazaire, Parsons left for Paris to enter the *Service Aeronautique*. He was told at the American Consulate that the only way he could enter into French flying was to first serve in the Foreign Legion and spend six months in the trenches. Then, if he was still alive, he could put in his application to his commanding officer for a transfer to Aviation. He was advised instead to go out to Neuilly and there join the American Ambulance Service. This he did.

Parsons first worked at the hospital at Neuilly-sur-Seine. In January of 1916 he began serving with S.S.U. 1 as an Ambulance driver with future Lafayette Flying

Parson's French mechanics beside his flipped Nieuport scout at Ham, March 1917.

Corpsmen Henry Sweet Jones and Frank Leaman Baylies. He was later given charge of a section of the Ambulance at Mrs. Whitney Warren's Hospital at Juilly, near the Soissons front.

In the winter of 1916 Parsons reapplied for entry into France's *Service Aeornautique*. Meanwhile he continued to serve with the American Ambulance Service. Through the help of Doctor Edmund Gros his acceptance papers arrived on 13 April and he was sworn into the Foreign Legion at Dijon with Ronald Hoskier.

The two Americans were mistakenly sent to the Pau aviation school which caused them one month's delay. On 15 May they began their preliminary training at Buc near Versailles on the Bleriot "Penguin." Both Americans had been pre-war flyers, and Parsons and Hoskier found the courses easy to complete.

Parsons received his *brevet militaire* on the Caudron on 23 August, receiving pilot badge no. 1212. Upon completion of his preliminary training he was transferred to the Nieuport school at Avord for pursuit training. One month of gunnery school followed at Cazeaux then on to Pau for two months of training in group flying and acrobacy. On 8 December 1916 he reached the G.D.E. at Plessis-Belleville for final training.

Most of the American pilots completing their aviation training had been sent to the *Lafayette Escadrille* in an effort to build it to the normal squadron complement of 15 to 18 pilots. On 24 January 1917 Parsons joined the Squadron at Cachy in the Somme sector just before its move to Ravenelle. In February he was promoted to *Sergent*.

Sergent Parsons flew many patrols and fought numerous combats but did not receive his first confirmed victory until 4 September 1917 when with ten rounds from his Vickers machine gun he downed a Rumpler C near Neuilly. For this action he was awarded the *Croix de Guerre*, with Palm.

Sergent Parsons was home on leave in Springfield, Massachusetts from November of 1917 until January of 1918. Upon his return to the Front, he found the *Lafayette* Squadron no longer under French command, but by then had become the U.S. 103rd Aero Pursuit Squadron. Parsons soon learned from his former squadron mates who had transferred to the U.S. Air Service that the conditions there were far from satisfactory and that long delays were anticipated before reassignment to flying status.

For this reason Parsons chose to remain in the French *Service Aeronautique*. He left the squadron on 26 February 1918 and requested assignment to *SPA 3*, *Capitaine* Georges Guynemer's old escadrille. On 24 April, in appreciation of his past service, Parsons was assigned to this famous Cigognes or "Stork" squadron.

While Parsons was with *SPA 3* he reached his full potential as a fighter pilot. On 6 May and 13 May 1918 he downed two enemy bi-places in the area above Montdidier. Following his third victory he received the *SPA 3* "silver stork" squadron pin. On 19 and 20 May he downed two more two-seaters near Montdidier and Gratibus.

On 17 June 1918 Parsons spotted Lieutenant Frank Baylies in the air above Montdidier as Baylies was returning from a two-hour combat patrol. At some distance Parsons watched his friend fall in a combat with four enemy Fokkers before Parsons could arrive to give aid. At the time of his death, Baylies had scored twelve victories, all as a pilot with *SPA.3*.

Parson's avenged Baylies's death with his sixth victory, this over a Fokker chasse near Morchain on 29 August. On 26 September he downed a two-seater south of Tahure, and his eighth and final victory came over a bi-place north of Somme-Py on 1 October 1918. He eventually received official credit for the destruction of eight enemy aircraft.

In the fall of 1918 Parsons was promoted to *Sous-Lieutenant*. Due to leaves, casualties, and illnesses, in October Parsons became the senior ranking officer with *SPA 3*. As a result, he was in command of the squadron for a period of two weeks. Then he too became ill later than same month and was granted sick leave at Biarritz until the signing of the November Armistice.

(Left) Ted Parsons (4th l.) in Hollywood, ironically playing a German pilot in the 1934 film, Hell in the Heavens. Warner Baxter stands on his immediate right. (Right) Rear Admiral Parsons, USN, during World War II.

In recognition of his war service, Parsons received France's *Croix de Guerre*, with eight Palms, the *Medaille Militaire*, and the *fourragere*. He also received Belgium's Cross of Leopold, *Croix de Guerre*, and the American Aero Club Medal.

One colleague said of Lieutenant Parsons: "Seeing him on leave one might easily have thought him an "aviateur des boulevards" who had never been nearer the Front than the "Camp Rentranche' de Paris." The resemblance was only superficial, for his keen eyes were never turned toward that promised land of all "aviateurs embusques." Permission over, he got into his well-worn flying clothes and a pair of sabots, and could always be seen, ten minutes before patrol time, clopping briskly out to the aerodrome."

Parsons returned to the United States in April of 1919 while still in the French Air Service. There, he served with the French High Commission into February of 1920 when he was finally discharged. Upon his return to civilian life, he graduated from the first class of special agents for the Federal Bureau of Investigation and was assigned to the Los Angeles office.

In 1923 he resigned from the Bureau and opened a private investigator agency in Los Angeles. That business failed and Parsons came out of it with only the clothes on his back. Fortunately he ran into a former Lafayette Flying Corpsman, William Augustus Wellman, who was building a solid reputation as a motion picture director. Wellman helped Parsons to secure work with Paramount Studios, and from 1926 to 1936 Parsons became a writer and technical director of aviation pictures in Hollywood. His most notable effort was his three year stint as aeronautical advisor on Howard Hughes's aviation epic, "Hell's Angels."

About this same time Parsons discovered a supplementary source of income when he began writing and selling aviation stories to pulp magazines. His stories were well received since they were written with an authenticity extracted from his own World War experience. In 1933 Parsons also scripted and narrated a live radio show over KFBK in Los Angeles. The fifteen-minute program was called "Heroes of the Lafayette Escadrille."

In 1937 he wrote a book titled *The Great Adventure* about his World War service. This book was later reprinted and retitled *I Flew With the Lafayette Escadrille.*

In 1934 Parsons joined the U.S. Naval Reserve. By 1939 he felt that the United States would soon be involved in another World War and he had asked for assignment to active duty. His first assignment as a Naval officer was aboard the tanker *U.S.S. Neches* off Alaska.

In 1939 when France was again threatened by a hostile German army Parsons, Paul Rockwell, and Harold B. Willis had attempted to do all they could to aid France. The three had begun to recruit young Americans with the intention of forming another *Lafayette Escadrille* until they were notified by U.S. officials that such activity was a violation of U.S. neutrality. At this point Parsons, still a U.S. Naval officer, withdrew from the effort while Rockwell and Willis sailed for France in an attempt to see it to reality. Their hopes were dashed by the sudden invasion of France by Germany on 10 May 1940, and the two fled Paris for their lives.

In June of 1940 Parsons began serving as an instructor at the Pensacola Naval Air Station. While there, he was appointed to the rank of Lieutenant Commander. In July of 1941 he was transferred to the Pacific combat zone where he eventually became commander of LST Group 41. He directed his unit's amphibious assaults upon the Philippine Islands and upon Okinawa.

Parsons eventually rose to the rank of Rear Admiral before his retirement from active duty in 1954. He settled in Osprey, Florida and in 1957 he married Catherine Gardner McKay, an actress and mother of actor Gardner McKay.

Parsons remained active in retirement, running unsuccessfully for local sheriff; maintaining a voluminous correspondence with aviation historians throughout the country; and forming with four other surviving Lafayette pilots the *Lafayette Escadrille N.124 Society* in 1960 to "preserve inviolate the good name of the Escadrille." In 1962 he received much belatedly France's *Legion d' Honneur*, and the *Medaille de la Marne*, which was awarded for his service with *SPA 3*.

Parsons died at age 75 of a lung ailment in Sarasota, Florida, on 2 May 1968. He is interred at Arlington National Cemetery, Washington, D.C.

PAUL PAVELKA

ESCADRILLE LAFAYETTE

"I will not forget the friends the Germans have deprived me of after I get back to the Front."

Paul Pavelka to Mrs. Alice Weeks, 17 January 1916, upon learning of the death of her son and his Legionnaire comrade, Kenneth.

Paul Pavelka, the 14th volunteer of the *Escadrille Americaine* was born in New York City on 26 October 1890, the son of immigrant parents Paul and Anna Pavelka. His father had come to the United States from Budapest, Hungary and had settled in the Wendover Avenue neighborhood of the Bronx, working as a foreman in a wire factory to support his family. Paul, Jr. spent a portion of his youth there, attending Bronx public schools until his father had saved the money to purchase a farm on Copse Road in Madison, Connecticut.

Tragedy struck the family soon after Paul's father had relocated to Madison. When Paul, Jr. was fifteen his mother, Anna, had injured herself on a hay fork and had died soon after at the age of thirty-six on 15 April 1907.

Several months after Anna's death, Paul's father married Barbara Balogh, a beautiful twenty-two year old woman from New York City. With his new stepmother only six years his senior, Paul, Jr. found himself strongly attracted to her beauty. Troubled by this attraction toward her, he departed the Connecticut farm at the age of sixteen and struck out alone.

After Paul had left home he worked on farms in Vermont and New Hampshire. He then moved into Canada and found employment at a lumber camp. The following summer he returned to New York State to take a position as a handyman at a sanitarium. The fall of 1908 saw him performing a series of odd jobs while riding empty boxcars through Ohio, Indiana, and Illinois.

In the spring Paul hopped a freight to North Dakota and took work on a sheep ranch. He then moved on to Glasgow and Malta, Montana, where he wrote the following to a friend:

"A few lines to let you know I am among the cow punchers. Here in Montana saloons are never closed and gambling is open. It's as lawless a country as any I have seen yet, always fighting among fellows in the saloons."

Paul earned his keep by sheering sheep at ten cents a head. When he had saved $56 he caught a freight to Washington State. Enroute, Pavelka was mistaken for a bandit while holed up in a baggage car between Wallace, Idaho and Spokane, Washington and was shot by an express guard. The guard soon realized his mistake and helped Paul to reach a hospital in Spokane. When Paul had recovered, the guard gave him the price of a fare to Seattle. There, in a nearby lumber camp, Paul served as a camp cook throughout the winter.

Paul spent the summer of 1909 living in a shack on a Washington river, surviving by fishing and hunting. He headed into California when the weather turned cold and found employment as a migrant worker. He eventually drifted south to the Panama Canal Zone and worked there some months before attempting a climb of the Andes which tragically cost the lives of his companions.

Pavelka again returned to Washington State and remained there a year and a half. Around 1911, at twenty years of age, Paul shipped out on a Pacific freighter headed for Japan. Next, he sailed for Australia and the South Sea Islands then continued westward to Europe and traveled the Continent before stopping in London where he spent some months.

Paul returned to the United States in 1912, landing in Norfolk, Virginia. It was here that he enlisted in the U.S. Navy and served as a seaman aboard the USS *Maryland*.

Following his discharge from the Navy, Paul spent the summer of 1914 in the Seamen's Home on South Street in New York City. When a recruiter from a Canadian transport ship approached him about work, Paul signed up. Paul joined the ship, the *Caribbean*, in Quebec. Its cargo were troops of the Princess Patricia's Canadian Light Infantry on their way to England and eventually to the trenches in France.

Not long after the ship docked in England, on 25 October 1914 Paul enlisted at Camp Wembley in the "1st Battalion De Tirailleurs," part of the 'Army' of the mythical South American 'Republic' of Counani led by one Adolph Beaufort. Pavelka wrote of this volunteer unit:

"It's a legion under the Belgian flag, and most of the boys have been in action during the sacking of Belgium and have been wounded, sent to British Hospitals, and then to camp here where we are reorganizing. We are expecting to leave for Dunkirk, France, in a fortnight's time."

Belfort's ragtag band of 300 men soon left Folkestone, England for France on the steamer *Sussex*. Paul was wary of entering combat with this disorganized Army of Counani, and he obtained his release and enlisted in the French Foreign Legion at La Rochelle on 28 November 1914. With a keen sense of foresight, he remarked "when this war is over I will be able to say that I was in one of the greatest wars in the world's history."

In February of 1915 Pavelka was sent to the Front with a detachment of reinforcements and was assigned to Company 2, Battalion B, of the *Premier Etranger*. Fellow American Kiffin Rockwell was also serving in that company, and the two soon became close friends. Of Pavelka, Rockwell later wrote to his brother, Paul that "he has never had any ambition, otherwise could probably have made a success as he has quite a lot of genius in several ways."

The Americans' unit suffered steady losses from enemy artillery and shell fire. Then on 8 May 1915 "Skipper" Pavelka, (so called once his comrades had learned of his seafaring background), moved with his battalion into the trenches near the village of Bethonsort, north of Arras, where the greatest offensive on the Western Front was being planned.

When the 9 May assault was launched against the German positions, Pavelka and Kiffin Rockwell charged across the rolling Artrois plain under heavy fire toward the German front-line trenches and quickly took prisoners. However, upon advancing, he later recorded: "I lost my American side-partner by the name of Rockwell, he went down with the remark of 'I got it in the hip; Paul go on and give them hell.'" That was the last I saw or heard of Rocky."

Enemy reinforcements stopped the Legionnaires' progress at the crest of a hill. Miraculously, Pavelka escaped injury although all his officers were soon either killed or wounded. Kiffin Rockwell later wrote of Pavelka that he "is recognized by all the officers in his company as being the best soldier in the company."

On 15 May 1915 Skipper wrote to his friend, Phil Kerner, in New York City that "I again take pleasure in informing you of my present welfare, which I cannot account for except that my day has not been intended to come quite yet. You see, I am somewhat of a fatalist."

At noon on 16 June Pavelka's Legionnaires launched an assault against the German positions around Souchez, including Cabaret Rouge, the 'Labyrinth' and Hill 119. On 25 June he wrote another letter to Kerner, this time from "Hopital Auxiliare No. 3." informing him that he had been "wounded on the 16th, by a bayonette."

Pavelka's unit had charged the enemy front line near Givenchy through a hail of rifle and artillery fire. Meeting little resistance, they quickly took the forward trench. But the enemy had set a trap and was waiting for them in the second line of trenches.

"It was in reaching this line that Fritz told me I was due for a little fun. I really was surprised by the way I came upon the trench, thinking it would be like the first line, where they offered a feeble resistance, ending up by 'Kamarade, Pas about,' and their dirty hands above their heads.

"Fritz saw me coming and was going to get me; he got out of sight until I had reached the edge of the trench and was about to get in, when up he sprang and made a great lunge at me with his bayonette. It was luck for me he hadn't a longer reach, or he may have done some real damage. As it was he only stuck it in my left leg.

"Phil, he won't stick any one else on this earth with a bayonette or anything else. Well, I stopped here, as the wound was bleeding badly and besides it was very painful."

When Kiffin Rockwell learned that his friend had been bayoneted, he wrote his brother Paul from the hospital where he was convalescing, asking him to do whatever he could for Pavelka.

"Skipper evidently never received my letter in which I enclosed the 'piece of change,' so I am sending him five francs today to keep him in tobacco until I can send him more, and I want you to do anything you can for him from Paris. Skipper's family didn't amount to anything and he never had any education to speak of, and he has been on his own lookout ever since a child, rambling about the world, but he has the sentiments of a gentleman. He did not have to enlist for France, he could have gotten a boat away from Europe and continued as a sailor. But he is a good lad and brave. He has no way of getting money from home and no friends in France. Now I want you to find someone in Paris who appreciates that a chap like that is fighting for France, and who will take him when he gets his convalescence, so that he can enjoy it before returning to the front. You and I can keep him in spending money, and his wants are simple and few."

On 3 July 1915 Pavelka again wrote his friend, Kerner from the hospital at Nogent-le-Rotrou that he was suffering from cramps in his mending leg muscles. But he went on to say he had the good fortune to recover in a hospital next door to a girls' boarding school and was enjoying the experience of giving them English lessons.

Upon his release, Pavelka spent several weeks in La Valbonne drilling with his unit. At the suggestion of Paul Rockwell he made out a demand on 10 August to transfer to the *Service Aeronautique*. The next day

Legionnaire Paul Pavelka on leave visiting Paul Rockwell at Place de la Concorde, Paris, November 1916.

Kiffin Rockwell learned that both he and Pavelka had been accepted into Aviation. However, there were delays. On 4 September Kiffin left for the Aviation Corps but Pavelka's transfer failed to come through. This puzzled Paul since both men had made their demands together.

On 21 September Paul joined *Battalion A.2, 1er Etranger* at Champagne to fight in the great battle there in September and October. Also participating were future *Lafayette Escadrille* pilots, William Dugan and Robert Soubiran.

The Legionnaires in Sector 109 launched an attack on 25 September. On the 28th they struck the Bois Sabot, a small wood strongly fortified by the enemy. The Germans had stretched barbed wire entanglements from tree to tree and had placed three rows of machine guns behind them. Despite their fortifications, Pavelka's commanding officer had ordered the *sergent* of his section to "Go over and wake them up."

Upon seeing the enemys' weapons above their trenches, none of the Legionnaires was anxious to advance. American Frank Musgrave said "Let's go over and stir them and see what kind of a show they put up." The forty-man section attacked. Only Musgrave and Pavelka made it back to their lines. Paul later wrote of his experiences in a 3 October letter to Mrs. Alice Weeks in Paris.

"Our company lost 160 men, and the most of these were killed. . . .We stayed on the field of battle exactly eight days, and were relieved yesterday morning, and now we are again reorganizing the regiment.

"I have not even a scratch on me only my clothes are torn to shreds by pieces of shrapnel. One shell burst about 2 metres in front of me, and I received the full charge of my steel cask, with which we are equipped. My rifle was demolished in my hands, and the overcoat torn to pieces. I wish you could see me, in my present state. Still I am happy, having had a dip in the river here and a good rest last night."

Paul and several other Americans, including William Dugan and Robert Soubiran, transferred to the 170th Infantry Regiment, hoping to find an easier time of it. Instead they found themselves ordered back to the very trenches they had just been holding as Legionnaires.

Paul became a runner, carrying messages between the different companies of his regiment. Soon winter set in and the men began to suffer frost-bite and frozen feet. Paul severely cut his hands one night stringing barbed wire entanglements. With his fingers bandaged, he could not hold and fire his rifle, and he prayed that the Germans would not attack. Physically spent, he requested and was granted a "permission" to Paris. He reached the capital on 20 November, one of just two survivors of the original American section of the Second Regiment *de Marche* of the First Foreign Regiment.

Paul Rockwell remembered how Pavelka had come to visit him in Paris when Pavelka had taken leave following the battle of Champagne with his tunic literally riddled with bullet holes. He had not had the opportunity to wash or change his uniform since before the Champagne offensive. Following a hot bath Pavelka and Rockwell together had picked almost a pound of spent rifle slugs and shrapnel out of the lining of his great coat.

Paul rejoined his regiment on 27 November 1915. To his great relief, orders awaited him to transfer to Aviation. The efforts of his friends, the Rockwell brothers, Mrs. Alice Weeks, and several prominent French officials had secured his transfer.

After being outfitted with a new uniform at Dijon, on 10 December 1915 Skipper Pavelka entered the Pau aviation school in southwestern France. He proved to be a most proficient student and received his *brevet militaire* on the Bleriot on 23 February. He remained at Pau through April of 1916 then advanced to Nieuport school before attending gunnery and acrobatic school at Cazeaux. While at Cazeaux in early June, future *Lafayette Escadrille* pilot Willis Haviland wrote this of Paul.

"Pavelka will return to Pau on June 10th for the ecole d' cauvat. Seems he has been eating 'em alive at Cazeaux. Is some little shooter. Busted everything they stuck in front of his gun including all school records and is now entitled to wear guns or something on his sleeve with that stripe for being wounded, one for a year at the front, aviation brassard, wings, corporal stripes and shoulder ornaments for being in a distinguished regiment. He will look like some young Christmas tree, won't he?"

With no vacancies at the Front Pavelka was returned to Pau on 15 July. He grew discouraged. However, his spirits were revived eleven days later when he received his flying wardrobe which consisted of "a great big fur overcoat, a fur-lined combination, overshoes, two pairs of gloves, goggles, a chart of the Front, and a nice new revolver."

Pavelka's time had finally arrived. While passing through Paris he hastily scribbled to a friend that "Skipper is at last on his way to the big noise."

Caporal Pavelka reached the *Escadrille Americaine* at Verdun on 11 August 1916 and became an immediate favorite with all the pilots there. On the eve of joining the Squadron, he had written the following to his boyhood chum, Kerner:

"Just received your letter, and it arrived what you may say in the nick of time; for in two days I am off again to the front in an esquadrille. I am more than delighted to have word from you at any time. You are the only person with whom I correspond in America. My father and brothers never having answered my letters to them. Well, Phil, this does not worry me. I have paddled my own canoe, so far, and at present am getting along splendidly. And I believe, I can continue to do so."

The day following his arrival at the Squadron, Paul made his maiden flight with former Legionnaire comrade Kiffin Rockwell and Squadron Adjutant deLaage to acquaint himself with the sector's landmarks. On 13 August he set out alone and strafed an enemy battery in the "hoodooed machine" that Bill Thaw had been wounded in and which Norman Prince had later wrecked. Continuing on with the patrol, at 9,000 feet over Verdun his Nieuport had caught fire.

A broken combination intake-exhaust valve in his Gnome rotary engine had poured raw petrol into the metal cowling which enclosed the red-hot whirling cylinders. The plane's nose immediately burst to flame, emitting a black oily cloud. Instinctively Pavelka closed his throttle and toggled off his ignition. He banked the Nieuport on its side. He then wing slipped to within 50 feet of the ground, keeping the flames out of his cockpit as fire consumed much of the skin on his fighter's wings and fuselage.

Barely ten feet from earth, Skipper reefed back on his stick and dropped the plane into a marsh near Verdun. Roaring heat filled his cockpit and seared his hands and face. Pavelka threw off his smoldering safety harness and leapt into the mud and fought his way clear of the fighter. He put fifty yards between himself and his ship before the fuel tank exploded.

In seconds nearby enemy gunners had sighted in on his smoking wreckage and began lobbing large-caliber shells in his direction. A veteran under fire from his many months with the Legion, Pavelka calmly but deliberately made his way to safety. The next day he was back in the air hunting Germans, his singed face and blistered hands painful mementoes of his perilous experience.

The Squadron moved on 15 September to Luxeuil in the Vosges Sector. It was there that Pavelka would lose his closest friend, Kiffin Rockwell, in combat on 23 September. "My heart is breaking over the death of Kiffin," Pavelka wrote to an acquaintance in Paris.

On 1 October 1916 Skipper Pavelka was promoted to *Sergent*. He was anxious to avenge Kiffin's death but had been without a machine since the Squadron had returned to Luxeuil. But when *Capitaine* Thenault had left on permission to Biarritz he had ordered Lieutenant deLaage to let Pavelka and Bert Hall use his fighter.

On 15 October Norman Prince was the next pilot to die before the Squadron moved to Cachy in the Somme sector. There *Sergent* Pavelka became a relentless hunter, often flying three patrols a day when planes were available. The bullet-riddled bus he brought in off patrol served as stark testimony to his bravery. Like Raoul Lufbery and Victor Chapman before him, he seemed a man without fear and appeared to thrive in a hostile atmosphere.

The Skipper's finest hour came in the first week of November 1916 when he was on patrol with William Thaw and James McConnell some ten miles inside German lines.

Thaw was the first to spot the flight of seven Albatros fighters patrolling menacingly beneath them. Upon Thaw's signal, all three Americans attacked. McConnell was the first to draw blood but caught two enemy fighters on his tail when coming out of his dive. Thaw had missed scoring in his first pass and immediately engaged the German flight leader. Pavelka had picked his target and closed to within seventy-five yards before firing his first burst. His patience paid off as he watched his adversary tumble from the fight and stagger across the sky out of control.

Seeing McConnell under attack, Pavelka immediately focused in on the two buses on his friend's tail. He full-throttled his engine and quickly closed on the Huns, patiently waiting until he was a mere fifty yards from one Albatros before releasing a *rafale* from his Lewis gun. The enemy ship swallowed up his tracers then immediately sideslipped and plummeted groundward in a vertical dive.

Skipper banked and climbed under the second German chasing McConnell. He hung on his prop, his engine whining, and fired four rounds from his Lewis gun into the enemy's vitals before his gun block jammed. A reddish flame glowed within the Albatros then erupted to life, consuming its belly as the bus flamed groundward trailing a stream of greasy smoke.

The enemy survivors disengaged. The three Americans also broke off and sped for their lines. Just as Pavelka made the forward allied trench his motor quit, but he had enough altitude to volplane to safety. Upon landing Pavelka went over his Nieuport with his mechanic and discovered sixteen bullet holes in his fighter. Three had struck his gas tank, draining all his fuel. Only a miracle had kept his bus from flaming.

Although Thaw and McConnell fought to get Pavelka's three victories recognized, French aerial observers were unable to confirm a single one since the combat had taken place deep behind enemy lines. Pavelka's name was posted in the order of the day as having successfully downed at least two enemy fighters, but he failed to receive even a citation.

While the Squadron was at Cachy during the winter of 1916–1917 it was subjected to frequent bombing attacks by enemy machines. *Sergent* Pavelka and French adjutant Lieutenant deLaage volunteered for a night mission, to rise up and intercept the German raiders.

They didn't wait long before the enemy struck the hangars of an adjoining escadrille. At 5 a.m. on 16 November Lieutenant deLaage and *Sergent* Pavelka hurriedly climbed into their 15 meter Nieuports and rose through the darkness to meet the threat.

Soon after becoming airborne, a small electric bulb powered by a dynamo blew out in the Skipper's plane. It had been fitted within Pavelka's cockpit to enable him to read his compass and other instruments. Im-

Sergent Paul Pavelka and his mechanics at Solonika, February 1917.

mersed in total darkness, Pavelka had only the moonlight on the Somme river to guide by. The battle front was defined by an inferno of cannon fire. Pavelka maintained his position by circling, always keeping the river in view. As the arc of his circle expanded he drifted over a large town appearing as a carpet of tiny lights. The sound of Pavelka's motor brought an immediate ground response as he was mistaken for a German raider. Shrapnel soon exploded on all sides of Skipper's Nieuport as he fell under intense allied antiaircraft fire. Pavelka later wrote this of his predicament:

"To come down certainly would have resulted disastrously for me. Every time I neared the earth, the machine guns on the ground would start in working on me. The gunners could not see me, but were shooting in the direction of the sound of my motors, with tracer bullets. Some came very close, I assure you."

Skipper Pavelka circled his aerodrome but realized he could never see to land there. He flew forty kilometers across country, exhausting his fuel supply, at the last minute finding sanctuary at daylight at Martainville.

"I spotted a promising looking chateau with a field near it and landed there. From the khaki uniformed troops about I knew I was in the British lines. Just as I was walking up to the chateau, out came a British colonel."

"'Hello, good morning, had breakfast?'" he asked. Pavelka remained with the British for three days, holed up in a chateau as he waited for the weather to clear. Undaunted by his heart-gripping experience, the fearless Pavelka and Adjutant deLaage went on to make twenty more night sorties, thus pioneering the concept of night pursuit which would not become routine with the allies for nearly a year.

In December of 1916 a request was sent to the French squadrons on the Western Front seeking volunteers to fly in the Army of the Orient. *Sergent* Pavelka volunteered, hoping a change of scenery might bolster his sagging spirits brought on by the death of Kiffin Rockwell and other comrades. Approximately one month later his request was approved, and he left the *Lafayette* Squadron on 24 January 1917, his destination, *Escadrille N.391* on the Salonica front.

Paul sailed from Marseille on 30 January 1917 and arrived in Salonika in early February. On 8 February he reached *Nieuport 391* following a four-day trip by tractor and immediately began flying his Nieuport five hours a day over the lines. The pilots of *N.391* were quartered in airplane cases. However, in a letter to a friend, Paul noted that the quality of the mess was good and the prices reasonable. He also noted the inexperience of the enemy pilots and the poor aim of their gunners. He went on to say:

"I am awfully glad that I came to the Orient. It is just what I desired, in the line of war. One hundred and eighty kilometres from nowhere, plenty of rough work, and some advancing to keep up the enthusiasm of every one concerned."

Paul flew missions with *N.391* until 15 June, 1917. That same day he transferred to *Escadrille N.507*, and in July he was mentioned in the Order for the Allied Armies fighting on the Macedonian front. Its Commander-in-Chief personally awarded him France's *Croix de Guerre*, with Palm. His citation, signed by General Sarrial, mentioned how: "he has given every proof of eagerness to fight and devotion; has had numerous combats, frequently with his machine riddled with bullets."

Pavelka wrote the following news on 30 July: "I have two machines now, a 120 H.P. Nieuport, and an A.R. 170 H.P., with a Greek officer as passenger, with whom I expect to make a few long-distance raids. He speaks perfect English, and is a charming sort of chap. However, he does not know much about handling a machine gun. This I will soon teach him. I expect to make my first raid in a few days.

"It has been very hot here these past few days. So hot that a good many of our boys are being sent to France, as they are not able to stand the heat. Malaria reigns supreme. We see many men who are "dango" from the heat; some become dangerous, and others quite amusing. Leaves are being accorded, and there are about ten thousand permissionnaires on their way to sunny France—or rather shady France, for God knows no one would want more sun than there is here in Macedonia."

In mid-August *Sergent* Pavelka and his driver were returning from Headquarter by automobile in pitch darkness of night to his flying field. The car was without headlights, and on a narrow bridge spanning a rocky ravine his driver lost control and crashed to the boulders below. Pavelka was pitched through the windshield, lacerating his face, breaking his jaw in two places, and dislocating both knees.

He was sent to a Salonika hospital to recover. As he lay in a ward in sweltering 120-degree heat, a fire broke out in the heart of the city which nearly consumed all the business and shopping district. The fire deprived Pavelka and the other flyers of the basic necessities they usually purchased from the open market to supplement their meager army rations.

Despite a wired jaw and bandaged knees, *Sergent* Pavelka continued to fly missions with *N.507* in his two-seater Nieuport through the late summer and early fall of 1917. But on 11 November 1917 an irony of fate would tragically end his life.

When Paul discovered that an English cavalrymen in a nearby regiment had been a former comrade from the Foreign Legion, he paid him a visit. That same day fresh horses had arrived at the regiment, and Pavelka, an accomplished horseman from his summers in the American West, had sought his friend's permission to ride one of the mounts. Always seeking a new challenge, Paul chose a horse of a vicious nature which was thought impossible to ride.

A crowd gathered to watch "the Yank broncobuster" climb onto the horse's back as a stable sergeant held tight the reins. Once freed, the horse reared and tossed and bucked violently but failed to throw Pavelka. Sensing that it could not throw its rider, the mare fell hard to the ground and rolled in the dust over the top of Pavelka, crushing him.

Onlookers rushed to Pavelka as others corralled the horse. Blood seeped from the Skipper's mouth into the dust where he lay. He was carefully placed on a stretcher and rushed to the hospital. He lingered on until the following day when he died on 12 November 1917 at age 26, his life ended by a freak accident after having survived so many months of combat with the Legion and so many aerial duels against enemy aircraft.

American Edgar Bouligny, a former Legion comrade of Pavelka's, was serving in Solonika with the French Air Service when his friend had been killed. Bouligny was granted permission to attend his funeral and described it in a 13 November 1917 letter.

"I did not know that poor Pavelka was near Salonika, and it was only this morning that I heard a pilot say that his funeral would take place at nine o'clock at the Zietenlick Cemetery at Salonika. . . . All the aviators of the Salonika sector were there; also a good many officers from the French and British camps.

"The piquet d'honneur was furnished by a French infantry regiment, and there was also an armed guard composed of Serbian solders. The funeral services were read by a chaplain from the British Army, and in the chapel a French almoner made a talk. He spoke of the brave and valuable services rendered to France by Pavelka, and also added a few words about the great friendship existing between the United States and France.

"In the crowd I noticed the United States Consul. On Pavelka's tomb were beautiful wreaths, sent by the pilots and observers of Escadrilles 502 and 507, and another from the mechanics and personnel of Escadrille 507."

Sergent Paul "Skipper" Pavelka had performed excellent service on behalf of France in nearly three years of duty with the French Foreign Legion and the *Service Aeronautique*. It was a fitting tribute that his body was removed from the Zietenlick Cemetery in Salonika, Greece to Marseilles, and in 1928 to a crypt in the *Lafayette Escadrille* Memorial near Paris where it rests today in the company of his fallen comrades.

ALFRED PELTON

Alfred Digby Pelton was born in 1888 in Montreal, Province of Quebec, Canada, the son of Godfrey Pelton, a hardware merchant, and Harriet Pelton. He received his primary education in Westmount, Province of Quebec, and later at Bishop College School, Lennoxville, Province of Quebec.

Pelton's older brother, George, was killed in France while serving with the Canadian Expeditionary Forces. About this time "Alf" Pelton married Miss Louise Moody of Terrebonne (north of Montreal). To avenge his brother's death in the War, Pelton tried to enlist in the Royal Flying Corps when their Canadian Training Plan was first established. However, Pelton was rejected because of his age.

Undaunted, Pelton sailed via New York to France where he was accepted into France's *Service Aeronautique* on 19 February 1917. From 27 February to 25 November 1917 Pelton attended the aviation schools at Avord, Pau, and the *G.D.E.* He received his *brevet* on the Caudron on 15 July 1917.

As Pelton progressed through the French aviation schools, he became close friends with his American comrades. At their suggestion, he applied for and was accepted into the Lafayette Flying Corps. Thus, he became that organization's only foreign member.

From 27 September to 1 December 1917 *Caporal* Pelton was assigned at the Front to *Escadrille N. 151*. Pelton was granted a three-month winter furlough to Canada. Upon his return to France, on 5 March 1918 *Sergent Pelton* was assigned to *N. 97*. He flew with that squadron until 31 May 1918, the day he was killed in aerial combat over enemy lines near Soissons. Pelton was 30 years old at the time of his death. Four months later word of his death reached his squadron through the International Red Cross. His remains are interred at the Lafayette Flying Corps Memorial near Paris.

Paul Baer, Alfred Pelton, Theodore De Kruijff, and Charles Kerwood at the Cafe d'Avord, April 1917.

DAVID PETERSON

ESCADRILLE LAFAYETTE

David McKelvy Peterson, the 36th volunteer of the *Escadrille Lafayette*, was born in Honesdale, Pennsylvania on 2 July 1894, the son of eye and ear specialist and Wayne County Coroner, Dr. Pierson B. Peterson, and Louise (Jadwin) Peterson, sister of Brigadier General Edgar Jadwin, commander during World War One of all construction work for the American Expeditionary Forces in France.

Following grammar and high school in Honesdale, young Peterson took a chemical engineering degree in 1915 from Lehigh University. Upon graduation, he went to work for the Dravo Construction Company of Pittsburgh. He left there to work at the Curtiss aviation plant in Buffalo, New York and in his spare time qualified as a pilot. He sailed for France in September of 1916 with the intention of enlisting in French aviation.

David Peterson in training at the Avord School of Aviation, April 1917.

Peterson enlisted as a private in the Foreign Legion and immediately transferred to French Aviation on 9 October 1916. He entered the aviation school at Buc on 16 October and received his *brevet militaire* on the Bleriot on 13 April 1917. Peterson progressed through the schools at Avord and Pau and was awaiting assignment at the *G.D.E.* pilot pool when he was called to the *Lafayette Escadrille* at Chaudun on 16 June 1917. He joined the Squadron the same day as James Norman Hall and Douglas MacMonagle.

Peterson soon proved a natural pilot. He had excellent air vision and was often the first to spot enemy activity and would immediately waggle his wings as a warning to his patrol. Because of his keenness of vision and unwavering courage, he soon became a patrol leader "without an equal," and by the time he had completed his service with the *Lafayette*, he had led more patrols in his SPAD emblazoned with its blue pennant markings than any other pilot in the Squadron.

The tall, angular Peterson was an immediate favorite with his fellow *Lafayette* pilots. But he proved an enigma to all his friends with the possible exception of Dudley Hill, whose own personality closely mirrored David Peterson's. Like Hill, nothing seemed to stir Peterson's "tranquil depths." And as one *Lafayette* colleague observed: "It may be said without any exaggeration that he (Peterson) is the only American who has never had a thrill from his adventures as an airman."

It wasn't that Peterson wasn't looking for adventure. He sought it with near recklessness, flying his SPAD ever deeper into German territory in pursuit of excitement. And more often than not it came in violent confrontations with the enemy or while hunting special targets such as observation balloons, a locomotive, a convoy of Boche trucks.

The conclusion of a combat patrol often saw the survivors huddled in animated excitement as they relived its highlights with tense voices and swooping gestures. Others would scrutinize the punctured skins of their fighters with their mechanics or would curse a bullet-jammed gun block, their legs weak with fatigue, their faces glistening from the sweat running out of their flying caps.

But not so, David Peterson. As James Norman Hall remembered him:

"Peterson would jump out of his bus, stretch his legs, wriggle the kinks out of his neck, and walk over to the *Bureau du Groupe* to make out the *Compte Rendu*—the pilot's report of the events of the patrol. Then, in the "Remarks" column he would write in a firm clear hand, *"Rein á signaler"*—nothing to report!

Sergent David Peterson beside his SPAD S.VII fighter, Chaudunm July 1917.

Some of us thought his idea of nothing to report a rather strange one."

Peterson had been fortunate enough to break into the Squadron routine in the early summer of 1917, flying fighter escort from the spacious Chaudun airfield. Next came a three-week interlude at Saint Pol-Sur-Mer near Dunkirk. The flight to the new aerodrome through dense fog proved more eventful than the sojourn there as a number of the pilots either lost their way or cracked up on the move. Peterson damaged his bus when he was forced to make an emergency landing on a British airfield.

On 11 August the Squadron moved to Senard in the Verdun area and the air work became intense as Peterson and the other pilots flew as many as three patrols a day. During its five-week stay there, Harold Willis was shot down and made a prisoner, and Steve Bigelow was so seriously wounded, he was forced to leave the Squadron. Both MacMonagle and Campbell would soon be killed.

On 18 September the Squadron was ordered back to Chaudun to support the planned 3d Aisne Offensive. On the 19th *Sergent* Peterson won France's *Croix de Guerre,* with Palm when he downed an Albatros scout between Trory and Montfaucon for his first official victory. To make certain his victory would be confirmed, in order to get an exact description of the crash site, he had followed the crippled fighter to such a low altitude before it had banged into the ground that his SPAD had been peppered with rifle fire from German troops.

On 7 December 1917 the Squadron was transferred to La Noblette. With the onset of winter, the drafty barracks became a sieve of discomfort. But not so, David Peterson's cubicle. As at past aerodromes, he would put his imagination and engineering skills to good use, chinking cracks, tacking up shelves, building furniture. As with Chaudun and Senard, "Pete's" quarters became the hangout for all the pilots wishing to avoid the draughty surroundings. There, it was "always snug and cozy, with a fire and a neat pile of fuel in the wood-box."

Peterson seemed to live a charmed life during his tenure with the *Lafayette* Squadron. He fought more combats than most of the other pilots. And often his SPAD showed damage from hostile ground and air fire.

One particularly violent confrontation took place over the forward trenches in the first week of December when Peterson and Ray Bridgman had flown one of their voluntary patrols. Together they had dropped out of the sun in a diving attack against an unsuspecting Boche two-seater. Peterson later described the action:

"At the first volley the Hun dropped like a shot-probably wasn't touched, only scared. Then things started happening for us.

"Seven Hun single-seaters dropped on our back out of nowhere; they, too, had evidently been using the sun. They were on top and were flying and shooting like wild men. Bridgie and I kept well together and attacked the Boche from underneath at every opportunity. About a dozen bullets had gone through my wings and Bridgie had his fuselage pumped full, but nothing serious had been done."

The enemy pack formed a deadly barrier between the Americans and the safety of French lines. Peterson could see no clear way through the gauntlet. He signaled Bridgman to turn 180 degrees and begin climbing into Germany. When they had climbed 1500 feet above the enemy formation, they turned back toward their lines and dove down on the enemy, firing their guns and scattering the Boche. They plunged through the gauntlet to safety while emptying their machine gun belts. Peterson concluded his report by saying:

"We had only been out half an hour, but what a lot of damage was done! Our machines were literally shot to pieces from the explosive bullets-but not a control cable or an engine was touched. Bridgie had four holes in his top plane a few inches from his head that you could stick your four fingers in."

On 19 January 1918 *Sergent* Peterson received his commissioned as Captain in the U.S. Air Service. The commission had been issued and dated 7 November 1917. On 18 February 1918 he joined the 103d U.S. Aero Pursuit Squadron and flew combat patrols until 31 March 1918. On 1 April he and fellow *Lafayette*

Carl Dolan, Kenneth Marr, William Thaw, David Peterson, Mr. George Ferry, and Dudley Hill at Chaudun, November 1917.

Escadrille pilot Captain James Norman Hall were assigned as flight commanders to Raoul Lufbery's 94th Aero Squadron at Epiez. Captain Edward Rickenbacker later wrote of their arrival:

"Both of them were experienced pilots who had become celebrated for their air work in the Escadrille Lafayette and who did much to enhearten us and instruct us during this forlorn period. We had all heard of these boys and idolized them before we had seen them. I cannot adequately describe the inspiration we all received from the coming of these two veteran airfighters to our camp."

Captain Peterson led the first All-American combat patrol accompanied by Lieutenants Rickenbacker and Reid Chambers on the morning of 14 April between Saint Mihiel and Pont-á-Mousson. Two weeks later, on 3 May 1918, he was awarded the U.S. Distinguished Cross for Extraordinary Heroism for action near Lunéville, France. His citation describes the engagement:

"Leading a patrol of three, Captain Peterson encountered five enemy planes at an altitude of 3500 meters and immediately gave battle. Notwithstanding the fact that he was attacked from all sides, this officer, by skillful maneuvering, succeeded in shooting down one of the enemy planes and dispersing the remaining four." The German scout that Peterson downed was later confirmed, raising his official victories to two.

Twelve days later, on 15 May, Captain Peterson added two more victories to his credit and received a Bronze Oak Leaf to his Distinguished Service Cross when he shot down two enemy fighters in five minutes in the Toul sector. Peterson later described the action:

"I was flying near Thiacourt at an altitude of 5,000 meters when I saw two German Albatros scouts going away at right angles in the distance. One was ahead of the other. I ascended 200 meters, circled, and stopped my engine and darted after them. I finally got close to the last plane. At a distance of 100 yards I opened fire with incendiary bullets and the German machine almost immediately burst into flame and fell.

"I continued straight on my course and within two minutes had the other enemy machine before me. He started a quick dive but I am certain that some of the bullets went home. One of the wings of his machine crumpled up at the same moment. As I circled about I saw one machine burning on the ground and the other, smoking in a downward plunge, crash to earth. That's all there was to it. He was a cinch."

Peterson was also awarded another Palm to his *Croix de Guerre* for this heroic action.. His citations called him "an excellent officer and a daring pilot, both exceptionally courageous and skillful."

Peterson flew with the 94th until 16 May 1918 when he was ordered to the 95th U.S. Aero Pursuit Squadron. On 17 May he brought down a German bi-place machine near St. Mihiel, his fifth victory, officially making him an ace. On 21 May near Chateau Salino he brought down another of a growing number of unconfirmed victories when he chased an enemy two-seater 40 kilometers along the battle lines then 12 kilometers behind German lines before attacking him from beneath. On 22 May he was officially made the commanding officer of the 95th. He was promoted to Major on 29 August 1918.

Major Peterson served as C.O. of the 95th until 8 October 1918, in the Toul sector, at Chateau Thierry, Saint Mihiel, and the Meuse Argonne offensives. His squadron fought 230 combats and was officially credited with the destruction of 47 enemy aircraft. And while Peterson's official tally of enemy aircraft has been placed at six planes destroyed, unofficially he may have downed as many as eighteen.

On 8 October 1918 Major Peterson was ordered back to the United States, to proceed to Gerstner Field, Louisiana, there to serve as an instructor. He and Lieu-

David Peterson standing near Dudley Hill, Chaudun, July 1917.

tenant Carl Dolan returned by transport together to the United States landing at Norfolk, Virginia around the end of October. Together they went on to Washington, D.C. and checked in with the Office of Air Service and were given the option of going to Carlstrom Field, Florida or to Rockwell Field, California. Both men chose Florida then proceeded home for one week's leave.

When Peterson and Dolan arrived at Arcadia, Florida Peterson was placed in command of aerial gunnery at Dorr and Carlstrom Fields while Dolan was made engineer officer there. In addition to this assignment, Peterson started the U.S. Air Service's first night flying school.

In early 1919 Major Peterson decided to remain in the regular army. In March of that year, in addition to his regular duties, he enrolled in a U.S. Military Command School in Florida in order to take the necessary training to receive a regular army commission.

On 16 March 1919 Major Peterson as pilot, with Lieutenant F.X. Paversick as passenger, was flying a D.9, a new type of ship, over Daytona Beach, near Sea Breeze, Florida. The D.9 had a dual control with a removable joystick which could be hung in a rack out of the way. While Peterson was piloting the plane during takeoff, Paversick accidentally jammed his joystick in the snap-lock socket of his cockpit in such a way that the plane continued to climb.

Peterson fought frantically to put the nose of the plane down as Paversick struggled to remove the joystick. The D.9 climbed steeper and steeper until it stalled and slid backwards and eventually hit on its tail. The engine broke loose from its mounts and crushed the twenty-four-year-old Peterson to death. Paversick was seriously injured and later died of his injuries.

Lieutenant Carl Dolan accompanied Major Peterson's body from Florida back to Honesdale, Pennsylvania. On 22 March 1919, the day of Peterson's burial, all the businesses in his hometown were closed in his honor. Dolan and Lieutenant Henry S. Jones escorted the coffin for burial at the Glen Dyberry Cemetery in Honesdale, Pennsylvania. Then Dolan and Jones and pilots from Peterson's 95th Aero Squadron carried his casket to the grave. At that time former *Lafayette* pilot Lieutenant Henry S. Jones, Peterson's friend and fellow Lehigh alum, paid him this tribute:

"Peterson will go down in history as one of the world's greatest aviators. He proved himself a master at fighting the Germans and was one of the four aces of our escadrille. He was able to think rapidly, to make prompt decisions and to act quickly. He was fearless to the last degree, and his ability to see everything in the air and on the ground while flying was uncanny at times."

On 16 May 1928 it was announced that Major Peterson's body would be disinterred from the cemetery in Honesdale for reburial at the *Lafayette Escadrille* memorial outside of Paris. However, the townspeople and the American Legion Post 254, which was named after Peterson, objected so strongly to the disinterment that the family decided against it. The tomb bearing his name at the *Lafayette Escadrille* Memorial outside Paris stands empty.

GRANVILLE POLLOCK

Granville Alexander Pollock was born near Fort Worth, Texas on 5 January 1889. In 1903 Pollock began work as a machine shop apprentice with the Southern Pacific Railway. He next worked as a mechanic for the Fairbanks-Morse agency in the U.S. and in Mexico.

In September 1914 Pollock sailed to England to enlist in the Royal Naval Air Service. But at that time British Aviation was not accepting Americans. He next attempted to enlist in the Belgian Army. When his services were refused, he became a European representative of the Pierce-Arrow Motor Company, taking charge of a number of trucks to be sold to the British Admiralty. Using this influence, he was able to enlist in the Royal Marine Artillery. He was made staff sergeant and was put in charge of mounting anti-aircraft guns in Admiralty vehicles.

Sergeant Pollock was sent to France where he took part in four battles. When his unit was returned to Britain, he was discharged on 23 February 1916.

On 24 December 1916, Pollock joined France's *Service Aeronautique*. From 1 January to 14 July 1917 he attended the aviation schools at Buc, Avord, Pau, and the *G.D.E.* He received his *brevet* on the Bleriot on 12 May 1917.

From 15 July to 14 October 1917 *Caporal* Pollock was assigned at the Front to *Escadrille Spa 102*. While with this unit, he made photo-reconnaissance missions in his SPAD deep within enemy lines. From 16 October 1917 to 8 January 1918 *Sergent* Pollock flew combat patrols with *Escadrille Spa 65*.

On 18 January 1918 Pollock left the French Air Service and was commissioned a First Lieutenant in the Aviation Section of the U.S. Signal Corps. He was returned to the United States and was attached to the Instrument Division, Technical Dept. USAS at Washington, D.C. He was ordered back to France where he was made officer in charge of flying at the American Acceptance Park, Orly. He remained in that position until the 1918 Armistice. For his outstanding War service, Lieutenant Pollock was awarded France's *Croix de Guerre*, with Star, along with British and American War Service decorations.

Following World War One Lieutenant Pollock planned a trans-Atlantic flight by way of the Azores in a two-seater Salmson bomber. At the last moment, General Pershing vetoed the plan, fearing for the young pilot's life. A short time later Pollock left the Air Service and returned to civilian life.

In 1921 Pollock established a chemical engineering business in Paris which he ran for many years. In 1925 he returned to French Service as a Major in the Riff War in Morocco. Following his discharge, he settled in Paris.

Pollock left his Paris-based business in 1941 and returned to the United States to enter World War Two. He served as a civilian employee for Air Force Intelligence. Following World War Two Pollock settled in Florida. He died of a heart attack in Carrizo Springs, Texas on 28 August 1950 at age 61.

WILLIAM PONDER

William Thomas Ponder was born in Llano, Texas on 16 March 1893, the son of William Perry Ponder, a Mangum, Oklahoma banker and rancher, and Mary Cornelius (Smith) Ponder. Ponder attended Mangum public schools before enrolling in Dartmouth College, Class of 1917. Following graduation Ponder sailed for France to join the American Field Service. Shortly after his arrival in Paris, on 4 June 1917 he enlisted in France's *Service Aeronautique*. From 19 June 1917 to 1 February 1918 he attended the aviation schools at Avord, Pau, Cazeaux, and the *G.D.E.* Ponder received his *brevet* on the Caudron on 7 November 1917.

From 3 February to 17 February 1918 *Caporal* Ponder served at the Front with *Escadrille Spa 67*. On 27 February 1918 he was commissioned a Second Lieutenant in the U.S. Air Service.

From 22 March to 12 May 1918 Lieutenant Ponder served as a ferry pilot at the American Acceptance Park, Orly. From 12 May to 1 September 1918 Ponder flew combat patrols with *Escadrille Spa 163*. He destroyed three enemy aircraft while with that squadron.

From 7 September 1918 to the 11 November Armistice Ponder was a pilot with the 103rd Aero Squadron, USAS. He was credited with the destruction of three additional enemy aircraft, bringing his total to six. For his heroic war service, Lt. Ponder was awarded France's *Croix de Guerre*, with four Palms. He was also awarded the U.S. Distinguished Service Cross for his extraordinary heroism on 23 October 1918. His citation describes his heroism.

"Having been separated from his patrol, Lt. Ponder observed and went to the assistance of an Allied plane, which was being attacked by thirteen of the enemy. Against great odds, Lieutenant Ponder destroyed one enemy plane, and so demoralized the rest that both he and his comrade were able to return to their lines."

On 14 May 1919 Ponder was promoted to Captain. While in Paris he married Mille. Jeanne Marie Dezerville. From this union two daughters were born, Colette in 1921, and Janine in 1923.

Following World War One Ponder was discharged and returned to Mangum, Oklahoma where he served as city manager in 1920. The Mangum airport was named Ponder Field in his honor. He left Mangum for Vernon, Texas then moved to Fort Worth where he headed a newspaper and magazine agency. While there, he became distributor for Lockheed aircraft sales in Texas and Oklahoma. He also became engaged in the oil contracting business. On 26 May 1932 he was arrested by U.S. Customs agents at the San Angelo, Texas airport for aerial liquor smuggling from Mexico. Ponder admitted to the charge and was released with a fine because of his exemplary war record.

During World War Two he joined consolidated Vultee as a trouble-shooter. Following the War Ponder again became an airplane distributor in San Angelo, Texas, handling Globe Swift aircraft with his brother, Rebis. He died of a heart attack on 27 February 1947 while on a business trip to Amarillo, Texas at age 53.

Captain William Ponder and his new bride, Jean Marie Dezerville Ponder, Paris, May 1918. (Photo courtesy Jack Moses)

THOMAS POTTER

Thomas Wyndeatte Potter was born in Westchester, New York, on 4 May 1891, the son of Edward Clarkson Potter. Thomas Potter attended St. Paul's School, Class of 1909. He worked as a stock broker before deciding to enter the European War.

Potter joined the American Ambulance Field Service in New York City on 14 December 1915. From late December 1915 to 1 June 1917 Potter served as a driver with SSU 3 in France. From January to October 1916 he served in Alsace, Verdun, and Pont-a-Mousson; and from October 1916 to May 1917 with the French Army of the Orient in Serbia, Albania, and Greece. He was awarded France's *Croix de Guerre* in Macedonia.

On 21 June 1917 Potter joined France's *Service Aeronautique.* He received his *brevet* on 10 November 1917. He was released that same month because of inaptitude.

Potter returned to the United States and on 12 January 1918 he joined U.S. Aviation. From April to August 1918 he attended the MIT ground school. On 2 August 1918 Potter entered U.S. Naval Aviation and served as a Chief Quartermaster until 4 June 1919 at the Naval Air Stations at Miami, Key West, and Pensacola, Florida. He was discharged from the U.S. Naval Reserve Forces on 1 August 1922.

On 18 April 1935 Potter married Elizabeth Hannah Ladenburg of New York City.

FREDERICK PRINCE

ESCADRILLE LAFAYETTE

Frederick Henry Prince, Jr., the 17th volunteer of the *Escadrille Americaine*, was born on 10 April 1885 at Boston, Massachusetts, the son of investment banker and industrialist Frederick Henry Prince and Abigail (Norman) Prince. He graduated from Groton and attended Harvard College, class of 1908.

Both Frederick and his younger brother, Norman, had received a portion of their education with private tutors in France while spending time on their father's horse-breeding estate at Pau in the Basses-Pyrenees. Frederick, in January of 1914, had shocked his friends and had incurred his father's displeasure by leaving Boston for Cincinnati to work as a freight brakeman on a railroad.

Norman had been living at his father's residences in Pau, France, and in Massachusetts when war had broken out in August of 1914. Having been a licensed aviator since 1912 and holder of Aero Club of American Aviator Certificate No. 55, Norman had sought to enlist in the French Air Service. He had sailed again for France in late December 1914 against his father's wishes and had soon begun using family influence and his own skills as an attorney to sell his idea at forming an all-American squadron of flyers.

Norman enlisted in French aviation on 4 March 1915. He returned home to Pride's Crossing, Massachusetts on Christmas furlough from his French squadron on 24 December 1915. Frederick was also home for the holidays, and with little persuasion on his younger brother's part, Frederick soon found himself bound for France in the company of Norman, Elliot Cowdin, and William Thaw.

Frederick enlisted in the Foreign Legion on 16 January 1916. He passed a cursory physical and was sent to Toul where he was outfitted in a French uniform. He was soon ordered to the Front to the "*Premiere Bataillon de Marche*" as a replacement.

Fortunately for the thirty-year-old American, a *sergent* appeared in the line calling for volunteers with previous cavalry experience. Prince, seeing a quick exit from the trenches, raised his hand and told the

sergent he had been a captain of a non-existent outfit, the Massachusetts cavalry. Next day he was transferred to Versailles, where he became a trooper in the "Vandasian Dragoons." What Prince hadn't realized when he had transferred was that the cavalry unit had been converted to infantry, and he soon found himself

(Left) Norman Prince, Frederick Henry Prince, Jr., Frederick Henry Prince at "Princemere," Massachusetts, circa 1905. (Above) Sergent Frederick Prince with his moniteur at Pau, May 1916.

back in the line serving where he had been in the Legion.

In early February Prince's transfer came through for the French Flying Corps. He was sent to the Aviation School at Pau, ironically beginning his first flight training on the very fields he had chased foxes over on horseback as a youth. Until recently the flying school had been part of his father's estate which the senior Prince had allocated for use by the French government after the outbreak of the war.

Prince began his flight training on the Bleriot "penguin." Then, as he later recalled "after about three hours circling the field at 100 feet, I started to do my triangles." He had just completed his first triangle when the school was shut down due to the accidental death of a young *mitrailleur* who happened to be the son of the Minister of War.

Frederick Prince was then ordered to the school at Buc with Willis Haviland, Robert Rockwell, and Robert Soubiran. After a ten-day wait for machines, the four Americans completed their altitude tests and triangles in one day, normally a two-week process. They received their military *brevets*—Prince on 21 May 1916—and returned to Pau for Nieuport instruction. That completed, Prince served a short time as *moniteur* of the Morane Saulnier. He was then sent for a week to the aerial gunnery school at Cazeaux then back to Pau for two days of acrobatics.

Sergent Prince was sent to the G.D.E. pilot pool following his aviation training to await assignment to *N.124*. As he later recalled:

"I remained there a fortnight flying worn-out machines. Since there were no casualties at *N.124*, I asked to go to another squadron. I was sent to the 3rd Squadron and remained less than a week when I received word that my brother had been seriously wounded in the Vosges. I was allowed to fly to Paris, to Fontaine, but arrived late in the afternoon in Fontaine and was told I would have to wait until next morning to fly into the Vosges where my brother was in hospital.

"Norman had hit telegraph wires while landing and had broken his right leg in five places and his left leg in two or three. He died at two in the morning with a blood clot on his brain. I arrived at five in the morning. He was lying in bed decorated with the Legion of Honor, *Medaille Militaire*, the *Croix de Guerre*, with Palm and Star, and we took his body to Luxeuil where the Squadron was quartered. My brother was buried there. The French gave him an imposing funeral. There were Australian and English flyers present. Thereupon that day I was taken into *N.124*."

On 22 October 1916 Sergent Prince, Willis Haviland, and Robert Soubiran together caught up with the Squadron at Cachy in the Somme sector. *N.124* had just made the transfer from Luxeuil. Prince flew with the *Escadrille* until some time after Christmas

Frederick Prince at Cachy beside his Nieuport 17 C.1 scout, October of 1916. LePrieur rocket launching tubes have been fitted to the struts of his fighter. Once armed, Prince will use the rockets to attack German observation balloons.

when he received word from his mother that his father was thought to be dying of typhoid fever in Pride's Crossing, Massachusetts. He secured emergency leave and sailed to the U.S. on the *Rochambeau*. When his father's crisis had passed, he returned to France on the *St. Louis* with Charles Choteau Johnson and William Thaw who had been on leave in the U.S. They arrived at the Squadron in late January, 1917.

Frederick Prince served dutifully with *N.124* until 15 February 1917. On that date he was removed from the Squadron by action of his father. The senior Prince wielded considerable influence with the French government through his gift of land for the flying school at Pau and by helping that government to secure war loans in the U.S. No longer wishing to risk the life of his sole surviving son, he brought that influence to bear and had Frederick removed from *N.124* and transferred to Pau as a flight instructor. The forced transfer was entirely against the younger Prince's wishes, one of the great disappointments of his life. And he never forgave his father's interference, as his father could not forgive his two sons' recklessness which had threatened to terminate the Prince name and legacy.

Following six weeks service as *monituer* at Pau, Prince was ordered to the French Aeronautical Mission at Washington, D.C. He returned to the U.S. in May aboard the *Rochambeau* with Laurence Rumsey, carrying on his person "all the secret information the French had on German motors and airplanes."

On 16 June 1917 Prince met Elizabeth Harding, daughter of William P.G. Harding, Governor of the Federal Reserve Board. They were married on 14 July in Harding's Washington, D.C. home.

In late September Prince was ordered back to France, to the G.D.E. at Plessis-Belleville. While awaiting reassignment to the Front, he was promoted to *Adjudant* and attached as a liaison officer to the 26th U.S. Division at Neufchateau where he remained until late January 1918. After a brief stint with *SPA 67*, Prince was then sent to Le Bourget in the capacity of ferry pilot until he was released at his own request from the French *Service Aeronautique* on 10 April 1918 in order to join the U.S. Air Service.

Prince could not meet the weight and eyesight restrictions for pursuit pilot, so he returned to the U.S. on 10 June. He was commissioned a First Lieutenant in the U.S. Quartermaster Corps on 8 August 1918. He was then assigned to the 549th Company, 428th Motor Transport Train. Wishing to return to France, he transferred into the infantry and was assigned to Headquarters 16th Infantry Brigade, 8th Division, and was appointed aide-de-camp to General R.E. Bradley.

Lieutenant Prince again sailed for France on 28 September 1918. He returned to the States following the Armistice. He was discharged on 6 January 1919, having participated while overseas in the Somme, Artois, Champagne, and 1918 Somme campaigns.

Following his return to civilian life, in 1922 Prince's only son, Frederick III was born. On 24 March 1923 Frederick's wife, Elizabeth obtained a Paris divorce from "Freddie" on grounds of abandonment, stating that her husband preferred to live in Pau with his parents rather than in America with her. The same day Mrs. Prince obtained her divorce, a Mrs. E.V.R. Thayer also secured a Paris divorce for abandonment from her husband, the director of New York's Chase National Bank. Three months later, on 1 July 1923 Mr. E.V.R. Thayer married the former Mrs. Prince at her father's Washington, D.C. home.

Perhaps it was due to the loss of his brother, Norman, combined with the intensity of his own World War service. Whatever the cause, Freddie lacked the enthusiasm his famous industrialist-operator and banker father demonstrated in the world of business. Hard driving, vain, possessive, the senior Prince was clearly disappointed that his dynamic son, Norman, lay dead in France, while Frederick, his namesake, shunned the glorious future he had mapped for him. That he would neglect his destiny blistered the old man's ego, and Frederick Prince made his son pay for it.

Frederick Prince, Sr. had been born in 1859 in Winchester, Massachusetts, the son of former Boston Mayor Frederick O. Prince. So anxious was he to immerse himself in business, he had left Harvard in his sophomore year and had soon begun increasing the already substantial family wealth. He joined the New York Stock Exchange on 10 December 1885, and by his death at age 93 on 2 February 1953 he was its oldest member.

Prince, in his lifetime, by his own assertion, controlled forty-six railroad companies. Little wonder, then, that he had questioned his older son's intelligence when Freddie had sought to be a freight brakeman on one of them.

Prince's major power base in the east had been his Frederick H. Prince & Co., one of New England's leading banking and brokerage houses. Using the wealth he generated from that firm, in 1882 he had directed his interests toward Chicago's stockyards where he purchased many failing businesses, including Armour & Co. He had been convinced of Chicago's potential as a major industrial and shipping center and had founded that city's industrial site, a move which years later would pay him enormous dividends.

Prince also founded the American Woolen Co. of New York City and eventually used that concern to take control over thirty of the largest woolen establishments in the Country. Then in 1922 he established the Los Angeles Central Manufacturing District.

Prince's business acumen saw him past the disastrous Crash of 1929 and the subsequent Depression. And when his contemporaries were jumping from window ledges he continued to build a $250,000,000 empire. In fact he had used the unfavorable climate to take advantage of low prices of securities. Then, as an ostentatious gesture to show he could succeed where others had failed, in 1932 he bought the famous Marble House at Newport, Rhode Island from Mrs. O.H.P. Belmont, formerly Mrs. William K. Vanderbilt.

Prince also owned homes in Boston, in Aiken, South Carolina, and a 994-acre estate, Princemere, in Essex County, Massachusetts. He also posessed a villa in Biarritz, France, and the estate at Pau where he had served as master of the foxhound hunt for twenty-five years.

Prince competed in sport as fiercely as he had in business and viewed both as one in the same. This was evidenced by the riding crop he always carried beneath his arm. His aggressiveness on the polo field lasted into his late 70's, and on 29 July 1929, during a match at the Myopia Hunt Club he had succeeded in getting himself banned from competition in Massachusetts when he had bludgeoned with his mallet one Arthur Mason, a groom and polo professional. The blow eventually cost Prince $50,000 to settle the damage suit filed by Mason. But In 1930 Prince paid $500,000 for the yacht, the *Weetamoe*, which he sailed many times and made an America's Cup contender through 1936.

Pursuing the ladies was another Frederick Prince sport. And he enjoyed bestowing lavish gifts upon his current interest. One, the dancer, Florence Walton, became a recipient from Prince of so many pearl necklaces that she finally suggested she would as readily accept diamonds or cash.

But there was one woman who was never allowed to handle Prince's money, and that was his dutiful wife, Abigail. The germ-conscious Prince feared that if she did so, she might contract some fatal disease. Abigail Prince soon came to share her husband's phobia, and to combat this threat she slept in a disinfected cradle with the walls of her bedroom draped with fresh sheets doused with antiseptics.

Freddie Prince tried to get along with his demanding and eccentric father, but he was too much his antithesis in personality and temperament. Unlike the old man, Freddie was a kind, decent man with many friends. The fact that he was well liked and a better polo player than the senior Prince infuriated his father to the degree to which Prince took every opportunity to demean his son, even supplying him with inferior polo ponies. In spite of this handicap, Freddie's team had still been able to win the 1919 National Open Championship.

Freddie mapped his own way in the business world, rejecting at once his father's assistance and interference, choosing instead to work at various Wall Street brokerage houses. One great success in his life was his marriage on 13 November 1935 to Virginia Mitchell Higginson, the widow of stockbroker James J. Higginson. Unlike his parent, Freddie proved a good father to his two stepsons and a loving husband until his death some twenty-seven years later.

The degree to which the senior Price would interfere with his son's fate was clearly demonstrated in 1937. Prince, some years earlier, had secured permission from the Bishop and board of governors of the National Cathedral of Washington, D.C. to erect the chapel of St. John in Norman's memory. The chapel was to hold burial vaults for Norman, and for himself and his wife, Abigail. But not for Freddie.

Following his death in 1916, Norman had first been buried at a chapel at Luxeuil, France in fulfillment of his final wish to be interred in French soil. Later, his body had been moved to the Procathedral in St. Cloud, near Paris, where it had been in repose with his comrades in the *Lafayette Escadrille* Monument. In the spring of 1937 the senior Prince had Norman's body removed from the Monument and brought to the U.S. on the French liner, *Normandie*, which arrived in the United States on 31 May 1937. Upon its docking, Prince vaguely explained the exhumation of his son to the waiting press:

"We thought it best to bring him home because you never know what is going to happen over there." Then, with an attentive audience before him, he used the opportunity to rail against the American Labor Party and to endorse President Roosevelt. He then departed Washington, D.C. without participating in the ceremonies for Norman's interment in the National Cathedral.

The true explanation for Norman's exhumation and re-interment is that the senior Prince had been foiled in his attempt at deifying Norman at the expense and memory of the other Lafayette pilots' sacrifice. And in anger he had lashed out by having his son's body removed from their hallowed ground.

Frederick Prince, Sr., the father who had vigorously objected to his sons' association with aviation, certainly championed the memory of Norman's participation within it and his ultimate supreme sacrifice. In 1917 Prince edited and financed the publication of the memorial edition *Norman Prince, A Volunteer Who Died For The Cause He Loved*. The book is filled with praise for his dead son. And nowhere within its pages is there a hint that it was assembled by a bitter, objecting parent.

In the early 1920s Georges Thenault, the former French commander of N.124, had turned his manuscript *L'Escadrille Lafayette* over to Prince for publication. Prince, before sending it to press, without Thenault's knowledge, had inserted additional paragraphs about Norman and his early role in the formation of the *Escadrille Americaine* which credited Norman in excess of what he deserved.

William Thaw seethed with anger when he saw Mr. Prince's attempts to take control of the Lafayette Memorial, to make it a "Prince Memorial." He used what influence he could bring to bear with Dr. Edmund Gros, Mr. H.C. Hoskier, and others, to break Prince's control, and to counteract the inaccuracies at the "instigation of his (Norman's) family" which built up Norman's role while ignoring the contribution of Thaw and other founding members.

When Thaw succeeded in breaking Prince's hold over the American Committee, Norman's father withdrew his own support from the Memorial and decided to remove his son's body from France. In 1927 he had petitioned and had received permission to build a chapel and a monument for Norman for $500,000 at Fort Meyer, Virginia. But when the National Commission of Fine Arts had objected to the chapel's location near the main entrance of Arlington Cemetery, Prince had withdrawn his funding and had abandoned the project on 8 April 1929. He then had approached the Bishop and board of governors of the National Cathedral.

Upon the burial of Norman at the National Cathedral, Charles Chouteau Johnson, a *Lafayette Escadrille* pilot who had flown with Norman, had admonished the senior Prince with this challenge:

"When are you going to leave Norman buried and stop digging him up every couple of years and re-burying him for your own grandisement? Well, I hope to God you leave him put now. You put him in the Capitol, and I don't know where the hell else you can put him."

In 1944 the senior Prince adopted his cousin, William Wood. In 1949 he made William Wood Prince president of the Union Stock Yard and Transit Company. Following his death four years later at his Biarritz home, Prince's will made William Wood sole guardian of the Prince fortune which was valued at $100,000,000. To his son, Freddie, Prince left a token stipend of $5,000 a year for twenty years.

Lafayette Escadrille pilot Carl Dolan recalled a visit by Freddie to his office at the Sperry Corporation shortly after the senior Prince's death. Freddie had lamented:

"That God damn old man of mine, he brought me up as the son of a millionaire and every time I ever tried to go into business he'd buy me off by increasing my remittance and telling me to get the hell out of it. I had an office down on Wall Street. I had several things but now that I'm in my sixties, about ready to quit, he disowns me. So I'm the type of fellow that has ridden the hounds. I was brought up as a millionaire's son, and I don't know how the hell to act, Carl."

Freddie's happiest years were spent in retirement participating in fox hunts at Orange Hill Farm in Marshall, Virginia. He died of heart failure on 5 October 1962 at age 77 in his home at old Westbury, Long Island, New York. He is buried at St. John's Memorial Cemetery, Cold Spring Harbor, Long Island, New York.

In death, as in life, people spoke of his kind and decent nature, and of the fact that he had been one of the thirty-eight pilots of the *Lafayette Escadrille.* And that he had a brother that he had loved whose name had been Norman.

Frederic Prince, Sr. typified the Eastern industrialist who flaunted his wealth. The jewel in his realm was the Newport, R.I. Marble House, which William K. Vanderbilt had built at the cost of $11,000,000 and had completed in 1892. Prince bought the magnificent mansion from Vanderbilt's former wife in August of 1932 for less than its $382,000 tax value. Shown in the top photo is the drawing room from his massive stone mansion located on his 900-acre "Princemere" estate in Essex County, Massachusetts. At the back of the room is a portrait of his dead son, Norman, which now hangs in a corridor in the National Air and Space Museum in Washington, D.C.

NORMAN PRINCE

ESCADRILLE LAFAYETTE

Norman "Nimmie" Prince, a founding member of the *Escadrille Americaine*, was born on 31 August 1887 at Prides Crossing, Massachusetts, the son of investment banker and industrialist Frederick Henry Prince and Abigail (Norman) Prince. Young Prince graduated from Groton and from Harvard College cum laude, class of 1908 with a Bachelor of Arts degree. He received his Bachelor of Laws degree from the Harvard Law School in 1911.

Following his admission to the bar, Norman began his practice of law in Chicago in the summer of 1911 with the firm of Winston, Payne, Strawn, and Shaw. However, he pursued his profession with little enthusiasm as he was consumed by an interest in the developing science of aviation. Prince took his training under the alias of George W. Manor to conceal his actions from his father who considered aviation a distraction "from the more serious concerns of life," and who called flying machines "stinking chicken cages," In 1912 Norman received his Aero Club of America flying license No. 55.

Norman remained with the Chicago law firm until the summer of 1913. By pleading troubled eyesight, he spent the next year between his father's residences in Massachusetts and in Pau, France.

At the outbreak of World War One, Norman had become anxious to serve the Allied cause. He considered France his "second country" since a good part of his youth had been spent on his father's estate at Pau. He had many friends in Paris, particularly among the old colony of Anglo-Americans.

In the fall of 1914 Norman had enrolled in the Starling Burgess flying school at Marblehead, Massachusetts in order to refine his aeronautical skills before volunteering his services to France. Also enrolled in the flying school was Frazier Curtis, a friend of Norman's from Boston and a fellow Harvard alum who had just returned from England where his application had been rejected for service with the Royal Naval Air Service. He had planned to train further at the Burgess school then reapply with the British.

Adjudant Norman Prince, one of the founders of the Escadrille Americaine.

It was to Curtis that Prince had first mentioned his concept of forming an all-American squadron of aviators who would fly as volunteers for France. Curtis, while receptive to the idea, felt he must first reapply with the British Flying Corps. In mid-December he left the Burgess school and sailed for England the day before Christmas.

Norman likewise left the flying school shortly after Curtis and set sail for France on 18 January 1915 aboard the *S.S. Rochambeau* arriving in Paris on the 29th. Edmond Genet, another future *Lafayette Escadrille* pilot, was also aboard the *Rochambeau* on his way to enlist in the French Foreign Legion. He wrote this of Prince to his mother in a 3 February 1915 letter:

"I also have quite a good friend in Norman Prince, one of the first Americans to try aviation. He is a rich young fellow—about 25 I should judge, and very quick and sensible. He came over on the *Rochambeau* and is entering the French aviation corps. He will have some exciting experiences I feel sure."

However, five weeks would pass before Prince would officially enlist in French aviation. In the meantime he would set himself up in a luxurious apartment and spend all his time in his effort to interest France in his cause for an all-American squadron of volunteers.

Norman's friend, Lieutenant Jacques deLesseps, a member of the Paris Air Guard, aided in getting the French War Department to listen to Prince's plans and ideas. Prince petitioned the aid of prominent members of the American colony in Paris and managed to secure the interest of some French military officials through Victor Chapman's father and uncle, John J. Chapman and Robert Chanler.

On 9 February 1915 Frazier Curtis arrived from Great Britain, having chosen not to enlist in the Royal Naval Air Service when he discovered the prerequisite, a renunciation of his American citizenship. Curtis joined forces with Norman and together they drafted a letter to Etienne Millerand, France's Minister of War, offering to raise a squadron of American volunteers to fly and fight for the Republic.

Their letter met with little support, with Millerand explaining France's position that she had more applicants for aviation than she had airplanes to fly. Temporarily rebuffed, Prince and Curtis enlisted as privates in the French Foreign Legion, Curtis on 2 March and Norman on 4 March 1915. Together they transferred to the Aviation Service and detailed to the School of Military Aviation at Pau.

Prince received his *brevet militaire* on the Voisin on 1 May. Frazier Curtis would not be so lucky. That spring he had suffered two serious accidents at Avord. Following the second accident Curtis suffered a nervous breakdown which required forty-five days' sick leave in Paris. While there, when he was able, he tried

Sergent pilot Norman Prince in the rear cockpit of a Voisin light bomber with his observer/gunner while assigned to V.B. 113, Fall of 1915.

to forward the two mens' cause of an American *Escadrille*. He also had attempted to recruit to French aviation future pilots from the American Ambulance Corps, the men needed to fill the roster of the proposed new squadron.

While recruiting from the Ambulance Corps, Curtis was able to make the acquaintance of its director, Dr. Edmund L. Gros, an American-born physician who had spent many years in Paris. Later on he would prove an invaluable contact in the fulfillment of their dream when they united forces to create the French/American committee with the objective of forming an American aero squadron.

On 1 May 1915 Norman Prince was assigned to *V.B. 108*. Two days later another American, Elliot C. Cowdin, joined the squadron. In June Norman was promoted to *Sergent*. On 1 July 1915 he was sent to *V.B.113* where he would remain until 15 February 1916.

Meanwhile Frazier Curtis had met with more bad luck. Following his convalescence, he had requested transfer from bombing to pursuit aircraft. In answer to his petition, because of his nervous state, on 8 August 1915 he was given an honorable discharge from the French War Office which had deemed him unfit for further flying. He departed France for England for four months' rest before returning to the United States.

Prince's dream of an all-American squadron of aviators had continued to germinate through the efforts of Frazier Curtis, Dr. Gros, and other members of the American colony. Prince's introduction through Robert Bliss, first secretary of the American Embassy in Paris, to Jarousse de Sillac, under secretary of the French Ministry of Foreign Affairs in February had begun to pay dividends. The French slowly saw the value in Prince's concept and came to realize how U.S. sympathy could be gained for their war effort through the formation of an American squadron.

In December of 1915 Prince, Elliot Cowdin, and a third American, *Sous-Lieutenant* William Thaw, were granted a thirty-day Christmas leave to the United States. Thaw, a pilot with *Escadrille C.42*, had begun his service in France in the Foreign Legion on 21 August 1914. In October of that same year, while in combat with the Second Regiment, he had become the first American to express the concept of "a squadron of American volunteers" to fellow Legionnaire Paul Ayres Rockwell, a full month prior to Prince's conversation with Frazier Curtis at Marblehead. That same month Thaw also had made the first attempt at transferring himself and his American comrades into a single French squadron and had continued to lobby for that effort whenever time from his duties allowed.

On the *SS Rotterdam* going home, Prince, Thaw, and Cowdin had agreed to unite their efforts. Once off the boat on 23 December 1915, the American press dogged their every move, playing them as heroes to a sympathetic public. However, not everyone was pleased with their volunteer service in France. Strident neutralists and pro-German voices called for their confinement for their "gross violation of American neutrality."

Norman, while home at Prides Crossing, had managed to recruit his older brother, Frederick, to his cause. Then shortly after the holidays the four Americans quietly boarded a ship bound for France.

The French government, now aware of the pilots' favorable reception in the United States, had seen the value of an all-American squadron. But the reality was still some weeks off. Back in Paris, Thaw lobbied his friend Lieutenant Brocard, later to become Commandant Brocard, Chief of French Air Service. Dr. Gros and de Sillac joined to form a Franco-American committee. When Thaw and Cowdin returned to their escadrilles, Norman had briefly returned to *V.B. 113* before transferring on 16 February to the R.G.A.(*Reserve General Aeronautique*) at Le Bourget for Nieuport training. Then on 14 March 1916 Colonel Henri Regnier, director of French military aeronautics, issued the following statement to the Franco-American Committee.

General Headquarters has just replied, informing me that an American squadron will be organized, with the pilots whose name follow: William Thaw, Elliot Cowdin, Kiffin Rockwell, Norman Prince, Charles C. Johnson, Clyde Balsley, Victor Chapman, Laurence Rumsey, and James R. McConnell.

On 16 April 1916 the *Escadrille Americaine* became a reality. Officially designated *N.124*, the roster was altered by the removal of Johnson, Balsley and Rumsey, with the addition of Weston Bert Hall, and two French officers, *Capitaine* Georges Thenault as squadron commander, and Lieutenant Alfred deLaage de Meux as his executive officer.

N.124 first began its operations at Luxeuil-les-Bains, the little resort town southeast of Paris. The pilots were quartered in a villa there awaiting the arrival of their Nieuport 11s. Meanwhile, a battle-weary Nieuport from *Escadrille N.49* was sent to their airfield for training purposes. Norman had the misfortune of being one of the first to fly it and had smashed it into a hangar, miraculously escaping injury but extensively damaging the plane. His poor eyesight may have been the cause.

Six Nieuport 11s arrived at Luxeuil in the first week of May. However, there was little activity with the exception of German night bombing attacks against the airfield, with the first combat patrol being flown on 13 May 1916. Five days later Kiffin Rockwell scored the Squadron's first victory when he downed a German reconnaissance plane between Mulhouse and Hartmannsweillerkopf.

The next day, 19 May, the Squadron was ordered to the Verdun sector. The war had suddenly become a serious business. The pilots occupied a villa between Bar-le-Duc and the airfield at Behonne and were soon flying patrols over the volcanic battle erupting beneath them.

On 19 June Prince was on patrol with Kiffin Rockwell and Clyde Balsely when they had been attacked by a large German formation north of Verdun. In the ensuing engagement, Balsley was seriously wounded and permanently crippled.

On 23 June Prince was again on patrol with *Capitaine* Thenault, Raoul Lufbery, and Victor Chapman. In combat with five enemy machines, Chapman was shot down and killed, becoming the Squadron's first fatality.

Throughout the months of July and August Norman Prince had been constantly on patrol. Obsessed with bringing down a German observation balloon, he had his Nieuport's v struts rigged with *Le Prieur* incendiary rockets. After a number of unsuccessful attempts, with the help of James McConnell he finally managed to incinerate an enemy *drachen* but did not receive official confirmation for it since its destruction took place five miles inside German lines.

On 23 August Prince claimed a single-handed victory over a German Aviatik bi-place during an aerial battle which he said took place six miles inside German lines. In a diving attack on the unsuspecting two-seater, Prince claimed to have killed the plane's observer before forcing its pilot down within French lines on a meadow outside of Verdun. When Prince had flown back to Behonne, his fellow pilots had met his story with skepticism. But *Capitaine* Thenault had allowed him the victory and had recommended him for France's *Medaille Militaire.*

It was this 23 August victory that had brought to a head a seething hostility which had existed for some time between Prince and Kiffin Rockwell. Rockwell had felt that Prince had been currying *Capitaine* Thenault's favor and was being unjustly cited while Kiffin's own efforts had been overlooked. Rockwell angrily said as much in a 1 September 1916 letter to his brother Paul.

"No one thinks X— (Prince) got a German, in fact, everyone is sure he didn't; yet the Captain proposed him for a citation, wanted to propose him for the *Medaille*, but everyone said if he did they would quit. I am going to have to call him out when he gets back (from Paris), as he talked awfully big about us behind my back when I was away. We have all agreed to try to run him out of the *Escadrille.*"

On 9 September Prince gained his second victory—the Squadron's thirteenth victory credit—when he had downed a German Fokker fighter. On 14 September the Squadron was ordered back to Luxeuil. On the 16th it was equipped with the larger, faster Nieuport 17. The first patrol from Luxeuil was flown on the 23rd by Kiffin Rockwell and Raoul Lufbery. Lufbery had retired to the airfield when his Vickers gun had jammed, with Rockwell continuing above Hartmannsweillerkopf where he was killed a short time later by a bullet to the chest fired from an Albatros two-seater.

On 26 September, three days after Rockwell's death, Norman Prince was awarded France's *Medaille Militaire* to accompany his *Croix de Guerre* for his 23 August and 9 September victories. On 10 October over Mulhouse he scored his third official victory but was nearly shot down and barely managed to land his bullet-riddled plane at Luxeuil.

A large British-French bombing raid had been planned for 12 October against the Mauser rifle works at Oberndorf, Germany. The bombers, combined with fighter escort, totalled nearly sixty machines. The *N.124* pilots assigned to fly fighter protection were deLaage deMeux, Lufbery, Masson, and Prince. The raid lasted five hours which forced the fighters with their two-hour fuel limit to refuel on forward fields before escorting the returning bombers home.

Prince and Lufbery, while awaiting the bombers, attempted to clear the air of hostile fighters. They found four German Fokker E.IIIs lying in wait for the allied bombers and immediately engaged them. Norman shot one down for his fourth confirmed victory and Lufbery another, making him *N.124*s first ace with five confirmed victories.

It was nearly dark by the time the last bombers had returned from the raid. Lufbery landed tentatively at the advanced field at Corcieux. Prince followed him in, but because of his poor vision and the closing darkness, he failed to see the telegraph cables above the tree tops surrounding the airfield. His wheels snagged on the cables and his Nieuport cartwheeled into the ground. His safety belt broke on impact and he was thrown from his plane, breaking both his legs and severely striking his head.

Pilots and mechanics rushed to the crash site. Lufbery was one of the first to reach Norman. However Prince's concern was not for the gravity of his own injuries, but for the pilot of a circling Nieuport seeking sanctuary on the airfield. Prince asked that flares be lit at once along the landing strip so that the pilot would not suffer his own fate.

It was a ten-mile journey by ambulance over unkind roads to the hospital at Gerardmer. Lufbery clutched Norman's hand and read the agony in his face. Prince sang to keep up his spirits. At first the doctors at the hospital had thought he would survive his injuries, but a blood clot formed on his brain. While lying unconscious on his cot he was promoted to *Sous-Lieutenant* and was awarded France's *Legion d'Honneur* by aviation group commander Felix Happe. Upon decorating Norman, Happe read the accompanying citation over him:

"For nineteen months in *escadrille* has distinguished himself by bravery and devotion to duty beyond compare in the execution of numerous bombardments and pursuit work. Was grievously injured on 12 October,

Adjudant Norman Prince in the cockpit of his Nieuport 17 C.1 fighter. This photo was taken on the Luxeuil airfield just days before the 12 October 1916 flying accident, which lead to his death.

1916, after having brought down one enemy plane. Already has the *Medaille Militaire*."

Norman Prince died at age twenty-nine on 15 October 1916 without regaining consciousness. Following an impressive military funeral at the Luxeuil aviation field, he was buried in a nearby chapel. Norman's Squadron mate, James McConnell, best summed up Prince's service when he wrote this of him in his book, *Flying For France*.

"It was hard to realize that poor old Norman had gone. He was the founder of the American *escadrille* and every one in it had come to rely on him. He never let his own spirits drop, and was always on hand with encouragement for the others. I do not think Prince minded going. He wanted to do his part before being killed, and he had more than done it. He had, day after day, freed the lines of Germans, making it impossible for them to do their work, and three of them he had shot to earth."

DAVID PUTNAM

David Endicott Putnam was born in Jamaica Plain, Massachusetts on 10 December 1898, the son of Frederick H. Putnam, a wool merchant, and descendant of American Revolutionary War General Israel Putnam, and English-born Janet (Hallowell) Putnam, who arrived in America at age four. Mr. Putnam died of rheumatic fever while David was a schoolboy.

David attended the Newton High School then entered Harvard College, class of 1920. Putnam was president of his class, class orator, and an excellent athlete, participating in four sports. While a college freshman, Putnam passed the U.S. Aviation service entrance exam. However, he was underage for enlistment, just beyond his 18th year. Determined to enter allied aviation, Putnam left Harvard after his mid-year examinations and found work on a cattle ship bound for France.

On 31 May 1917 Putnam enlisted in France's *Service Aeronautique*. From 10 June to 10 December 1917 he attended the aviation schools at Avord, Pau, and the *G.D.E.*. He received his *brevet* (# 9328) on the Caudron on 17 October 1917.

On 12 December 1917 *Caporal* Putnam was assigned to *Escadrille SPA 94* and flew with that squadron until 1 January 1918. *Sergent* Putnam flew with *Escadrille SPA and (M.S.P.) 156* from 7 February to 1 June 1918, gaining four official victories. From 1 June to 14 June 1918 Putnam flew with *SPA 38*, gaining five more victories. On 8 June he had been commissioned a First Lieutenant in the US Air Service. From 24 June to 12 September 1918, the day of his death when he was shot twice through the heart during aerial combat, Lieutenant Putnam was flight commander of the 139th Aero Squadron, USAS. While with that unit he scored four more victories, bringing his official total to 13.

Since many of Putnam's combats took place deep within enemy lines, his actual victories were undoubtedly much higher. Some sources credit him with the destruction of 34 aircraft. On one single day, 5 June 1918, he probably shot down five enemy machines but was credited with just one victory.

For his heroic war service, Putnam was awarded France's *Croix de Guerre*, with 6 Palms, and was posthumously awarded the *Legion of Honor*, and the U.S. Distinguished Service Cross. He was also recommended for the Medal of Honor. His DSC citation, dated the day of his death, reads: "For extraordinary heroism in action near La Chaussee, France, September 12, 1918. After destroying one of the eight German planes which had attacked him, he was turning to our lines when he saw seven Fokkers attack an allied biplane. He attacked the Germans and saved the biplane, but was himself driven down, shot through the heart." Putnam was the eighth victim of German ace Georg von Handtleman, of Jasta 15. He was 19 years of age at the time of his death, the Lafayette Flying Corps's youngest and second-ranking ace.

The night before Putnam was killed, his sister, Dorothy dreamed of his death from her bedroom at the family's Massachussetts home. In her dream, her brother, who was dead in his cockpit, came down in a skeletonized fighter in a field of blood. The next day Putnam's SPAD (#15137) landed at Limey in a field of blood-red poppies. It was later skeletonized by souvenir-seeking soldiers.

Lieutenant Putnam was interred with full military honors in the Forest of Sebastopol, near Toul, beside the grave of Raoul Lufbery. In 1928 his remains were moved to the Lafayette Flying Corps monument near Paris.

RUFUS RAND

Rufus Randall Rand, Jr. was born in Minneapolis, Minnesota on 25 May 1892, the son of Rufus Rand, President of the Minneapolis Gaslight Company, and Susan (Nealey) Rand. Rand attended Minneapolis public schools and St. Paul's School, Concord, New Hampshire. He enrolled in Yale University before entering Williams College, Class of 1916, where he helped to found the School's Aero Club.

Upon receiving his BA degree from Williams College, Rand joined his father's firm. He left Minneapolis to enter the European War and enlisted in the Norton-Harjes Ambulance Service. He sailed for France in April 1917 and served with Section 21 near Ham. On 26 July 1917 he enlisted in France's *Service Aeronautique*. From 28 July to 4 December 1917 Rand attended the aviation schools at Avord, Tours, Pau, and the *G.D.E.* He received his *brevet* on the Caudron on 14 September 1917.

On 6 December 1917 *Caporal* Rand was assigned at the Front to *Escadrille SPA 158* where he flew many patrols and fought a number of combats, receiving several unconfirmed victories. He remained with that squadron until the 11 November Armistice. *Adjudant* Rand received France's *Croix de Guerre*, with Palm and Star for his heroic war service.

Rand returned to the United States in April 1919. That same year he married Helen Chase of Waterbury, Connecticut. Rand had met his wife while both were serving in France, she as an auxiliary nurse.

The Rands settled in Minneapolis, Minnesota where Rufus became a prominent business and civic leader. He served as vice-president for the family-owned Minneapolis Gaslight Company. An amateur architect, Rand designed and constructed in Minneapolis a copy of a French Norman castle where he lived with his wife and six daughters. In 1929 Rand designed and built the Rand tower in Minneapolis. Because he built the work with non-union labor, union officials tried to damage his home and threatened to kidnap his children. However, Rand managed to finish the project. He also designed airplanes and helped to found Northwest Airlines.

During World War Two Rand served as Major then Colonel in the U.S. Army Air Corps. He was in charge of security at a bomber base in England and also took part in the invasion of Italy from North Africa.

Following World War Two, Rand served as Mayor of the Minneapolis suburb of Wayzata. He was also a regent of the University of Minnesota, and State Commander of the American Legion. He died in Minneapolis on 15 October 1971 at age 79.

JOHN RANDALL

John Francis Randall was born in Meriden, Connecticut on 10 August 1897, the son of Reverend Arthur T. Randall, an Episcopal minister, and Mary Brainard (Deshon) Randall. John Randall attended Meriden public schools before attending Rutgers University, New Brunswick, New Jersey. He worked briefly as a journalist for the Hartford *Times* before joining the American Field Service on 20 June 1917 and sailing for France.

Shortly after arriving in Paris, on 20 July 1917 Randall enlisted in France's *Service Aeronautique*. From 28 July to 8 December 1917 he attended the aviation schools at Avord, Tours, Pau, and the *G.D.E.* He received his *brevet* on the Caudron on 1 October 1917.

From 11 December 1917 to 18 April 1918 *Caporal* Randall served at the Front with *Escadrille SPA 158*. On 10 February 1918 Randall was severely burned when he fell in the barracks kitchen and overturned a kettle of boiling water onto his legs. Randall required skin grafts at the French hospital at Villers Cotterets. On 23 March, while *Sergent* Randall was in hospital, he was commissioned a First Lieutenant in the U.S. Aviation Signal Corps.

Following his recovery Lieutenant Randall served as an instructor at the machine gun school at Cazeaux, then as a ferry pilot flying fighters to the Front. On 6 June 1918 Randall was assigned at Dunkirk to the 103rd Aero Pursuit Squadron, USAS. On 12 June 1918 Lieutenant Randall was severely injured in a landing accident, fracturing a leg. Because of the severity of his injury, he was sent to hospital in England for convalescence. On 13 October 1918 Randall returned to duty at the American Aviation Instruction Center at Issoudun where he remained until the 11 November Armistice. Lieutenant Randall was discharged in Washington, DC on 6 August 1919.

Following World War One Randall worked in the United States and in Canada for the Pratt and Whitney Aircraft Co. He married Agnes Clark shortly after World War Two. Randall died in New Paltz, New York on 26 November 1987 at age 90.

ROBERT READ

Robert Emery Read was born in Franklin, Pennsylvania on 26 November 1892, the son of Emery C. Read, a surveyor, and Ellen (McClellan) Read. Read attended Franklin public schools before attending Beaver College in Jenkintown, Pennsylvania. Read then spent some time in the US Naval Reserve Forces before enrolling in the Columbia University pre-medical school, New York City, New York. Read left Columbia University after one year to join the American Ambulance Field Service. The city of Watertown, New York raised money for an Ambulance, and Read volunteered to drive it. In 1917 he served in France for five months with SSU 2 as a dresser of wounded.

On 13 June 1917 "Doc" Read enlisted in France's *Service Aeronautique*. From 23 June 1917 to 20 January 1918 he attended the aviation schools at Avord, Pau, Cazeaux, and the *G.D.E.* He received his *brevet* on the Caudron on 18 November 1917.

On 24 January 1918 *Caporal* Read enrolled as an Ensign in the U.S. Naval Reserve Flying Corps at Paris, France. On 20 February 1918 he was made Naval Aviator #564(HTA) at Moutchic, France. Read served at the Front at the Dunkirk Naval Air Station, first as a pilot, and later as a commanding officer.

Alvin Cushman, George Dock, and Robert Read at Avord, July 1917

For his diligent war service, Lieutenant Read received the following Letter of Commendation from the U.S. Navy. "U.S.N.R.F. LTJG Robert E. Read performed meritorious service when called upon to take command of the U.S. Hydroplane Base at Dunkirk after his chief had disappeared and maintained this unit up to a high standard of activity and military efficiency." Read also received from France the Legion of Honor and the *Croix de Guerre*, with Palm.

When the Dunkirk base was decommissioned in January 1919 Lieutenant Read's final orders by his commanding admiral were to burn all the hydroplanes at the Dunkirk Station. Wishing to preserve one example, he had one hydroplane dismantled and crated for shipment to the United States. But when Read learned that the Admiral was going to attend the destruction, he had to uncrate the plane and reassemble it, just in time for the admiral to witness it being burned with the others.

Following World War One Read was stationed at Washington, DC. He married Helen Smith, and on 16 June 1920 a son, Robert, was born, followed by a daughter, Jacqueline. The Reads would eventually divorce, and Read married Adelia Carpenter. From this union a second son, Alan, was born.

On 23 January 1922 Read was honorably discharged as a Lieutenant (JG). Following his Naval Service Read worked as a developer and engineer at the International Paper Company designing and building paper machines which produced molded pulp for containers such as egg cartons. He then worked for the Union Bag Company. In 1950 he established a consulting firm that specialized in package-machinery design and pulp-molding machinery. He retired from business in 1962. Read died of a malignant brain tumor in Glens Falls, New York on 6 September 1976 at age 83.

LEONARD RENO

Leonard Minor Reno was born in Little Rock, Arkansas on 23 September 1897, the son of Harry O. Reno, a Chicago publisher and owner of *Furniture Age Journal*, and L.D. Reno. Reno attended Wisconsin's Kuwatin Academy and the Kentucky Military Institute where he excelled in all sports. Following school, "Len" began work as an office manager for his father's H.O. Reno company. He left that firm on 19 February 1917, following an argument with his father, and joined the Franco-American Flying Corps.

Reno sailed for France, and on 20 March 1917 he enlisted in the *Service Aeronautique*. From 23 March to 20 July 1917 he attended the aviation schools at Avord, Pau, and the *G.D.E.* He received his *brevet* on the Bleriot on 10 May 1917.

On 23 July 1917 *Caporal* Reno was assigned at the Front to *Escadrille SPA 103*, Rene Fonk's squadron. In early September, while on a scouting expedition within enemy lines over Fort Holthurst, Reno's SPAD was disabled by anti-aircraft fire as he attempted to flee four enemy fighters. However, he managed to volplane into allied lines before piling his machine into a clump of trees on Belgian soil.

Reno was taken to a Dunkirk hospital where he underwent two operations to repair his injuries. During his hospitalization, through a mixup of records, he was reported killed in action, and dispatches in French and American newspapers carried his obituary, causing grief among his comrades and relatives. Reno was transferred to a Paris hospital before the report had been corrected. He attempted to cash a check at the W.F. Bonbright & Co. and was startled to learn that he could not, as the teller believed him dead.

Reno separated from *SPA 103* on 18 September 1917. He remained hospitalized in Paris until November then was granted convalescent leave in December to the United States. Upon his return to France, on 4 June 1918 *Sergent* Reno was assigned to *Breguet 134*, a day bombardment squadron. He flew numerous patrols over the Montdidier-Noyon Sector. On 18 July 1918, his last day with the French Air Service, Reno knocked down an enemy fighter on his return from a bombing raid upon Oulchy-le-Chateau. Twenty kilometers inside enemy lines, Reno was attacked by a four Albatross scouts. His bus was shot to pieces, but he was still able to land safely within Allied lines. For his heroic actions that day, Reno was awarded France's *Croix de Guerre*, with Palm.

Reno was honorably discharged from the French Army to accept an appointment as an Ensign in the Naval Reserve Flying Corps at Paris on 23 July 1918. He received his Navy training at Pauillac and Moutchic. On 26 September 1918 he was designated Naval Aviator #1352, later #1377 (HTA) at Moutchic. Ensign Reno served at Ile Tudy, France and at Porto Corsini, Italy. For his War Service in Italy, Reno was awarded the Italian War Cross.

Ensign Reno returned to the United States and served at Rockaway Island, Long Island, New York until his honorable discharge on 30 September 1921. That same year he married Muriel Reed of New York. From this union a daughter, Muriel, was born.

Following World War One Reno returned to Chicago where he was employed as an advertizing salesman for his father's publication. The Reno's were eventually divorced, and in 1928 Reno married Jessica Reed, a famous Ziegfield Follies show girl. They divorced in 1932.

At the outbreak of World War II Reno attempted to re-enlist in the U S. Navy but was refused for medical reasons. Reno left Chicago and relocated to Provo, Utah where he was employed as a benzol operator by the Midland Steel Company. He died of a heart ailment in the Provo Valley Hospital on 24 December 1944 at age 47.

WALTER RHENO

Walter Davis Rheno was born in Vinyard Haven, Massachusetts on 26 February 1895, the son of Simeon L. and Clara (Pratt) Rheno. Rheno was educated at Vinyard Haven public schools. As a young man, Rheno entered the automobile business in New York City. In late 1916 Rheno left his business to sail for France to enter the European war.

On 24 December 1916 he enlisted in France's *Service Aeronautique*. From 31 January to 16 July 1917 he attended the aviation schools at Buc, Avord, Pau, and the *G.D.E.* He received his *brevet* on the Bleriot on 10 May 1917.

From 18 July to 15 September 1917 *Caporal* Rheno was assigned at the Front to *Escadrille SPA 80*. On 18 August, and on 6 September, he was credited with the destruction of an enemy aircraft. For these actions, *Sergent* Rheno was awarded France's *Croix de Guerre*, with two palms.

On 17 October 1917 Rheno was granted a three-week leave to the United States. While at home, he asked for a transfer to the U.S. Air Service. After a futile wait of nearly a year for his commission, Rheno returned to France in early October 1918, hoping to be reassigned to a French Squadron. While returning to the Front, Rheno received his commission as First Lieutenant in the U.S. Air Service. However, Rheno became ill with pneumonia a few days after his arrival in France. He died of this illness on 10 October 1918 at the American Hospital at Neuilly-sur-Seine. He was 23 years of age at the time of his death. His remains were returned to the United States for burial.

HUGH RIDLON

Hugh Owen Thomas Ridlon was born in Ohio on 24 August 1897, the son of Dr. John Ridlon, an orthopedic surgeon, and Emily (Robinson) Ridlon. Ridlon attended the Chicago Latin School and was a student when he decided to enter the European War. Ridlon enlisted in the American Ambulance Field Service in order to secure passage to France.

On 14 June 1917 "Hot" Ridlon enlisted in France's *Service Aeronautique*. On 13 August 1917 he was released upon his own request. Ridlon returned to the United States and served in the U.S. Navy until the end of hostilities.

Following World War One Ridlon worked as a salesman of welding equipment for National Cylinder Gas Co. and other companies. Ridlon married Katherine Kemp in Chicago, and from this union two sons, Hugh, Jr, and Robert were born.

As an interesting side-note, Hugh Owen's sister, Jeannette Piccard, became the first woman in history to enter the stratosphere when she accompanied her husband, Jean, in a balloon from Dearborn, Michigan, to a height of 57,579 feet. Her husband's purpose was to study cosmic rays. By accompanying him, Jeannette set the altitude record for women, a record which she held for many years.

It is believed that Hugh Ridlon was the first person to successfully weld aluminum. In World War Two Ridlon served in the U.S. Navy as a Chief Petty Officer. Because of his extensive welding knowledge, he was the first man to receive the Chief Machinist Mate-Gas rating. Following the War, Ridlon returned to work for National Cylinder Gas and managed convention hospitality for that company. He died of a heart attack in Hudson, Florida on 5 May 1966 at age 68.

GEORGE ROCKWELL

George J. Rockwell, Jr. was born in Meridan, Connecticut on 16 February 1884, the son of George Rockwell, Secretary of the International Silver Company of Waterbury, and Minnie (Battles) Rockwell. George was working as a silver plater at the International Silver Company when he decided to enter the European War. From 7 February 1915 to early May 1915 he served at the Front as a Section *Sous Chef* with Sections 1 & 4 of the American Ambulance Field Service. While under fire at Dunkirk, Rockwell's knee was seriously injured. He was awarded France's *Croix de Guerre* for heroism. He spent several months recuperating at the American Ambulance hospital at Neuilly.

Following his extended hospitalization, on 8 August 1916 Rockwell was released from the Ambulance service and enlisted in France's *Service Aeronautique*. In November of 1916, while still in training at Buc, he was released from aviation at his own request.

He returned to the United States and on 4 September 1917 he joined the Enlisted Reserve Corps at MIT, Cambridge, Massachusetts. On 23 October he was classified as Special Mechanic-Aviation. On 23 November 1917 he was assigned to the Foreign Detachment Air Service Signal Corps. On 22 February 1918 he enrolled at the 8th Aviation Instruction Center, Foggia, Italy. On 18 March 1918 he was appointed Second Lieutenant, Air Service. On 27 October 1918 he joined the 3rd Aviation Instruction Center. AEF. On 13 November 1918 Rockwell was appointed First Lieutenant. He sailed to the United States in late December 1918 and was honorably discharged at Garden City Long Island, New York on 7 January 1919.

Following World War One Rockwell worked for the General Electric Corporation in New York. In the mid-1920s he set up a flying school in that state. In 1926 he married Edith Hickson, an author and editor of the Girl Scouts Magazine. The Rockwells relocated to New Jersey where George Rockwell worked as an inventer and self-employed merchant in various small businesses. The Rockwells adopted a son, Eaton, in 1930. George Rockwell died in the All Souls Hospital, Morristown, New Jersey, on 26 February 1967 at age 83.

KIFFIN ROCKWELL

ESCADRILLE LAFAYETTE

"The man who enters this war should consider himself dead from that moment. Every day thereafter that he lives should be accounted as so much good luck—as so much to be grateful for."

Kiffin Yates Rockwell, a founding member of the *Escadrille Americaine*, was born in Newport, Tennessee on 20 September 1892, the son of the Reverend James Chester Rockwell, a Baptist minister, and Loula (Ayres) Rockwell. Before Kiffin reached his first year, his father died at age 25 of typhoid fever. Kiffin spent his youth on the South Carolina plantation of his maternal grandfather, Major Enoch Shaw Ayres. It was there that he became an outdoors man and an excellent horseman, and first demonstrated a fierce determination that later manifested itself in his adulthood as a fighter pilot in France.

When Rockwell was not yet four years old, his grandfather presented him with an untamed Shetland pony. Every time young Rockwell climbed up onto the Shetland, he was bucked to the ground. But despite repeated discouragements, by day's end he had tamed the pony and was able to ride it with ease.

Both of Rockwell's grandfathers had served with the Confederacy in the Civil War. And with his young imagination stirred by Major Ayres's chivalrous tales of battle, Kiffin decided to become a soldier.

Kiffin had a keen love of the outdoors and spent much of his youth hunting and fishing in the woods and swamps of South Carolina and East Tennessee. But when Kiffin turned fourteen, his mother moved to Asheville, North Carolina in order that her sons might take advantage of the educational opportunities there. To support her family, Mrs. Rockwell opened a practice in osteopathy, which soon proved successful.

Kiffin attended Asheville's Orange Street school from 1906–1908 then enrolled in the Virginia Military Institute for the 1908 fall term. He was able to secure an appointment to the Naval Academy at Annapolis in 1909 but soon resigned the appointment and that same year entered Washington and Lee University where his brother, Paul, had enrolled.

In 1911 Kiffin left Washington and Lee and found employment with a firm which published advertising editions for newspapers. His work took him up into Western Canada and down along the Pacific Coast, to San Francisco, where he took a position with an advertising agency.

In the fall of 1913 he returned to Asheville then joined his brother, Paul on New Years Day 1914 in Atlanta, where Paul had been working as a reporter for the *Atlanta Constitution.* They shared an apartment on West Peachtree Street, surviving on Paul's salary until Kiffin was able to secure a position with the Massengale Advertising Agency. During their spare time they collaborated on an "unsatisfactory" newspaper called "The Commerce of Greater Atlanta."

Kiffin Rockwell as a student at Washington and Lee University, 1911.

As war grew imminent between France and Germany, the Rockwells hastened to the side of the Republic. They were spurred on by their French Huguenot heritage, and by a feeling that the cause of France was "the cause of humanity." On 3 August 1914, Kiffin wrote the following to the French Consul-General at New Orleans:

"I wish to volunteer to serve in case of actual warfare and wish to know whether I can report to you at New Orleans and go over with the French Reservists, or would it be wiser to go to France before enlisting? I am twenty-one years old and have had military training at the Virginia Military Institute. I am very anxious to see military service and would rather fight under the French flag than any other, as I greatly admire your nation. If my services can be used, I will bring my brother who also desires to fight under the French flag."

Shortly after writing this letter, Kiffin learned that he could secure passage from New York to France on 7 August aboard the American liner *Saint Paul.* He and Paul caught the next train to New York and immediately embarked for Europe, not waiting for the French Consul-General's reply.

Mrs. Rockwell did not understand her sons' desire to serve France in a European conflict which America knew little about. She immediately began using all means at her disposal to prevent them from doing so. Her efforts included many entreaties to the Department of State in Washington and later to the French government asking that her sons be released from military service. Only later, after the supreme sacrifice of one of her sons, and through their letters to her, did Mrs. Rockwell come to see the high motivation behind her boys' enlistment.

Upon their arrival in France, the Rockwell brothers set out for Paris, for the Hotel des Invalides. There, on 21 August 1914, they enlisted in the French Foreign Legion with forty-one other Americans. On the morning of 25 August they marched out of Paris behind the American flag for the training camp at Rouen.

The American volunteers remained in Rouen only a few days before being transported by freight cars to Toulouse on 1 September. There, they shot target-practice and began drilling, and executed field exercises and forced marches. On 12 September a call was made for volunteers with previous military experience to accompany a group of veterans to the Front. The Rockwell brothers claimed to have served in the Mexican Army in order to be chosen.

Upon the issuance of new uniforms and field equipment, the Rockwells assembled with one thousand other men of Battalion C of the Second Marching Regiment of the Second Foreign Regiment on the morning of 30 September to entrain for the Front. A two-day cramped ride in freight cars ensued which ended on 2 October at Camp de Mailly, in the War Zone. The Rockwell brothers, with fellow Americans Bert Hall

Kiffin Rockwell (left) in the Aisne trenches, January 1915.

and William Thaw, were soon in the line below the Craonne plateau, virtually in the center of the battlefront. There they worked ten hours a day, widening the trenches and erecting their dugouts.

In December of 1914 near Chemin-des-Dames Paul Rockwell was wounded by an exploding shell which knocked him back into the trenches and smashed his collar bone. He was sent to the hospital but never regained his strength and was invalided out of the Legion.

On 18 March 1915 Kiffin—now called "Rocky" by his American comrades—applied for transfer and was sent to the 1st Regiment of the Moroccan Division near Reims. During the first week of May this unit was ordered 90 miles northeast to Artois to become part of the first major offensive of the War. On 9 May his division became the second wave that made a bayonet charge at La Targette against the German positions across the rolling Artois plain.

That morning Rockwell and fellow American Paul Pavelka struggled up out of the trenches and charged across the brilliantly sunlit plain into the fury of German machine gun emplacements. Sixty of their comrades were cut down around them, and as Rockwell later recalled it, "as fast as men fell it seemed new men sprang from the ground to take their places. To think of the horror of the thing was impossible. All I could think of was what a wonderful advance it was and how everyone was going against that stream of lead as if he loved it."

The Legionnaires cleared out the German trenches then stopped to catch their breath. The French assault penetrated nearly three miles into German territory before it met stiff resistance. Soon Rockwell and Pavelka found themselves pinned down by machine gun fire. Their lieutenant stood up and ordered his men forward. Rockwell and Pavelka followed him and had gone but twenty yards when a machine gun slug slapped into Kiffin's thigh, knocking him to the ground. When Pavelka stopped to help, Rockwell ordered him forward before tumbling into a shell hole to examine his wound.

Kiffin had been hit in the upper right thigh. Fortunately, the bullet had missed the bone. He bandaged his wound then pulled himself to the rear across an open field by clutching tufts of grass. It would be four days before he would have his wound treated at a hospital near Rennes. Still, he was one of the lucky survivors, as over half his regiment had been wiped out in the assault on La Targette and Neuville St. Vaast.

Rockwell showed his fearlessness and disdain for death when he later related this description of the attack to an American comrade:

"There is nothing like it; you float across the field, you drop, you rise again. The sack, the three hundred and twenty-five extra rounds, the gun, have no weight. And a ball in the head and it is all over—no pain."

Kiffin spent six weeks at *Hopital Auxiliaire No. 101* at Rennes. He further convalesced at the Paris apartment of Mrs. Alice Weeks where his brother, Paul, had been staying. Since leaving the Legion Paul had served in the Ministry of Public Information, France's propaganda office, and also had worked as a correspondent for the *Chicago Daily News*.

While convalescing in Paris, in July Kiffin ran into William Thaw, his former Legion comrade, who was then flying at the Front with *Escadrille Caudron 42*. He had come to Paris to pick up an aircraft. When Thaw learned of the condition of Kiffin's wound, that he would probably be unfit for further field service, he recommended that Kiffin attempt a transfer to aviation. Kiffin soon put in his request.

On 1 August 1915 Kiffin arrived at the Depot at Lyon. On the 9th he was sent to the Legion's training camp at La Valbonne. There, he was reunited with his comrade, Paul Pavelka, who was also just reentering active service after being wounded. Kiffin told Pavelka of his own attempt to enter French aviation and suggested Pavelka do the same. He then helped Pavelka to compose his request for transfer since Pavelka couldn't write in French. On 12 August both men took their preliminary physicals which were required prior to entering French aviation.

On 4 September Kiffin was ordered back to Lyon and was told to leave for Dijon to enter the *Service Aeronautique* This had been accomplished through the help of his brother Paul, and with the assistance of Mrs. Alice Weeks, and William Thaw. That same day he drew a new uniform at Dijon then headed for the school at Avord.

On the very first day of his arrival at Avord Kiffin took to the air with his instructor in a double-command Maurice Farman. He progressed rapidly through the course and had hopes of quickly receiving his *brevet* but was delayed by inclement weather and lack of aircraft. During a lull he traveled to Bourget and ordered a tailor-made uniform: black kepi, tunic and trousers "with the broad red artillery stripes" running down the legs.

Twenty-six September saw the arrival of another American at Avord. Before the day was out, Rockwell

Jim McConnell, Victor Chapman, and Kiffin Rockwell shooting pool at the Hotel Pomme d'Or, Luxeuil-les-Bains, May 1916. Madame Voge, the proprietress, looks on.

had discovered in young Victor Chapman a comrade who shared his own intense love for France and its cause. Upon first meeting Kiffin at Avord, Chapman wrote the following to his father: "I find a compatriot I am proud to own here. A tall, lanky Kentuckian, called Rockwell." The two immediately became the closest of friends, and a bond grew between them not unlike the bond between Kiffin and his brother, Paul.

While at Avord, on 22 October 1915 Kiffin received his *brevet militaire* on the Maurice Farman. He then entered the Nieuport school at Avord, living with Victor Chapman in a little house in the nearby village. Again bad weather slowed Rockwell's progress, and he chafed at the delays. He was ever anxious to reach the Front upon receiving word of the loss of many of his Legionnaire comrades in combat. In a November 1915 letter to a friend he wrote that "I have many scores to settle, and there is going to be more than one "Boche" aviator to settle them, or I will not live to tell the tale."

Kiffin graduated from Nieuport school ranked at the top of his class. On 14 December he had arrived at the R.G.A.(*Reserve General Aeronautique*) depot just outside of Paris. The depot supplied the escadrilles at the Front with pilots and planes.

On the 19th the engine of Kiffin's Nieuport stopped as he was attempting a landing on the field. He barely cleared a hut before he struck a pole and sheared off one wing.

For Kiffin Rockwell the weeks crawled into months, and he had still had not reached the Front. On 6 April he wrote the following to a friend:

"The fact is I have been more or less demoralized by never doing anything and waiting each day to go to the front. I ask each day to be attached to a regular French escadrille, but they continue to keep us all back, waiting until the Escadrille Americaine is formed. Why they don't form it, I don't know. We have enough men ready, know who will be the captain, that we are to go to Belfort, and that our number is "N.124." Yet we are not sent out. In the meantime, I fly a little and spend much time in Paris."

Finally on 20 April 1916 Kiffin Rockwell and his friend, Victor Chapman became two of the first seven members of the *Escadrille Americaine* when they arrived at the Squadron's quarters at the health resort at Luxeuil-les-Bains. Rockwell's first discovery upon his arrival at Luxeuil was that the Squadron had no aircraft. And he was both pleased and baffled by the idyllic setting of the Luxeuil resort which seemed far removed from the war. On the same day of his arrival there, he wrote this to his mother:

"We all eat together at an hotel where wonderfully good meals are served. We occupy a villa that has been requisitioned for us, with orderlies to wait on us. We go down each day about one hundred yards from here to bathe in a bath-house that is over two hundred years old. The scenery around the town is wonderful and we ride over the surrounding country with our captain and are planning to do a little fishing and hunting, so you can see that it is not much like being at war. . . .

"I am sitting by my window now with a good warm sun coming in and a wonderful view of the birds singing. If it were not for looking in the glass and seeing myself in uniform I should not be able to believe that I am at war, or that there is such a thing as war." But enemy planes soon passed overhead bombing the Luxeuil airfield, a jarring reminder that the war was still on.

On 29 April 1916 Kiffin took Bert Hall's unarmed fighter on a two-hour patrol over the lines. He ran into three German machines and quickly retreated into allied airspace.

On 5 May Rockwell was finally able to take his own Nieuport out on its first flight. However, he could not cross the lines, as he still had no machine gun.

On 9 May Rockwell and Victor Chapman took to the air and flew a total of five hours at 300 meters in a reckless attempt to lure enemy fighters skyward. Rockwell complained afterward that the enemy wouldn't rise to meet the challenge but "beat it as soon as they see us, and as soon as we are sighted, not a Bosche will leave the ground, not even their famous Fokkers."

On 13 May Rockwell led the Squadron's first regular patrol over the lines accompanied by *Capitaine*

Sgt. Kiffin Rockwell beside his Nieuport fighter, Bar-le-Duc, July 1916.

Thenault, Victor Chapman, James McConnell, and William Thaw. No enemy ships were encountered and the pilots returned to the Luxeuil aerodrome without incident.

On the morning of 18 May 1916 *Caporal* Rockwell, on his fourth patrol, flew a history-making sortie over the Front. Upon crossing the enemy lines near Thann, his engine began to cough and sputter. As he turned back toward his aerodrome he spotted a German L.V.G. bi-place 2000 feet beneath him. Spurning fancy tactics, Kiffin prepared to fight his first combat, and he immediately pushed in his throttle and nosed his Nieuport downward.

As he swooped down on the enemy bus, he watched the gunner quickly warn the pilot. The pilot rushed toward his lines as the gunner trained a burst of fire on Rockwell's Nieuport.

As he closed the distance, Kiffin felt his fighter shudder. His main wing spar had been struck by an enemy slug. But as he later described it, "I didn't pay any attention to that and kept going straight for him, until I got within twenty five or thirty meters of him."

Only when Kiffin thought he would ram the enemy plane did he fire four rounds from his Lewis gun and swerve sharply at the last instant to avoid a collision. Incredibly, Rockwell's shots had done mortal damage. Glancing back, he saw the German gunner fall backward toward the pilot who was already crumpled up in the forward cockpit.

The L.V.G. fell off on one wing then dropped into a vertical dive. Three minutes later it hit the ground in flames, billowing a plume of black greasy smoke just inside enemy lines. The victory, the first for Rockwell and the first for the American Squadron, was confirmed by a French observation post which quickly telephoned the news back to Luxeuil.

By the time Kiffin had returned to the aerodrome, his comrades were on the tarmac waiting to congratulate him. They rushed his Nieuport, and cheering, they lifted him from his plane. And as his friend, James McConnell, noted with envy, "all Luxeuil smiled upon him—particularly the girls." Word of his victory soon reached Paris where it also caused "a tremendous wave of excitement."

Kiffin's brother, Paul, was working in Paris at the time of the announcement. He rushed to Luxeuil to congratulate his brother and presented him with a bottle of vintage bourbon. Kiffin poured himself a shot then handed the bottle to Victor Chapman. Chapman declined and instead suggested that the bourbon be saved "for rare occasions." He went on to explain that each pilot who shot down a German plane should be "entitled to one slug." Thus, a ceremony was instituted which saw a number of the pilots drink from what was later called the "Bottle of Death" until the bourbon was finally consumed.

Rockwell continued to prove an excellent aviator. His comrades saw him as the soul of the Squadron and greatly admired his idealism, his enthusiasm, and the fearlessness he had demonstrated in the air on numerous occasions as he closed to within feet of his adversary before he tripped his machine gun.

On 20 May 1916 the Squadron left Luxeuil for Bar-le-Duc in the Verdun sector. On the 22nd a French counteroffensive was launched and *N.124* was assigned to fly morning air cover. In the early hours of the 24th Rockwell flew a patrol with Thaw which saw Thaw shoot down a Fokker. Following the mission, they landed at Behonne to rearm and refuel.

Capitaine Thenault was preparing to lead the regular morning patrol with Victor Chapman and Lieutenant deLaage. As Rockwell and Thaw joined Thenault he commanded all pilots only to attack the enemy on his signal.

Over Etien they spotted a formation of a dozen German biplaces. Victor Chapman, ignoring Thenault's command not to attack broke from the group and dove on the enemy, dragging Rockwell and Thaw into the fray. A fierce combat ensued which saw all three pilots wounded. A bullet struck the windscreen of Rockwell's Nieuport, peppering his face with glass and bullet fragments. Wiping blood from his goggles, he attacked and shot down one of the biplaces (which was never confirmed) before being forced to return to Behonne after exhausting his fuel supply.

Despite their wounds, he and Chapman both refused hospitalization. Chapman's wounds were less severe and he was back flying patrols within twenty-four hours. However, Rockwell's face was badly disfigured, and at first he had thought he had lost his nose. At the aid hospital while fragments were removed from his face Rockwell's mind reeled from the events of that morning, having fought a total of eight combats inside German lines.

Thenault ordered Rockwell to hospital for fifteen days. Instead Kiffin took an eight-day leave to Paris to recuperate at the apartment of Mrs. Alice Weeks. She dressed his wounds and later described his appearance in a letter to a friend:

"He looks terribly with those dreadful eyes men always have after going through heavy firing. I can not describe them. They are sunken and yet have a sharp look."

By 5 June Kiffin had returned to the Squadron still weakened from his wounds and suffering from fatigue. He had begun to feel that his previous effort with the *Escadrille* had gone unrecognized, and he blamed *Capitaine* Thenault for this, as he indicated in a letter to his brother, Paul.

"My citation hasn't yet gone through, so can't send you a copy yet. Don't think there is much doubt of the Medaille (*Medaille Militaire*), but don't expect two citations. There is no reason why I shouldn't have them, except that we are very unlucky in having a captain who is a nice fellow and brave, but doesn't know how to look after his men, and doesn't try to. I have been fighting with him ever since being back, mainly about the fact that I have no machine, he having given my old one to Prince and not managing right about getting me a new one. I think that in a few weeks I will be pretty sick with the outfit."

But by 10 June Kiffin was again flying missions from the Bar-le-Duc airfield. However, the end results were often unsatisfactory, as he indicated in a letter to Alice Weeks:

"For a wonder the weather changed and there has been a lot of activity, but I didn't have much success and was very tired and disgusted last night. Instead of bringing down anyone, I came close to being brought down myself a couple of times. Yesterday morning Chapman and I had a mixup with a number of them and I felt good when both of us got out of it safe and sound. I was coming in yesterday evening about dark, thoroughly tired and disgusted, not paying any attention and passed right beside a number that had been bombarding Bar-le-Duc, without seeing them. So you can see that I am not in good humor."

On 18 June Rockwell, Thenault, Prince, and Balsley fought an air battle with a large number of Germans east of Verdun which saw Balsley shot down and seriously wounded. On 23 June the death of Victor Chapman in combat near Verdun dealt a crushing blow to Kiffin and the Squadron. Kiffin wrote this of Victor the day following his death:

"He and I had roomed together and flown together a great deal, and I had grown very fond of him. I am afraid it is going to rain to-morrow, but if not, Prince and I are going to fly about ten hours and will do our best to kill one or two Germans for Victor."

As the war ground on, as many faces of dead comrades filled Rockwell's memory, he seemed driven to personally avenge each death with a victory over the enemy. He spoke of this driving determination in a letter to his brother, Paul:

"I had thought beforehand that yesterday and today I would try my damndest to kill one or two Germans for the boys (comrades in the Legion) who got it this time last year—but I had no luck. Am tired out now; have been out four different times to-day, all the time going up and down."

On 28 June during a ceremony at Bar-le-Duc, Kiffin was promoted to *Sergent* and was awarded the France's *Medaille Militaire* and *Croix de Guerre*, with Palm for his 18 May combat while on patrol with Chapman and Thaw. His citation described him as a "courageous and skillful pilot."

In July on the Verdun Front Rockwell fought forty officially reported combats. Still feeling that his efforts had not been appreciated, he wrote his brother, Paul, that he planned to go to another escadrille. "...to tell you the truth,: he said, "I want the Legion d'Honneur and a Sous-Lieut.'s grade. I don't give a d--- how conceited it may appear, but I think I have well earned the two."

Friction continued between himself and *Capitaine* Thenault, the latest row having been a dispute over how to run the *popote*. (mess). "Everyone here is scrapping and discontented," Rockwell wrote to Paul, "and I am about the worst of any."

The discontent continued, as Kiffin again indicated in a 31 July letter to Paul:

"I want to be changed to a French Escadrille unless certain conditions change here, and several others will follow my example.

"I think I have the most hours of flight and the most fights for the month of July on the Verdun front of any Nieuport pilot, but am not sure. I don't think, however, that a full report of my work has gone out of this office, and a number of times my report on a fight has been changed. The machine that (Bert) Hall brought down X. did his best to prove that it wasn't brought down, and so far Bert hasn't even been proposed for a citation."

The "X" that Rockwell had referred to was Norman Prince, whom Rockwell had viewed as self serving and had felt he had been currying Thenault's favor. He questioned Prince's sincerity and believed Prince's motive for helping to found and serve with the Squadron had been for his own personal glory.

In the month of August Kiffin Rockwell fought thirty-four air duels with the enemy. James McConnell described Rockwell as a pilot who "attacked so often that he has lost all count and who shoves his machine gun fairly in the face of the Germans."

By 4 September, the day Norman Prince returned to the Squadron following eight days *permission*, the hostility between Rockwell and Prince threatened to bring them to blows. Rockwell had felt that Prince had been padding his score of victories, particularly Prince's claim of 23 August, and he had stated as much in a 1 September letter to Paul.

"No one thinks X—got a German, in fact, everyone is sure he didn't; yet the Captain proposed him for a citation, wanted to propose him for the Medaille, but everyone said if he did they would quit. I am going to have to call him out when he gets back (from Paris), as he talked awfully big about us behind my back when I was away. We have all agreed to try to run him out of the Escadrille."

The dispute between the two young Americans temporarily dissipated under the pressure of flying combat patrols. On 9 September near Vauquois Rockwell came upon a German two-seater just inside enemy lines. He killed the observer with his first burst then relentlessly pursued the pilot down to 4000 feet, firing steadily. Only when two Fokker fighters jumped him from above did Rockwell break off the combat. But by then the bi-place was finished, and two French ground observers witnessed its crash just inside German lines. It was Rockwell's second confirmed victory.

On 14 September 1916 the Squadron returned to Luxeuil. They had been ordered to fly fighter cover for combined allied bombing raids against Germany.

On 23 September 1916, three days after his twenty-fourth birthday, *Sergent* Rockwell and Raoul Lufbery prepared to take the 8 a.m. patrol to challenge the enemy Fokker fighters from Colmar and Habsheim which had been harassing the French and British bombers. Four days earlier the Squadron had been supplied with new Nieuport 17s, and they were anxious to try out the speedier, better armed fighters. Once airborne, Lufbery had trouble with his Vickers gun, and was forced to land at the airfield at Fontaine to make adjustments to the weapon. Rockwell chose to continue on alone.

Kiffin soon found himself near Rodern, over the area where he had scored his first victory some four months earlier. There, he spotted an Albatros two-seater observation plane 11,000 feet beneath him, heading toward allied lines, and he plunged to the attack. The German gunner in the rear cockpit quickly fired a salvo from his machine gun. Undaunted, Rockwell maintained his usual pattern of attack and only when a collision appeared imminent did he fire his own gun.

A French infantry captain observing the combat through field glasses apparently thought that the Albatros had been struck and seemed to fall. But it was Rockwell's Nieuport which had suffered the mortal blow. Suddenly, it faltered, and the nose dropped rapidly and the wings on one side of the fuselage ripped away and fluttered through space. Rockwell's Nieuport augured groundward at terrific speed and crashed in a field of flowers.

Poilu in nearby trenches rushed to the crash site and pulled Rockwell's lifeless body from the wreckage while enemy gunners zeroed in on the smashed Nieuport. Once under cover, the French soldiers examined his broken body and found a gaping wound in his chest where he had been struck by a disintegrating bullet.

The news of Rockwell's death was quickly relayed to the Squadron. The pilots were struck with grief. Lufbery had learned of it when he had dropped down at the airfield at Fontaine to rearm and refuel following a combat with an enemy bus. He immediately took to the air and flew directly above the enemy airfield at Habsheim in an attempt to bait the Germans to avenge his friend's death. But no one would accept his challenge.

Capitaine Thenault and several of the pilots rushed in a staff car to where Kiffin had fallen. Upon their return to Luxeuil Thenault gathered the other Ameri-

Sgt. Kiffin Rockwell with his mechanics test-firing his Lewis gun.

can pilots around him as they returned from patrol, With tears in his eyes, he told them: "We are all saddened. When Rockwell was in the air, no German passed. The best and bravest of us is no more."

Rockwell had had a premonition that he would not survive the war. He had written of this to his mother: "Now, I don't want you to worry about me. If I die, I want you to know that I have died as every man ought to die fighting for what is right. I do not feel that I am fighting for France alone, but for the cause of all humanity, the greatest of all causes."

The night before he was killed Kiffin Rockwell had told Paul Pavelka that, in the event of his death, he should be buried where he might fall, and further "to take whatever money you happen to find on me and drink to the destruction of the damned Boche." The location of his crash site ruled out his first request. Instead his body was brought back to Luxeuil for burial, his final resting place. His friend, James McConnell would later write of his loss:

"No greater blow could have befallen the escadrille. Kiffin was its soul. He was loved and looked up to by not only every man in our flying corps but by every one who knew him. Kiffin was imbued with the spirit of the cause for which he fought and gave his heart and soul to the performance of his duty. He said: "I pay my part for Lafayette and Rochambeau," and he gave the fullest measure. The old flame of chivalry burned brightly in this boy's fine and sensitive being. With his death France lost one of her most valuable pilots. When he was over the lines the Germans did not pass—and he was over them most of the time. He brought down four enemy planes that were credited to him officially,* and Lieutenant de Laage, who was his fighting partner, says he is convinced that Rockwell accounted for many others which fell too far within the German lines to be observed."

Sgt. Rockwell's smashed body in the wreckage of his Nieuport 17 C.1 fighter, Rodern, Alsace, 23 September 1916.

On 25 September, at ten in the morning, a funeral "worthy of a general" was held in Kiffin's honor at Luxeuil. His coffin was draped with a French flag and was strewn with flowers as it passed between the ranks of many allied pilots and their ground crewmen. Also present were French officers and men from regiments of the French Territorial and Colonials. Airplanes circled above the procession dropping flowers. Rockwell was posthumously promoted to the rank of *Sous-Lieutenant* and was made a *Chevalier* of the *Legion d'Honneur*. An accompanying citation, signed by General Joffre, praised "his courage and his daring."

Eventually the procession reached the Luxeuil cemetery two blocks off the main street and Kiffin Rockwell was placed in the ground, this his final resting place.

In a 24 September 1916 letter to his mother, Kiffin's friend and comrade, Charles Chouteau Johnson penned these words which could well have served as Kiffin's epitaph.

"There isn't space to tell you what a man he was. A man all the way through, hard to know at first but once your friend he would stick all the way through. Brave to the core, never flinching."

**Kiffin Rockwell had a total of two confirmed victories.*

ROBERT ROCKWELL

ESCADRILLE LAFAYETTE

Robert Lockerbie Rockwell, the 15th volunteer of the *Escadrille Americaine*, was born in Cincinnati, Ohio on 18 March 1892, the son of C.L.& N. Railroad executive Marion Clay Rockwell and Mary Foote (McQuat) Rockwell. Rockwell received his early education with a private tutor. His formal education began at Howe Military School, Howe, Indiana. Following graduation from Howe, he attended Kenyon College, Gambier Ohio from 1911–1913. He then studied medicine at the College of Physicians and Surgeons at New York University from 1913 to early 1915 when he left school to join the Anglo-American Ambulance Service.

Rockwell sailed for France on 13 March 1915. He began his service as an *Interne* on 29 February 1915 at the Anglo-American Hospital at Paris. He worked there, and at the auxiliary hospitals at St. Valery-en-Caux and at Yvetot as a dresser of wounded with charge of three hospital wards until 3 February 1916. A distant cousin of Paul and Kiffin Rockwell, he followed Kiffin's lead into French aviation via the Foreign Legion on 7 February 1916.

Rockwell began his aviation schooling at Pau on 15 February. He received his *brevet militaire* while at the school at Buc on the Bleriot on 20 May 1916 at approximately the same time as his American classmates Willis Haviland, Frederick Prince, and Robert Soubiran. Unlike Rockwell, they *breveted* on the Caudron. Following a ten-day hiatus awaiting machines, Rockwell and the three other Americans completed their altitude tests and triangles in one day, normally a two-week process. After receiving their *brevets* they returned to Pau for Nieuport instruction. Rockwell then proceeded to gunnery and acrobatic school at Cazeaux then on to the *G.D.E.* pilot pool for final instruction. His pre-war flying experiences would propel him to the Front a full month ahead of the other three Americans.

On 17 September 1916, while Rockwell was in Paris, he was the American chosen by *Capitaine* Thenault, squadron commander of *N.124*, to replace Victor Chapman who had been killed in combat the

previous June 23. He accompanied Thenault to the Squadron's quarters at Luxeuil in the Vosges sector. Before long Rockwell had picked up the nickname "Doc" when the other American pilots had learned of his background in medicine.

Caporal Rockwell and the other pilots soon began flying patrols and fighter escort for French and British bombing groups stationed near Luxeuil. On 23 September Kiffin Rockwell was killed while attacking an Albatros two-seater near Hartmannsweillerkopf. On 15 October Norman Prince died of injuries he sustained while attempting a landing following a bombing raid. The Squadron was then ordered to Cachy on the Somme front. As it passed through Paris on 22 October 1916 it picked up three American replacements: Willis Haviland, Frederick Prince, and Robert Soubiran.

The Squadron was assigned to French *Groupe de Combat 13* and was housed in the sieve-like barracks

Sergent Robert Rockwell posing with a British nurse while on leave. (Below) Adjudant Rockwell in Paris visiting with former Ambulance Service comrades.

at the Cachy airfield. By mid-November, inclement weather had settled in, making patrols less and less frequent. Fog closed the air and rain pounded the field to quagmire. In January, as air activity became almost non-existent, *Sergents* "Doc" Rockwell and Dudley Hill took a three-week leave to the United States.

On 26 January 1917 the *Escadrille Lafayette* moved to Ravenal, near Saint Juste-en-Chausse. The pilots were ordered to lie low and await the planned spring offensive. February's weather allowed just twelve days fit for flying.

Sergent Rockwell spent his leisure time catching up on the voluminous correspondence he maintained with giddy young maidens in the United States. His method was to hook them by sending them "a shamefully idealized" photo of himself before breaking their hearts by ceasing all correspondence with them, allowing them to "pine away in hopeless longing."

March proved February's antithesis. By the 19th, the day James McConnell was killed in combat near Jussy, the Squadron had flown nearly 150 sorties.

On 7 April 1917 the Squadron moved again, this time to the airfield at Ham. The pilots took to the air almost daily flying protective sorties and offensive patrols. Edmond Genet was killed on 16 April in the air south of Saint Quentin when he was struck by fragments from an anti-aircraft shell and rode his fighter full-throttle into the ground. On 23 April Ronald Hoskier was also killed with his French gunner when his Morane Parasol was sent earthward minus its wing beneath the rattling guns of swarming Albatros fighters.

Throughout this intense period of combat by now *Adjudant* Rockwell was also being pounded beneath the gunsights of German aircraft. On 6 May 1917, the day he would win his *Croix de Guerre*, with Palm, he suffered serious damage to his Nieuport while combating German fighters. The engagement began on a solo patrol when he had spotted a formation of seven German monoplaces at 15,000 feet. Rockwell, seeing that they were flying into the sun, had hoped they would not spot him as he dropped onto the rear plane and fired. The other six fighters scattered, climbed, and swooped madly down on Rockwell's Nieuport.

A single Albatros latched onto his tail and fired bursts into his fuselage. Bullets riddled his plane. He frantically kicked his rudder bar to escape the path of deadly fire. The pursuit continued to 6,000 feet. There, in desperation, Rockwell put his nose down and dived away from the German who by then had emptied his machine gun magazines into Rockwell's aircraft.

Half and hour later Rockwell landed on the Ham airfield and began counting the bullet holes in his Nieuport. There were twenty-seven on either side of his cockpit with numerous other holes in his wings, tires, aileron controls, and landing-gear braces. He would later salvage the plane's propeller, making one-half into a walking stick while retaining the bullet-splin-

tered half as a memento of the fierce engagement which had nearly cost him his life.

Twenty-three May 1917 saw the death of the Squadron's second-in-command, French Lieutenant Alfred deLaage de Meux, who crashed a new SPAD S.VII fighter shortly after takeoff from the Ham airfield. He had been a favorite with the American pilots, and his death was deeply mourned by Rockwell and the others.

Captain Robert Rockwell, commanding officer of the U.S. 93rd Aero Pursuit Squadron, November 1918. The Squadron's Indian head insignia is clearly visible on the side of Rockwell's SPAD S.XIII fighter.

On 3 June the *Escadrille* relocated to Chaudun. A short time later it was redesignated *SPA 124* since the pilots now flew primarily SPAD S.VII fighters. On 17 July the Squadron was again moved, this time to St. Pol-sur-Mer, a small village approximately two miles from Dunkirk. While awaiting the soon-to-be launched British offensive, the pilots struck enemy aircraft sitting on the ground. When the British offensive was launched on 31 July, *SPA 124* was unable to partake because of fog and rain.

Senard became the site of the Squadron's next move, some 250 miles away near the Argonne Forest. This took place on 11 August. Once the pilots had settled in, the weather cleared, and each man flew three times daily for twenty-four days without interruption. Harold Willis was shot down and captured on 18 August while flying fighter escort with the other pilots on a bombing raid against Dun-sur-Meuse, forty kilometers within enemy lines.

On 24 September Douglas MacMonagle was shot down and killed on a morning patrol with Doc Rockwell, Raoul Lufbery, and Edwin Parsons. Rockwell nearly suffered the same fate later that day. The motor of his SPAD fighter had failed him while flying above a German formation, forcing him to plunge through the heart of the enemy patrol. During his rapid descent, his oil radiator had burst, spewing oil over his windscreen and goggles. With an enemy formation now on his tail, Rockwell continued an extended nose-dive, hoping his pursuers would think him dead and break away. The ruse worked, and Rockwell left the vrille 300 meters above the ground, hugging the landscape back to the aerodrome. He was awarded a second Palm to his *Croix de Guerre* for bringing home his disabled plane.

By the end of September it was back to Chaudun for the Squadron. Shortly after the move Andrew Courtney Campbell became the next Squadron fatality when he was killed in combat on 1 October 1917 in the air north of Soissons. Not long after his death, several pilots transferred out of *SPA 124* either to other French squadrons, or to the aviation instruction centers at Tours and Issoudun.

On 7 December 1917 the Squadron made its final move, this time to the aviation field at La Noblette in the Champagne sector. Shortly after the move, the *Lafayette Escadrille* pilots remaining on the roster were told to seek their releases from French aviation in order to receive commissions in the U.S. Air Service. *Adjudant* Robert Rockwell was commissioned a Captain on 31 January 1918. On 18 February, as the *Escadrille Lafayette* passed out of existence, Rockwell and the remaining *SPA 124* pilots formed the nucleus for the new U.S. 103rd Aero Pursuit Squadron.

Captain Rockwell served as pilot and flight commander with the 103rd under commanding officer Major William Thaw from 18 February until 10 August 1918. At that time Thaw was promoted to command the newly formed 3d Pursuit Group and Captain Rockwell assumed command of the 103rd. Rockwell, while squadron commander, continued to fly regularly as patrol leader and to serve his squadron diligently until he was relieved of command for a rules infraction on 18 October 1918. Captain Robert Soubiran took over as squadron commander that same day and served in that capacity until the November Armistice.

Captain Rockwell flew for more than two years at the Front. His final overseas assignment, performed after the November Armistice, was as commanding officer of the U.S. 93rd Aero Pursuit Squadron. In recognition of his many months of sacrifice to the Republic which had begun in February of 1915, France made him a *Chevalier de la Legion d'Honneur* on 17 May 1919 by issuing the following citation:

"American citizen enlisted at the beginning of the War in the Foreign Legion. Has been distinguished in all the operations in which he took part, and has given proof of the highest qualities as a pilot of L'Escadrille Lafayette."

Following the war, Rockwell returned briefly to the United States. Upon his discharge from the service, he became associated with the Texas Company, working in their export department in New York City. He later did overseas investigation work for that company in Europe, Algeria, Morocco, Tunisia, Gibraltar, and along Africa's west coast. He resigned from the company in 1921 and returned to Cincinnati, Ohio, where he became engaged in farming and the raising of pedigreed dogs and prize poultry.

Rockwell married Mary Dilatush Van Note, the widow of Dewitt Van Note, on 4 April 1923, in Cincinnati. They eventually had three sons: William Andrew, Richard Lockerbie, and Frank Allen. In 1925 Rockwell obtained his commercial pilot's license. In 1928 he became associated with the Embry-Riddle Company, a commercial aviation firm, directing its flying school at Cincinnati. In 1929 he left Embry-Riddle and moved to Dayton, founding the Dayton School of Aviation that same year and operating it until 1934. He offered a correspondence course in flying which he had written himself to some 5000 students. Rockwell then became a tester of engines for designer Prothro and worked in a civilian capacity at Wright Field.

From 1921 on Rockwell remained on flying status with the Army Air Corps Reserve, flying from Lunken Field, Cincinnati, Ohio. He also attended officers school at Fort Leavenworth, Kansas. He returned to active duty in the military in 1939 at the rank of full colonel in the Air Corps. His first assignment was head of the Aviation Cadet Examination Board which included parts of Kentucky, Ohio, and West Virginia, with headquarters in Lexington, Kentucky.

Colonel Robert Lockerbie Rockwell while serving in the U.S. Army Air Corps during World War Two.

In March of 1942 Rockwell became commanding officer of the Newark, New Jersey airport. In the spring of 1943 he was given command of Air Depot Group 344 and served as station technical advisor at the Air Depot Training Station, Albuquerque, New Mexico. In December of 1943 he then became technical inspector at the Ogden Air Depot, Hill Field, Ogden, Utah. Additional assignments were at San Bernardino Army Air Station, San Bernardino, California; Williams Field at Chandler, Arizona; Tonopah Air Base, Tonopah, Nevada; and Muroc Air Base (now Edwards Air Force Base), California.

In 1946 Colonel Rockwell retired from active duty. In 1949 he enrolled in the University of Redlands, Redlands, California under the G.I. Bill, and received a degree in Geology in 1953. He passed away from a heart attack in San Bernardino, California on 25 January 1958 at age 65.

MARIUS ROCLE

Marius Romain Rocle was born in Brussels, Belgium on 6 June 1897 of French parents, Etienne Rocle, an engineer, and Marie Louis (Ferron) Rocle. Marius spent his youth in Paris where his father was editor of the *Figaro*. The senior Rocle relocated to New York City to work for the phone company, and Marius attended the Harlem High School. Shortly after the outbreak of World War One, Rocle, then a youth of seventeen, left school for France to enter the European War.

On 26 September 1914 Rocle enlisted in the French Foreign Legion in Paris. He served with the Infantry in the trenches for one year and eight months. He was wounded in combat and was awarded France's *Croix de Guerre*, with Star for the courage he demonstrated on 28 September 1915.

On 5 June 1916 Rocle enlisted in France's *Service Aeronautique*. From 5 June to 28 January 1917 he attended the aviation schools at Buc, Cazeaux, and Pau. Rocle was one of three Lafayette Flying Corpsmen who served as machine-gunner /observer in the French Air Service, the other two being William McKerness, and Fred Zinn.

From 1 February 1917 to 19 February 1918 *Caporal* Rocle served at the Front with *Escadrilles N.84*, *C.46*, and *Br.213*. On 19 February Rocle was commissioned a Second Lieutenant in U.S. Aviation. From 19 February to 15 March 1918 Lieutenant Rocle served with the 13th Aero Squadron, USAS. From 15 April to the 11 November 1918 Armistice he served with the 644th Aero Squadron, USAS. By the end of the War Rocle had served more than four years on active duty, and had been involved in eight plane crashes as an observer-gunner.

While in Paris on 1 October 1918 Lieutenant Rocle met and married Margot King, a portrait and landscape artist who had left her studies in New York City to serve as a Red Cross worker in France. From this union a son and two daughters would be born.

Following Marius's discharge in 1919, the Rocles settled briefly in New York before again returning to France where Marius represented an American machinery firm. They eventually returned to New York City then moved to California in 1922 where they bought a Chula Vista citrus ranch.

Rocle was introduced to oil painting by his wife, an accomplished artist. He grew adept at the medium and produced a number of landscapes and startling surrealistic works which were exhibited jointly with Mrs. Rocle's works in various museums, including the San Francisco Museum of Art.

The Rocles later moved to the Echo Hills Farm south of Ramona where Marius raised Arabian and saddle-bred show horses and Shetland ponies. Rocle died of cancer in a San Diego hospital on 16 November 1967 at age 70.

JOHN ROLF

John Fletcher Rolf, Jr. was born in Centerville, Maryland, on 19 April 1894, the son of John F. Rolf. Rolf attended Harvard University, Class of 1914, 1920. He was a student at that institution when he decided to enter the European War.

Rolf joined the Lafayette Flying Corps in New York City on 9 June 1917. He sailed for France aboard the *Chicago* on 19 June.

On 31 July 1917 Rolf enlisted in France's *Service Aeronautique*. He received his *brevet* on 28 September 1917. On 11 November 1917 he was released upon his own request.

Rolf died on 21 January 1979.

William Rodgers

William Blackstock Rodgers, Jr. was born in Pittsburgh, Pennsylvania on 27 January 1895, the son of William Blackstock Rodgers and Adilyn (McVay) Rodgers. He attended the Culver Military Academy Summer Naval School at Culver, Indiana in 1914. Rodgers then attended Princeton University, Class of 1917. He left that institution in his freshman year and entered Cornell University. While a student there, Rodgers decided to enter the European war.

On 23 February 1917 Rodgers entered the U.S. military service at Key West Barracks, Florida as a Sergeant in the Aviation Section. On 6 March he was called to active duty and was assigned to the Curtiss Aviation School at Miami. On 26 May 1917 Rodgers received an honorable discharge in order to sail for France to enlist in the Lafayette Flying Corps.

On 25 June 1917 Rodgers enlisted in France's *Service Aeronautique* in Paris. From 29 June 1917 to 1918 he attended the aviation schools at Avord, Juvisy, Chateauroux, and the *G.D.E.* He received his *brevet* on the Caudron on 1 October 1917.

Caporal Rodgers left the French Air Service to accept a commission in U.S. Naval Aviation. On 10 April 1918 he was commissioned an Ensign in Paris. From 10 April to 1 June 1918 he attended the U.S. Naval Instructional Center at Moutchic-Lacanau. He was made Naval Aviator #2160 (HTA).

From 1 June 1918 to 20 December 1918 Ensign Rodgers served as chief pilot for seaplanes at the U.S. Naval Air Station at Lake Bolsena, Italy. On that date he was transferred to the Naval Aviation Office at Paris. On 9 January 1919 he was released from active service at Paris.

Following his World War Service Rodgers remained abroad for 15 years as a Paris representative of an American firm. He returned to the United States in 1934 and settled on Florida's West Coast. He married Penelope Hoopes. From this union one son was born. Rodgers died in the Sarasota Memorial Hospital on 20 December 1954 at age 59.

CLIFFORD DEROODE

Clifford deRoode was born on 9 September 1886 in Chicago, Illinois, the son of Holger de Roode, a Chicago insurance broker, and Cornelia Marie (Volrider) deRoode. DeRoode attended Chicago schools and Benton Harbor College. He left school to work as a dining-car conductor and as a reporter for the Chicago *Herald Tribune*. In 1913 he went to Europe as an escort for an older lady. At the outbreak of World War One he was in Germany studying hotel management and was interned there. When he secured his release, he left through Switzerland to Paris, France.

From 1915–1916 deRoode served in the American Field Service with SSU 1 for 9 months as an ambulance driver. He joined the *Service Aeronautique* and was *breveted* on 28 September 1917. de Roode was later commissioned a *Sous-Lieutenant* and was appointed interpreter and drillmaster at the aviation school at Avord. When he left French service he was commissioned a First Lieutenant in the U.S. Air Service.

Following World War One deRoode returned to the United States and entered the export-import business, purchasing antique furnishings in Europe for U.S. hotels. He married Louise Jagsch, a dress designer, from Columbus, Ohio. From this union two sons were born.

DeRoode and his family were living in Europe when World War Two broke out, effectively ending his export business. DeRoode remained in Vichy, France, but sent his two sons to the United States to live with relatives. deRoode again joined the French Air Force and became the first American to fight in World War Two. He became the commanding officer of the 2nd Lafayette Escadrille.

In the fall of 1942 deRoode had to flee France and escaped through Spain and Portugal to the United States. After the war he worked with travel agencies and with stock brokerage firms. He lived in New York and New Jersey then moved to Nashville, Tennessee. He died in a Brentwood, Tennessee nursing home on 25 August 1973 at age 86.

RAYMOND ROSS

Raymond Thomas Ross was born in Crawfordsville, Indiana on 23 July 1895, the son of Charles W. Ross, a banker and realtor, and Alice (Green) Ross. Ross attended Crawfordville High School before entering Amherst College, Amherst, Massachusetts, Class of 1917. Ross left school and on 5 April 1917 he joined the Norton Harjes Ambulance Service, and was issued Red Cross #302. On 15 April 1917 he enlisted in France's *Service Aeronautique*. On 12 August 1917 he was released because of defective eyesight.

Ross returned to the United States where he became involved in the lecture circuit. On 1 January 1918 he returned to France and enlisted in the French Tank Corps where he served as a *sous lieutenant* with 365th Company, 19th Battalion, 507th Regiment. According to the *Amherst Memorial Volume*:

"Ross became a fighter pilot in the Air Service. He was injured in a combat with a German aviator and sent back to the United States on a furlough of three months; but he returned to France in January, 1918, and was commissioned a First Lieutenant in Aircraft Artillery on July 12, 1918. He received France's *Croix de Guerre*, with two Stars and Palm. He was wounded several times, finally having his arm badly shattered just before the armistice was signed."

Following World War One Ross returned to the United States and entered the University of Chicago School of Business. He returned to Crawfordsville where he worked for his father's firm. On 27 February 1924 he married Dorothy Green in Crawfordsville. From this union two daughters, Mary and Alice, were born.

From 1927 to 1941 Ross served as assistant director of the U.N.Y. Community Trust, an organization of banks administering charity trusts. In 1941 he settled in West Lafayette, Indiana where he engaged in the real estate business and became a manager of a publishing house. He died in Lafayette after a five-week illness on October 18, 1957 at age 62.

KENNETH ROTHARMEL

Kenneth Rotharmel was born in Bluffton, Indiana on 5 September 1892, the son of Albert C. Rotharmel, and Eva (Martz) Rotharmel. Rotharmel attended Wabash College in Crawfordsville, Indiana for 3 years from 1911–1914. He left school to work as a reporter for the Birmingham, Alabama *Age Herald*. In the spring of 1916 he was working as a copy writer for the Miami, Florida *Metropolis* when he decided to join the American Ambulance Service. On 9 December 1916 he left New York on the *Espagne* for Bordeaux. Rotharmel served in France with SSU 4 for 7 months from 23 December 1916 to 9 July 1917.

On 10 July 1917 Rotharmel joined France's *Service Aeronautique*. From 12 July to 24 February 1918 he served at the aviation schools at Avord, Tours, Pau, Cazeaux, and the *G.D.E.* He received his *brevet* on the Caudron on 23 October 1917.

From 25 February to 26 April 1918 *Caporal* Rotharmel served at the Front with *Escadrille SPA 112*. On 4 April he was commissioned a Second Lieutenant in U.S. Aviation. He remained with his French squadron on detached service, flying combat patrols until the 11 November Armistice. He received France's *Croix de Guerre*, with Star for his courageous service and numerous combats between March and November of 1918.

Following World War One Rotharmel worked briefly in Birmingham, Alabama as a news reporter before joining the staff of the Miami *Herald*. He then moved to Chicago to work for an advertizing agency, then for the Portland Cement Association. In 1921 he married Evelyn Clarke of Chicago. From this union a daughter and son were born. Rotharmel then moved to Cincinnati, Ohio with the Portland Cement Association.

In 1926 Rotharmel moved to New Orleans, Louisiana where he worked as advertizing manager and Eastern Sales Manager for Pan-American Petroleum Corporation (Standard Oil of Indiana) for 26 years. He died of cancer at his New Orleans home on 17 October 1952 at age 60.

LELAND ROUNDS

Leland Lasell Rounds was born in New York City, New York on 22 August 1891, the son of Dr. William Rounds, and Dr. Eirene King (Lasell) Rounds, New York City physicians. As a youth he resided with relatives in Orange, New Jersey, and attended the Montclair, New Jersey Military Academy, Class of 1911. Following graduation, he spent some time in Europe before returning to the United States.

On 14 September 1916 Rounds sailed from New York for France aboard *LaTourainne*. On 16 October 1916 Rounds enlisted in France's *Service Aeronautique*. From 16 October 1916 to 1 August 1917 he attended the aviation schools at Avord, Pau, and the *G.D.E.* Rounds received his *brevet* on the Bleriot on 8 May 1917.

From 3 August to 22 December 1917 *Caporal* Rounds served at the Front with *Escadrille SPA 112*. Rounds flew numerous combat patrols with this squadron and was awarded France's *Croix de Guerre*, with Palm for his official destruction of an enemy fighter.

Rounds was forced to discontinue flying combat missions because he would black out while patrolling above 3500 meters, an altitude above which most pursuit work took place. On 3 January 1918 *Sergent* Rounds was commissioned a First Lieutenant in U.S. Aviation. From 1 January to 1 May 1918 he was Chief Pilot at the American Aviation Instruction Center at Tours. From 1 May to the 11 November Armistice he was on duty at U.S. Aviation Headquarters, Paris. On 4 March 1919 Rounds was discharged from the U.S. Air Service.

Following World War One Rounds spent much of his life in France working for the Diplomatic Service of the U.S. Government. He married Marcel Julianne in the 1930's aboard ship on the Atlantic Ocean.

During World War Two Rounds served as U.S. Vice Consul and Technical Adviser at Oran, Algeria. He was decorated by the United States, France, and Algeria for his clandestine operations on behalf of the Allied Cause. Rounds retired from the Diplomatic Service in New York City. It was there that he died of cancer on 2 July 1965 at age 73.

LAURENCE RUMSEY

ESCADRILLE LAFAYETTE

Laurence Dana Rumsey, Jr., the 11th volunteer of the *Escadrille Americaine*, was born on 2 September 1886 at Buffalo, New York, the son of tannery tycoon, railroader, and banker Laurence Dana Rumsey and Jennie (Cary) Rumsey.

Young Rumsey attended Harvard University, Class of '04–'08. Following school, he played polo professionally until the outbreak of World War One when he sailed for France in late 1914 to serve as an ambulance driver with the American Ambulance Field Service. He was with S.S.U.I from January to July 1915, attached to the French Army in France and in Belgium. Seeking the more active roll of a combatant, he left the Ambulance Service and enlisted in the French Air Service via the Foreign Legion in Paris on 9 September 1915. Rumsey picked up his uniform and flight gear at Dijon and arrived at the French School of Military Aviation at Pau on 11 September.

Rumsey had little trouble completing his penguin course and his "tour de piste" and had anticipated receiving his *brevet* before Christmas 1915. Two unfortunate accidents set him back and ultimately prevented his becoming one of the *Escadrille Americaine's* original squadron members. A letter home, dated 16 January 1916, records his unfortunate luck.

"I started on my triangles . . . I had almost done two sides of it when my engine went wrong and I had to come down. It was a very poor landing, and I smashed the machine to splinters. I was out in the country, and when one comes down one has to stay with the machine until the mechanicans come and cart it back to Pau. I hung around in a small village called Grenade, north of Pau, for six days. I started again on the 10th, and had almost finished my triangle and my two heights—one 2,700 metres, the other 3,000 metres—when I got lost in the clouds. I went clean into the mountains across the border of Spain, but when I saw snow I decided it was time to start north again, so came out. I had been flying four hours, and as my gas was giving out I had to come down again. Well, I had to land on the side of a hill, and the result was I smashed up entirely the second time."

Rumsey received his *brevet militaire* on the Caudron on 2 February 1916. From Pau he went to the aviation school at Avord where he continued his flight training on the double-command Voisin. While at Avord, he put in a request to fly Nieuports at the Front. He was sent back to Pau for chasse training in early March then was sent on to the *G.D.E.* (*Groupe des Division d'Entrainement* at Plessis-Belleville. He was assigned to *Nieuport 124* on 4 June 1916. While he was waiting to go out on patrol, he had his personal insignia, "RUM" painted on his Nieuport 11, on the sides of the fuselage and the top surface of the upper wing.

Rumsey wrote a series of eight letters home while he was with the *Escadrille Americaine*. The first one is dated *June 14, 1916*.

"I have my own Nieuport machine now and everything is on it, machine gun, clock, altimetre, compass, and map. I have been out several times looking over the country and studying the lay of the land, etc., and

(Left) Laurence Rumsey beside his Nieuport 11 Scout showing his personal insignia "RUM" at Bar-le-Duc, July 1916.
(Above) Another view of Rumsey's Nieuport 11 Scout (serial number 1290) at the Behonne airfield, Bar-le-Duc. Note: Lewis gun covered with protective case.

yesterday I went to the line beyond Verdun. It was quite misty so did not see any Germans. Just as well, as I want some more practice, because the gun is stationary, and you have to aim by steering your machine. The machine is so small, light, and fast that you have only to move your feet a hair's breath to manoeuvre. From high up the country beyond Verdun looks just like a plowed field. It is all brown, and all the trees have been swept away by shells and fire. They are still fighting. I could see the blast from the cannon. They took a crack at me, but I was at 4,000 metres and a darn small mark, especially going at 190 kilometres(about 120 miles) an hour. We are the only escadrille in which all the pilots have these Bebes, and we are going to try and deliver the goods."

June 27, 1916. "Bar-le-Duc is in a valley, quite a pretty town. We live in a villa overlooking the town. It belongs to an artist, as it has a studio with skylight, etc., is surrounded by gardens, and it is very pretty and pleasant. We all eat together, two majors, two captains, two lieutenants, and about ten of us. We are most careful now about attacking escadrilles over the German lines. Also we are flying together better than at first, and have some sort of system, and are not flitting all over the heavens."

June 28, 1916.(Note: This day Rumsey made *Sergent.*) "Well, we have been seeing and tasting something of real fighting, and it is very exciting indeed. The artillery at the forts around Verdun is very powerful and numerous. It kicks up the dust for over two miles and it is dangerous to get into it, because we might lose each other and also get our carburetors full of dust, which means motor trouble. We travel at about 3,500–4,000 metres. When they are shelling each other on the ground, the Germans don't trouble much about shelling us in the air. We have to be careful about a very fast new machine they have out. It is faster than our fastest, carries two men, two machine guns, has two planes and a 220-horsepower Mercedes Benz stationary engine. That is why we are very careful now about staying together, as the two men we have lost got separated from the bunch and had no show. When it is good weather, we go out twice a day to the lines for two and a half hours, making five hours a day. It is pretty tiring, so now they have brought another escadrille to help us, and we do not go out so often. It is a great nerve strain.

July 12, 1916. "We are still here at Bar-le-Duc, flitting up and down and trying to entice a non-suspecting German near enough to our lines so that we can go after him. They very seldom come across, except in swarms, and then they beat it back quick before we are up or else to some other part of the line. We go up twice a day officially, and whenever we want at other times, but that is not often, as we usually sleep or relax as much as possible in between times."

The usual means of relaxation for the pilots was a night in the cafes of Bar-le-Duc. Bert Hall recorded the aftermath of one such journey in his diary entry for 20 July 1916:

"Last night when I came home from town I had Rumsey in tow. He was in excellent spirits, singing little snatches of a song about an "English king a thousand years ago, who ruled his land with an iron hand, but his mind was weak and low." When we got to the villa I piloted Laurence up the winding stone stairs.

Just as I got to the top, I relaxed my attention a moment and down went Mr. Rumsey of Buffalo, New York, right to the bottom of the stairway.

"I followed him down and found him sitting on the bottom step fanning himself with a little branch of a tree. He looked up and said:

"That you, Bertie, you old bozo!"

"Yes," I said, "this is me. Get up now, goddamit, we're goin' in the house."

"Why, say, Bertie, I nearly fell, didn't I?"

So I gathered him up and dragged him up the steps and threw him in the villa. Rumsey and I are very good friends. I like him very much."

Rumsey's next letter home was dated *July 29, 1916*. In it he describes an air action that has just taken place.

"Two Frenchmen brought down a German 30 kilometres (19 miles) within our lines yesterday. The pilot had been wounded but landed without breaking his machine a bit. It is a new kind of Albatross bombardment machine. We all went over and it was interesting looking over. A big stationary Mercedes engine, two men, two machine guns, and very well finished."

September 20, 1916. "We are at Luxeuil-les-Bains, near Switzerland. We did not go to the Somme after all, and all are disappointed. We had to fly over the Vosges to get to the line here, and it is quite far. The captain took us along the line yesterday, showing all the landing places. It was very interesting, and the Vosges are very beautiful. They are not so very high, and are covered with large pines, much like the Adirondacks. We will probably be here all winter."

October 20, 1916. "We had hard luck in Luxeuil; we only went there for the raid. I was not in it. Only four machines went over. We have now changed to the Somme. We are out in the country about 20 k.m.(12° miles) from Amiens, sleeping on the floor of barracks. I think we are in for the real thing now. The English and Canadians gave us a good-bye party at Luxeuil, and all seemed to be sorry that we were going. They tried to get us attached to them. From here you can hear the big guns, just the way we could when in Belgium, only more so."

November 1, 1916. "We are getting settled here very slowly, but we are not really comfortable yet. The barracks leaked on top, sides, and floor. It does nothing but blow and rain, so first we were uncomfortable, but now we have fixed the roof, papered the walls, and plugged the floor. We also bought some oil stoves, so it is much better. We are 17 k.m. east of Amiens, which is the English headquarters. I went down to the general reserve the other day to get myself a new machine, latest type, 110 horsepower, 200 k.m.(124 miles) an hour; climbs to 4,000 metres in less than ten minutes. It seems like a bear-cat."

Laurence Rumsey would later describe his most harrowing mission to reporters when he had returned to the family mansion in Lewiston near Buffalo, New York in 1917.

"I started out with five other flyers one morning. It was pretty misty and we got in a lot of clouds. When I got clear I thought I saw the other five fellows so I flew over amongst them. Pretty soon I heard the 'put-put-put' of German machine guns and saw the whole five taking cracks at me. I had mistaken a German flying squad for my own fellows. Say, I dropped about 1,000 meters and got away from there."

As Rumsey had noted in a letter home, by mid-October 1916 the pilots of *N.124* had been forced to relinquish their pampered quarters at Luxeuil and to transfer to Cachy in the Somme Sector. The American aviators must have felt they had fallen from grace and had been sent to a frigid hell for an egregious breech of duty. As pilot James McConnell noted in his book *Flying For France*:

Laurence Rumsey (left) and Dudley Hill playing chess in the Cachy barracks, November 1916.

"Before its arrival in the Somme the escadrille had always been quartered in towns and the life of the pilots was all that could be desired in the way of comforts. We had, as a result, come to believe that we would wage only a deluxe war, and were unprepared for any other sort of campaign. The introduction to the Somme was a rude awakening. Instead of being quartered in a villa or hotel, the pilots were directed to a portable barracks newly erected in a sea of mud.

"It was set in a cluster of similar barns nine miles from the nearest town. A sieve was a watertight compartment in comparison with that elongated shed. The damp cold penetrated through every crack, chilling one to the bone. There were no blankets and until they were procured the pilots had to curl up in their flying clothes. There were no arrangements for cooking and the Americans depended on the other escadrilles for food. . . . The thick mist, for which the Somme is famous, hung like a pall over the birdmen's nest dampening both the clothes and spirits of the men."

The condition of the field and the air above it made missions nearly impossible from Cachy. The pilots had little to do to pass the time but to write letters, play cards, and drink cognac. On one frigid evening in November, "Whiskey", the Squadron's lion cub mascot, entertained himself by chewing on Laurence Rumsey's new uniform cap. Heedless of Rumsey's commands to stop, "Whiskey" simply growled and tugged harder at the cap between his paws. Enraged, Rumsey released his bottle of cognac and grabbed a nearby walking stick, and clubbed the lion about the head. "Whiskey" recoiled, dropped the cap, and shambled uncertainly to the table where other pilots were playing poker. Looking down, they examined their mascot and noticed that Rumsey's blow had blinded his right eye.

The use of alcohol played a major role in the lives of World War One pilots of all nationalities. It was no different with the Americans of *Nieuport 124*. Edwin Parsons later recalled this of the Squadron: "They were a pretty hard drinking kind, some of them." And Edmond Genet noted in his diary entry of 25 February 1917:

"It seems a mighty difficult and quite impossible proposition to keep entirely away from drink with this Escadrille. If one goes into town any day with any of the fellows it's impossible to keep from going in and drinking without absolutely being discourteous and uncomradelike. Perhaps I'm a fool but I don't like it one bit."

The drinking habits the young Americans developed in the cafes and barracks of France they took home with them, those who survived the war. And in a number of cases there is little doubt that alcohol contributed to their untimely deaths. Still, at the time its use seemed a necessity, a tonic with which to unwind between missions; a means to forget for a time the faces of dead comrades.

With Laurence Rumsey it became more. While he was considered a capable pilot, he sought in alcohol the courage to fly, and too often he consumed so much of it, he was unable to leave the airfield. His sense of duty in the ambulance corps and his perseverance through flight training in spite of the growing size of the flying school cemetery testified to his good intentions. But at thirty years of age, the recklessness of youth had since deserted him, and in the five months since he had joined the Squadron he had seen three comrades wounded and three killed. The blinding of "Whiskey" completely unnerved him, and in a matter of days had become manifest in boils which had broken out over his body and had sent him to hospital. By 25 November 1916 he was dropped from the Squadron, having flown less than half a dozen missions during his entire tenure with *N.124*.

Rumsey's lowest moment had come when the Squadron had made its "mass flight" from Luxeuil to Cachy. Former *N.124* pilot Edwin Parsons related the incident as it was told to him by Frederick Prince.

"They (the Squadron) were going to leave from some field outside of Paris and meet over the field and fly together to Cachy. During the time they were preparing at Paris to go up, the morning when they were to leave, Rumsey took on too much of the spirits fermenti and when the boys left to get the planes everyone tried to discourage Rumsey from going because they didn't think he was in quite a fit position to go. However, he tried to leave with them and they all took off simultaneously.

"Rumsey didn't join the group and they waited a while for him and then decided he had landed and wasn't quite capable of flying. They flew on up to Cachy and landed in the expectation that Rumsey would come tooling along after them. They began to get worried as it grew dark. They went out and put flares on the field hoping that nothing had happened to him. A short while later they received a telephone call over the communications system asking if the escadrille had a pilot named Rumsey. They admitted that they did. They said 'your pilot landed at Delouze' which was directly opposite from the field at Cachy far behind the lines, and when he landed there he didn't quite know where he was and remembering his instructions if he thought he was landing in German territory he was to set fire to his machine. And that's what he did, and the machine burned up on the field."

In May of 1917 *Sergent* Rumsey returned to the United States on board the *Rochambeau* with fellow Lafayette pilot Frederick Prince. Prince had secured emergency leave home to visit his ailing father. Laurence Rumsey reached the family mansion in Lewiston, New York on 18 May.

Despite *Sergent* Rumsey's ill health, and his more than two years' service in France, following a brief recuperation he enlisted in the U.S. Army as a private and was assigned to the 4th Field Artillery. The value of his war experience was soon recognized and he was detailed to Field Artillery Central Officers' Training School, Camp Zachary Taylor, Kentucky. He was commissioned a First Lieutenant in the Field Artillery on 1 March 1918. He was discharged in January 1919 and was commissioned a First Lieutenant in the Field Artillery Officers' Reserve Corps on 20 January 1919.

Following the War, Rumsey again pursued his interest in polo with his brother, Charles Cary Rumsey, the famous sculptor. As their father had before them, they played together for the Buffalo Country Club team and on the Orange County team with Averell Harriman. And together they vied for a spot on Tommy Hitchcock's 1921 Westchester Cup team with former *Lafayette Escadrille* pilot Frederick Prince.

Charles "Pad" Rumsey's dual careers in art and in polo were tragically snuffed out by his untimely death in an auto accident in 1922. Laurence continued to compete in polo and like fellow Lafayette Flying Corpsman Tommy Hitchcock won many awards while riding on the great Eastern teams. He sustained himself with inherited family wealth and never sought active employment.

Unlike several of his Squadron mates, Rumsey never used his Lafayette service to thrust himself into public view. And he avoided all contact with the other survivors of his Squadron. When the last "survivor's reunion" was held in Asheville, North Carolina in 1960 he did not attend. He continued to live in relative obscurity in Youngstown, New York surrounded by his polo trophies in an apartment building which had been converted from an ancient horse stable. He died on 11 May 1967 in Buffalo, New York at age 80.

Joseph Saul

Joseph Saul was born in Lancaster, New Hampshire, the son of Fred Saul. On 10 June 1917 Saul enlisted in France's *Service Aeronautique*. On 12 July 1917 he was released for disciplinary reasons. He returned to the United States that same month.

HAROLD SAXON

Harold Young Saxon was born in Atlanta, Georgia on 5 December 1894, the son of Thaddeus Glenn Saxon, a furniture merchant, and Alla (Pollard) Saxon. Harold's parents separated when he was an infant, and Alla Saxon moved with her two small children to the Indian reservation at Carlisle, Pennsylvania where she took a teaching job. Alla's brother, Eldon Pollard, was a U.S. Cavalry trooper there. When trooper Pollard was transferred out West, Mrs. Saxon and her children accompanied him. When she contracted tuberculosis, she left her job and moved to New York City. She died in Washington, D.C. in 1918.

Harold attended the Cathedral School of St. Paul on Long Island. Also known as St. Paul's, it was an Episcopal prep school. He later enrolled in Bowdoin College, Brunswick, Maine.

"Hally" Saxon left Bowdoin College to join the American Ambulance Field Service. He sailed for France, and on 10 June 1917 he enlisted in France's *Service Aeronautique*. From 21 June 1917 to 18 January 1918 he attended the aviation schools at Avord, Pau, and the *G.D.E.* He received his *brevet* on the Caudron on 14 November 1917.

From 21 January 1918 to 17 June 1918 *Caporal* Saxon served at the Front with *Escadrille SPA 31*. From 17 June to the 11 November Armistice *Sergent* Saxon flew combat patrols with *SPA 12*. Of his war service, one colleague said: "Saxon was in the air at every opportunity, and fought through every important battle on the Western Front during 1918." He was credited with the destruction of two enemy aircraft. For his dedicated and diligent service, Saxon was awarded France's *Croix De Guerre*, with two Palms.

Following World War One Saxon journeyed West to Hollywood, California where he worked in the movies. He eventually returned to New York City to work for the Washington Dean Company. He later relocated to Pittsburgh, Pennsylvania where he worked as a door-to-door salesman selling carpet sweepers and vacuum cleaners. He married Dorothy Cannon of that city. Saxon died of cancer in Pittsburgh on 26 October 1965 at age 70.

LAWRENCE SCANLAN

Lawrence Scanlan was born in Cedarhurt, Long Island, New York, the son of Mr. & Mrs. Michael Scanlan. Scanlan attended Seton Hall College, South Orange, New Jersey, to study electrical engineering. He left school in the fall semester and sailed for France where he enlisted in the Foreign Legion in November 1914. Scanlan was wounded at Souches on 16 June 1915. He later described the incident:

"We all left the advanced trench together, but I soon found myself in the Greek battalion, and with it I reached the first German trench. Some distance away I saw my captain and other Americans and I started over to them. Then machine gun bullets hit me, smashing my thigh and an ankle. After lying fifty-six hours on the battlefield, I was found by stretcher-bearers and carried to the rear."

While in hospital, "Red" Scanlan was awarded France's *Croix de Guerre*, being the first American volunteer to receive the honor. He was cited for his exceptionally bravery and for the gallant conduct he demonstrated on 9 May and 16 June 1915.

Scanlan spent the next year and a half in the hospital. On 1 January 1917 he was invalided out of the Foreign Legion, with one leg six inches shorter than the other. Scanlan headed directly to Paris to enlist in France's *Service Aeronautique*.

Scanlan tried valiantly to become an aviator, but he was severely hampered by the war injuries to his leg. While he was making his trial flights for his *brevet* he was involved in a spectacular accident at the Avord training camp. Scanlan lost control of his aircraft and crashed it through the Avord bakery. The soldiers who had been mixing bread inside the building fled outdoors. Scanlan, uninjured, extracted himself from his wrecked plane and left the building covered with flour.

The commander of the aviation school mistook Scanlan for a baker and ordered him back into the building, shouting:

"Come into the bakery and help get the body of the dead pilot there!"

Replied Scanlan: "I am the dead pilot, Captain."

Scanlan's wounded leg caused him much pain. In September 1917 Scanlan was forced to give up aviation and return to the United States.

Following World War One Scanlan operated a wholesale slaughter-house in New York State. In late October 1921 his war wounds became inflamed and infected. He was rushed to the hospital and his leg was amputated. Eight successive operations were performed on the leg in a period of a month in an effort to stop the infection. Finally, too weakened to resist, Scanlan died in the hospital on 25 November 1920.

EDWIN SCHREIBER

Edwin Booth Schreiber was born in Zurich, Switzerland on 22 January 1892, the son of Dr. H.W. Schreiber of Hopewell Junction, New York, and Ellen (Booth) Schreiber. Edwin Schreiber attended schools in Switzerland. He was working at the mining smelter at Anaconda, Montana when he decided to enter the European War.

On 29 June 1917 Schreiber enlisted in France's *Service Aeronautique*. On 18 February 1918 he was released upon his own request for the purpose of taking a 2nd Lieutenant's commission in the U.S. Air Service. Schreiber was sent to the U.S. 8th Aviation Instruction Center at Foggia, Italy before being sent to the 3rd Aviation Instruction Center, Issoudun, France. On 8 August 1918 he was killed in a flying accident at Issoudun. He was 26 years of age at his death.

Schreiber was buried that same day at the 3rd A.I.C. Cemetery. In November of 1921 that cemetery was abandoned, and his remains were removed to the St. Mihiel American Cemetery at Thiaucourt, Grave 21-A-9. Schreiber had requested, if he should be killed, that his body remain where it was first buried. Thus, his remains were not transferred to the Lafayette Flying Corps Memorial in 1928. His name does not appear on the list of the Corps's honored dead.

Edwin Schreiber (seated front row) on his way to France.(Back l.r) Tommy Hitchcock, Walter York, Wally Winter, David Guest, William Rodgers.

Horace Seaver

Horace Seaver was born in Hartford, Connecticut. Seaver attended Hartford Schools, and the Cooper Union School, New York, New York, Class of 1914, where he studied electrical and mechanical engineering. Seaver left school before graduating and worked as an elevator constructor.

On 10 March 1917 Seaver enlisted in France's *Service Aeronautique*. He was released on 13 August 1917 because of injuries he received in a flying accident.

WALTER SHAFFER

Walter John Shaffer was born in Dauphin, Pennsylvania on 25 February 1892, the son of Charles E. and Clara May (Davidson) Shaffer. Shaffer graduated from Dauphin High School in 1909 and worked in the printing trade until 1917. In his spare time Shaffer took civilian flight training and learned to fly in a pusher seaplane at Essington, Pennsylvania in 1915. When War broke out in Europe he lacked the funds to travel to France. He borrowed the passage money from Lewis Gross, his fiance's father.

On 1 August 1917 Shaffer enlisted in France's *Service Aeronautique* in Paris. From 1 August to 28 December 1917 he attended the aviation schools at Avord, Tours, Pau, and the *G.D.E.* He received his *brevet* on the Caudron on 7 October 1917.

From 1 January to 1 June 1918 *Caporal* Shaffer flew with *Escadrille SPA 156*. From 1 June to 3 October 1918 *Sergent* Shaffer flew with *SPA 38*. While flying with that squadron he was credited with the destruction of two enemy aircraft: a balloon on 4 August 1918, and a pursuit plane on 21 August 1918. On one patrol, during an encounter with Richthofen's squadron, Shaffer had the back of his airplane seat shot away.

On 10 October 1918 *Sergent* Shaffer was himself shot down southeast of Laon in occupied Belgium while attacking an enemy balloon. He was captured and held a prisoner in Germany. He managed to escape but was recaptured three days later and was held in a prison hospital in Namur, Belgium until the Armistice. For his heroic war service, *Sergent* Shaffer was awarded France's *Medaille Militaire*, and the *Croix de Guerre*, with three Palms. On 8 January 1918 he was discharged from the French Army.

Following World War One Shaffer continued in aviation and barnstormed the country in Curtiss Jennies from 1919-1926 as a test pilot and flying instructor. In 1926 he began his career as an airmail pilot. On 19 March 1927 he was commissioned a 2nd Lieutenant in the U.S. Army Air Corps Reserve. On 6 August 1927 he married Gertrude Gross of Brooklyn, New York. From this union two daughters were born.

From 1927–1929 Shaffer served as night airmail pilot for Pitcairn Aviation of Philadelphia, Pennsylvania. In this capacity he pioneered night flying with instruments on the New York to Washington run.

In 1929 Shaffer became one of Eastern Air Lines' original pilots, then called Eastern Air Transport Inc. of Brooklyn, New York. He flew night mail and also an 18-passenger plane.

Shaffer returned to active duty in World War Two, serving with the U.S. Army Air Corps in Australia and in the Philippines as chief pilot of the Southwest Pacific Wing of the Air Transport Command.

Following the War Major Shaffer returned to Eastern Air Lines until his retirement in 1952 after 24 years and 4 million miles without an accident as an airline pilot. Shaffer died in Atlanta, Georgia on 27 February 1974 at age 82.

WALTER SHIPLEY

Walter B. Shipley was born in Page, West Virginia, the son of Douglas S. Shipley. Shipley attended the University of Virginia, Charlottesville. He was involved in business when he decided to enter the European War.

On 9 June 1917 Shipley enlisted in France's *Service Aeronautique*. He received his *brevet* on 10 December 1917.

On 18 March 1918 Shipley was assigned at the Front to *Escadrille SPA 67*. On 22 March 1918 he was released at the Front.

CLARENCE SHONINGER

Clarence Bernard Shoninger was born in Chicago, Illinois on 9 October 1892, the son of Joseph Shoninger, Vice-President of the B. Shoninger Company, the piano manufacturing firm, and Tillie (May) Shoninger. Mrs. Shoninger died in 1897, when Clarence was five-years-old. His father died in 1908.

Clarence Shoninger attended the New York Military Academy and the Columbia Grammar School. He entered Yale University in 1912 to study Law. He graduated from that institution in 1915. Shortly after graduation he took a position in the credit office of B. Altman & Company. He left that firm in June of 1916 to serve in France with the American Ambulance Field Service.

Shoninger served as an ambulance driver with SSU 8 for six months at Verdun while attached to the 16th French Army. On 24 May 1917 he enlisted in France's *Service Aeronautique*. From 5 June 1917 to 20 February 1918 he attended the aviation schools at Avord, Juvisy, Pau, and the *G.D.E.* He received his *brevet* on the Caudron on 26 November 1917.

From 22 February to 29 May 1918 *Caporal* Shoninger served at the Front with *Escadrille N.99*, patrolling with *Groupe de Combat 20* above Lorraine. On 29 May 1918, while trying to drive off enemy *reglage* aircraft, he was jumped by several Albatross fighters and was shot down over German lines. Shoninger crashed into a German antiaircraft battery near Crugny. Wounded, he was taken prisoner and was evacuated to an enemy hospital. *Caporal* Shoninger was eventually sent to the prison camp at Karlsruhe where he was held until 29 November 1918.

Shoninger made his way back to Paris, and on 19 January 1919 he was discharged at the rank of *Sergent* from the French Army. He was awarded France's *Croix de Guerre* for his diligent war service.

On 20 October 1920 Shoninger married Ruth Dorn of Flushing, New York. From this union a daughter, Ruth, was born. Shoninger was employed by a number of banking firms, including the Manufacturer's Trust Co., working in New York State and in Connecticut. He also worked as a freelance writer doing stories for such aviation magazines as *Flying Aces*.

Shoninger entered the U.S. military during World War Two. From 12 April 1942 to 16 June 1944 Shoninger served in the Army Air Corps in England as an intelligence officer, briefing American pilots on how to act if taken a prisoner of war. Major Shoninger retired to Coral Gables, Florida. He died of heart disease at the University Hospital in New York City, New York on 1 February 1956 at age 62. He is buried at the military cemetery at Farmingdale, Long Island, New York.

REGINALD SINCLAIRE

Reginald Sinclaire was born in Corning, New York on 30 August 1893, the son of Henry Purdon Sinclaire, co-founder of the Corning Glass Co., and Annie (Watson) Sinclaire. As a youth he attended St. Martin's prep school and St. Paul's School, Concord, New Hampshire, before enrolling in Yale University in 1912. Sinclaire left Yale in 1916 and returned to Corning, New York to work in the family glass works there.

In April of 1917, following America's declaration of War with Germany, Sinclaire attempted to join the aviation branch of the U.S. Signal Corps. When he was denied a commission because he had not completed his studies, Sinclaire decided to join the Lafayette Flying Corps.

On 15 June 1917 Sinclaire enlisted in France's *Service Aeronautique*. From 20 June to 2 December 1917 he attended the aviation schools at Avord, Pau, and the *G.D.E.* He received his *brevet* on the Caudron on 2 October 1917.

From 7 December 1917 to 4 October 1918 Sinclaire flew at the Front with *Escadrille SPA 68*, *Groupe de Combat 20*. Sinclaire flew 167 combat sorties and was involved in 30 aerial combats. He had 3 confirmed victories, and unofficially shot down 8 other enemy aircraft. He never lost a plane. For his courage in combat, *Adjudant* Sinclaire was awarded France's *Croix de Guerre*, with three Palms.

Following World War I, Sinclaire returned to New York. He married Jean Hamilton. A son and daughter were born of this union. In 1921, when his wife contracted tuberculosis, Sinclaire moved her to the healthful atmosphere of Colorado Springs, Colorado where he purchased the 6000-acre Plum Creek Ranch in order to raise thoroughbred horses.

When not running his ranch, Sinclaire played polo and collected and flew aircraft. He participated in the 1928 National Air Races in Los Angeles, California, finishing second. He continued to fly twin-engine Beechcraft airplanes, holding his pilot's license until 1973.

During World War Two Sinclaire served in the U.S. Navy from April 1942 to October 1945 as Officer in Charge of Free Gunnery Training in Hawaii, and in the continental U.S. He retired at the rank of Commander. Sinclaire died in Colorado Springs, Colorado on 15 February 1989 at age 95.

Glenn Sitterly

Glenn Nelson Sitterly was born in Spring Valley, Illinois on 22 August 1897, the son of J.C. Sitterly, owner of a Spring Valley gravel company, and Mrs. Sitterly. Sitterly graduated from Hall Township High and Vocational School in Spring Valley on 31 May 1915. In September 1916 he entered Beloit College, Beloit, Wisconsin, a member of the Class of 1920. Sitterly left Beloit after one year and enlisted at 17 years of age in the American Ambulance Service. Sitterly enlisted in France's *Service Aeronautique* on 31 May 1917. From 12 June to 24 March 1917 he attended the aviation schools at Avord and the *G.D.E.* He received his *brevet* on the Caudron on 22 September 1917.

On 26 March 1918 *Caporal* Sitterly was assigned at the Front to *Escadrille C. 46*. He flew a number of intense combat missions with that squadron until 20 August 1918. In July, his machine was brought down in flames near Villers-Cotterets, but Sitterly and his two *mitrailleurs* managed to escape the flames uninjured.

On 15 October 1918 *Sergent* Sitterly was assigned to *Capitaine* Madon's chasse squadron *Escadrille SPA 38*. He flew numerous combat sorties and scored his second confirmed victory, over an enemy two-seater. *Adjudant* Sitterly was awarded France's *Croix de Guerre*, with Palm for his heroic war service. Of that service, one colleague has written:

"Sitterly is reticent about recounting his experiences, and it is only by seeing his uniform, torn by shrapnel and bullets, that one can realize the narrow escapes he has had."

Following World War I Sitterly served as Sales Manager then President for the Western Sand & Gravel Company in Spring Valley, Illinois. He married, and from this union a son and daughter were born. In early June 1944 Sitterly was erecting a swing set for his children. While testing it, he fell and struck his head. This resulted in his untimely death on 8 June 1944 at the age of 46.

Samuel Skinner

Samuel Wiggins Skinner, Jr. was born in Cincinnati, Ohio on 17 June 1891, the son of Samuel Wiggins Skinner, and Elizabeth (Jones) Skinner. Skinner attended Cincinnati public schools and the Groton School. He entered Harvard College in 1911 and graduated in 1915. Skinner then entered business in Cincinnati with the Central Trust and Deposit Company.

In early 1917 Skinner prepared to enter the European War. He received his pilot's license at the Curtiss Flying School, Newport News, Virginia. In May of 1917 he sailed for France.

On 31 May 1917 Skinner enlisted in France's *Service Aeronautique*. He received his *brevet* on 26 July 1917. On 16 October 1917, while awaiting assignment at the Front, *Caporal* Skinner shot and killed himself at the *G.D.E.* He was 26 years of age at the time of his death.

ROBERT SOUBIRAN

ESCADRILLE LAFAYETTE

Robert Soubiran, the 18th volunteer of the *Escadrille Americiane*, was born in Avallon, France on 16 March 1886, the son of Theodore Soubiran and Clementine (Malapris) Soubiran. Robert's mother died shortly after his birth. His father brought him to the United States in 1890 when he was four years of age.

Soubiran had no opportunity for formal education, as he was forced to begin work in his youth selling newspapers on New York City street corners to supplement the family income. He became a naturalized U.S. citizen on 10 August 1898 through his father's naturalization on that same date. In early adulthood he demonstrated an uncanny ability toward mechanics and began maintaining and racing automobiles with Ralph de Palma.

Robert's ties to France drew him there in its time of peril. As early as 7 August 1914, three days after Germany's declaration of war against France, Soubiran embarked for Europe to fight for the Republic.

Soubiran enlisted in the Foreign Legion in Paris on 21 August 1914 with the first group of forty-three Americans. On 25 August he left that city to begin training at Rouen. Following further training at Camp de Mailly he served in the Chemin des Dames sector with James Bach, Bert Hall, Kiffin and Paul Rockwell, and William Thaw. However, his initial front-line service was brief. When the French became aware of his mechanical abilities, he was ordered to drive a threshing machine to help to harvest the fall wheat still standing in the war zone.

Through his warm congenial nature and his fluent command of French, Soubiran made many friends among the villagers where his unit was on *repos*. He whittled bread and drank *pinard* "with the natural ease of an old *poilu* reservist."

Beginning on 25 September 1915 Soubiran served with the 2nd Foreign Regiment in the Battle of Champagne. The 2nd Regiment was continually under bombardment or in the throes of battle. When the opportunity arose for transfer to another regiment thirty-five kilometers behind the line, Soubiran, Paul Pavelka, William Dugan, and a number of his American comrades seized the opportunity to join the 170th Line Infantry Regiment.

The Americans soon realized their mistake. Their new regiment, an attacking regiment, had participated in every major engagement since the declaration of war and was only briefly recuperating from the awful losses it had suffered in the Champagne offensive. As a result of its service and great courage, the unit had been continually decimated and was in constant need of replacements. The American Legionnaires helped to fill the void then were marched back to the very trenches they had just left, there serving as replacements for

Sergent Robert Soubiran in the cockpit of a 50 h.p. Morane-Saulnier scout.

their fellow Legionnaires who were then being withdrawn from the line.

Shortly after occupying the forward trench, on 19 October 1915 Soubiran caught a shell splinter in his knee which sent him to the hospital. He spent the next four months recuperating. While he was there, he made his application to join the *Service Aeronautique* which was furthered along through the efforts of the Rockwell brothers and others. Kiffin wrote the following to his brother, Paul, in a 6 November 1915 letter.

"Give Bob S.(Soubiran) my regards. I would like to have him with me, but if it is worked it will have to be done from Paris. The thing to do is to state that Bob is an expert mechanic and that as such he wants to go in the aviation, and that he wants to go to the front as my mechanic, we both being Americans and friends and my having absolute confidence in his ability."

Following Soubiran's release from hospital, on 27 February 1916 he entered French Aviation as an *eleve pilote*. Soubiran received his first aviation training at the school at Pau. He received his *brevet militaire* on the Caudron on 27 May and attended further aviation schools at Buc and the G.D.E., completing his instruction on 20 October 1916. Two days later, on 22 October, he joined the *Escadrille Americaine* at Cachy in the Somme sector with former flying-school comrades Willis Haviland and Frederick Prince.

Soubiran brought along in his kit a folding Kodak camera and film. This proved historically invaluable, the significance of which fellow *Lafayette* pilot James Norman Hall would later note:

"The pilots in the Escadrille Lafayette have Soubiran to thank for the only complete photographic record which they have of their life at the Front. He was present with his camera at every ceremonial in which the Squadron took part. He snapped every crash, every bizarre accident at the aerodrome, from one sector to another, squadron parties at Bar-le-Duc, Villers-Cotteretts, Nancy, Soissons, Dunkirk, Chalons, Epernay-photographs which are priceless to them now."

"Bobby" Soubiran quickly proved a dedicated, willing pilot who flew more than his share of patrols. In his eagerness to do his part he sometimes paid a price; such as when he left Cachy field into a heavy mist in the winter of 1916 only to be forced groundward. Upon landing on Cachy's frozen furrowed surface, he groundlooped his Nieuport with such force that he snapped his collar bone.

Harold B. Willis, who served nearly six months with Soubiran in the Squadron, and who flew a number of patrols with him, later remembered him with fondness:

"Robert Soubiran was a great fella. A cheerful and a very valuable fella, and a lot of fun. He was a great flyer. And he stuck by you in the air."

The Lafayette pilots sought various methods of staying warm during the bitter winter of 1917 at Ravenel. Soubiran developed one which also provided the Squadron with food. As *Capitaine* Thenault later recalled:

"To warm himself Soubiran used to go hunting the numerous partridges and hares of the neighbourhood and it was not uncommon to see him suddenly dash breathless into the shed, hastily hide his gun, plunge into the bed clothes and appear to be sound asleep. That meant that the forest rangers were after him hot and heavy, for hunting was forbidden. But our Chef did make such delicious jugged hare!"

American trainees at Pau Aviation School, February 1916. (l.r.) Soubiran, Rumsey, Pavelka, Prince, McConnell, Haviland, Robert Rockwell.

Sergent Soubrian before his Nieuport fighter preparing for a patrol over the Somme. Cachy, December 1916. Notice the first style Seminole Indian head which was used as the Squadron's insignia from October of 1916 to April of 1917.

Another service that Soubiran performed well was in helping *Capitaine* Thenault recover the Squadron's downed aircraft. With his mechanical abilities, and his knowledge of the Front, he was able to locate and often repair his colleagues planes so that they could be recovered to the their aerodrome.

On 4 July 1917 Soubiran was chosen to convey to Paris the United States flag which had been carried in August 1914 by the American volunteers in the Foreign Legion. These men had written their names upon the banner and it was to be presented to the French government. It had been entrusted in his care in the spring of 1915 upon the death of Rene Phelizot. Soubiran, in the presence of those American Legionnaires and *Lafayette* pilots who had been granted special leave for the occasion, transmitted the flag to the officials of France in the court of honor of the *Hotel des Invalides*. It was then placed in the Hall of Honor among other battle-stained Legion banners.

On 9 November 1917 *Adjudant* Soubiran received France's *Croix de Guerre*, with Palm for his action on 17 October when he received his first and only confirmed victory over an enemy aircraft. His citation states:

"An American enlisted since the beginning of the war in the Foreign Legion, where he took part in the combats in the Aisne in 1914 and in the Champagne attacks in 1915. Wounded 19 October 1915. Passed into the Aviation, he showed himself an excellent pilot, fulfilling with remarkable ardor the missions confided to him. 17 October 1917, while protecting an attack on Drachens, forced an enemy to land out of control."

Adjudant Soubiran was commissioned a captain when the *Lafayette Escadrille* was taken over by U.S. Aviation on 18 February 1918 to become the U.S. 103rd Pursuit Squadron. He flew frequent combat patrols in his SPAD S.XIII fighter which displayed his personal insignia, a diamond-framed Indian head.*

While Captain Soubiran was on leave in Cravant visiting his aunt, he met his future bride-to-be, Mademoiselle Ann-Marie Choudey. Ann-Marie had been orphaned at age nine and had been adopted by Robert's aunt. At the time of their meeting, she had been teaching school in France. She and Captain Soubiran were married on 20 August 1918 in Langres, France with many of his Legionnaire and *Lafayette* comrades in attendance.

Prior to his wedding, on 29 July 1918 Captain Soubiran was transferred from the U.S.103rd to duty as the Operations Officer of the 3rd Pursuit Group. He remained there until ordered back to the squadron as pilot and commanding officer on 18 October 1918, there to succeed fellow *Lafayette Escadrille* pilot, Captain Robert L. Rockwell. Soubiran served in that capacity until the November 11 Armistice.

On 9 April 1919 France made Captain Soubiran a *Chevalier de la Legion d'Honneur*. He was also awarded a second Palm to his *Croix de Guerre*. His 17 May citation reads:

"American citizen enlisted in August 1914 in the Foreign Legion. Was distinguished in the infantry (wounded in September 1915), then as a pilot in l'Escadrille Lafayette where he showed the highest qualities of courage and audacity."

Captain Soubiran accumulated over four-hundred hours of combat flying in his twenty-three months with the French Air Service and in ten months with the U.S. Air Service. Few Americans would match his war record and sacrifice. He was mentioned in the General Orders of the Commander of the First U.S. Army Air Service. General Orders No. 2 spoke of his "meritorious service", of how he had "courageously and ably fulfilled the duties of Flight Commander, Squadron and

---

** Soubiran's personal insignia evolved as he changed aircraft. It began as a dark serpentine S on the fuselage of his Nieuport 17 C.1 preceded by a dark band and the Seminole-type Indian head. His SPAD S.VII first displayed a dark RS within a dark diamond frame followed by a Sioux-type Indian head. This was changed to a white S on a dark diamond background followed by the Sioux Indian head. Finally his SPAD S.XIII carried no initials, but a diamond-framed Sioux Indian head with squadron commander's stripes running fore and aft of the diamond.*

The wedding of Captain Robert Soubiran and Ann-Marie Choudey, Langres, France, 20 August 1918. (Below) Sergent Soubiran with his mechanics before his Nieuport scout, Cachy, October 1916.

Group Operations Officer, and Commanding Officer of the 103rd Aero Squadron."

G.O. No. 17 expressed his Wing Commander's congratulations to him and his personnel, stating that "no other organization in the American Army had a right to such a high measure of satisfaction in feeling its difficult task had been performed."

Following his overseas service, by then Major Soubiran was stationed at Langley Field, Virginia. Unable to adjust to peacetime service, he left the Army after one year. His daughter, Elizabeth, was born in Washington, D.C. in 1920, the year he returned to Paris, the birthplace of his daughter, Rosemary, to sell refrigerators for the Delco Manufacturing Company. Soubiran proved a natural salesman and traveled extensively in this capacity in Eastern Europe and in Morocco.

Soubiran moved back to Paris, his mother's city, in 1928, to work for General Motors. In 1929 a third daughter, Jacqueline, was born in Auxerre. Shortly after that date Robert lost his position with General Motors when the company was forced to reduce its overseas staff. With few opportunities open in France, Soubiran returned to the United States in 1931. For a time he managed Frederick H. Prince's "Princemere" estate in Essex County, Massachusetts.

From 1939–1940 Soubiran worked for the British Purchasing Commission in New York. He was then hired by Lafayette comrade Carl Dolan and placed in the inspection department of his Wissahickon Tool Works. When the U.S. entered World War Two Soubiran became an aircraft inspector for aviation parts plants in Norwich and Hartford, Connecticut. He then worked for the War Assets Administration for two years and finally for Republic Aircraft in Farmingdale, New York until he developed signs of a heart condition and was forced to take medical retirement.

Soubiran spent his final days in Baldwin, New York showing his famous *Lafayette Escadrille* photographs to interested historians. He died after a brief illness in St. Albans Naval Hospital, Queens, New York on 4 February 1949 at age 62.

Mrs. Robert Soubiran passed away on 10 March 1982. Prior to her death she transferred her husband's historic photo collection to the National Air and Space Museum at Washington, D.C. so that all researchers and historians might have the pleasure of studying their content.

WALLACE SPEERS

Wallace Carter Speers was born in Bloomfield, New Jersey on 7 November 1896, the son of James M. Speers, co-founder of New York's James McCutcheon department store. Speers graduated from the Hill School, Pottstown, Pennsylvania, and attended Princeton University, Class of 1918. He left school while in his junior year, and on 26 April 1917 he joined the American Ambulance Field Service in New York City. On 5 May he sailed for France aboard the *Espagne*.

On 19 July 1917 Speers enlisted in France's *Service Aeronautique*. On 1 August 1917 he was released because of ill health. He returned to the United States to recuperate. On 2 May 1918 he enlisted as a private with the 102nd Ammunition Train, 27th Division at Fort Slocum, New York. In June of 1918 he sailed for France and served at the Front with the 33rd and 79th U.S. Divisions. He returned to the United States in March 1919 and was discharged on 3 April.

Following World War One Speers married Lila Kelsey. From this union four children were born. Speers worked for the Crucible Steel Co. of Harrison, New Jersey. In 1923 he joined his father's company in New York City as a buyer of table linens. He eventually became a department manager and director of industrial relations. When the firm closed in 1957, Speers became an account executive for the Norm Advertizing Agency of New York.

Speers was also active in the Presbyterian Church. He was involved in fund raising for that institution, and was the founder in 1942 of the Laymen's Movement for a Christian World. He also served as a member of the Montclair Board of Education. He died of a heart attack in his Montclair, New Jersey home on 31 March 1963 at age 66.

DUMARESQ SPENCER

Dumaresq "Stuffy" Spencer was born in Chicago, Illinois on 4 December 1895, the son of Earl W. Spencer, and Agnes (Hughes) Spencer. He attended the Phillips Academy, Andover, Massachusetts. He graduated from that institution in 1913. He then enrolled in Yale University, Class of 1917.

During the summer of 1916 Spencer served with the Yale Battalion at Tobyhanna, Pennsylvania. On 6 April 1917 he enrolled with his good friend, George Moseley, as a Carpenter's Mate 3d Class in the Aviation Section, First Battalion, New York Naval Militia. He took aviation training at Bay Shore, Long Island, New York. He was discharged in June and on the 20th of the month he sailed for France to join the Lafayette Flying Corps.

On 10 July 1917 he enlisted in France's *Service Aeronautique*. From 18 July to 24 December 1917 he attended the aviation schools at Avord, Tours, Pau, and the *G.D.E.*, receiving his *brevet* on the Caudron on 21 October 1917.

From 27 December 1917 to 22 January 1918 *Caporal* Spencer served at the Front with Escadrille *N. 150*, the day he was killed in an aeroplane accident near Belfort. Returning to his airfield from a combat patrol, he had left his formation to empty his guns into a target on a lake near the aerodrome. While making a sharp turn near his target, he lost control of his plane and crashed to earth. The hood of his machine struck him just below his eyebrows, instantly killing him. He was 23 years of age at the time of his death.

Spencer was buried at Belfort on 25 January 1918, the same day his commission as an Ensign in U.S. Naval Aviation had arrived from Paris. He was posthumously awarded France's *Croix de Guerre*, with Star for his 19 January combat against a group of enemy aircraft. In 1928 his remains were transferred to the Lafayette Flying Corps Monument at St. Cloud.

ALFRED STANLEY

Alfred Holt Stanley was born in Elmira, New York on 23 December 1896, the son of Rufus Clark Stanley, a social reformer and co-founder of the 4H Club, and Charlotte Ellsworth (Rose) Stanley. Stanley attended Elmira public schools and the McKenzie School in New Jersey. In 1914 Stanley became involved in commercial aviation, working as an engine tester at the Curtiss Aeroplane Company at Hammondsport, New York.

When the United States entered the European War, Stanley tried to join the U.S. Air Service. When he was unsuccessful, he heard of the Ambulance Service in France, and he enlisted in that organization to secure overseas passage. Stanley joined France's *Service Aeronautique* in Paris on 12 April 1917. From 2 June 1917 to 22 February 1918 he attended the aviation schools at Avord, Pau, Cazeaux, and the *G.D.E.* He received his *brevet* on the Caudron on 13 November 1917.

From 24 February 1918 to the November Armistice *Caporal* Stanley was assigned at the Front to *Escadrille SPA 23*. Stanley flew heroically and brilliantly with *Capitaine* Pinsard's squadron and was credited with the official destruction of two enemy aircraft. On one patrol near Verdun his small *nourrice* petrol tank was set on fire by a stream of incendiary bullets from the rear guns of an enemy bi-place. Stanley managed to extinguish the blaze and bring his SPAD safely home in spite of a pierced crank case and a broken connecting rod. *Adjudant* Stanley, in recognition of his outstanding service to France, was awarded its *Medaille Militaire,*, and *Croix de Guerre*, with four Palms and two Stars.

While in Paris, during a zeppelin bombardment, *Adjudant* Stanley and Miss Laura Melvin had sought cover in the same doorway. Miss Melvin was working in Paris as an interpreter and ambulance driver for a YWCA medical unit. Several months after this chance meeting, Stanley and Miss Melvin were married in Paris on 14 August 1918.

Following World War One the Stanley's returned to New York State where a son and daughter were born. Stanley started a local airline in Elmira and also ran an auto agency. But his businesses were wiped out by the depression, and he barnstormed, did contract flying, and aerial photography. In the late 1920s Stanley began working for Edward Link, the inventer of the Link trainer. During their decade of association, Stanley taught Link to fly and helped him with instrument testing.

Alfred and Laura Stanley separated in 1938 and were divorced in 1942. Stanley continued to support himself in aviation-related positions. He died of a heart attack in Kirkwood, New York on 24 June 1961 at age 64. He is buried in Elmira, New York.

FRANK STARRET

Frank Elmer Starret, Jr. was born in Athol, Massachusetts on 16 July 1893, the son of Frank Elmer Starret, and Lena Dora (Hoehm) Starret. Starret graduated from Athol High School, and attended Worcester Academy, Worcester, Massachusetts, Class of 1912. He graduated from Brown University, Class of 1916.

Starret joined the Red Cross Ambulance Service on 1 November 1916. He served in the Verdun sector with Red Cross Section 5-646 and was awarded France's *Croix de Guerre*. On 19 July 1917 Starret enlisted in France's *Service Aeronautique*.

From 1 August to 3 January 1918 Starret attended the aviation schools at Avord and Tours. On that date he was making his *brevet* flight in a Caudron when his plane fell near Pontlevoy, and he was killed in the subsequent crash. Starret was 24 years of age at the time of his death. He was buried at the American cemetery at Tours. His remains now rest at the Lafayette Flying Corps Monument near Paris.

RUSSELL STEARNS

Russell Falconer Stearns was born in Pawtucket, Rhode Island on 4 January 1896, the son of Walter Henry Stearns, a successful textile manufacturer, and Abbie Harris (Razee) Stearns. Stearns was educated at the Moses Brown School, Providence, Rhode Island, and at the Lake Placid School, Lake Placid, New York, before enrolling at the University of Virginia, Charlottsville, Class of 1918.

On 8 January 1917 Stearns left the University of Virginia to join the American Ambulance Service. Following two months service at the Front, on 12 April 1917 he enlisted in France's *Service Aeronautique.* From 26 April to 24 December 1917 he attended the aviation schools at Avord, Juvisy, Pau, and the *G.D.E.* He received his *brevet* on the Caudron on 21 October 1917.

From 27 December 1917 to 24 February 1918 *Caporal* Stearns flew at the Front with *Escadrille SPA 150, Group de Combat 16.* Stearns had a high-strung personality. He hated flying and dreaded going into the air. Still, he flew a number of combat patrols and escorted bombers over the Alsace-Lorraine sector.

In May 1918 Stearns was granted a furlough and returned to the United States where he entered the U.S. Marine Flying Corps. He was sent to Florida as a 2nd Lieutenant where he served for two months. Shortly thereafter, Lieutenant Stearns was given an honorable discharge after suffering a nervous breakdown.

Following World War One Stearns's mental state continued to deteriorate. Upon the death of his father, he and his sister, Margaret, inherited the family fortune. Stearns was unable to care for himself, and he was assigned a male companion. Using some of his inheritance, Stearns bought a Buick touring car. He had a siren installed in the back seat. As his chauffeur drove him around Providence, he would sit in the back seat with a pet monkey and run the siren. Russell's sister, Margaret Stearns, devoted much of her life taking Russell throughout the European continent in an attempt to find a cure for his illness.

He died in a mental institution in Aiken, South Carolina on 10 January 1938 at age 42.

JOSEPH STEHLIN

Joseph Charles Stehlin was born in Sheeps Head Bay, Brooklyn, New York, on 29 January 1897, the son of Charles Stehlin, a beer wagon driver, and Sarah (Burr) Stehlin of England. As a youth he attended St. John's Academy.

Stehlin enlisted in the Lafayette Flying Corps on 19 February 1917. He received his *brevet* on the Bleriot on 23 June 1917. From 18 August to 2 October 1917 he flew at the Front as a *Sergent* with *Escadrille SPA 95*. He is credited with the destruction of one enemy aircraft on 7 September 1917 for which he was awarded France's *Croix de Guerre*, with Palm.

In January of 1918 Stehlin was to be released from the *Service Aeronautique* in order to join the U.S. Air Service. His papers were not completed when he was granted leave by France to return to the United States. Delays in the process of his paperwork caused him to remain in the United States where he became a speaker for the Liberty Loan Bond drives. Stehlin joined the U.S. Air Service on 1 July 1918. He subsequently received a commission and sailed for France on 30 October 1918. He reached Paris at the signing of the Armistice.

Following the end of World War One Stehlin fought with the Polish Army against the Bolsheviks. He then served in the White Russian Army, was awarded the St. Stanislas Cross, and flew in the Latvian Air Force. While he was in Europe he married a countess. From this union a son, Charles, and a daughter were born.

Stehlin and the Countess divorced, and Stehlin returned to the United States where he married the actress, Pearl White. They divorced, and he married Elizabeth Morton of St. Louis, Missouri. They had two children, Joseph, Jr., and Patricia. Stehlin was working for the National Lead Co. at the time of their marriage. The Stehlins moved to Florida which was the source of the sand used in processing lead. While there, he lost his position with his company and began developing real estate. He and Mrs. Stehlin named the Florida community of Ponte Vedra.

The Stehlins divorced and Stehlin married Dorothy Hoover of the Hoover Vacuum Industry. They also divorced and Stehlin married Jeanette Nelson and adopted her son, James.

During World War Two Stehlin served both in the Atlantic and Pacific Theaters, attaining the rank of Colonel in the Army Air Forces. Following World War Two he continued in the development of real estate and also ran a casino in Florida. He passed away in Miami, Florida on 4 May 1967 at age 70.

HENRY STICKNEY

Henry Elmer Stickney was born in Greenwich Village, New York on 8 January 1895, the son of Henry E. Stickney, a well-known Brooklyn cartoonist, and Mary (Sullivan) Stickney. Henry attended New York public schools. While still a youth, his parents died, and he was placed in the care of his uncle, William Stickney, former Governor of Vermont. He was educated at the Vermont Academy, Saxtons Rivers.

Stickney's interest in aviation dated to 1911 when he built and flew a 40-foot glider. In 1914 he flew at Turland, Maine in a plane he had assembled from Curtiss bi-plane parts. He studied aviation in Detroit, Michigan then returned to Rutland, Vermont to work as a hotel bellboy. At the outbreak of World War One Stickney attempted to enlist in the U.S. Air Service but was rejected because of his small stature. He enlisted in the Lafayette Flying Corps at Mineola, Long Island, New York. He sailed to France and On 21 July 1917 he enlisted in France's *Service Aeronautique* in Paris, becoming the smallest man in the Lafayette Flying Corps.

From 31 July to 2 December 1917 Stickney attended the aviation schools at Avord, Tours, Pau, and the *G.D.E.* He received his *brevet* on the Caudron on 3 October 1917. On 4 December 1917 *Caporal* Stickney was assigned at the Front to *Escadrille SPA 150* and flew with that squadron until 20 June 1918. On 3 June 1917 *Sergent* Stickney had been commissioned a First Lieutenant in the U.S. Air Service. As an American officer, he was reassigned to his old French Squadron and continued to fly with *SPA 150* until the 11 November Armistice. He was awarded France's *Croix de Guerre*, with Palm for his courage and aerial combats with the enemy. He was credited with the destruction of one German aircraft. Stickney was promoted to Captain at the time of his discharge.

Following World War One Stickney returned to the United States and worked as a flying instructor before joining the Canadian Allied Flying Corp., doing exhibition flying and carrying passengers at various fairs

Sergent Henry Stickney (right) with Caporal Russell Stearns at the Front with SPA 150, January 1918.

and celebrations throughout New York and New England. In 1920 he married Delia Black at Bellows Falls, Vermont. For a time he worked as a millwright for the Moore & Thomson paper company of Bellows Falls. Stickney eventually established his own flying school and an airport at Rutland, Vermont. He also established an airport at Springfield, Vermont, and ran the Osgood & Stickney Curtiss Aeroplane agency there. In 1923 he added another airport at Bellows Falls.

In October 1927 Stickney joined the New England Aircraft Corp. as manager and chief instructor at their Worcester, Massachusetts airport. He was seriously injured in an airplane accident in Massachusetts on 10 January 1928 when the student he was instructing stalled the plane then banked at an altitude of 50 feet. George Knowlton was killed outright. Stickney suffered numerous broken bones, and was incapacitated for two years. When he returned to flying he became a maintenance superintendent for Pan American Airways. Stickney spent two years with that firm in Texas before being transferred to Colon, Panama where he died of a stomach hemorrhage on 18 September 1934 at age 37.

DONALD STONE

Donald Edward Stone was born in New York City, New York, the son of Mr. & Mrs. Edward T. Stone. As a young man, Stone worked as a cattle-rancher in Mexico. In 1916 he sailed for Europe and joined the Norton-Harjes Ambulance Service. He served with Section 11 at the Front into 1917 and was awarded France's *Croix De Guerre*, with Star for his service.

On 8 April 1917 Stone enlisted in France's *Service Aeronautique*. From 15 June 1917 to 16 March 1918 he attended the aviation schools at Avord, Pau, Cazeaux, and the *G.D.E.* He received his *brevet* on the Caudron on 22 October 1917.

On 18 March 1918 *Caporal* Stone was assigned at the Front to *Escadrille* SPAD 12. He flew with that squadron until 21 April 1918, the day he was killed in aerial combat in a fight involving thirteen enemy aircraft. His remains were never found. He was posthumously awarded France's *Croix de Guerre*, with Star for the action which led to his death.

Fellow squadron mate, George Dock, later recorded the following after learning of Stone's death: "I went to our room, where Don's cot remained as he had left it that morning, and beside it on a table were his books, cigaret case and camera. In one corner of the damp room hung his clothes, among them a uniform that Don had bought the day before. I thought of the night, less than a month ago, when he had said "In a day or two maybe we'll think we deserve all this. Just now I feel like a four-flusher." On a chair was his diary, open at the last page, where I read this sentence. "I think the highest ambition of my life is to get a Boche."

GERALD STONE

Gerald Starr Stone was born in Spencer, Massachusetts on 9 March 1897, the son of Ralph Stone, a lumber company owner, and Sarah (Starr) Stone. Stone attended Worcester Academy, Worcester, Massachusetts, and Dartmouth College, Hanover, New Hampshire, Class of 1920. In the summer of 1916 Stone served with the 2nd Regiment at Columbus, New Mexico during the Mexican Border Campaign. Stone left Dartmouth College to join the American Ambulance Field Service. He served near Soissons from May to July 1917 with TMU 526 for 3 months, hauling ammunition to the Front.

On 2 August 1917 Stone enlisted in France's *Service Aeronautique*. He received his *brevet* on 29 October 1917. On 13 January 1918 he was released upon his own request while at the *G.D.E.* and returned to the United States. Stone joined the U.S. Naval Reserve Forces in June of 1918 and served at Hingham and Wakefield, Massachusetts until his discharge on 18 December 1918.

Following World War One Stone returned to Dartmouth College to complete his studies. He received a B.S. in chemistry in 1920 then worked in the Southwest as a roustabout and tool dresser in the oil fields. He married Frances Cooper of Atlanta, Georgia. From this union a daughter and son were born.

In 1927 Stone joined the R.H. Macy Co, of New York City as a buyer/manager of the china glass department. He left that company in 1955 to become the U.S. manager for Johnson Bros, Ltd. of England, a manufacturer of china and crystalware. He retired from that firm in 1962 and pursued his hobbies of fishing and travelling. Stone died in New York City on 18 November 1963 at age 66.

UPTON SULLIVAN

Upton Supple Sullivan was born in Philadelphia, Pennsylvania on 4 June 1891, the son of Jeremiah J. Sullivan, a banker and founder of the Philadelphia Electric Company, and Anna (Patterson) Sullivan. Sullivan graduated from Phillips Academy, Andover, Massachusetts, and from Harvard University, AB 1913–14.

On 13 June 1917 "Sully" Sullivan enlisted in France's *Service Aeronautique*. From 25 June 1917 to 6 January 1918 he attended the aviation schools at Avord, Pau, and the *G.D.E.*, receiving his *brevet* on the Caudron on 16 November 1917.

On 8 January 1918 *Caporal* Sullivan was assigned at the Front to *Escadrille N.90* and served with that squadron at Nancy until 8 April.

On 8 April 1918 *Sergent* Sullivan received an honorable discharge from the French Army in order to join the U.S. Naval Aviation. On 12 April he was appointed an Ensign in the Naval Reserve Flying Corps at Paris. On 1 July 1918 he was made Naval Aviator #1031(HTA) at Moutchic and was assigned to the Naval Air Station at Stonehenge, England. He served with the British Royal Air Force for six months until the Handley-Page bomber he was piloting was involved in a severe accident. Sullivan spent several weeks on sick leave in hospital before being assigned to the Northern Bombing Group. He flew bombing missions with that unit until the 11 November 1918 Armistice. Sullivan returned to the United States and was discharged an Ensign in the Naval Reserve Flying Corps on 30 September 1921.

Following World War One Sullivan returned to Philadelphia. In 1919 he joined the staff of Cassatt & Co., a banking and brokerage firm. He married Hannah Wright in Philadelphia on 15 October 1921. From this union a daughter was born. Sullivan retired from the investment house in 1943 to devote himself to his horse farm in Malvern where he raised thoroughbred hunters and beagle hounds for the hunt. He was also involved in farming. Sullivan died at the Lankenau Hospital, Philadelphia on 24 December 1981 at age 91.

Leslie Taber

Leslie Ray Taber was born in Syracuse, New York on 15 April 1893, the son of William R. Taber, a railroad engineer, and Elizabeth (Morrow) Taber. Taber attended Brown University, Class of 1917 but left that institution his senior year to join the American Ambulance Service. Taber drove ambulances in France for 5 months with SSU 4. On 25 June 1917 he enlisted in France's *Service Aeronautique*.

From 19 July 1917 to 9 March 1918 Taber attended the aviation schools at Avord, Tours, Sacy-le-Grand, and the *G.D.E.* He received his *brevet* on the Caudron on 26 October 1917. From 11 March to 17 March 1918 *Caporal* Taber served at the Front with *Escadrille BR. 29*. He resigned from the French Army to join U.S. Naval Aviation.

On 19 March 1918 Taber enrolled as an Ensign in the USNR at Paris, France. On 23 May 1918 at Paris he became Naval Aviator #620(HTA). He trained at Moutchic, France, and Stonehenge, England. Ensign Taber was assigned to the Northern Bombing Group and made numerous bombing runs from Dunkirk and Eastleigh in his Handley-Page bomber against the German bases on the Belgian coast.

On 15 August 1918 Ensign Taber made the only successful U.S. Naval night-bombing raid of World War One when he flew a single-plane raid on the submarine repair docks at Ostend, Belgium. While a member of Air Squadron One based at St. Inglevert, France, he piloted a Caproni bomber and dropped one 550 and two 250 pound bombs, destroying a railroad factory. Taber was also credited with the Navy's longest flight when he flew a Caproni Bomber from Milan, Italy to Calais, France.

Ensign Taber was awarded the U.S. Navy Cross "For distinguished and heroic service as a seaplane pilot in which capacity he made flights for patrolling the sea and bombing the enemy coasts, showing at all times courage and a high spirit of duty."

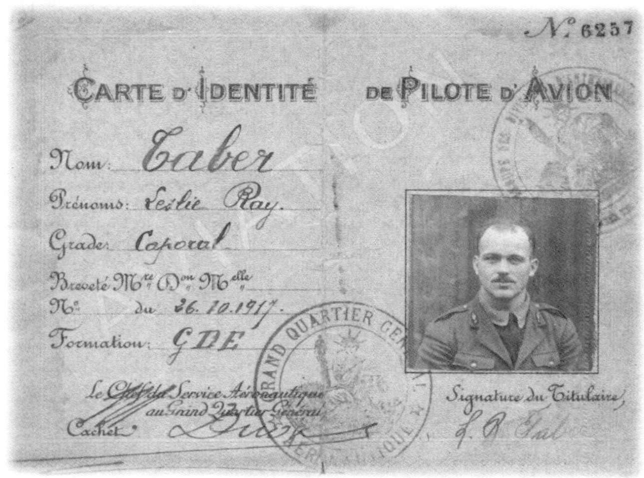

Taber concluded his war service as a test pilot in England. Following the Armistice he was stationed briefly at the air training station at Hampton Roads, Virginia. He married Orrell Bond in New Rochelle, New York on 22 December 1920. From this union three sons and a daughter were born.

Taber returned to Brown University to finish his studies for his Bachelor of Philosophy degree. He then enrolled in the Cornell Medical College and became an MD in 1924. He interned as a physician and surgeon at the Paterson, New Jersey General Hospital where he co-founded the Lendrim Tumor Clinic, an early user of radium treatment for cancer. Taber became Chief of Surgery at the hospital, retiring in 1958. He died on 21 October 1963 of arteriosclerosis at the Valley Hospital, Ridgewood, New Jersey at age 70.

WILLIAM TAILER

William Hallett Tailer was born in New York City, New York on 3 February 1895, the son of Henry Pennington Tailer, president of Banker's Trust Company of New York City, and Clara (Wright) Tailer. Tailer attended the Newman school and grew up at the family estate on the shore at Hempstead Harbor in Roslyn, Long Island. The Tailer property was considered one of the showplaces of that community.

As a young man "Billy" Tailer worked for his father's Wall Street brokerage firm and in his free time built a reputation as a tennis player, golfer, and horseman among the sporting colony of Nassau County. His carefree days were interrupted in 1916 when Tailer served as a sergent with New York's 7th Regiment at McAllen, Texas on the Mexican Border. While with that unit he won a silver marksmanship trophy.

Following his Mexican Border service, Tailer returned to New York City to a position with his father's Banker's Trust Company In March of 1917 when the United States entered into World War One Tailer took aviation training at the U.S. Army flying field at Mineola, Long Island in the hope of gaining a commission in U.S. aviation. Having completed the course, he left for Washington, D.C. to enlist but grew impatient while waiting for his commission and sailed for France in the hope of joining the Lafayette Flying Corps.

On 21 July 1917 Tailer enlisted in France's *Service Aeronautique.* From 31 July to 12 December 1917 he attended the aviation schools at Avord, Tours, Pau, and the *G.D.E.* He received his *brevet* on the Caudron on 10 October 1917.

On 14 December 1917 *Caporal* Tailer was assigned to *Escadrille SPA 67.* He flew with that squadron until 5 February 1918, the day he was killed by enemy antiaircraft fire while flying a patrol near Verdun. Those who witnessed Tailer's death saw his machine nose down and go into a *vrille.* At a thousand meters his plane shed its wings and crashed three kilometers inside Allied lines, instantly killing Tailer. He was 23 years of age at the time of his death.

On Memorial Day, 1918, a bronze tablet was erected in Tailor's honor at the clock tower in his home town of Roslyn. Eight-hundred of his friends and neighbors participated in the memorial ceremony. In 1928 his remains were re-interred at the Lafayette Flying Corps Memorial near Paris.

ELMER TAYLOR

Elmer Bowden Taylor was born in Cedar Grove, New Jersey on 21 November 1892, the son of Mr. & Mrs. Edwin Taylor. He attended Bloomfield New Jersey High School.

Shortly after the U.S. entry into World War One Taylor sailed for France. On 21 July 1917 he joined the *Service Aeronautique* at Paris. From 31 July 1917 to 31 January 1918 he attended the aviation schools at Avord, Tours, Pau, and the *G.D.E.* He received his *brevet* on the Caudron on 2 November 1917.

On 1 February 1918 *Caporal* Taylor was assigned at the Front to *Escadrille C.74*. He flew with that squadron until 1 April 1918, the date he was assigned to *SPA 102*. *Sergent* Taylor flew with that squadron until 6 April, when his transfer came through for U.S. Naval Aviation.

On 12 June 1918 Taylor enrolled at Paris as an Ensign in the Naval Reserve Flying Corps. On 29 July 1918 he was made Naval Aviator # 1037(HTA) at Moutchic, France. Tailor flew at the Front with the 9th Squadron, Northern Bombing Group from 15 June to 27 October 1918. On that date Ensign Taylor died of bronchial pneumonia at British Hospital #30, Calais, France. He was 25 years of age at the time of his death.

Taylor was buried in the British Military Cemetery, Les Baracques, Calais, France. He was posthumously awarded the U.S. Navy Cross "for distinguished and heroic service as a seaplane pilot in which capacity he made flights for patrolling the sea and bombing the enemy coasts, showing at all times courage and a high spirit of duty." In 1928 Taylor's remains were re-interred at the Lafayette Flying Corps Memorial outside Paris.

HUGH TERRES

Hugh Terres was born in Diquini, Santo Domingo on 6 September 1896, the son of John Terres, U.S. Consul at Port-au-Prince, Haiti, and Mrs. Terres. He attended St. Peter's College, Westminster, England. On 15 October 1915 he entered the University of Oxford. Terres left that institution before completing the examinations necessary to take a degree to join French aviation.

On 15 May 1917 Terres enlisted in the *Service Aeronautique*. From 26 May 1917 to 30 March 1918 he attended the aviation schools at Avord, Crotoy, and the *G.D.E.* While in aviation schools, Terres served as interpreter and intermediary between his American comrades and the French authorities. He received his *brevet* on the Caudron on 20 November 1917.

On 16 April 1918 *Caporal* Terres transferred to the U.S. Naval Reserve Flying Corps as an Ensign at Paris. On 25 May 1918 he was made Naval Aviator #632(HTA). Ensign Terres trained at Stonehenge, England, before being assigned to the seaplane patrol station at Le Crotie, France. He was later assigned to the Northern Bombing Group at Dunkirk.

In mid-August 1918 Ensign Terres was ordered to Italy to ferry a Caproni bomber from the factory at Turin to France. On 17 August, in the late afternoon, Ensign Terres and his two-man crew took off from Mirafiore Field in Caproni B-13 (#11523) for their flight across the Alps. At an altitude of 150 feet the port engine quit. As Terres attempted to return to the field, the right engine also quit. The bomber sideslipped and crashed nose-down at 5:15 p.m., killing Terres and his crew. Ensign Terres was 22 years of age at the time of his death. He was buried in a Milan cemetery. In 1928 his remains were re-interred at the Lafayette Flying Corps Monument near Paris.

William Thaw

ESCADRILLE LAFAYETTE

William Thaw, a founder of the *Escadrille Americaine*, was born in Pittsburgh, Pennsylvania on 12 August 1893, the son of wealthy business executive, Benjamin Thaw, and Elma Ellsworth (Dows) Thaw, an heiress and a prominent art collector. In his youth William Thaw attended the Browning School in New York City and the Hill School in Pottstown, Pennsylvania. He enrolled in Yale University in 1911 but left there following his sophomore year to pursue a career in aviation.

Thaw first learned to fly at the Curtiss School of Aviation at Hammondsport, New York under the instruction of Francis "Doc" Wildman in the summer of 1913. From a generous allowance bestowed upon him by his father he purchased a Fiat runabout which he used to shuffle himself and his fellow students from the boarding house where they stayed to the flying school grounds.

Thaw was an excellent student and completed the course of instruction on 8 August 1913. On 1 October he became a member of the Aero Club of America but evidently did not obtain a flying license. With funds from his father he purchased a Curtiss Model E Hydro flying boat, and upon delivery from the company, he had it shipped to Newport, Rhode Island to "Beachmound," the senior Thaw's summer home which overlooked the beach clubs there.

This large structure with a portico supported by Grecian columns had been named "the cottage" by Mrs. Thaw, and had been designed for Mr. Thaw by architect Stanford White who was later shot dead by Harry K. Thaw. Soon young William Thaw was carrying Newport's social elite on flights in his Curtiss aircraft out over Narragansett Bay.

Stephen MacGordon, Thaw's Curtiss School classmate, joined his friend in Newport. On 5 October Thaw and MacGordon flew the first of several trips from Newport to New York City, this time by way of New Haven, a distance of 230 miles in 220 minutes. Once there, they began transporting the social set as passengers above the city and made headlines by flying under the four East River bridges before circling the Statue of Liberty. They also attempted one of the first shore-to-ship deliveries by dropping a bundle of newspapers from Thaw's aircraft onto the German ship *Imperator* as she entered New York Harbor.

On 13 October the two aviators participated in the Round-Manhattan Race at Oakwood Height, Staten Island before heading for the warmer climate of Palm Beach, Florida in November. There, MacGordon purchased a flying boat and he and Thaw carried the resort's vacationers out over the sea, taking them "as high as they wanted to go" for $20. They also operated an aerial ferry, carrying visitors across Lake Worth.

In the spring of 1914 Thaw and MacGordon left the United States for Great Britain where they pur-

Thaw family at the Beachmound "cottage," Newport, R.I., July 1912. (Front row l.r.) William, Henrietta, Benjamin Sr., Elma, Alex Blair. (Back row) Stephen and Benjamin.

chased a Deperdussin monoplane which they intended to enter in the Schneider Trophy competition to be held at Monaco on the French Riviera. Thaw joined a field with seven other planes there on 20 April. But after watching Howard Pixton's Sopwith "Tabloid" perform over the course, Thaw and two other participants withdrew from the competition, knowing their aircraft would be unable to challenge the pace Pixton had set.

Thaw left for Paris with MacGordon and his younger brother Alexander Blair, where they intended to enter his 100-horsepower Curtiss Model R Hydro airplane in the *Securitie en Aeroplane* Competition. Blair Thaw had invented an automatic stabilizer for the Curtiss in his New York machine shop, and they hoped to win a portion of the $100,000 safety competition prize offered by the French Government. While in France Thaw secured Aero Club of France F.I. Hydro License No. 2 then took his plane in a tour along the Mediterranean, to Jaun-les-Pins and other resorts, fine-testing the stabilizer. Following this stint, he thereafter became known to his friends as "the playboy of the Riviera."

By the time the Thaws and MacGordon had returned to Paris for the pending competition, hostilities between France and Germany had forced its cancellation. Thereupon, Thaw had offered his services to the French *Service Aeronautique* but was turned down. In the early stages of World War One the French military had far less aircraft than men needed to fly them. However, Thaw and MacGordon were determined to serve France.

Thaw turned his Curtiss Hydro airplane over to the French Government for its use, then on 21 August 1914 he and MacGordon and forty-one other Americans marched to the Hotel des Invalides in Paris and enlisted in the French Foreign Legion as common infantry privates. Thaw was given L.M. no. 5,503.

On 30 September 1914 following four weeks of drill at Rouen and Toulouse Thaw and his American comrades left in boxcars for the Front. Thaw's first assignment in the Chemin des Dames Sector was to act as scout for his seventeen-man squad. This meant that with his company on the march Thaw had to walk three times as far and as fast as the moving ranks over rough terrain in order to warn of the nearness of any enemy concentrations. His American comrade Bert Hall remembered how Thaw's feet would swell up "like sausage balloons."

Thaw soon realized he could not maintain such a pace. And he still retained his hope and desire to enter French aviation. In early October 1914 while in combat with the Second Regiment he related to Paul Ayres Rockwell his hope of forming "a squadron of American volunteers" to fly for France. This marked the earliest date that any American had spoken of such a concept, a full month prior to Norman Prince's conversation with Frazier Curtis at Marblehead, Massachusetts when Prince had told Curtis of his dream of forming an all-American squadron of aviators.

Thaw, to this end, that same month of October had secured permission to hike the thirty-two kilometers to the aerodrome of *Escadrille D.6* to see if he could arrange a transfer to that same French squadron for himself, Bert Hall, and James Bach. It was there that Thaw made the acquaintance of senior pilot, Lieuten-

William Thaw with Stephen MacGordon, summer of 1913.

ant Brocard, later Commandant Brocard, Chief of French Air Service, who would play a key role in Thaw and Prince's efforts to form the all-American squadron.

In November James Bach received his orders for transfer to French aviation and on the 15th he left for Saint-Cyr to start his training. Thaw, however, did not receive his transfer and again trekked the thirty-two kilometers to *Escadrille D.6*. There, Lieutenant Brocard assured him his transfer orders were on the way and on 24 December 1914 he was ordered to join the squadron. Bert Hall left the trenches four days later, for the Pau flying school.

Before Thaw departed from his Legionnaire comrades, he assured fellow Americans Kiffin Rockwell and Charles Trinkard, among others, that he would work hard for their transfers into aviation. He then departed again for *Escadrille D.6* where he joined the roster as a *soldat-mitrailleur*, thus becoming the first American to actually serve in a French squadron at the Front.

Armed with a pistol and carbine, Thaw flew as an observer-gunner in a Deperdussin bi-plane on raids and reconnaissance missions over German territory until 1 February 1915. On that date he was sent to the aviation school at Saint-Cyr, having convinced the French *Service Aeronautique* that he was a qualified pilot.

Thaw trained and received his *brevet militaire* #714 on the Caudron G.2 on 15 March 1915, the first American to be so licensed by the French Government. He received further training at Buc and the R.G.A. (*Reserve General Aeronautique*) before being assigned to *C.42* on 26 March.

Thaw had run into Norman Prince in Paris on 20 March, six days prior to Thaw's assignment to the Front. Thaw had been using the period between the end of his training and forward assignment to press the cause of his American Legionnaire comrades in the trenches, among them Kiffin Rockwell and Victor Chapman, in order that they might join him in French aviation.

Prince had been in Paris since early January of 1915, using his influence with prominent members of the American colony in Paris to push his concept of an all-American squadron of aviators. This same concept had long been in Thaw's mind, and while he and Prince did not unite in purpose, the two recognized this shared desire.

Thaw served with distinction as a pilot with *C.42* at Nancy and Luneville. He was promoted to *Sergent* on 18 May and was cited and awarded France's *Croix de Guerre*, with Palm on three occasions. Meanwhile Norman Prince had completed his aviation training and had arrived at the Front, joining *V.B. 108* on 20 May 1915. Another American, Elliot Christopher Cowdin, had been serving with that squadron since 3 May.

Whenever free time from their duties allowed, Thaw and Prince had separately pressed their efforts at forming an all-American squadron. Then in December of 1915 they and Elliot Cowdin were granted a thirty-day Christmas leave to the United States. Sailing home aboard the *Rotterdam*, the three Americans had agreed to unite their efforts.

As soon as the Americans had exited the ship at New York on 23 December 1915 the press fell on their heels. The pilots were reluctant to give specifics on their service in French aviation, but reporters pursued them and built them to heroic proportions for their reading audience.

While this caught the interest and sympathy of the American public, it simultaneously brought out the ire of both neutralist and pro-German factions who jointly called for the confinement of the three Americans for their "gross violation of American neutrality." In fact Count von Bernstorff, Germany's Ambassador to the United States, personally warned Thaw in a New York barbershop that it would be best if he and his two American comrades remained in the United States. Thaw scoffed at the ambassador's veiled threat and walked out of the establishment.

Following an eight-day stay in the United States, the three Americans with Norman Prince's older brother, Frederick, whom Norman had recruited, returned to France to rejoin their squadrons. The French government had closely observed the press reports on the three Americans while visiting their homeland, and had noted the positive reaction by a potentially valuable ally. They also recognized that considerable propaganda value and a bonding between the two nations might arise through the formation of an all-American squadron of aviators.

Following their trip to America, Thaw, Prince and Cowdin now discovered a favorable atmosphere existing in Paris. They redoubled their efforts. Thaw's friend, Lieutenant Brocard, had left the old Deperdussin squadron for a position of influence in Paris which Thaw could now exploit. Dr. Edmond Gros of the American Ambulance Service and Jarousse de Sillac, under secretary of the French Ministry of Foreign Affairs, formed a Franco-American committee to further promote the Americans' concept.

Lieutenant William Thaw upon his return to Bar-le-Duc from hospital after having been wounded in the left arm in aerial combat north of Verdun.

When *Sergent* Thaw had returned to France following his visit to the United States, he continued to serve with *C.42* at the Front until 29 January 1916. After that date he returned to Paris and stayed with his sister, Mrs. Lawrence Slade, at his mother's apartment at 148 Rue de Longchamp. The Slade home soon became a favorite hangout for Thaw's American comrades, among them Victor Chapman, Elliot Cowdin, Norman Prince, and Bert Hall. Out of gratitude for the Slade's hospitality, Bert Hall later dedicated his book "*En l'air!*" to them.

While Thaw stayed at his sister's apartment, during the day he attended the R.G.A. at nearby Plessis-Belleville. There, through February and most of March he was retrained from bi-place to *chasse* pilot and was frequently in Paris lobbying for the cause of an all-American squadron. Twice he approached Colonel Barres, Chief of French Aeronautics in the *Zone des Armees* and was finally able to gain his official sanction for the formation of this new squadron and to win for *Capitaine* Georges Thenault, his old squadron commander with *C.42*, the position as commanding officer.

On 14 March 1916 Colonel Henri Regnier, director of French Military Aeronautics, issued a statement to the Franco-American Committee that an American squadron was to be organized. William Thaw, Norman Prince, and Elliot Cowdin were to be included in its roster. On 28 March Thaw joined Cowdin at *Escadrille N.65* while Norman Prince completed *chasse* school. Thaw remained there until 15 April 1916 when he and Cowdin were ordered to Luxeuil-les-Bains, the little resort town southeast of Paris to join the roster of *N.124*, the designation assigned the new American squadron. Americans Victor Chapman, Weston Bert Hall, James McConnell, and Kiffin Rockwell soon joined Thaw, Prince, Cowdin, squadron commander *Capitaine* Thenault, and his adjutant, French Lieutenant deLaage.

On 13 May *Sergent* Kiffin Rockwell had led the Squadron's first regular patrol over the lines southwest of Mulhouse. *Capitaine* Thenault, *Sous-Lieutenant* Thaw, *Sergent* Victor Chapman, and *Sergent* James McConnell had accompanied Rockwell.

On 24 May 1916 William Thaw led the morning patrol over the lines north of Verdun and shot down a Fokker E fighter. He made a second patrol at 9:30 a.m. and ran into three enemy aircraft near the lines. He immediately engaged and shot down a second Fokker fighter before his Lewis gun jammed and he was forced to turn back toward allied lines with two Aviatiks in hot pursuit. They immediately opened fire upon Thaw's Nieuport. A round struck him in the left elbow and continued on into his gasoline tank, making a hole there the size of a fist. Petrol poured out over Thaw's legs and feet. When he reached forward to throw the engine switch, he realized that his arm had been broken.

Thaw volplaned down between the second and third allied trenches near Fort Travennes and pancaked to earth. Dizzy from loss of blood, he threw off his harness and staggered to safety. Ambulance men removed his gas soaked combination before placing him on a stretcher. The gasoline in his clothes had evaporated so quickly that it had burned and blistered his body. He was immediately evacuated to the American Hospital at Nieully where his wounds were treated. He finished his recuperation at his sister's residence in Paris.

The history and service of the *Lafayette Escadrille* can serve as a record of the history and service of William Thaw. The two are inseparable, bonded in time and history. Thaw alone of its thirty-eight pilots served from the Squadron's inception on 20 April 1916 until its men were transferred to the 103rd Aero Pursuit Squadron, the American Lafayette, on 18 February 1918.

Promoted to full Lieutenant on 22 May 1916, two days before he was wounded in combat, Thaw soon became the American commander of the *Escadrille* under French *Capitaine* Thenault. As Edwin Parsons, his *Lafayette* colleague, would later write of Thaw, his American C.O.:

"Most of the time he was in practical, if not nominal, command and handled with unfailing tact and good humor one of the toughest jobs ever thrown on the shoulders of one man. It is reasonable to suppose, since he seemed to be doing all right before there was any such thing, that he could have gotten along without the Escadrille. But most certainly the Escadrille could never have gotten along without Bill. Many times he must have been seething inside, but he never showed it. He made a brilliant success of a job where most men would have gone mad."

William Thaw was revered by the pilots who served with him. And for good reason. He had an innate ability to lead men. This was complemented by his diplomatic abilities which he fully utilized when serving as liaison and buffer between his roisterous American pilots and their stoical French commander. He led the best way possible, by winning the pilots' devotion and respect. Thaw never asked them to do anything which he was unwilling to do himself. And his heroic record bore this out.

During the course of his French and U.S. service Thaw was seriously wounded through the left arm in aerial combat. He was officially credited with the destruction of five enemy aircraft. He was awarded France's *Croix de Guerre*, with nine citations (one bronze Star, one Silver star, seven Palms). He was awarded the *Chevalier* and Officer grade of the *Legion d'Honneur*. He was awarded the Montenegro Military Medal First Class, a special AEF citation, and the U.S. Distinguished Service Cross with Bronze oak leaf, indicating two citations for extraordinary heroism. He was also awarded the Medals of Valor from the Aero Club of France and America and the Harmon Trophy medal.

But the welfare of his pilots was always first and foremost in William Thaw's mind, whether he was commanding the *Lafayette Escadrille*, the U.S. 103rd Aero Pursuit Squadron, or the Third U.S. Pursuit Group. He had an unflappable personality. He ignored what stood in his way and used a disarming smile and a soft-spoken demeanor to win his point rather than resort to his intimidating physical attributes. Only twenty-one years old when he had joined the Squadron, he nevertheless won the loyalty of pilots older and more worldly than he, and carried his duties with a maturity far beyond his years.

By walking a delicate line between the disciplinary restrictions of the French and the unruly antics of his brash but often heroic American pilots, he was thus able to preserve their unique identity and *elan*. But Thaw himself was hardly the grim stoic. Beneath his French tunic beat the heart of a fun-loving young American. He worked hard at commanding his men but played just as hard when the opportunity allowed. Edwin Parsons again noted this side of Thaw's character:

"If there was any drinking or gambling, he wanted to be in on it. If there wasn't anything doing, he started something. He didn't care where you went or what you did or how much you drank, just so you could show up for patrols and at least make a pretense of getting into the air.

"There was never a dull moment in his company. No matter where we were sent, Bill had generally been in that sector before. He always knew a houseful of more or less alluring young wenches in the next village or an attractive widow who owned a chateau with a grand wine cellar, and he was most generous about letting his friends in on his soft spots.

"Many times, through a dim haze, I have seen Bill led off to bed leaning heavily on Percy's* shoulder at two o'clock in the morning when he was scheduled for a show at four. In the gray dawn, eyes red rimmed, and weaving a bit on unsteady legs, he'd be on his way to the hangars with a cheerful grin and an encouraging word for everybody, all set to hop in and lead a patrol. He would never ask anyone to do what he wouldn't do himself, and he'd do two men's work if necessary."

Attending the funerals of his fallen pilots was Thaw's hardest task. No fewer than eleven of his *Escadrille* comrades would die in the Great War, including French officer deLaage and *Sous-Lieutenant* Norman Prince. But the most painful funeral he had to attend was that of his nineteen-year-old brother, First Lieutenant Alexander Blair Thaw, killed in a plane crash en route to Orly Field, Paris. At the time of his

**Robert Percy, Thaw's orderly. He was a New Orleans Negro who had served with Thaw in the Legion Etrangere.*

William Thaw and Dudley Hill (standing left) picnicking with the ladies.

Mr. and Mrs. Benjamin Thaw with "Bronco" at the grave of their son, Lt. Blair Thaw.

While on the flight to Orly the DeHavilland had either run low on fuel or had developed motor trouble. Thaw had attempted to put the plane down at Fere-Champenois but had snagged his wheels on telephone wires and was instantly killed in the crash. Meyer escaped with a broken leg but was so severely injured he would spend the rest of his overseas service in hospital. Bronco was unhurt, but fled the accident scene. He was found a few days later near the wrecked plane and was turned over to Mrs. Thaw. Again William Thaw found himself in the sad position of attending another funeral.

Prior to his brother's death, on 26 January 1918, William Thaw had left the French Air Service to accept a commission as Major in the U.S. Air Service. However his many months of hard service had taken their toll, and he required a personal waiver from General Pershing before his commission was granted.

On 5 November 1917 Pershing recommended "William Thaw as Major, waiving defective vision left eye, 20/80 ophthalmoscopic left shows atrophy plus pigmentation in focal area, hearing defective 15/20, and recurrent knee injury with limitation of motion." In death young Thaw was the commanding officer of the U.S. 135th Aero Observation Squadron.

Lieutenant Alexander Blair Thaw had been considered "an ideal commanding officer" despite his youth. Tall, handsome, his youthful features hidden beneath a mustache, he flew more than his share of missions to prove himself to his older officers.

On the day of his death Lieutenant Thaw was flying a Dehavilland DH 4 to Orly Field where it was to be turned over on loan to the 55th Royal Air Force Squadron at Ozelot for test purposes. The day prior to Thaw's flight a pilot who had flown the DH 4 had warned of a loose and vibrating throttle and a low-slung pilot's seat. But Thaw evidently had ignored the warnings in an eagerness to make the trip to Paris which would also allow him the opportunity to visit his mother and sister.

Accompanying Thaw in the observer's cockpit was First Lieutenant Cord Meyer who had been a pilot with the U.S. 103rd and 93rd Pursuit Squadrons. Sharing Thaw's pilot cockpit was an Irish Terrier," Bronco", the squadron mascot. "Bronco" had been Lieutenant Thaw's pet prior to his overseas service. Thaw's mother had brought "Bronco" overseas with her on one of her trips to Paris. Since that time "Bronco" had flown regularly with Lieutenant Thaw on missions.

Colonel William Thaw with Major Carl Spatz at the 3d Aviation Instruction Center, France.

spite of these defects, Major Thaw took command of the U.S. 103rd Pursuit Squadron on 18 February and served in that capacity while actively flying missions until 10 August 1918. On that date he turned over command to his former *Lafayette* colleague, Captain Robert "Doc" Rockwell.

Major Thaw left the 103rd to become Group commander in the 1st Army Air Service for the newly formed 3rd Pursuit Group which included the 103rd, 28th, 93rd, and 213th U.S. Pursuit Squadrons. An active commander who always got the most from his pilots, by war's end his group had been credited with the official destruction of 96 enemy aircraft. In recognition of his exemplary service he was promoted to Lieutenant-Colonel three months after his twenty-fifth birthday on 12 November 1918.

Thaw had been blessed with an ability to extract the best from his pilots. Because they idolized him they performed what he asked and worked hard to please him. One pilot stated unequivocally that he was "the most striking and popular figure on the front," and that his warm personality and generous nature were recognized by one and all from Dunkirk to the Vosges.

Following his overseas service, from January to July 1919, Lieutenant Colonel Thaw was made executive officer for Rockwell Field, San Diego, California. On 17 July 1919 he received an honorable discharge from the U.S. Air Service and shortly thereafter became engaged in prospecting in lower California and in Mexico. In 1921 he spearheaded lighter-than-air promotion in California. On 2 March 1921, following a brief courtship, Thaw married Marjorie Everts Priest, a St. Louis divorcee, in Coronado.

In 1923 Thaw returned to Pittsburgh where he became an insurance agent for his father's General Insurance Company and involved himself in commercial aviation. He returned periodically to California where his wife maintained a residence, and where Thaw moored a racing schooner.

In February of 1924 Thaw received an announcement from war acquaintance, Dr. Edmund Gros, one of the champions of the *Lafayette Escadrille*, regarding the proposed construction of a monument near Paris to honor the pilots of the *Escadrille* and the Lafayette Flying Corps. Gros, in his letter, also solicited Thaw's assistance toward the raising of funds.

Thaw, in his 1 March 1924 reply to Dr. Gros, for the first time revealed a smoldering animosity he felt toward Frederick Prince, Sr., father of *Lafayette Escadrille* pilots Norman and Frederick Prince. Upon offering Gros his assistance, Thaw wrote:

Marjorie Thaw, Katherine Thaw Meyer, and William Thaw near Wisdom, Montana, 1930.

"I am enclosing an article which appeared recently in the New York Herald, and I must admit that its numerous inaccuracies were far from pleasing to one of the survivors of the original Lafayette Escadrille, of which this article says "There is none." Such clumsy publicity (I have no idea of its origin) is almost certain to dampen the ardor and lessen the cooperation of the "survivors," both of the Escadrille and of the Flying Corps. . . .

"Again, in referring to the history of the Escadrille, Prince is repeatedly referred to as the only founder of the Escadrille, the originator of the idea. I distinctly recall hiking 32 kilometers in October, '14 to endeavor to arrange for the transfer of Hall, Bach, and myself to the same French Squadron, and when we were finally transferred from the trenches, our parting words to Kiffin Rockwell, Charles Trinkard and others were that we would do our best to get them with us in the aviation. This I later and always did, and with the cooperation of yourself and Norman Prince the plan finally worked out, but I do not feel that Prince's activities in Paris had any more to do with the ultimate organization of an American Unit than did, for example, the meeting with Besnard, arranged, as I recall it, by you, or my personal friendship with Lieut. Brocard, later Commander Brocard, and Chief of the French Air Service. Please do not think I am endeavoring to detract from what Prince did, but I do maintain that he did no more than Elliot Cowdin, you or myself, and the insistent references to him (undoubtedly at the instigation of his family) in newspaper articles as *the* founder of the Squadron, frankly has finally gotten my goat. An excellent example of what I mean by the "instigation of his family" is a page and a half of drool, some of which is untrue, that was inserted by Prince's family in Thenault's book without his knowledge, when it was given to them for publication. . . . Also, please do not think that I am in any way seeking any further publicity merited or otherwise; in fact, I hope you will consider this letter confidential.

"Again, the article says the Escadrille was formed on April 18, 1916, when "a handful of Americans, only one of whom could really fly, left Paris for Luxeuil" and later speaks of "the original four" being killed in action. As a matter of fact, as you will recall, the original squadron was composed of Thenault, De Laage, myself, Prince, Cowdin, Hall, K. Rockwell, Chapman and McConnell. Of these, three, Rockwell, Chapman and McConnell, were killed in action, and Prince hit some high tension wires well within our lines. De Laage crashed testing a motor, and incidentally, was the finest chap, French or American, I ever knew. All of these men were finished pilots. . . .

"In closing, may I personally request that if you have any way of reaching the source of this and any similar attempts at publicity, will you please make sure that the articles are corrected before publication by someone such as yourself, who has the necessary records available."

On 4 March 1924 Mr. H.C. Hoskier, father of slain *Escadrille* pilot Ronald Hoskier, approached Colonel Thaw to ask his assistance in the same fund-raising effort to culminate in the building of the monument near Paris to the *Escadrille* and Flying Corps pilots. A number of the slain Americans were to be buried there in crypts near the central monument.

Thaw immediately responded that he would assist Hoskier in whatever manner possible. However, he objected to "the undue prominence given to Norman Prince" in the Memorial Association's fund-raising circular.

Hoskier wrote back suggesting that Thaw cable Dr. Gros about changing the information, as Hoskier did "not want to quarrel with the 'P' family."

Thaw had clearly seen that the flurry of press releases in New York and Paris newspapers had come from Frederick Prince, Sr. in his attempt to alter history and to glorify Norman's participation in the *Escadrille* to the degree to which it diminished Thaw's role and the role of the other pilots. Thaw saw that the senior Prince was attempting to take control of the American Committee for the Memorial to champion the memory of his dead son, when, ironically, he had been a vigorous objector to Norman's entry into aviation. Thaw eventually proved a formidable foe to Prince, then considered one of the wealthiest and most powerful men in America. On 12 March 1924 Thaw wrote the following letter to Mr. H.C. Hoskier.

"As to the publicity article marked 'Received from Paris,' it is just this sort of thing that I wish to keep out of print, because of its unfairness and untruthfulness, and I know that I am backed up by the sentiments of the members of the old Lafayette, both those who were killed and the few who still survive, with the possible exception of Elliot Cowdin, who was Prince's particular friend, and who, as you will no doubt recall, developed heart trouble a few weeks after the forming of the Squadron. As an example of what I mean by unfairness, why shout about Prince being the first to join the French Air Service, etc., when I was a breveted pilot, waiting to be assigned to a French Squadron, in March, '15, before Prince had even enlisted or learned to fly, and at the same time Hall and Bach had already been assigned to aviation, after five months of infantry, and were waiting to be sent to flying school, and Chapman, Rockwell, Dugan, Trinkard and others were in the trenches and ready and anxious to join me in the aviation. I see no mention whatever of Lufbery. . . . If any names are to be mentioned, let them speak of the entire original outfit as "the founders" and if any particular one, let them at least choose a man who "got a Boche" single handed, or who died fighting.

"I have recently and very thoroughly talked over the Prince situation with Captain Thenault, who is now French Air Attache in Washington, and he feels the same as I do. He commanded the outfit from its formation until October, '17, when I took charge, and was in close touch with every step of its organization, and will tell you that it was my two conferences with Commandant Barres, then Chief of the French Air Service, that finally secured official sanction for the formation of the Squadron, and Thenault's assignment as C.O.

"In view of the fact that this Lafayette Memorial is now putting the situation on the basis of historical record, I feel it my duty, both to myself and the boys who served with me, to prevent, as much as possible, any inaccuracies or misplaced credit. Until I have received word from Dr. Gros in answer to my letter, and definite assurance from Paris that this is to be a Lafayette, and not a Prince Memorial, I am sorry to say that I can not conscientiously offer my cooperation in raising the funds, or in any other way.

"I hope you will forgive me for boring you with this little airing of my personal feelings in the matter, but when I look back on those four years, and realize how much more your son, and half a dozen others I could name, did for the morale and the reputation of the Lafayette Escadrille than did Norman Prince, I am sure that you will understand."

Hoskier, in a 14 March 1924 reply to Thaw stated that he would cable Dr. Gros in Paris, asking that he "suppress mention of Norman Prince as the principal founder" of the *Lafayette Escadrille* in the circular announcing the opening of the Memorial fund drive. He assured Thaw that it would not become a "Prince Memorial."

On 18 March, following his cable to Dr. Gros, Hoskier again wrote to Colonel Thaw mentioning how the circular had been modified "to mention the founders of the Escadrille and not Mr. Norman Prince solely." He then brought up the fact that he had planned to use in facsimile a tribute by Marshal Foch on the back of the Memorial folder which contained the sentence: "In January 1915 Mr. Norman Prince and later his 267 aviator compatriots flew to the rescue of France and that of the liberty of the World." He asked if Thaw would object to his using that statement. On 20 March Thaw fired back the following reply.

"I was very glad, indeed, to get your letter of the 18th, and to hear that Dr. Gros is cooperating with us to some extent in preventing any further inaccuracies.

"As regards the facsimile tribute of Marshal Foch . . . I think the Marshal, if he knew that in January, 1915, Prince did not know how to fly, would agree with me that both France and the world would not have been saved quite so promptly. His use of the Prince's name merely demonstrates with what thoroughness and almost intelligence the propaganda for the publicity of the Prince family has always been handled. If the Marshal had been a little more thorough in his search for data to use in his tribute, and for the names of those who first flew to the rescue of the world, he would find that Bert Hall, Steve MacGordon and I enlisted on August 7, 1914. Mac was later forced to withdraw. He would also find that at the time of my enlistment I presented a $5000 aeroplane to the French Government, and I have never received either the plane, thanks, or reimbursement therefor."

On 24 March Hoskier informed Thaw that Frederick Prince, Sr. had accompanied the architect of the proposed Monument to the building site to examine it following removal of underbrush and stumps in preparation for evacuation. Thaw, meanwhile, wrote his friend and former *Lafayette* comrade, Robert Soubiran, explaining Prince's direction, and soliciting his aid. He also suggested that if any one pilot should be singled out for praise above the rest, it should be Raoul Lufbery.

Soubiran responded that he agreed with Thaw's position. He also shared Thaw's opinion of Lufbery, that "his name should top the list." He then suggested that Thaw organize the living members of the Squadron as a counter force "to decide whom the hero is, etc., and not a few Parasites who by their eagerness to be in the lime light are misdirecting some well meaning and benevolent citizens."

H.C. Hoskier informed Thaw in a 31 March letter that he was soon to meet with Frederick Henry Prince in New York City. He planned to confront him to "clear away any misunderstanding. In this effort he would use a letter he had received from Prince's cousin, Charles Prince, which had Charles Prince stating that the information presented by Frederick Prince regarding Norman was "irrelevant and inexact."

It is likely that Prince came out of this confrontation angered and further resolved to continue the fight. Prince wielded considerable power in France as one who had not only donated one of his estates to the French Republic for military use in World War One, but also had secured substantial war loans in the U.S. for France's War effort.

But Thaw received additional support for his cause when his brother-in-law, Lawrence Slade, an executive with the National City Bank of Paris, was named a member of the Paris Lafayette Memorial Committee. Thaw immediately sent Slade copies of all his correspondence on that subject. Slade soon replied from Paris that he felt as Thaw did, "that there was entirely too much Prince in this affair." He then promised to contact Dr. Gros "to change the point of view."

The skirmishing between Thaw and Prince continued for several years. Thaw would ultimately win the battle. And while Prince would allow his son's remains to be placed in the crypt of the completed *Lafayette* Monument, as early as 1927 he was seeking an alternate resting place for Norman in the United States. Thaw made reference to this in an 8 May 1929 response

William Thaw at his desk in Pittsburg, 1931.

to an editorial published in the *Pittsburgh Post Gazette*.

"I am taking the liberty of writing you relative to your editorial of Thursday, April 25th on the subject of the service of Norman Prince with the Lafayette Escadrille, and his interment in a private chapel in Washington. Without wishing in any way to detract from the good work done by Prince while with the Lafayette Escadrille, I do wish to correct your figures. You state that he "fought 122 engagements and was credited with five enemy planes, and with four others not officially recorded."

"I was with the Lafayette Escadrille as second in command under Captain Thenault...during the entire period (except for a few weeks in the hospital) of Prince's sojourn therewith, and during that time the entire squadron had 156 aerial combats, and destroyed officially 17 enemy planes, for three of which Prince was given credit. . . .

"It is illogical to assume that any one pilot actively participated in 80 per cent of the squadron's combats, and won 53 per cent of its victories, especially when it is remembered that the Escadrille at that time counted among its members such daring and successful pilots as Lufbery, Chapman, Rockwell, Nungesser and a few others, who, to my knowledge, took the air oftener than Prince did.

"It is my personal opinion that a private chapel is being purchased near the remains of President Wilson and Admiral Dewey because the surviving Lafayette pilots and the committee in charge of erecting our memorial in France refused to permit any partiality in connection with our memorial at Paris, or any exaggeration of the services of any one associated with the Lafayette Escadrille, either in connection with its founding, operation or administration."

On 4 July 1928, when the *Lafayette* Monument was officially dedicated at Villeneuve L'Etang, near Paris, Thaw's voice had been heard and heeded. No one pilot would be singled out for praise above the others. And the *Lafayette* Memorial booklet was just as impartial with one small exception. Inside is a photo showing the "first four pilots of the Escadrille Lafayette to be killed." They are McConnell, Rockwell, Prince, and Chapman. In parenthesis beneath Norman Prince's name are the two words "the Founder." Evidently one small concession to Frederick Prince, Sr. which, when seen by Thaw, nevertheless must have made his blood boil.

While Colonel Thaw had been battling Frederick Prince, he had continued to pursue an interest in aviation. In September of 1928 he purchased a specially built 525 h.p. Lockheed Vega Hornet for $30,000 from the firm's Burbank factory which he planned to enter in the National Air Races' non-stop flight from New York City to Los Angeles. Thaw would act as navigator on the flight and he chose Captain John P. Morris, a Pittsburgh flying instructor, to pilot the Vega.

Days prior to the flight Thaw sold the Vega to publisher Bernarr Macfadden for $35,000 with the proviso that he and Morris still fly the ship. Macfadden agreed, and on 13 September 1928 the two left Roosevelt Field, New York for Los Angeles in the *True Story*, the name Macfadden had painted on the plane.

That same night an oil line snapped on the Vega, and the aircraft crashed into a rail fence on a farm near Geneva, Indiana. Thaw suffered a dislocated hip and Morris a cracked pelvis. Both men lay injured in the wreckage from midnight until dawn when they were finally rescued. Thaw gave up flying following this race in which not a single entry finished.

In 1931 Thaw's mother, Elma Dows Thaw, passed away in her Paris residence of cancer. Upon the death of her father, Stephen Leland Dows, an Iowa banker and railroad executive, Mrs. Thaw had become the recipient of his wealth. Her husband, Benjamin, following the death of their son, Stephen, of a brain tumor in 1923, had journeyed to Paris to meet with his wife and suggested they combine their funds to form a children's fund. Its purpose was to guarantee the security of their grandchildren. This was mutually agreed upon and the fund was established.

Benjamin Thaw, in setting up this fund, was reflecting the action of his own father, William, who had established the William Thaw "Coke Trust" with his three sons serving as trustees. Thaw had assembled large tracts of coal acreage in Fayette and Westmoreland counties, Pennsylvania and by the terms of his will, his sons were to lease the properties to others to mine, but not to mine coal from the tracts themselves. The revenues generated from the mining were to be distributed among his direct descendants. William's father, Benjamin, had been managing trustee throughout the most profitable time of the "Coke Trust", and had substantially increased its value.

John Thaw, William Thaw's great-grandfather, had been born in Philadelphia in 1779 and had gone to Pittsburgh when it was only a hamlet to open the first bank there. His son, William, born in 1818, had built the family fortune. Of William's three sons, William II, Alexander Blair, and Benjamin, Benjamin showed the greatest business acumen.

Benjamin had been born in Pennsylvania on 14 March 1859. He took an A.B. from the Western University of Pennsylvania in 1878 followed by a year of graduate study at Yale University. In 1889, upon the death of his father, he became managing trustee for the "Coke Trust" as well as a director of several banks, railroads, and utilities. He was also an avid book and document collector, and assembled a significant and valuable historical collection which he later donated to the Historical Society of Western Pennsylvania.

William Thaw following a day of fishing in Alexandria, Minnesota, 1932.

The Great Depression brought many profound changes to American business and industry. It also had a deleterious effect on the Thaw holdings. In 1932, Benjamin, in ill health, turned over his position of managing trustee of the Thaw "Coke Trust" to his son, William. But when Benjamin died in Pittsburgh on 19 August 1933 at age 74, it came as no surprise to William that his father had appointed him only nominal director of his empire: the Iowa Railway and Light Corporation, the Pittsburgh, Youngstown, and Ashtabula Railroad, the Enterprise Transit Co., and the Hurd Coal and Iron Company.

In August of 1914 when William had entered into the service of France he had done so against his father's will. And it was with a mixture of pride and envy that Benjamin Thaw had followed his young son's brilliant and meteoric military career which had indicated a tremendous future potential. However, once the war had ended Benjamin had been able to again control and direct William's life by using the bestowal of a periodic remittance upon him to stifle his growth and potential.

By 1933 William Thaw was clearly not the same man who had served with such vitality in World War One, who had commanded fighter squadrons, who later had done battle with Frederick Prince. Stifled by a domineering father and spiritually ravaged by his war experience, he had increasingly sought release through drink. He appeared dispirited, much older than his thirty-nine years and lacked the dynamism and business interest his father had demonstrated.

William had no offspring. And in his lifetime he had buried many old comrades, friends, and family. He had lost his younger brother, Alexander Blair, in the Great War; his brother, Stephen, in 1923; his mother in 1931; and his father in 1933. Only his sister, Henrietta, and his brother, Benjamin, survived. Benjamin would die tragically in 1937 from a strep infection resulting from a razor cut.

As a sign of shifting times and fortunes, the windows of Morewood Place, the Thaw family's Pittsburgh mansion, were boarded over not long after William's father had passed away. William, its last occupant, had left to take up residence in the Schenley Apartments.

Sergent Elliot Cowdin, Lieutenant deLaage, Capitaine Thenault, and Sous-Lieutenant William Thaw at Luxeuil-les-Bains, May 1916.

In April of 1934 he had gone to the mansion to sort out his war mementoes which had lain dusty and neglected in the attic. A lethargy permeated him, the result of chronic Bright's disease. Of late he had been feverish, and when he returned to his apartment, he collapsed into unconsciousness.

Thaw's wife, Marjorie, immediately called in two doctors. Thaw was found to have pneumonia, and the doctors treated him as best they could. He remained comatose, and yet his fever dropped and he seemed to rally.

But on the morning of 22 April 1934, around 8:30 a.m. Thaw regained consciousness and told his wife the end had come. He then thanked the doctor for trying to save his life and expired. He was forty years of age.

Thaw's remains were buried in the family plot in the Allegheny Cemetery in Pittsburgh. Among those attending were his *Lafayette Escadrille* comrades Charles Chouteau Johnson and Robert Soubiran. A squad of reserve corps aviators had been assembled in Thaw's honor, and four buglers "sounded the last post."

Several months after his death an effort was made to remove Thaw's remains from the Allegheny Cemetery to the *Lafayette* Memorial in France. There, he was to rest with Norman Prince and the other fallen pilots of the squadron.

William Nelson Cromwell, one of the officers of the *Lafayette* Monument, offered to pay all expenses for the transfer of Thaw's remains to a special chapel he would privately finance and have constructed in the center of the *Lafayette* Memorial. And while Thaw's surviving family were in agreement with his proposal, Thaw's widow, Marjorie, remained undecided and seemed reluctant to give her consent. Before she could arrive at a decision she died of an epileptic seizure in Los Angeles on 15 March 1936.

Approximately one year later, in the spring of 1937, Frederick Prince, Sr. had Norman's body removed from the *Lafayette* Monument and had brought it to Washington, D.C. for burial at the National Cathedral. This had been the final expression of his anger for having been thwarted in his attempts to make the *Lafayette* Memorial a tribute to Norman. William Thaw had led the opposing forces which had stopped this action. Now both founders, William Thaw and Norman Prince, were destined for eternity to lie outside the hallowed ground which held their fallen comrades.

Clifton Thompson

Clifton Badlam Thompson, Jr. was born in Hyde Park, Massachusetts on 9 November 1892, the son of Mr. & Mrs Clifton Thompson. Thompson entered Dartmouth College, Class of 1917, where he captained the cross-country team. He left that institution in the spring of 1917 and sailed for France on 4 May 1917 with the First Dartmouth Ambulance Unit.

Upon his arrival in Paris, on 10 June 1917 "Tommy" Thompson enlisted in France's *Service Aeronautique*. From 18 June 1917 to 13 January 1918 he attended the aviation schools at Avord, Pau, and the *G.D.E.* He received his *brevet* on the Caudron on 30 October 1917.

On 15 January 1918 *Caporal* Thompson was assigned at the Front to *Escadrille SPA 99*. Thompson flew with that squadron as a *Caporal* until 28 May 1918, the date he was commissioned a Second Lieutenant in U.S. Aviation. Lieutenant Thompson continued to fly combat missions on detached service with his old French squadron until the 11 November Armistice.

Thompson proved a diligent, dedicated pilot, and one well-liked by his French comrades. He was an asset to his squadron both on and off duty. On one occasion, at the request of *SPA 99* commandant, Captain Rougevin, Thompson ran a cross-country endurance race in which his entire squadron staked their fortune. Using his running skills, Thompson easily won the race in record time and forever won a place in the hearts of his comrades. Lieutenant Thompson was awarded France's *Croix de Guerre*, with Star and Palm for his courageous World War service.

In 1919, while still in France, Thompson married Mlle. Adeline Leger. Eventually, three sons and a daughter were born. Upon his return to the States, Thompson began work for the Lewis-Shepard Co, of Boston in their platform department. He moved to Providence, Rhode Island as an agent for the Lovejoy shock absorber Company. Thompson maintained an interest in aviation and served as instructor and general manager for the Providence Flying Club. He also actively promoted an airport for Providence.

On the evening of 26 July 1928 Thompson's wife had driven from their summer cottage to the Pothier aviation field near Warwick, Rhode Island. As she parked her car and started toward his plane which sat on the ground, she stopped to watch an aircraft stunting above her. The plane, a black Arlington monocoupe, suddenly fell into a tail spin from a height of 100 feet, crashed, and burst into flame. She soon learned that her husband had been a passenger in the plane. Neither Tommy Thompson nor Osmond Mather, the pilot, survived the crash. Mather, a salesman for the Arlington aircraft firm, was trying to interest Thompson in buying a plane for the Seekonk, Massachusetts airport. When the plane crashed to earth, the fuel tank burst into flame. It was two hours before the remains of the occupants could be extracted from the wreck. Both men had been burned beyond recognition. Thompson was 35 years old at the time of his death.

CHARLES TRINKARD

Charles L. Trinkard was born in Ozone Park, New York, the son of Mr. & Mrs. Trinkard. Trinkard was orphaned at an early age and was raised by his older sister, Josephine. He attended New York public schools before enrolling in the Pratt's Institute, New York, then worked in New York City as a jeweller's engraver. Trinkard sailed for France aboard a cattle boat, and on 24 August 1914 he joined the French Foreign Legion. Following training at Rouen, Trinkard was sent to the Front with the Second Foreign Regiment. Trinkard fell ill with fever and spent several months in hospital.

Upon his recovery, Trinkard was shifted from the Second to the First Foreign Regiment at La Valbonne. In September 1915 "Tiny Trink" Trinkard was twice wounded in the right shoulder by machine gun fire during the first Champagne offensive. The first bullet struck him just over the heart but was deflected upward into his shoulder by the safety razor he carried in his left pocket. He spent months in the hospital but returned to his unit in time to participate in the Battle of the Somme where he served as a signalman for his unit. He was awarded France's *Croix de Guerre*, with Star for his heroic service with the Legion.

On 1 March 1917, through the aid of William Thaw, Trinkard was withdrawn from the trenches and sent to the French Flying Corps. On 13 March he enlisted in the *Service Aeronautique*. From 20 March to 30 August 1917 he attended the aviation schools at Avord, Pau, and the *G.D.E.* He received his *brevet* on the Caudron on 24 July 1917.

From 1 September to 29 November 1917 *Caporal* Trinkard flew on the Lorraine Front with *Escadrille N.68*. On 29 November 1917, Thanksgiving Day, he was returning from a patrol when he decided to "buzz" his Legionnaire comrades billeted near Toul. While doing loops and nose-dives, in a vertical turn at low altitude he wingslipped and crashed to the ground and was instantly killed, his spine and both legs broken. Christopher Charles, one of Trinkard's Legion comrades later wrote of his death:

Charles Trinkard (right) with Wainwright Abbott at Avord, April 1917.

"We did not know who had fallen, but when we saw the khaki uniform and the red *fourragere* of the Legion, we were mightily grieved, and the Americans were especially sad, for we all knew Trink and the splendid work he had done as an infantryman. He did more than his duty in this war and did it cheerfully."

Trinkard's body was removed to the Toul aviation field. With eight of his former Legionnaire comrades in attendance, he was buried in the De Choloy cemetery there. In 1919 his remains were moved to the St. Mihiel American Cemetery, Thiaucourt. In 1928 his remains were again moved to the Lafayette Flying Corps Monument near Paris.

DUDLEY TUCKER

Dudley Gilman Tucker was born in New York City, New York on 7 April 1887, the son of Gilman Henry Tucker, secretary of the American Book Company, and Caroline Low (Kimball) Tucker. Tucker prepared for college at Dr. Ray's School in New York, and the Hackley School, at Tarrytown. He entered Harvard University, Class of 1907, and proved to be a good student and athlete. Following graduation, he made his fourth trip to Europe then enrolled in Columbia Law School. Finding he was not fitted for the legal profession, in 1909 he joined his father's American Book Company. Five years later Tucker left that firm to become business manager of the Washington Square Players in New York.

It was there that he met Austen "Billy" Parker, the theater's manager. The two quickly became friends and decided to leave New York and sail to the Orient to study theater there. When their steamer reached Panama, they left the ship to visit the ruins and mines at Darien. They lost their way in the jungles and were held prisoner by a German plantation owner who thought they were U.S. Army Intelligence agents. They made their escape and paddled a dugout canoe 150 miles to Panama. While there, they learned that America had entered the European War. They cancelled their Orient trip and bought steamer passage to Bordeaux, France.

On 9 May 1917 Tucker enlisted in France's *Service Aeronautique*. From 22 May 1917 to 26 January 1918 he attended the aviation schools at Avord, Pau, and the *G.D.E.* He received his *brevet* on the Caudron on 30 September 1917.

On 28 January 1918 Tucker was assigned at the Front to *Escadrille SPA 74*. He soon transferred to *SPA 15* and flew numerous offensive patrols from Rheims to Montdidier. On 7 July 1918 Tucker was preparing to take his first leave in eight months. He was weak from the recurrence of jungle malaria, and his mind was fogged from the consumption of drugs. However, at the last minute, despite a strong premonition of doom, Tucker surrendered his leave to fellow American Henry Forster.

On 8 July 1918 *Sergent* Tucker and four comrades were patrolling the Marne salient when they dove to rescue two French reconnaissance planes under attack by a dozen enemy Fokkers. Tucker disappeared in the ensuing fierce combat. Several months later the Red Cross at Berne, Switzerland received word from Germany that Tucker had been wounded and was being held prisoner. However, he failed to turn up after the Armistice.

Tucker's mother sailed to France and searched for his grave. In August 1920 the Red Cross examined the German records at Berlin. It was learned that he had been removed, unconscious, from his wrecked Spad, to a field hospital at the Checrise church. He was buried in the adjoining churchyard. He was 31 years of age at this death. Tucker's remains were removed to the U.S. Military Cemetery at Seringes-et-Nesle. His body now rests in the Lafayette Memorial near St. Cloud. In 1922 *Sergent* Tucker was posthumously awarded France's *Medaille Militaire*, and *Croix de Guerre*.

GEORGE TURNURE

George Evans Turnure, Jr. was born in Lenox, Massachusetts on 13 July 1896, the son of George Evans Turnure, a banker and stockbroker, and Elizabeth Gardiner (Lanier) Turnure. George Turnure attended the Fay and Groton Schools, and entered Harvard University, Class of 1920. Turnure left Harvard in 1916 for military training at Plattsburg, New York. In 1917 he left Plattsburg to join the American Ambulance Field Service and sailed for France.

Turnure enlisted in France's *Service Aeronautique* on 16 February 1917. From 25 February to 25 July 1917 he attended the aviation schools at Avord, Pau,

and the *G.D.E.* On 16 June 1918 he received his *brevet* on the Bleriot. On 27 July 1917 he was assigned at the Front to *Escadrille SPA 103*. He flew combat patrols with that squadron until 16 December 1918. On one memorable patrol with *Adjudant* Rene Fonk, he was credited with assisting Fonk in bringing down a German two-seater which was piloted by Captain Weissman, the pilot who had been credited two weeks earlier for having downed the famous Guynemer. Two weeks prior to this victory, Turnure had shot down a two-seater near Ypres.

On 2 January 1918 *Sergent* Turnure was commissioned a First Lieutenant in U.S. Aviation. From 12 February to 18 February 1918 Lieutenant Turnure was assigned to *SPA 124*.

From 18 February to 28 August 1918 he flew with the 103rd Aero Squadron, USAS. While with that squadron, Turnure was credited with destroying two enemy observation balloons. From 28 August to the November Armistice, Lieutenant Turnure served as flight commander with the 28th Aero Squadron, USAS. As a result of his heroic war record, Lieutenant Turnure was awarded France's *Legion of Honor*, and *Croix de Guerre*, with three Palms and a Star.

Following World War One Turnure returned to New York City and was engaged in the offices of White, Weld & Co. investment securities. The Turnure's had a summer home at Lenox, Massachusetts. As a member of the party scene there, Turnure was called "the champion motorcyclist of the Berkshires" because of his fondness for daredevil riding. Turnure had gone to Lenox from New York for the Thanksgiving holiday. Following a party, he had returned to the Lenox home too intoxicated to get into the house. He fell asleep on the porch. The next day Turnure returned to New York to his parent's home at 115 East Thirty-sixth street. The following day he fell ill with pneumonia. He died there of the illness on 30 November 1920 at age 24.

STEPHEN TYSON

Stuart Mitchell Stephen Tyson was born in Milwaukee, Wisconsin on 12 March 1898, the son of Reverend Stuart L. Tyson, a clergyman, and Katharine (Rosengarten) Tyson. Tyson grew up in Princeton, New Jersey. He was educated in Oxford, England, and at the Haverford School, Pennsylvania. In 1915 he was employed by the Midvale Steel Company.

On 14 October 1916 Tyson joined the American Ambulance Field Service. He served as an ambulance driver with Section One at Verdun until 14 April 1917. Tyson left the Ambulance Service to join French aviation. Finding himself too overweight to be accepted, Tyson went on a month-long weight reduction regimen and lost fifteen pounds.

On 25 May 1917 Tyson enlisted in France's *Service Aeronautique*. From 6 June to 17 December 1917 he attended the aviation schools at Avord, Pau, and the *G.D.E.* He received his *brevet* on the Caudron on 16 October 1917.

On 6 June 1917 *Caporal* Tyson was assigned at the Front to *Escadrille SPA 85*. Of his work there, Tyson wrote this to his father:

"Our group has been assigned to shooting up the German column as they march up from the rear. We fly very low, so you can imagine what two machine guns on each aeroplane, flying full in the face of the enemy, can do. It is very exciting work. We are in the trajectory of shells from both sides, with anti-aircraft guns shooting up. I have had awfully good luck. Not been touched yet, although my machine has been badly hit twice."

On 19 July 1918 *Sergent* Tyson was flying protection for a photographic two-seater between Dormans and Chateau-Thierry. Spotting eight enemy chasses beneath him, he broke away from the patrol and attacked. The enemy turned to meet him, and Tyson's SPAD was caught in their concentrated fire. His fighter veered wildly, fell downward in a *vrille*, burst into flames, and exploded at 6,000 feet. Dead at age 20, Tyson was posthumously awarded France's *Croix de Guerre*, with Palm in recognition of his extraordinary heroism.

Following his death, Tyson's squadron commander wrote of him: "Stephen Tyson was a brave and capable pilot, always ready to do more than his duty, and was beloved by all his comrades in the Esquadrille."

WILLIAM VAN FLEET

William Carey Van Fleet, Jr. was born in Sacramento, California on 25 December 1891, the son of William Carey Van Fleet, a Federal District Judge, and Lizzie (Crocker) Van Fleet. He attended the Pacific Heights School and the Lowell High School, San Francisco, before attending the University of California, 1909–11.

Van Fleet left university to work as a rancher in Nevada and in the oil business in California. He also assisted in the experimental development of M.J. Trumble's small one-man submarine which the U.S. Government showed no interest in. However, Japan's military industry incorporated its design into the miniature submarines it employed during World War Two.

On 21 July 1917 Van Fleet enlisted in France's *Service Aeronautique*. From 31 July 1917 to 30 June 1918 he attended the aviation schools at Avord, Tours, Pau, Cazeaux, and the *G.D.E.* He received his *brevet* on the Caudron on 7 October 1917.

From 1 July to 28 August 1918 *Caporal* Van Fleet flew at the Front with *Escadrille SPA 78* where he proved to be a "skillful and aggressive pilot." On 24 August 1918 *Sergent* Van Fleet, at the request of his father, enrolled in the Naval Reserve Flying Corps as an Ensign in Paris. He was made Naval Aviator #1731 (HTA).

From 1 September to 15 September 1918 Ensign Van Fleet served at the U.S. Naval Air Station at Dunkirk. From 15 September to 30 October 1918 he flew *chasse* aircraft at the Front with *Escadrille de Saint-Pol* and was awarded France's *Croix de Guerre*, with Palm for "his great initiative and courage during numerous pursuit patrols and combats." He was also awarded the U.S. Navy Cross for his "combats with enemy airplanes, and in attacks upon enemy batteries."

Ensign Van Fleet served with the Northern Bombing Group from 31 October to the 11 November Armistice. Part of his service was spent aboard the U.S.S. *Texas*, 6th Battle Squadron. He and fellow Lafayette Flying Corps pilot Lieutenant Willis Haviland introduced the U.S. Fleet to the English method of piloting land scout planes from the ship's deck.

Following his World War service, Van Fleet was stationed at the seaplane training station at Bay Shore, Long Island, New York. Ensign Van Fleet was honorably discharged on 30 December 1921.

In the early 1920's Van Fleet settled in Mexico and worked in the oil business. In 1925 he married Margaret Clare MacNachtan, of Lake Ontario, Canada. In the late 20's Van Fleet returned to California and established the Van Fleet-Durkee chain of auto filling stations. In 1932 he sold the chain to Shell Oil. Eventually, he became the third senior Shell officer in the United States and president of Shell Services, Inc.

Van Fleet retired from Shell Oil in the late 1930s and pursued his fly-fishing hobby. He died in Greenbrae, California on 11 April 1976 at age 84.

CHARLES VEIL

Charles Herbert Veil was born on 9 June 1896 in Big Run, Pennsylvania, the son of C. Frank Veil, a factory storekeeper, and Annie (Moot) Veil. As a youth, Charles's parents moved to East Palestine, Ohio where he attended local schools. In the fall of 1915 Veil enrolled in Wesleyan College, Middletown, Connecticut but left school for New York City in January 1916. Several days later he signed on as a crewman aboard the *Chipana* and sailed for South America. In Buenos Aires, Argentina he found employment with Swift & Co.

Veil returned to the United States and that fall enrolled in Columbia University, New York City, to study mining. He left School the 1917 winter semester and sailed for France aboard the French liner *SS Chicago* to join French aviation.

Veil enlisted in France's *Service Aeronautique* on 12 April 1917. From 17 May to 16 December 1917 he attended the aviation schools at Avord, Pau, and the *G.D.E.* Veil received his *brevet* on the Caudron on 20 October 1917.

From 18 December 1917 to 9 October 1918 Veil served at the Front as a *Sergent* with *Escadrille SPA 150*. On 9 October he was commissioned a First Lieutenant in U.S. Aviation and continued to fly with his French Squadron until the November Armistice. He was awarded France's *Croix de Guerre*, with three Palms, and the *Medaille Militaire* for his frequent combats and his official destruction of three enemy aircraft. While still in Paris, Veil managed to fly a fighter through the *Arc de Triumphe*, a feat he accomplished by turning his plane on its side just before passing through the arc. While in France, Veil married Mille Lola Fridell of Moline. From this union a daughter, Annie, was born.

Lieutenant Veil was honorably discharged on 8 September 1919 at Garden City, New York. Following his World War service, in 1920 Veil went to Poland to organize the Paris-Warsaw airline. He returned to Paris in 1921 and engaged in the aviation industry there. That same year he went to Russia with the American Relief Administration as a field superintendent. In 1924 he engaged in the oil business, in real estate, and in aviation in California. He also gold-mined in the Sonora area. He and Lola Veil divorced and Veil married Ruth Turpin of San Diego. From this union a daughter, Shirley, was born. Veil moved to Los Angeles and became vice-president and manager of Airco, Inc.

A self-proclaimed adventurer and world-traveler, Veil documented his exploits in two books: *Adventure's a Wench*, and *Ninety-Nine Lives*, both published in 1934. That same year he began working as a mining engineer in Central and South America. Shortly after the outbreak of World War Two, Veil returned to the United States and joined the U.S. Army Air Corps as a Captain. He resigned his commission at the end of the war. Veil was killed in an automobile collision near San Francisco, California on 8 September 1949 at age 53.

NEAL WAINWRIGHT

Neal Wainwright was born in Milton, Massachusetts on 18 January 1898, the son of Robert Wainwright and Elsie (Nourse) Wainwright. Wainwright attended Boston's Noble and Greenough School, and Harvard University, Class of 1919.

Wainwright was in his sophomore year at that institution when he decided to enter the European War. From February to July 1917 he served in France as an ambulance driver with the American Ambulance Field Service in SSU9. On 6 July 1917 Wainwright enlisted in France's *Service Aeronautique*. He was detailed to the aviation school at Avord, but in November of 1917, after wrecking an aircraft, he was radiated out of aviation by the French military before completing his training.

On 2 February 1918 Wainwright enlisted as a private in the U.S. Army in Paris. He was assigned to Battery A, 102d Field Artillery, 26th Division. He served in five engagements and was cited as follows in general orders, Headquarters, 26th Division, AEF.:

"For gallant conduct and devotion to duty in the field on April 20–21, 1918, while repelling a fierce enemy attack on the Toul sector."

On 31 October during the Meuse-Argonne Offensive Wainwright suffered the loss of his left hand and forearm when the 75-mm gun he was serving on exploded while conducting night harassment fire. During the next three or four years Wainwright would endure several amputations, and throughout the remainder of his life he suffered unceasing pain and nervous anxiety from his wounds.

Following a period of recuperation in an allied hospital, Wainwright returned to the United States on 22 February 1919. He was discharged on 12 June 1919.

After the War Wainwright became employed by the Autocar Company of Ardmore, Pennsylvania. On 29 December 1921 he married Mona Crozer of Philadelphia and New York. When they later divorced, Wainwright sailed for France in 1922. He settled in Nice where he worked at translating French books into English. It was there in 1930 that he met and married Marie Antoinette Koehler of Nancy. While in Nice Wainwright spent eight years translating the novels of Maurice de Kobra into English. Wainwright died in Nice of cerebral congestion on 19 March 1935 at age 37. His remains were shipped to the United States for burial. He was interred at the Sleepy Hollow Cemetery, Concord, Massachusetts.

Benjamin Walcott

Benjamin Stuart Walcott was born in Washington, D.C. on 2 July 1896, the son of Dr. Charles D. Walcott, a world-renown paleontologist and author, and Helena (Stevens) Walcott. As a youth, Stuart accompanied his father, then with the U.S. Geological Survey, on expeditions to Montana and British Columbia to gather Cambrian fossils.

Stuart prepared for college at the Taft School then entered Princeton University, Class of 1917. Anxious to participate in the European War, he left Princeton in the spring semester and offered his services to the U.S. Signal Corps.

In April of 1917 Walcott entered the Army and took flight training at Newport News, Virginia. In May he resigned from the Army and sailed for France aboard the *S.S. Chicago* with the intention of joining the Lafayette Flying Corps. He arrived at Bordeaux on 31 May. On 3 June 1917 he joined France's *Service Aeronautique*.

From 3 June to 27 October 1917 "Toots" Walcott attended the Aviation Schools at Avord, Pau, and the *G.D.E.* Considered a brilliant pilot by his *monituers*, he was *breveted* on 6 September 1917, six weeks ahead of his class.

On 29 October 1917 *Caporal* Walcott was assigned at the Front to *Escadrille SPA 84*. Following numerous delays, Walcott flew his first combat patrol on 12 December 1917. He engaged a German bi-place above enemy lines over the Champagne sector and shot it down. He started back for French lines when he was overtaken by three Albatross scouts and was himself shot down and killed near Saint Souplet. He was buried by German troops at Leffincourt on 14 December 1918. He was 21 years of age at the time of his death.

In early December 1917 *Caporal* Walcott had passed his examinations for transfer to American Aviation. His commission for First Lieutenant in the U.S. Army arrived at his squadron shortly after his death.

Walcott was posthumously awarded France's *Croix de Guerre*, with Palm for his heroic combat. In 1923 France presented his family with the *Medaille Militaire*. In 1928 his remains were placed in the Lafayette Flying Corps Memorial near St. Cloud.

WILLIAM WASS

William Ethelbert Van Wass was born in Brunswick, Maine on 24 August 1897, the son of Edward Van Wass, Professor of Music at Bowdoin College, and Mrs. Van Wass. Wass attended Brunswick high school. In 1916 he joined the American Ambulance Field Service and sailed for France where he served six months with SSU 2.

On 3 June 1917 Wass enlisted in France's *Service Aeronautique.* From 3 June to 12 February 1918 he attended the aviation schools at Avord, Pau, Cazeaux, and the *G.D.E.* Wass received his *brevet* on the Caudron on 30 October 1917.

On 15 February 1918 *Caporal* Wass was assigned to the Front to *Escadrille SPA 91*. He flew numerous combat patrols with that squadron until 4 November 1918 and was shot down three times by enemy aircraft. During one engagement with a group of enemy Fokkers, Wass's guns jammed and his engine failed just above a forest. Wass crash-landed and was thrown from his machine, badly disfiguring his face.

On 5 November 1918 *Sergent* Wass was commissioned a First Lieutenant in the U.S. Air Service. He was awarded France's *Croix de Guerre*, with Palm and Star for his courageous war service.

On 22 March 1919 Lieutenant Wass was discharged from the U.S. Military at Garden City, Long Island, New York. He spent some time in a New England Veterans' hospital where his features were restored. He then headed for California and worked briefly as a movie-director, was married, and divorced. Plagued by ill health from his war wounds, Wass fell on hard times. He was reduced to pan-handling for a living and began "riding the rails" from city to city, checking himself into various veteran's hospitals. His life became a series of brief stays at various veterans' institutions as he traveled by rail from San Francisco to Los Angeles, to Reno, Nevada, to Salt Lake City, Utah.

In mid-September 1938 Wass "hopped a freight" for Denver's Fitzsimons Hospital to seek medical attention. On 19 September he was arrested in Colorado Springs for "begging and drunkenness." Unable to pay the $10 fine, he faced a five-day jail sentence. Wass explained his condition to the judge, that he was seeking attention for his old war wounds. He was ordered to spend that night in jail and was released the next morning.

On September 20th Wass returned to the Santa Fe rail yards to continue his journey to Denver. While attempting to hop a moving boxcar, Wass slipped beneath its wheels, and his left foot was crushed. He was taken to the hospital where his foot was amputated at the ankle.

Former Lafayette Flying Corps comrade Reginald Sinclaire, who owned a ranch near Colorado Springs, came to Wass's aid. But once Wass's injury had healed, he continued his odyssey. It eventually ended at the V.A. Hospital at Yountville, California. Wass died there of pneumonia on 18 November 1963 at age 66. He is buried at the Golden Gate National cemetery, San Bruno, California.

WILLIAM WELLMAN

William Augustus Wellman was born in Brookline, Massachusetts, on 29 February 1896, the son of Arthur and Celia (McCarthy) Wellman. He attended Newton High School where he won the nickname "Wild Bill" for his athletic prowess. Following graduation, he entered his brother's wool business in Boston and played hockey on the side. In the spring of 1917, with America's entry into the European War imminent, Wellman attempted to enter U.S. Aviation but was rejected because of his youth. He then enlisted in the Norton-Harjes Ambulance Service in April 1917 in New York. On 22 May he sailed for France aboard the *Rochambeau*.

Shortly after arriving in France, Wellman decided to join the Lafayette Flying Corps. On 13 June 1917 Wellman enlisted in France's *Service Aeronautique*. From 29 June to 1 December 1917 he attended the aviation schools at Avord, Pau, and the *G.D.E.* He received his *brevet* on the Caudron on 29 September 1917.

From 3 December 1917 to 21 March 1918 Wellman was assigned at the Front with *Escadrille SPA 87*. On 21 March *Sergent* Wellman's Nieuport was struck by antiaircraft fire which severed his rear control wires. He crashed into treetops and was thrown from his fighter into the tree. Clinging to a bow, he slid safely to the ground. However, he had badly injured his back, and on 29 March he was discharged from the French Army as unfit for further aviation service. *Sergent* Wellman received France's *Croix de Guerre*, with two Palms for his "enthusiasm and courage," and for his official destruction of two enemy aircraft.

Wellman returned to the United States and wrote a book about his war service titled "Go Get 'Em!" During treatment and recuperation, he participated in Liberty Loan bond drives before enlisting in the U.S. Aviation. On 16 September he was commissioned a First Lieutenant. He was assigned to Rockwell Field, San Diego, California as an Advisory Instructor of Combat Flying.

Following his World War service Wellman became a wing-walking barnstormer. While pursuing this career, he met Douglas Fairbanks, Sr. when he landed his plane on Fairbank's Hollywood polo field. Through Fairbanks's aid, he broke into the motion pictures, first as actor, then as director.

In his long and distinguished career Wellman directed such classics as "Wings," "The Ox-Bow Incident," and "Battleground." Wellman directed 82 films in all, but one he probably wished he had not directed was his last effort titled "The Lafayette Escadrille." Based loosely on his own service in France, the film's hero, as played by Tab Hunter, was a troublemaker who deserted the French Air Service and holed up in Paris with a prostitute before eventually redeeming himself. The film incurred upon Wellman the enduring wrath of the surviving Lafayette pilots and their families.

Wellman died at his Brentwood, California home on 9 December 1975 at age 79, still considered an outcast by the pilots with whom he had served. He was survived by his wife, the former aviatrix, Marjore Crawford, whom he married in 1930, and eight children.

FRANK WELLS

Frank Willard Wells was born in Syracuse, New York on 18 January 1890, the son of Fred Wells and Minnie (Albertson) Wells. Wells attended Syracuse High School. In late 1916 he sailed for France to participate in the European War.

On 6 January 1917 Wells enlisted in France's *Service Aeronautique*. From 12 January to 4 August 1917 he attended the aviation schools at Buc, Avord, Pau, and the *G.D.E.* He received his *brevet* on the Caudron on 10 May 1917.

From 6 August to 23 December 1917 *Caporal* Wells flew at the Front with *Escadrille SPA 93* on the Verdun sector. On 3 January 1918 *Sergent* Wells was commissioned a First Lieutenant in U.S. Aviation. From 10 January to 17 June 1918 Lieutenant Wells became a *Brevet Moniteur* at the American Aviation Instruction Center at Tours, France. On 25 June he was ordered to the United States, to Wilbur Wright Field, Dayton, Ohio. From 15 July to the November Armistice Lieutenant Wells served there as Supply Officer and Test Pilot.

Following World War One Wells was employed by the Ford Motor Company. He left Ford and worked two years at the House of Corrections, City of Detroit, Michigan. He was then employed by the Chrysler Corporation in Canada, South America, and in Detroit, working in the export department, and as a production engineer. He married Nellie Livingston, and from this union a daughter was born. Wells died in a Bloomington, Indiana hospital of a heart attack on 27 December 1950 at age 60.

HERMAN WHITMORE

Herman K. Whitmore was born in Brunswick, Maine on 12 June 1895, the son of John Amherst Whitmore, a Brunswick coal merchant, and Anna (Crehore) Whitmore. Whitmore's father died before he was born. His mother died five years later. Whitmore's earlier years were spent with relatives as he attended Brunswick schools, and the Haverhill High School where he captained the football team. The school yearbook called him "the most popular fellow in the entire school." Whitmore graduated from Haverhill High in 1914 and enrolled in Dartmouth College, Class of 1918. He left school in his third year, and on 5 May 1917 he joined the American Field Service and sailed for France.

"Hick" Whitmore served briefly in the Field Service in Paris before joining the *Service Aeronautique* on 13 June 1917. From 21 June 1917 to 22 March 1918 he attended the aviation schools at Avord, Pau, Cazeaux, and the *G.D.E.* He received his *brevet* on the Caudron on 30 October 1917.

From 24 March to 6 April 1918 *Caporal* Whitmore flew at the Front with *Escadrille SPA 77*. On that date he confronted a patrol of Albatros fighters and scored an unconfirmed victory over one of them before being shot down near Montdidier. Whitmore was captured and held prisoner in Germany until the November Armistice in spite of several attempts to escape.

Following World War One Whitmore completed his studies at Dartmouth and received a post-war liberal arts degree. On 4 June 1921 he married Cora Fraser whom he had met in France where she was secretary to the head of the YMCA. From this union a daughter, Jean, was born.

Whitmore worked on the staff of the Robert M. Gair Co., a box-board firm. When the company was bought out by Continental Can, he became a vice-president for that firm. Whitmore retired in 1958. He died in Laguna Beach, California on 3 December 1983 at age 88.

JOHN WHITMORE

John Joyce Whitmore was born in New York City, the son of John Whitmore. Whitmore attended St. Paul's School, Concord, New Hampshire from 1904–1911, repeating at least one grade. He had academic problems, and a number of illnesses, including pneumonia. His father withdrew him from St. Paul's School in 1911.

On 8 February 1917 Whitmore enlisted in France's *Service Aeronautique*. From 25 February to 22 November he attended the aviation schools at Avord, Pau, and the *G.D.E.* Soon after beginning his training on the Bleriot he was seriously injured in an accident, breaking a leg and injuring an eye. He spent several months in the hospital and when he returned to school, Bleriot training had been discontinued. He received his *brevet* on the Caudron on 3 September 1917.

On 24 November 1917 *Caporal* Whitmore was assigned at the Front to *Escadrille SPA 315*. He flew with that squadron until 5 February 1918 when he was again hospitalized from his training injuries. On 13 May 1918 *Sergent* Whitmore was reassigned at the Front, to *SPA 314*. On 23 May he was sent to the rear again due to complications from his injuries. On 19 June 1918 Whitmore was released from French service because of the injures he had received in training.

CHARLES WILCOX

Charles Herbert Wilcox was born in Minneapolis, Minnesota on 22 May 1891, the son of Herbert A. Wilcox, a mining engineer, and Jennie (Best) Wilcox. From 1906–1907 Wilcox attended the Staunton Military Academy, Virginia. He attended high school in Aspen, Colorado and in San Diego, California. From 1910–11 he attended the Throop Academy, Pasadena, California. From 1911-1915 he attended the California Institute of Technology, receiving a B.S. degree in mechanical engineering. Following graduation, from June 1915 to May 1916 Wilcox worked as a machinist with the Craig Ship Building Co. at Long Beach. From November 1915 to May 1916 Wilcox was a member of the Aviation Section of the California Naval Militia.

From May 1916 to January 1917 Wilcox worked as a student engineer at the General Electric Co. at Schenectady, New York. While there, he read an ad seeking pilots for the Lafayette Flying Corps. Against his father's wishes, he applied and was accepted. His supervisor secured him a passport to France, ostensibly to work there for the company. Wilcox sailed for France shortly before the U.S. entered the European War and enlisted in France's *Service Aeronautique* on 8 February 1917.

From 10 February to 16 July 1917 Wilcox attended the aviation schools at Avord, Pau, and the *G.D.E.* He received his *brevet* on the Bleriot on 31 May 1917. From 18 July 1917 to 18 January 1918 *Caporal* Wilcox served at the Front with *Escadrille SPA 80*. From 25 January to 18 February 1918 *Sergent* Wilcox was assigned to the *SPA 124*, a period when its American pilots had been removed from the French Air Service.

On 9 January 1918 Wilcox transferred to the U.S. Air Service at the rank of First Lieutenant. From 18 February to 18 June 1918 he served with the 103rd Aero Squadron, USAS. He was officially credited with the destruction of three enemy aircraft. For his war service, Wilcox was awarded France's *Croix de Guerre*, with two Palms and a Star.

From July to October 1918 Lieutenant Wilcox served as test pilot at Wilbur Wright Field, Dayton, Ohio. He sailed to England on 19 October to test aircraft and remained there until after the 11 November Armistice. From December 1918 to 30 September 1920, the date of his discharge, he was test pilot, armament designer, and student at the Officers' School of Application, McCook Field, Dayton, Ohio. While still in the Service at Dayton, on 23 January 1919, Lieutenant Wilcox married Maria Binder, a Zigfeld girl.

Following his military service Wilcox became a test-pilot in Ohio. A crash-landing ended his career. He returned to California to his parent's Pasadena mansion with Mrs. Wilcox. He was employed by the Los Angeles Gas and Electric Co. at Los Angeles. He and Mrs. Wilcox were divorced in 1924. Wilcox eventually became chief steam engineer for L.A. Gas & Electric. On 19 July 1929 Wilcox married Lu Verne Beal of Los Angeles, a talented singer and piano player. Wilcox died on 8 September 1959 in Beverly Hills, California at age 68.

MARCELLUS WILD

Marcellus Edward Wild was born in Stockport, New York on 3 October 1890, the son Mr. & Mrs. Wild. Marcellus Wild attended Stockport public schools. From 1911–1914 he attended Harvard University, Class of 1915. During the 1916 Winter semester Wild left Harvard to join the Norton-Harjes Ambulance Corps. On 29 February 1916 he sailed on the S.S. *Lafayette* for Paris. He served 12 months in France as an ambulance driver with Red Cross Section 7 and was awarded France's *Croix de Guerre*, with Star for his heroic evacuation work.

On 29 March 1917 Wild enlisted in France's *Service Aeronautique*. From 13 April to 18 October 1917 he attended the aviation schools at Avord, Pau, and the *G.D.E.* He received his *brevet* on the Caudron on 24 August 1917.

From 20 October 1917 to 30 March 1918 *Caporal* Wild was assigned at the Front to *Escadrille SPA 15*. He served on the Champagne Front, patrolling the lines from Rheims to the Argonne Forest. In May 1918 *Sergent* Wild returned to the United States on sick-leave.

Upon recovering from his illness, Wild enrolled in U.S. Naval Aviation and was assigned to the Naval Air Station at Pensacola, Florida where he served until the end of the war. On 16 December 1918 Ensign Wild was made Naval Aviator #1893(HTA) at Pensacola. He served at the seaplane training and patrolling station at Miami until his honorable discharge as an Ensign on 30 September 1921.

Following World War One Wild was involved in interior decorating. He was also an investor in stocks and bonds. He married Winifred Graustein, and from this union a son, Mark, was born. In the Spring of 1938 the Wilds separated. Mrs. Wild sought a divorce in Reno, Nevada. On the night of 12 May 1938 Wild entered the apartment of Alice Tripp, a 21-year-old acquaintance. Wild told her "I am sorry to muss up your place." He then dropped to the floor and shot himself in the temple. He died in a Boston hospital that same night. Wild was 47 years of age.

GEORGE WILLARD

George Gale Willard was born in Chicago, Illinois on 13 November 1894, the son of Charles E. Willard and Ella (Felton) Willard. He attended Chicago's public schools before enrolling in the University of Chicago, Class of 1917.

Willard spent the summer of 1914 in Berlin as a correspondent for the Chicago *Daily News*. That same August war broke out between France and Germany, and Willard volunteered his services to the American embassy there. He was arrested in Munich in September by the German police and charged with spying for Russia. His arrest had come about when the police had discovered a pencil diagram his father had made for him prior to his trip to Germany to help him locate his aunt's residence in Boston. Once Willard had convinced the police of the purpose of the sketch he was released from custody.

When Willard left Germany, concealed in his shoe was the first description of the big German Krupp guns which had been written up by Raymond Swing, the Berlin bureau manager for the Chicago Daily *News*. Following that exciting experience, Willard returned to his studies at the University of Chicago.

By the autumn of 1917 Willard, in his last term at the University, had decided to sail for France to serve in the American Ambulance Field Service. When he told the school of his intentions, university officials informed he that since he was leaving school early to serve in France, he would still be conferred his bachelor of philosophy degree during the June convocation.

On 29 April 1917 Willard left New York aboard *La Touraine* for Bordeaux, France for service in the American Field Service. Willard's ship docked in Bordeaux on 9 May. Upon his arrival in Paris, Willard discovered that there were 150 other young Americans waiting to join the Field Service. On 27 May he sent the following cable to his mother from Paris. "Ambulance Service still overcrowded. Believe my greatest service to France in Aviation. Our cause needs aviators not ambulanciers."

On 11 June 1917 Willard enlisted in France's *Service Aeronautique*. From 12 June to 24 December 1917 he attended the aviation schools at Avord, Juvisy, Pau, and the *G.D.E.* He received his *brevet* on the Caudron on 1 November 1917.

From 26 December 1917 to 13 January 1918 *Caporal* Willard was assigned to the Front to *Escadrille SPA 157* at Belfort. Because of a continuing lack of aircraft, Willard flew no missions with *N. 157* and left that squadron for Plessis-Belleville on 3 February.

Once back at the *G.D.E*, *Caporal* Willard began flying a variety of aircraft. In early February he also put in a request for transfer to the U.S. Air Service and was assured a first lieutenancy. On 6 February 1918 he was ordered to report to Paris for transfer to the American Army as a second lieutenant, date of rank 26 January 1918. While in Paris he was attached to *SPA 471*, the Defense of Paris Squadron, and flew a number of sorties over that city.

On 9 March *Caporal* Willard was ordered to the gunnery school at Chantilly. While he was at Chantilly, the school was a frequent target for German raiders.

On 28 March Willard was ordered to Paris. He had been commissioned a second lieutenant in the U.S. Air Service and received orders to return to *SPA 471*. On 8 April Willard reported to LeBourget. On his first flight from that field he was nearly killed when he collided with another aircraft in heavy fog.

On 1 June Second Lieutenant Willard began flying combat patrols over enemy lines across the Marne and the Seine rivers. On 18 July Willard was relieved from further duty with *N. 471* and was ordered to Orley Field, to American Acceptance Park No. 1.

On 24 July 1918 Lieutenant Willard was assigned to the 147th Aero Squadron, USAS, at Saints. He flew numerous combat and ground straffing missions, and protection patrols. He would remain with that squadron until 1 December 1918, serving at Chateau Thierry, St. Mihiel, and in the Argonne Offensive.

On 1 December Lieutenant Willard was removed from duty with the 147th Aero Squadron and was ordered to the 3rd Aviation Instruction Center at Issoudun. On 23 December he was ordered to the Port of Embarkation for return to the United States.

On 4 February 1919 Willard arrived in New York aboard the steamer *Cedric*. On 7 February he was discharged at the rank of Second Lieutenant at Garden City, New York.

Following his World War service Willard went into investment banking and worked for a bank in Minneapolis. He also worked for Paine, Webber & Co. in Chicago and in New York. In 1929 he joined Brown Brothers Harriman & Company Bankers of Chicago and eventually became a manager of that firm. He married Grace Van Evers in June of 1920. They had two daughters, Linda and Phoebe. Mrs. Willard died in 1934.

When Willard retired from Brown Brothers he moved to Florida. He died of cancer at St. Francis Hospital, Miami Beach, on 12 June 1968 at age 73.

HAROLD WILLIS

ESCADRILLE LAFAYETTE

Harold Buckley Willis, the 25th volunteer of the *Escadrille Lafayette*, was born on 9 February 1890 in Boston, Massachusetts, the son of John B. Willis and Myrta (Gale) Willis. He attended Newton High School and graduated from Harvard Architectural College in 1912. Willis worked for the firm of Allen and Collins until 1914, at the same time serving with Battery A of the Massachusetts Volunteer Militia.

In 1914 Willis sailed for France to serve with the American Ambulance Field Service, becoming an ambulance driver in Section 2 from February 1915 to June 1916 on the Bois le Pretre and Verdun Fronts. Four other future *Lafayette Escadrille* pilots also served in Section 2: Willis Haviland, Walter Lovell, James McConnell, and Kenneth Marr. On 4 July 1915 during the Bois le Pretre offensive, Willis was cited "for rescue of wounded under fire" and was subsequently awarded France's *Croix de Guerre*, with Star while with the Sanitary Service of the 73rd Division. His citation related the action.

"Willis, Harold, driver, of American Automobile Sanitary Section No. 2: Has always given proof of courage and fearlessness worthy of the highest praise, notable during the assault of July 4, in volunteering to go after the wounded in a very dangerous spot; his car was riddled with fragments of shell."

Willis eventually grew "fed up" with what he saw as "America's non-combatant role in a cause which we felt was the right one." When he had run into the founding members of the *Escadrille Americaine* at Verdun in the spring of 1916, he had then and there decided to enter into French aviation.

Willis requested his release from the Ambulance Service. When it was denied him, he decided to take "French leave." He appropriated a car at Pont-au-Mousson and drove the back roads to Paris with neither a commission nor the necessary papers to do so.

Once in Paris, on 1 June 1916 Willis enlisted in the French Foreign Legion. He then transferred to the French Aviation Service and on 30 June he arrived at the aviation school at Buc, joining his Ambulance Corps comrade Walter Lovell who had arrived the previous day. In August Willis attended the flying school at Avord and received his *brevet militaire* on the Bleriot on 20 October. In November he attended the machine gun school at Cazeaux and in January of 1917 he attended the school of combat and acrobatics at Pau.

Sergent Harold Willis beside his Nieuport scout, Ravenal, March 1917.

Willis arrived at the G.D.E. at Plessis-Belleville in February and remained there until the end of the month. On 1 March 1917 he joined the *Escadrille Lafayette* at Ravenal, arriving the same day as Edward F. Hinkle and three days after his friend, Walter Lovell.

On one of his initial flights over the Front, Willis had flown one of the first Nieuport 17's to be armed with a Vickers machine gun which had been synchronized to fire through the propeller. On a flight deep into enemy territory he had attacked a German scout. In the ensuing combat, his interrupter gear had failed and he shot several holes in his propeller.

The cold, bitter air of early March 1917 rushed into the Nieuports' open cockpits. Willis later recalled how the pilots would come back from patrol "with frozen hands and frozen faces. It was an awful fight to keep your hands from freezing," he remembered. "You'd fly with one hand, beat the other on your knee, then change hands, trying to keep the blood circulating while you flew."

On 19 March 1917 *Sergent* James McConnell, Willis's former Ambulance Corps comrade, was killed in combat near Jussy. On 7 April the Squadron moved to Ham in pursuit of the retreating German lines. Willis remembered it was a "pretty dismal world that we moved into. The Germans had blown up all railroads and all crossroads. They had cut down all the fruit trees and poisoned all the wells."

On 16 April *Caporal* Edmond Genet was killed south of Saint Quentin when his fighter was struck by antiaircraft fire. Seriously wounded, he plunged full motor into the ground.

Sergent Willis led the patrol on 23 April that witnessed the death of *Sergent* Ronald Hoskier, shot down with his French gunner while flying a Morane Parasol two-seater.

"I was very close to him at that time," Willis later recalled. "And he confided to me that he thought he could really do some fighting with the Morane, carrying (Jean) Dressy as a machine-gunner. He was not flying the plane over the lines for the fun of it, but was flying it as a combat machine.

"Over Saint Quentin we ran into about a 60% cloud cover. . . . As we were flying in between these clouds we suddenly came upon several German fighters and some brushes resulted. All of us were flying in and out of clouds, doing some firing and chasing the enemy into the clouds. The fight eventually broke off and we climbed above these small cumulus clouds and then tried to find the enemy planes between and below the clouds. Hoskier did not join us, and fell close to our front lines at Saint Quentin. Some of us went up the lines to recover his body later that day."

Sergent Willis and Edward F. Hinkle, both architects by profession, had looked with a critical eye at the Seminole Indian head serving as the squadron's insignia. They had found it bland and unwarlike and had set about designing a new insignia. They came up with a yelling Sioux brave.

(Top) Harold Willis (3.1, at the Buc School of Aviation August 1916. Edward Hinkle is on Willis's right. French ace Jean Navarre is on his left. Walter Lovell stands far right. (Below) Willis while in training in a 110 h.p. Nieuport.

"We thought we could improve on the Indian head insignia which the Squadron was then using. It was a rather round Indian head, and Hinkle and I thought we could do one that would be a little more distinctive from a distance. Also we wanted to achieve an insignia that, from extreme distances, would give out a blotch of red, white, and blue, even if you couldn't make out the Indian head. Together we drew up the insignia which the Squadron later used and which the French Lafayette Squadron uses to this day."

Willis's skills at the piano keyboard rivaled his skills with a stylus. And in off-duty hours he and fellow *Lafayette* pilot Stephen Bigelow took turns entertaining their Squadron mates with a variety of lively tunes.

The Squadron's next move was on 3 June 1917 to the Chaudun airfield. The pilots remained there until 17 July when they flew their SPAD S.VII fighters to

Harold Willis and Walter Lovell on repos with an American Ambulance driver and an unidentified lady. (Below) Willis in the cockpit of his SPAD S.VII fighter, Senard airfield, August 1917.

Pol-sur-Mer to fly support for the 3rd Ypres Offensive.

"We moved to Dunkerque, on the British front," Willis remembered. "We utilized the field on the beach, on the west side of the Dunkerque Harbor, behind the lighthouse. There was a great mass of German aviation there and we were helping out the British. A lot of fog too. Very often when we would be over the lines, the fog would sweep in from the North Sea and cover the field.

"I tried photographic work with a SPAD. The French were very interested in it because I could go deep into the lines very quickly. I took stereoscopic pictures by flying at a given altitude and a given speed and taking pictures by a stop watch."

French Squadron commander Thenault later wrote highly of Willis's contribution in his book *l'Escadrille Lafayette*:

"With his trained and intelligent brain he always brought back from trips information that was greatly prized by the High Command. It is very hard to make good observations from a single seater. Willis had made a specialty of it and thanks to the speed of his machine, he was able to go places where the slower two-seaters could not have ventured without being brought down."

However, Willis's aerial observations weren't always of a military nature. Edwin Parsons had recalled how his comrade had disappeared following a mission over Compiegne. When he had failed to show up at the aerodrome more than an hour later, Parsons had been ready to give him up as lost.

Willis did return as the second hour approached, and when queried about where he had been, he allowed that he had taken "the splendid opportunity to fly to Chartres-nearly a hundred miles to the south - to get an aerial view of the cathedral there. He walked across the field to the barracks to warm up, still expounding on the glories of Gothic architecture."

On 11 August *SPA 124* moved to the Senard airfield at the edge of the Argonne Forest. A French offensive had been planned for Verdun and the Squadron was to fly support. The weather, unlike that at Pol-sur-Mer, was ideal for combat patrols, and the Americans flew as many as three-a-day against menacing Albatros D.III and D.V fighters.

On 18 August 1917 *Sergent* Willis was ordered to fly a protective mission for a bomber strike deep into German territory. This penetration over a narrow salient allowed the enemy fighters to "sweep back and forth" behind the American flyers. Willis later recalled the mission:

"The commanding officer of the bombing group seemed to have charge of us that day. They split us into two groups, one on each side of the bombers. Instead of giving us freedom for maneuvering, as they should have, they told us to stay with the slower, bigger planes. I forgot who was flight leader, but he obeyed his orders. He stayed right alongside the bombers, zig-zagging to keep in position.

"We were jumped by a lot of fighters after we got well into Germany and got quite badly shot up. A German Albatros came right down in front of me to shoot up the flight commander of my flight. I think it was Didier Masson. Anyway, I shot up this fellow, but another Albatros came down behind me and started shooting me up. I did an Immelmann to get out of that situation, and thereby lost my flight and had an individual fight for a short time. The fellow got another burst into me and got some bullets in my engine, which finished me.

"He followed me all the way down, this fellow and some of his friends. I suppose we were flying at 12,000 feet, or something like that, and I had quite a time to keep out of his fire, without an engine. He very properly tried to shoot me down, because he didn't know whether or not I'd get my engine going again and get back to the lines.

"He shot up my windshield and he finally shot my glasses off, which was close enough. Eventually I saw the ground coming up and it began to look like I might get out of it. I never expected to. I made a landing right on top of a hill near Dun-sur-Meuse, which is downstream from Verdun on the Meuse River.

"Then this German who shot me down landed right beside me, which was good luck, because a group of German soldiers came running up, and he was able to protect me. Sometimes pilots were badly treated by soldiers on both sides. This pilot who shot me down proved to be completely correct and considerate. We waited around and pretty soon some other officers came up in a car, and we all went down to the German pilot's own field and had breakfast together.

"I was only at this German squadron until after lunch. In the early afternoon I was taken and put in the old French fortress at Montmedy which is now abandoned. They put me in an old medieval prison with a lot of French officers. I was there three or four weeks.

"I wasn't very proud of being a prisoner. In prison one has the feeling that perhaps if he had put up a better fight he wouldn't have been taken. The only way to clear one's self of that feeling was to damned well escape. I did, eventually, quite a long time afterwards."

Edwin Parsons rather than Didier Masson had been the flight commander whose aid Willis had come to. He remembered how Willis had picked a German Fokker off his tail before "the buzzard knew what hit him, for he went spinning off like a gigantic top into the wide-open spaces."

Willis had begun the mission wearing only green-striped pajamas and two oil-stained sweaters beneath his flight suit. No Squadron member had ever anticipated being taken alive. In fact, Willis was the only *Lafayette Escadrille* pilot with that dubious distinction. And when he was, he found himself ill-equipped for his confinement, even lacking identity papers.

Once Parsons had learned of his friend's fate, that he had been captured, he had bundled up Willis's "uniform, boots, cigarettes, and money" and had dropped them from his fighter behind German lines where he had hoped they would eventually reach Willis.

Sergent Harold Willis standing before the stererscopic camera he carried in his SPAD scout while pioneering aerial photography from a fighter plane.

Parsons had addressed the bundle to Lieutenant Willis, knowing he would receive better treatment as an officer than as a non-com. Willis, without the papers to contradict himself, had already claimed officer status. And Parsons's parcel added veracity to the claim. Until the ruse was discovered, Willis received the treatment accorded a commissioned officer.

Sergent Willis was taken from Montmedy to the prison at Karlsruhe, Germany. From there he was sent to the aviation distribution camp at Landshut, Bavaria. He remained there one month before he was sent to Gutersloh near the Dutch border. He wasn't there long before he was moved with all the French prisoners to the camp at Holstein, north of Lubeck.

Conditions at the new camp were severe. Willis endured them through December of 1917, and January and February of 1918, before attempting an escape in the middle of March. He made it to the outer barbed wire barrier before he was recaptured.

Following his unsuccessful escape attempt, *Sergent* Willis was sent to the small camp at Bad Stuer in Mecklenburg. He remained there until 1 July 1918 when he was transferred to the prison camp at Villingen. It was there that he had his first contact with other Americans since being shot down.

Harold Willis in France in 1940, again serving in the American Field Service, this time at age 50.

Willis found the conditions at Villingen appalling. The prisoners were kept in a "pen of huts" where the sanitary conditions were "indescribable." He noted while imprisoned there that "the whole camp is alive with fleas and vermin of all sorts." But the Americans managed to endure with the help of Red Cross food boxes.

Several unsuccessful attempts at escape were made under the direction of U.S. Navy Lieutenant Victor M. Isaacs whose ship had been torpedoed off the French coast. Another escape was planned for 5 October. The prisoners were divided into escaping teams. Willis's team dressed as best they could to appear as German guards, utilizing faked German caps and overcoats and carrying carved wooded guns. When the German guards poured out of their barracks to pursue another team outside the prison compound, Willis's team mingled with them in the darkness.

"With the guards I rushed around to the southwest side of the barracks where three teams of our men were escaping. But as we ran, I edged off more and more into the darkness. An under-officer saw me edging off and shouted something at me. When he shouted again, I dropped all pretense, let my gun fall, and ran off at top speed.

"The squad began shooting at me then, but their aim was poor. I had a hard run uphill and was much distressed by the time I got to the top of the hill. My speed wasn't very great for my heavy prison shoes were loaded with mud and each seemed as if it weighed ten pounds."

Willis rendezvoused with Lieutenant Issacs. They set off together into the darkness. The first day of their flight was spent in dense thicket avoiding German school children. The next night was spent mired in a swamp, swimming and wading streams as they followed the crude maps they had brought with them.

They crossed the Schwarzwald valley the second night and spent the following day huddled together under pine boughs. By the fourth night Willis had lost track of the many mountain streams they had crossed.

The two Americans finally reached the Rhine river following days of hiding and nights of bitter cold. Wearing only trousers, their money and papers strung about their necks, they entered the river's fierce current, with Switzerland and freedom just 600 feet away on the opposite bank.

Both men were nearly swept away by the icy water. Willis managed to reach freedom but lost track of his friend, Issacs. Fortunately the young Lieutenant had landed safely up river and the two were soon reunited. He and Issacs remained in Berne, Switzerland until 17 October when they left for France.

When Harold Willis reached Paris he found the War nearly over. In his absence he had been promoted to *Sous-Lieutenant* and had been proposed for a majority in the U.S. Air Service pending his transfer from French aviation. He was awarded France's *Medaille Militaire* before returning to the Argonne front. By the time Willis left Paris for the United States shortly after the Armistice, he had logged more than 800 hours as a pursuit pilot.

Upon his return home, Lieutenant Willis took part in War bond drives until his honorable discharge that same year. He then returned to civilian life as a partner in the architectural firm of Allen, Collins, and Willis. On 7 May 1920 he married Cornelia Fiske, sister of Lieutenant Gardner Fiske, a pilot with the 20th U.S. Aero Observation Squadron. The Willis's eventually had three children: Harold B, Jr., Andrew, and Hannah.

Willis served as a Captain in the Massachusetts National Guard from 1922 to 1926. In 1939, when

Germany invaded France, Willis, Paul Rockwell, and Edwin Parsons initiated an effort to recruit Americans for another *Lafayette Escadrille* until they were discouraged by the F.B.I. They then took their efforts to Paris, and while trying to bring the second Lafayette squadron to fruition, they organized another American Field Service. Willis served in this unit with the 10th French Army from 1939–1940 until the Republic fell to the Germans and he barely escaped into Spain.

In 1940 Willis was commissioned a Major in the U.S. Army Air Corps Reserve. In 1941 he entered active service and designed air bases for the Ferry Command at Ascension Island, Ankara, Khartoum, and Trinidad. He later joined the staff of General Spaatz, attaining the rank of Colonel while serving as Chief Tactical Liaison Officer in North Africa. While he was stationed there he spearheaded and helped to create another *Lafayette Escadrille*, but with French pilots and under French control. Colonel Willis's military awards included France's *Legion d' Honnuer*, the *Medaille Militaire*, the *Croix de Guerre*, (with five citations), and the *Medaille des Evades*.

Willis returned to the practice of architecture following his World War Two service, designing churches in the Gothic style. He was elevated to Fellow in the American Institute of Architects in recognition of his standing as one of this country's leading ecclesiastical architects. Among his many works are the Riverside Church at New York City, Mount Holyoke College, the Union Theological School, and Springfield College. Harold B. Willis died of cancer at age 73 on 17 April 1962 at Weston, Massachusetts.

WESTEL WILLOUGHBY

Westel Robinson Willoughby was born in Washington, D.C., on 1 November 1895, the son of Westel W. Willoughby, a political scientist and author, and Grace (Robinson) Willoughby. Willoughby attended Baltimore public schools, and Princeton University, Class of 1916. Following graduation, Willoughby joined the American Ambulance Service in August 1916 and served at American Ambulance Hospital at Neuilly-sur-Seine until January 1917.

On 9 July 1917 Willoughby enlisted in France's *Service Aeronautique*. He received his *brevet* on 4 December 1917. He was released in mid-December at his own request to join the American Expeditionary Forces.

On 16 December 1917 Willoughby entered the U.S. Army in Paris as a Second Lieutenant. From December to March 1918 he attended the Field Artillery School, Saumur. From April to June 1918 he was attached to Battery A, 59th Artillery, 32nd Artillery Brigade.

In July 1918 he was sent to the Aerial Observers' School, 2nd Aviation Instruction Center. From August to September 1918 he trained as a pilot at the 3rd Aviation Instruction Center, Tours. In October 1918 he attended the Aerial Gunnery School, St. Jean-des-Monts. From 23 November 1918 to February 1919 he was assigned as a pilot to the 90th Aero Squadron, USAS, at Battelainville and Belrain. From February to May 1919 he served at the 1st Air Depot, Columbey-les-Belles. On 1 May 1919 Willoughby was returned to the United States. He was discharged on 2 June 1919.

Following World War One Willoughby sold ladies' garments in New York City. In 1922 he moved to Washington, D.C., and joined the Department of State's Latin American Division. He married Kathleen Beatrice Gibson in 1923. From this union two daughters and a son were born. From 1930–35, he was an assistant to the president of the District National Bank. From 1935 until he retired in 1948, he was with the National Archives, at one time serving as the chief of its treasury division. His wife Kathleen passed away and he married again in 1979. Willoughby died at a Bethesda, Maryland, hospital on 3 July 1989 at age 93.

JOSEPH WILSON

Joseph Volney Wilson was born in Grove City, Pennsylvania, on 16 September 1895, the son of Reverend Gil Irwin Wilson, a Presbyterian minister, and Amanda (Rob) Wilson. In 1910 Rev. Wilson was appointed minister of the Presbyterian church at Wheeling, West Virginia. Volney attended Wheeling schools before entering Washington and Jefferson College, Washington, Pennsylvania, where he undertook ministerial studies. He left that institution in the winter of 1917 to join the Norton-Harjes Ambulance Service.

On 17 February 1917 Wilson left New York City for France aboard the *Chicago*. He arrived in Bordeaux on 5 March and entered the Ambulance Service on 1 June. On 21 June Wilson joined France's *Service Aeronautique*. From 3 August to 19 November 1917 he attended the aviation schools at Avord, Tours, and the *G.D.E.* He received his *brevet* #8483 on the Caudron on 25 September 1917.

On 21 November 1917 Wilson was assigned at the Front to *Escadrille Br. 117*. He flew with that Squadron as a *Caporal* until he was commissioned a First Lieutenant in U.S. Aviation on 16 January 1918. Lieutenant Wilson continued to fly with his old French Squadron until 1 July 1918. On that date he was assigned as a bombing instructor at the American Aviation Instructional Center at Clermont-Ferrand until 30 September 1918. From 30 September to 23 October 1918 Lieutenant Wilson flew as senior flight commander with the 163rd Day Bombardment Squadron, USAS on bombing and strafing runs.

Caporal Joseph Wilson (left) with Sergent Charles Kinsolving while attached to Escadrille Br. 117, December 1917.

On the afternoon of 23 October Wilson and his observer were testing a D.H. 4. He had just left the ground when the plane's motor stopped and the plane fell. In the ensuing crash, the observer was not injured, but Wilson struck his head on the top wing and was instantly killed. He was 23 years of age at the time of his death.

Wilson was credited with the destruction of one enemy aircraft. He was awarded France's *Croix de Guerre*, with Star for his "remarkable courage." He was also recommended for the U.S. Distinguished Service Cross although the award was never made. Lieutenant Wilson was buried at Gondrecourt on 26 October 1918. On 30 May 1921 his remains were re-interred at Parkersburg, West Virginia.

PIERRE WILSON

Pierre Marie Wilson was born in New York City, New York, the son of M.L. Wilson, of Fifth Avenue, New York City. Wilson attended New York schools and the Universities of Nancy and Marseilles, France. He was a student when he decided to enter the European War.

On 27 February 1917 Wilson enlisted in France's *Service Aeronautique*. He received his *brevet* on 11 July 1917. He was released in November 1917 because of ill health.

Pierre Wilson (center row, 4th from left) with American trainees at Avord, March 1917.

ALAN WINSLOW

Alan Francis Winslow was born in Chicago, Illinois, on 2 October 1895, the son of William H. Winslow, a foundry owner, and Edith (Henry) Winslow. He was educated at the Latin School, Chicago, and the Coyt School, Munich, Germany. He enrolled in Yale University, Class of 1918 but left there in the spring of 1917. That June he sailed for France aboard the S.S. *Chicago*.

On 10 July 1917 Winslow enlisted in France's *Service Aeronautique*. From 19 July to 24 December 1917 he attended the aviation schools at Avord, Juvisy, Pau, and the *G.D.E.* Winslow received his *brevet* on the Caudron on 12 October 1917.

From 24 December 1917 to 12 February 1918 *Caporal* Winslow served at the Front with *Escadrille SPA 152*. On 20 February *Sergent* Winslow was commissioned a Second Lieutenant in U.S. Aviation.

On 1 April 1918 Lt. Winslow joined the 94th Aero Squadron, USAS, at Toul. On 14 April, while with that Squadron, he was credited with being the first U.S. Army aviator in a newly formed squadron to bring down an enemy aircraft when he downed an Albatross D.V.A. On 6 June Winslow received a second official confirmation by downing a Hannover CL.III above St. Mihiel.

On 31 July 1918 Winslow was seriously wounded, his left arm shattered, during an aerial combat that saw him shot down and taken prisoner. His injuries resulted in the amputation of his left arm while he was held in captivity. He was held in Germany until after the Armistice, arriving in Paris on 9 January 1919.

For his heroic war service Lieutenant Winslow was awarded France's *Croix de Guerre*, with Palm, and the U.S. Distinguished Service Cross for Extraordinary Heroism for his 6 June 1918 destruction of an enemy aircraft. His citation states:

"While on patrol, consisting of himself and 2 other pilots, he encountered an enemy biplane at an altitude of 4,000 meters near St. Mihiel, France. He promptly and vigorously attacked, and after a running fight extending far beyond the enemy lines shot his foe down in flames near Thiacourt."

Following his World War service, Winslow served in the U.S. Diplomatic Service from 1919 to 1929. On 20 October 1923, in Washington, D.C., he married Rosamond Castle, daughter of Under Secretary of State W.R. Castle, Jr. From this union three sons were born. In 1930 he joined Pan American Airways, Inc., as assistant-chief of the Board of Directors of Uraba, Medellin Central Airways, Inc., a Pan American subsidiary.

On 26 February 1932 Rosamond Winslow died of pneumonia in New York City. Alan, in ill health, but with three small children to care for, continued working for Pan-American. On 12 August 1933, while conducting negotiations in Ottawa, Canada, between his firm and the Canadian government, Winslow fell 50 feet from his hotel window. He was transported, unconscious, to the Water Street Hospital, where he died on 15 August 1933 at age 37. He was buried at Arlington National Cemetery.

CARROLL WINSLOW

Carroll Dana Winslow was born in New York City, New York, on 30 May 1889, the son of Francis Winslow, a stockbroker, and Emma (Carroll) Winslow. He attended the Taft School, Watertown, Connecticut, and Yale University, New Haven, receiving his degree in 1910. On 17 February 1912 he married Rose O'Neill of Tuxedo Park. From this union three children were born.

In 1913 Winslow joined Mondex Shock Absorber Company and became involved in auto racing. In June of 1915 he sailed for France with his wife and infant daughter to work at the American Ambulance Hospital in Paris. On 8 July he resigned his position there to enter France's *Service Aeronautique.*

From 25 October 1915 to 1 May 1916 Winslow attended aviation schools at Pau, Chartres, Chateauroux, and the *G.D.E.* On 13 March 1916 he received his *brevet* #2940 on the Maurice Farman.

From 3 May to 13 July 1916 *Sergent* Winslow served with *Escadrille M.F. 44* in the Verdun sector. On that date Winslow returned to Pau for Nieuport pursuit training. On 10 August he was granted sick leave when his daughter became ill with diphtheria.

Winslow arrived in New York harbor on 19 August 1916. While waiting for his daughter to recover, Winslow finished his manuscript *With The French Flying Corps.* Published by Charles Scribner's Sons in 1917, the book gave the American public its first glimpse of an American volunteer in France's *Aeronautique Militaire.* Winslow described without embellishment the steps a student aviator took from enlistment through training on to his assignment to a forward squadron.

On 1 January 1917 Winslow returned to France to complete his combat training. On 22 March he was assigned to *Escadrille N.112* at LaChappe near Chalons. He flew with this squadron until 30 April 1917 when he was released from the French Air Service to join the U.S. Air Service. Winslow secured his release on 16 May and returned to the United States.

In June of 1917 Winslow was commissioned a First Lieutenant in the U.S. Air Service and served as a flight instructor at Dayton, Ohio. On 21 June 1918 Winslow transferred to Marine Aviation and became a gunnery sergeant. He was made an instructor with the Marine Aviation Detachment at the Marine Flying Field, Miami, Florida. He served there until his 21 December 1918 discharge.

Winslow resumed civilian life by joining his father's brokerage firm of Munds, Winslow, and Stackpole. On 3 March 1923 he became a first lieutenant in the Air Service Officer Reserve Corps. He was discharged a captain on 4 September 1924. That same year Winslow resigned from his father's firm.

In 1925 Winslow founded the Electric Ferries, Inc., which ran from New York City to Weehauken, New Jersey, and from Port Washington, Long Island to Rye, across the Sound. Winslow died of pneumonia in New York City on 27 December 1932 at age 43.

WALLACE WINTER

Wallace Charles Winter, Jr. was born in St. Paul, Minnesota, on 4 May 1896, the son of Wallace Charles Winter, and Florence Lillian "Lottie" (Robbins) Winter. When Wallace was in his eighth year, his father relocated to Chicago where he became a partner in the brokerage firm of King, Farnum and Company.

"Wallie" Winter attended Chicago schools, and the Hotchkiss School in Lakeville, Connecticut, before entering Yale University's Sheffield Scientific School, Class of 1918. Winter was a member of the U.S. Naval Reserve at the time America declared war on Germany. He requested and was granted an honorable discharge in order to join the Lafayette Flying Corps. He left Yale in May of 1917 and in early June 1917 he sailed for France.

On 25 June 1917 "Wint" Winter enlisted in France's *Service Aeronautique* at Paris. From 29 June to 28 November 1917 he attended the aviation schools at Avord, Pau, and the *G.D.E.* He received his *brevet* on the Caudron on 15 September 1917.

From 1 December 1917 to 1 January 1918 *Caporal* Winter flew at the Front with *Escadrille SPA 94.* On that date he was assigned to *Escadrille M.S. 156.* He was awarded France's *Croix de Guerre,* with Star for his 21 February destruction of an enemy aircraft.

On 1 February 1918 *Caporal* Winter was commissioned a First Lieutenant in the U.S. Signal Reserve Corps. However, he had not received a formal discharge from the French Army, so he continued to serve with his squadron.

On 8 March 1918, the day before his transfer to the U.S. Air Service as a First Lieutenant, Winter left the squadron's aerodrome at Chalons and headed for the lines in a Morane Saulnier. Winter and another pilot came upon four German two-seater machines above the Champagne sector. Winter immediately attacked one bi-place from behind. When the distance between the two ships reached ten yards, the wings of Winter's Morane monoplane folded up and the fighter plunged earthward, its bracing wires evidently shot away. His plane fell a mile and a half inside German lines and crashed at Pont-Faverger, a small village 12 miles northeast of Rheims. He was buried by the enemy in the German Terre Rouge Cemetery near Aussonce.

During the October Allied advance a Yale friend came upon Winter's remains. On Thanksgiving Day 1918 his remains were removed to the Aisne-Marne American Cemetery, Belleau, Aisne. In 1930 they were transferred to the Lafayette Flying Corps Memorial near Paris. In June 1919 Yale University awarded Winter a Ph.B. degree, *post obitum, honoris causa.*

Henry Woodward

Henry H. Houston Woodward was born in Chestnut Hill, Pennsylvania, on 27 February 1896, the son of Dr. George Woodward, a physician, and Gertrude (Houston) Woodward. Henry prepared for college at Chestnut Hill Academy and at the Taft School, Connecticut. He spent one and one-half years at Sheffield Scientific School, Yale University, Class of 1920.

In June 1916, Woodward served with the Yale Battery at Tobyhanna, Pennsylvania. Woodward resigned from Yale while in his sophomore year to join the American Ambulance Field Service. On 19 February 1917 he sailed for France aboard the S.S. *Chicago* and was assigned to Section 13 as an ambulance driver. He was attached to a French division and served in the Champagne offensive until 23 July 1917.

"Woody" Woodward enlisted in France's *Service Aeronautique* on 24 July. From 24 July to 14 December 1917 he attended the aviation schools at Avord, Juvisy, and Pau. He received his *brevet* on the Caudron on 1 October 1917.

On 16 December *Caporal* Woodward was assigned at the Front to *Escadrille SPA 94.* He flew with that squadron until 1 April 1918, the day he was killed in aerial combat near Montdidier. While on an afternoon patrol *Caporal* Woodward was seen by his comrades to engage several enemy aircraft just before he disappeared.

Approximately one year later the charred remains of his SPAD fighter were discovered three kilometers south of Montdidier. Woodward's grave was found close by, marked with pieces of his aircraft. He was posthumously awarded France's *Croix de Guerre,* with Palm for his 6 January and 1 April 1918 combats with enemy aircraft. He was officially credited with the destruction of one enemy aircraft. Woodward was 22 years of age at the time of his death. His remains are buried in the Suresnes Cemetery.

WARWICK WORTHINGTON

Warwick Worthington was born in New York City, New York, the son of L.N. Worthington. Warwick attended New York Schools and Cambridge University.

On 9 March 1916 Worthington enlisted in France's *Service Aeronautique*. From 15 March to 28 February 1917 he attended the aviation schools at Buc, Etampes, and the *G.D.E.* Worthington received his *brevet* on the Farman on 24 September 1916.

From 3 March 1917 to 13 February 1918 *Caporal* Worthington flew at the Front with *Escadrille C. 53*. While serving as an observation pilot, he was involved in numerous combats and crashes. He was awarded France's *Croix de Guerre* for his dutiful war service.

On 18 February 1918 *Sergent* Worthington was commissioned a First Lieutenant in the U.S. Air Service. From the date of his commission until the 11 November 1918 Armistice, Lieutenant Worthington served as an instructor at the American Aviation Instructional Center at Tours.

It is believed that Worthington died in 1931.

Buc school of Aviation, 6 June 1916. (standing l.r.) Hoskier, Rocle, Frey, Dowd, Bigelow, Lt. Behiere-Chef Pilote, Worthington, moniteur, Terres, Chatkoff. (Front row) Genet, Parsons, Boal, Littauer.

HAROLD WRIGHT

Harold Everett Wright was born in Brooklyn, New York. He attended Brooklyn schools before enlisting in France's *Service Aeronautique* on 20 March 1917. From 25 March to 8 September 1917 he attended the aviation schools at Avord, Pau, and the *G.D.E.*, receiving his *brevet* on the Caudron on 18 July 1917.

From 11 September to 23 December 1917 *Caporal* Wright was posted at the Front to *Escadrille SPA 155*. On that date *Sergent* Wright left his squadron for the United States. From 19 January 1918 on he was posted as a deserter on French military records. While in the U.S. Wright applied for a commission in the U.S. Air Service but was refused because of a negative French Army report.

In the summer of 1918 Wright submitted a series of articles to the *Saturday Evening Post* titled "Aces High," which portrayed his alleged aerial-combat experiences. These articles produced a deluge of protest-letters from his comrades, which revealed the truth of his war service. One such rebuttal by *Sergent* Austen Crehore appeared on 18 January 1919 issue of the *Post:*

"Dear Sirs: Due to the fact that I am the only American now left in Groupe de Combat 18 of the French Army, I have been requested by Capitaine George, as well as by the other French officers and pilots of Harold E. Wright's former Escadrille, to write the American press and have the true facts concerning former Pilot Wright's performance brought to light.

"The articles appearing in your widely circulated magazine have caused a good deal of comment among the pilots on the Front here, who knew Harold E. Wright and his history while connected with the French aviation.

"This letter is addressed to you in the hope that you will correct the impressions given by Wright in his articles . . . you will observe the facts and statements made herein are vouched for by Capitaine George, of Wright's old Escadrille.

"Besides the many untruths in the articles, the insulting fact is that a pilot, who did not enjoy the confidence and esteem of his comrades, would go to America and, by means of cleverly written articles, appearing in a widely circulated magazine, set himself up as the performer and hero of many adventures which were not lived by him at all, but were really achievements of other pilots.

"In the first place 'Sergeant' Wright was a corporal while in the Escadrille and left that squadron during December, 1917.

"While with the Escadrille he never flew over the enemy lines for more than a total of ten hours and was a participant in but one combat during his ten hours.

"We also declare as false the statement in one morning he had seven fights or combats. No wonder he can't remember the details. Why a fellow with one combat's experience should say he once held the American record for having fought seven in two hours is also far from being clear to us.

"Now he is and apparently has been in America for about three or four months, writing, we admit, interesting articles but ones that contain many large and small absolute falsehoods."

Following World War One Wright continued his career in fiction writing. Wright is believed to have died in Brooklyn, New York, in August 1924.

WALTER YORK

Walter Raymond York was born in Somerville, Massachusetts, on 13 January 1894, the son of James M. York, owner of a food and wholesale company, and Nellie Eunice York. As a youth, York was a prominent football player at the Somerville High School.

York enrolled in Dartmouth College, Hanover, New Hampshire, Class of 1916. He left his studies at Dartmouth College in February 1916 and sailed for France. On 2 March 1916 he joined the Norton-Harjes Ambulance Corps. He served with Section 7, attached to the 21st Division of the French Army. For three months he drove an ambulance on the Champagne front, and for three months at the battle of Verdun. He was discharged on 7 September 1916 and returned to the United States and enrolled in the fall term at Dartmouth College.

York again withdrew from Dartmouth College before the end of the term. On 9 December 1916 he sailed aboard the *Espagne* from New York to Bordeaux and re-enlisted in the Norton-Harjes Ambulance Service. He again served as an ambulance driver until 14 March 1917.

York was on his way back to the United States when America entered World War One. Upon his arrival in the States, York immediately enlisted in the U.S. Naval Flying Corps. He trained at Squantum, Massachusetts, until 31 May 1917 when he secured his discharge to join the Lafayette Flying Corps in France.

On 28 June 1917 York enlisted in France's *Service Aeronautique* in Paris. From 28 June to 1 March 1918 he attended the aviation schools at Avord, Pau, and the *G.D.E.* He received his *brevet* on the Caudron on 2 December 1917.

From 3 March 1918 to the 11 November Armistice *Caporal* York was assigned at the Front to *Escadrille SPAD 97*. York flew more than 100 patrols over enemy lines. On 17 September 1918 he was credited with the destruction of an enemy aircraft for which he was awarded France's *Croix de Guerre,* with Palm.

Sergent York was commissioned a *Sous-Lieutenant* on 11 October 1918. On 19 November 1918 York was chosen by General Petain as one of the aviators designated to fly over the surrendered fortress at Metz as the French Army entered there.

York left France around Christmas 1918 for a furlough in the United States. He became critically ill with the flu and pneumonia while aboard ship. On 1 January 1919 he arrived in New York and was placed in a New York hospital. York never fully regained his strength, although he left the hospital to work in Lincoln, New Hampshire, for the Parker, Young Company, a large lumber and pulp concern. On 26 April 1919 York married Helen Ireland of Somerville, Massachusetts. At the end of July 1920 York was forced to give up work when he was diagnosed with tuberculosis.

In September 1920 York entered the sanitorium at Saranac Lake, New York, for treatment. It was learned he had intestinal tuberculosis, and an operation was performed on him in late October. The operation proved unsuccessful. York became gravely ill. With his wife at his side, he died of heart failure at Saranac Lake on 6 January 1921 at age 27.

FREDERICK ZINN

Frederick Wilhelm Zinn was born in Galesburg, Michigan, on 14 February 1892, the son of A.K. Zinn, owner of a poultry-grain firm. Zinn attended Galesburg public schools then enrolled in the University of Michigan and received a civil engineering degree in 1914. Following graduation, Zinn planned an around-the-world trip. He began his sea cruise from Montreal, but at the outbreak of World War One his ship was diverted to Liverpool, England. Zinn attempted to join the British Army but was rejected because he was not a British subject. He crossed the Channel to France and enlisted in the French Foreign Legion (Infantry) on 24 August 1914 at the *Hotel Invalides* in Paris.

Following training at Rouen Zinn served at the Front until 1 February 1916. While with the Legion, Zinn carried his personal camera and photographed the American legionnaires' experiences at the Front. His photographs were widely published in magazines and newspapers.

Zinn was severely wounded in October 1915 during the battle of Champagne when he was hit in the side by a shell-splinter. Physically unfit for further infantry service, Zinn enlisted in France's *Service Aeronautique* on 14 February 1916. From 17 February to 10 December 1916 Zinn attended the aviation schools at Etampes, Pau, Cazeaux, and the *G.D.E.* He received his *brevet* as a *mitrailleur-bombardier* on 29 August 1916.

From 12 December 1916 to 21 October 1917 Zinn served with *Escadrille F. 24* as an observer/gunner. He was an expert cameraman and photo-interpreter and accomplished numerous photo-reconnaissance patrols, often fighting his way back from dangerous missions. He was twice awarded France's *Croix de Guerre*, with Palm and Star for his bravery and daring work as a photographic observer.

Sergent Zinn was the first man the AEF U.S. mission requested. On 16 November 1917 he was commissioned a Captain in U.S. Aviation and was attached to the First Air Depot at the American G.H.Q. at Chaumont where he directed the transfer of Lafayette Flying Corps pilots from their *Escadrilles* to U.S. Squadrons. He also assigned newly brevetted American pilots to French *Escadrilles* in order that they might gain practical experience.

Following the November 1918 Armistice, Captain Zinn was commissioned a Major and became chief of the Graves commission at Berlin for locating graves of missing American airmen. At the time of his discharge he had spent five years overseas serving the Allied cause.

Upon his return home from France, Zinn joined his father's A.K. Zinn feed firm in Battle Creek, Michigan. On 2 August 1919 he married Ruth Hitt. Zinn served in World War Two in Europe as a Lieutenant Colonel in the Army Air Corps's Strategic Services.

Following the War he formed a real estate management firm, ran for public office, and was involved in many charitable organizations. Zinn died of cancer in Battle Creek, Michigan, on 4 October 1960 at age 68.

6 • Lafayette Escadrille Memorial

"We have weathered successfully many storms, such as the attempt of the Prince family to take away from us the Memorial at Garches to all our Dead, and turn it into a memorial to one pilot only."

—Paul Ayres Rockwell

In the beautiful park of Villeneuve L'Etang eight miles outside of Paris and midway to Versailles stands the "Memorial de l'Escadrille Lafayette," a striking monument erected to the memory of the 209 American volunteer pilots of the *Escadrille Lafayette* and the Lafayette Flying Corps.

The monument is composed of a central Arch of Triumph and the sanctuary crypt. The arch is one-half the size of the *Arc de Triomphe de l'Etoile*, and upon its stone facades are inscribed the names of the dead and the surviving pilots of the Lafayette Flying Corps in order of enlistment.

The initial idea for a monument was conceived by former Lafayette Flying Corps pilot Edgar Guerard Hamilton who had assisted the Allies in locating the bodies of his fallen comrades following the Armistice. He felt it appropriate that they should all be brought to rest in one memorial which would also serve to perpetuate the spirit which inspired their volunteer service to France. To this end he enlisted the aid of prominent American and French officials and in March of 1923 an association called the "Memorial de l'Escadrille Lafayette" was formed to bring Hamilton's dream to fruition.

Almost from the moment of its conception the monument was fraught with controversy. Even before its construction began a war broke out between former pilot William Thaw and Norman Prince's father, Frederick Henry Prince, which lasted through the years of the memorial's construction and which ultimately saw the embittered Prince refusing to allow Norman's remains to rest in the memorial. On the very eve before the monument's dedication a further dispute erupted between the former pilots it was intended to honor which threatened to disrupt the dedication ceremonies.

An effort was begun on 15 May 1928 to remove the bodies of the 67 members of the Lafayette Flying Corps from their resting places in France, Belgium, Italy, and the United States and re-inter them at the memorial in time for the July 4th dedication ceremonies. There were 68 sarcophagi in the sanctuary, one of which was to be left empty. In truth, a number of the tombs lie empty today because of the impossibility in locating or removing several of the Lafayette pilots' remains.

Only six of the twelve dead pilots of the *Escadrille Lafayette* whose names appear on tombs in the Memorial's crypt are truly buried there: Edmond Genet, James Ralph Doolittle, Ronald Hoskier, Raoul Lufbery, Douglas MacMonagle, and Paul Pavelka. The crypt for Andrew Courtney Campbell may or may not contain his remains.

On the evening of 3 July 1928 a reunion dinner was held at the Hotel Chatham in Paris reuniting all the Americans who had served as pilots in France's *Service Aeronautique*. During their war service the Chatham had served as a favorite gathering point where the American pilots had come together on leave in a spirit of comradeship. However, little comradeship prevailed on the eve of the dedication.

The friction began when a *Lafayette Escadrille* pilot had objected to the inclusion of the names of the Lafayette Flying Corps dead with the names of the *Lafayette Escadrille* dead for which the monument had been intended. The former pilot further complained that the Lafayette Flying Corps, which had been established in 1917, was attempting to "steal the thunder" of the *Lafayette Escadrille*.

Other *Lafayette Escadrille* members stood to object to the sharing of the Lafayette Flying Corps in the monument's inscriptions and in the official ceremonies and further complained that the names of living

The dedication of the "Memorial de l'Escadrille Lafayette" on 4 July 1928. The monument at Villeneuve L'Etang, eight miles outside of Paris, was intended to honor the service and memory of the "209 American volunteer aviators" of the Lafayette Flying Corps. From its conception to the day of its dedication, the monument was a source of friction and controversy.

Corpsmen should not be inscribed on the monument as had been planned.

At the most heated point of the argument former Lafayette Flying Corps pilot Austen Crehore took the floor and made an impassioned speech which he concluded by saying: "We have brought you our dead. Don't exclude them. We all fought for the same cause." Following Crehore's speech a committee was selected from both groups, and they worked through the night, finally "harmonizing the conflicting views."

So in the end, the memorial which had been constructed to house and honor the dead of the Lafayette Escadrille, contained the remains of only six or seven of that squadron's pilots but held the remains of forty-two Lafayette Flying Corpsmen who had never served in the Lafayette Escadrille.

There is much about the monument and the publicity that accompanied its dedication that is confusing and historically inaccurate. The dedication booklet which was distributed to surviving pilots, surviving family members, and donors for the memorial's construction is misleading and often incorrect.

Sixty volunteers were completely overlooked by the Board of Directors of the Memorial Association. The most glaring omission was founding Lafayette Escadrille pilot Bert Hall. His name was not inscribed among the surviving pilots of the Lafayette Flying Corps on the memorial's stone facades, nor was it included in the dedication booklet. Memorial directors Dr. Edmund Gros and Paul Rockwell deemed Hall's service unworthy despite his four confirmed aerial victories, so they made no mention of him and removed him from history. Lafayette Flying Corps pilot Harold Wright suffered the same fate as Hall (as did Eugene Bullard, America's first Black combat aviator), when both men were judged unfit to be included in the list of surviving members.

The authors of the booklet listed among the names of the Honor Roll of the Dead, the "67 (who) were killed" in France. Yet, several Corpsmen died of illness, two committed suicide, and four pilots died in the United States. The booklet's authors also further muddied the distinction between those thirty-eight men of the Lafayette Escadrille and the rest of the volunteers of the Lafayette Flying Corps buried in the monument by calling the edifice "the last resting place of the sixty-seven dead of the Escadrille Lafayette."

Ten thousand people, including the relatives of many dead airmen, attended the July 4th, 1928 dedication of the monument which had been made possible by the private donations of many French and American citizens. Also observing the ceremonies were Ferdinand Foch, Marshal of France, France's Minister of War, Painleve, U.S. Ambassador Myron T. Herrick, and a number of former French and American pilots.

The memorial was formally presented to France by Dr. Edmond L. Gros, President of the Board of the Memorial Association. This was followed by a speech in which Georges Thenault, France's air attache to Washington and the *Lafayette Escadrille's* former commander, related the history of the Squadron. Then Ambassador Herrick took the podium and in an impassioned speech said this of the pilots:

"During three terrible years, when the sting of criticism cut into every American soul, they (the pilots) were showing the world how their countrymen could fight if they were only allowed the opportunity. To many of us they seemed to be the saviors of our national honor, giving the lie to current sneers upon the courage of our nation."

A benediction followed which was led by Bishop William T. Manning of New York. Planes droned low overhead. Then the French honor guard stood at attention while three guns were fired as a salute and a bugler sounded taps. Thus ended the moving and harmonious ceremonies, much to the relief of the monument's board of directors.

William Nelson Cromwell, a Paris-based American lawyer and the Founder of the Lafayette Escadrille Foundation who had seen to the financing and building of the monument, had contributed 600,000 francs to assure its perpetual maintenance. However, his sizeable donation eventually proved inadequate, and the passage of time took its toll on this perpetual symbol of America's first combat aviators. However, it has since been restored to the beautiful edifice it once was.

The Lafayette Flying Corps Memorial booklet which was distributed to surviving pilots, family members, and donors to the monument's construction.

Corpsmen Who Are Entombed at the Monument

Leif Barclay
James Baugham
Frank Baylies
James Bayne
Philip Benny
Vernon Booth, Jr.
Andrew Campbell, Jr.
Oliver Chadwick
Cyrus Chamberlain
Charles Chapman, Jr.
Roger Clapp
Phelps Collins
Philip Davis
James Doolittle
Meredith Dowd
Sidney Drew, Jr.
Stuart Edgar
Dinsmore Ely
Eric Fowler
Edmond Genet
Norman Grieb
Andre Gundelach
Robert Hanford
Warren Hobbs
Ronald Hoskier
Harry Johnson
Theodore de Kruijff
Edward Loughran
Raoul Lufbery
James McConnell
Douglas MacMonagle

French and American dignitaries, former pilots and their families, have assembled for the dedication of the Memorial Monument. Originally intended to honor the "memory of the heroes of the Escadrille Lafayette, its facades bear the names of the fallen airmen and the survivors of the Lafayette Flying Corps. In its sanctuary are 68 sarcophagi intended to hold the remains of the dead American aviators. But because of the difficulty in locating and identifying their remains, a number of the tombs stand empty.

William Meeker
Walter Miller
Alan Nichols
Henry Palmer
Paul Pavelka, Jr.
Alfred Pelton
David Putnam
Samuel Skinner
Dumaresq Spencer
Frank Starret
William Tailer
Elmer Taylor
Hugh Terres
Charles Trinkard
Dudley Tucker
Stephen Tyson
Wallace Winter, Jr.
Benjamin Walcott

Corpsmen Buried Elsewhere in France
Stafford Brown-Suresnes 23-B-4
Dennis Dowd-St. Germain Cemetery
Kiffin Rockwell-Luxeuil
Henry Woodward-Suresnes 27-B-16

Corpsmen Buried in the United States
Julian Biddle
Arthur Bluethenthal
Linn Cookson
William McKerness
David Peterson
Norman Prince
Walter Rheno
Lawrence Scanlan
Joseph Wilson

Corpsmen Whose Remains Were Not Found
Alan Ash
Victor Chapman
Schuyler Lee
Manderson Lehr
Carter Ovington
Donald Stone

7 • Lafayette Escadrille Reunion

"They shall grow not old, as we that are left grow old:
Age shall not weary them, nor the years condemn.
At the going down of the sun and in the morning
We will remember them." Laurence Binyon, "For the Fallen."

Reunion of Lafayette Flying Corps pilots at Hotel Lafayette, New York City 11 March 1949.

Reunion of the Escadrille Lafayette *pilots at Asheville, North Carolina, June 1960. (l.r.)Harold Willis, Frederick Prince, Carl Dolan, Henry Jones, Edwin Parsons, Paul Rockwell, Capt. Philip Flammer (USAF representative), and Capt. Louis Richard (French Air Force and current* Escadrille Lafayette *representative.)*

Sources

NOTE: THIS TITLE IS TO BE INCLUDED AS A REFERENCE IN ALL THE FOLLOWING SOURCES

Hall, James Norman, and Charles B. Nordhoff. *The Lafayette Flying Corps*. In two volumes. Boston: Houghton Mifflin, 1920.

•*LAFAYETTE FLYING CORPS*

Above The French Lines. Letters of Stuart Walcott American Aviator. Princeton: Princeton University Press, 1918.

American Legionnaire in France. Compiled by James Howard Gore. Washington: W.F. Roberts Co.

Banister, George T. *Friends of France The Story Of The American Field Service (1914–17) And The American Field Service Medal, Insignia, and Documents* C.1985, St. Petersburg, FL

Dunbar, Ruth. *The Swallow*. New York: Boni and Liveright, 1919.

Gordon, Dennis. *Lafayette Escadrille Pilot Biographies*. Doughboy Historical Society, 1991

Memoirs of Harvard Dead in The War Against Germany.

Unpublished letters of George Dock, Jr.

Author conversations and correspondence with Bill Foley

• *AVION DE CHASSE TRAINING*

An American for Lafayette: The Diaries of E.C.C. Genet, Lafayette Escadrille. Charlottesville: University Press of Virginia, 1981.

Forster, Henry. *Over There 1917–1918: Recollections of Wartime*. Unpublished memoirs.

Hall, James Norman. *High Adventure: A Narrative of Air Fighting In France*. Boston: Houghton Mifflin Company, 1918.

McConnell, James Rogers. *Flying For France*. Garden City, New York: Doubleday, Page & Company, 1917

Molter, Bennett A. *Knights of the Air*. New York: D. Appleton and Company, 1918

Parsons, Edwin C. *I Flew With the Lafayette Escadrille*. Indianapolis: E.C. Seale & Co., 1963.

Edwin Parsons taped interview conducted by Dr. S.H. Kleiser November 1962

Unpublished war memoirs of George Dock, Jr.

• *ENEMY*

Bishop, Major William A. *Winged Warfare*. New York: George H. Doran Company, 1918.

Jackson, Robert. *Aerial Combat:"The World's Great Air Battles*." New York City: Galahad Books,1976.

Munson, Kenneth. *Aircraft of World War I*. New York: Doubleday and Company, Inc., 1968.

Springs, Elliot White. *War Birds: Diary Of An Unknown Aviator*. New York: Grosset & Dunlap, 1926

Thenault, *Capitaine* Georges. *l'Escadrille Lafayette*. Paris: Librairie Hachette, 1939

Wallace, Melville. *Pilots of World War 1*. London: Flammarion, 1978

Winter, Denis. *The First Of The Few." Fighter Pilots of the First World War*." Athens: University of Georgia Press, 1983.

Unpublished war memoirs of George Dock, Jr.

• *FRONT*

Hall James Norman & Charles B. Nordhoff. *Falcons of France*. Boston: Little, Brown, and Company, 1929

Hall, James Norman. *High Adventure*. Boston: Houghton Mifflin Company, 1918

Kahnert, M.E. *Jagdstaffel 356*. London: John Hamilton, LTD.

Munson, Kenneth. *Fighters 1914–1919: Attack and Training Aircraft*. London: Blandford Press, 1968

Parsons, Edwin C. *I Flew With The Lafayette Escadrille*.

Thenault, *Capitaine* Georges. *l'Escadrille Lafayette*.

War Letters of Edmond Genet. NY : Charles Scribners Sons, 1918.

Winter, Denis *The First Of The Few*. "Fighter Pilots Of The First World War." Athens: University of Georgia Press, 1983.

Connell, Dennis & Frank W. Baily. "Lafayette Escadrille and Lafayette Flying Corps Victory Logs." *Cross & Cockade Journal*, 21, No. 4 (1980), 363–368.

Sjordal, Capt. Paul E., USAF. "Carl Dolan, Last Survivor of the Lafayette Escadrille." *Air Force Magazine*, February 1974.

Carl Dolan interview conducted by Ken Boege

Unpublished memoirs of George Dock

Fred Andrews interview with Henry Jones September 1969

• *PARIS*

Marshall, General S.L.A. *The American Heritage History of World War One*. New York: American Heritage Publishing Co., 1982

History of World War I AJP Taylor, Editor. England: Octopus Books, 1974

Parsons, Edwin. *I Flew With The Lafayette Escadrille*.

Weeks, Alice. *Greater Love Hath No Man*. Boston: Bruce Humphries, Inc.,1939

Wren, Jack. *The Great Battles of World War I*. New York: Grosset & Dunlap, Inc.,1971

Soltan, Arthur M. & James Davilla, M.D. "The Bombardment Of Paris: 1914–1918." *Over The Front*, 4.,No. 3 (1989), 250–261

Henry Jones interview conducted by Fred Andrews September 1969.

Carl Dolan interview conducted by Ken Boege, 1974

Herman Whitmore war diary.

• *ACES*

Franks, Norman & Bailey, Frank *Over The Front*. London: Grub Street, 1992

Connell, Dennis and Frank W. Bailey. "Lafayette Escadrille and Lafayette Flying Corps Victory Logs." *Cross & Cockade Journal*, 21, No. 4 (1980), 351–368.

• *ABBOTT, WAINWRIGHT*

American Aviators In The Great War 1914–1918, Glasebrook Foundation, 1984.

Yale in the World War. In two volumes. George Henry Nettleton, ed. New Haven: Yale University Press, 1925.

The New York *Times* 16 January 1985

Author interview with Mrs. Augustus S. (Nancy V. Abbott) Ballard 8 December 1990

Author correspondence with Bill Foley 4 November 1995

• *ADAMS, JOHN RUSSELL*

Author correspondence with the Department Of Veterans Affairs 21 November 1990.

• *ALLEN, SIDNEY*

History of the American Field Service In France. Houghton Mifflin, 1920.

Sloan, Jr, James J. *Wings Of Honor*. Schiffer Publishing Ltd., 1994.

Sidney Allen passport application via the National Archives.

• *APPLETON, WALTER*

Rockwell, Paul Ayres. *American Fighters in the Foreign Legion*. Boston: Houghton Mifflin Co, 1930.

Walter K. Appleton record—New York Adjutant General's Office via Marilyn Van Dyke

• *ASH, ALAN*

Letters of Alan Ash via the State Historical Society, Madison, WI

author correspondence with Marjorie E. Haws, Magnolia, Illinois 17 June 1992.

author interview with Alan Ash 2 November 1992.

Author interview with Lawrence Ash 10 October 1992.

• *ATEN, ARTHUR*
records of Dr. Gros from Crehore collection via Dr. Jim Parks
The New York *Times* 19 May 1918 via Marilyn Van Dyke.
• *BACH, JAMES*
Salt Lake *Tribune* 24 February 1960.
Author correspondence with Mrs. Adrienne Buck 14 April 1982; 19 May 1982
Author correspondence with Alumni Records, the University of Utah, 15 January 1982
• *BAER, PAUL*
The Bluebook of Aviation. C. 1932, The Hoagland Co., Los Angeles.
Old Fort News, Vol. XXX1, No. 4. "Paul Baer Scrapbook." Fall 1968
• *BAIRD, BENJAMIN HESTER*
Arthur, Reginald Wright. *Contact: Naval Aviator Careers 1–2000*. Naval Aviator Register, Washington, D.C., 1967.
Benjamin Baird passport application via the National Archives.
Benjamin Baird California death certificate
• *BALSLEY, CLYDE*
War Letters of Kiffin Yates Rockwell. Garden City, New York: Country Life Press, 1925
Balsley, Horace Clyde. "Severely Wounded." *Century Magazine*. February 1919.
"The Diary of H. Clyde Balsley." *Cross & Cockade Journal*, 18, No. 2 (1977), 97–123.
San Antonio *Express* July 15, 1917
Clyde Balsley letter to Austen Crehore 5 March 1928 via Dr. Jim Parks
Author interview with Mrs. Bernice Balsley 1/29/84; 1/30/84
Author interview with Mrs. Mary Craig 11/17/85; 11/24/85
• *BARCLAY, LEIF*
Memorial Volume of the American Field Service in France. James W. D. Seymore, ed. American Field Service, 1921.
• *BASSETT, CHARLES CHESTER JR.*
Arthur, Reginald Wright. *Contact: Naval Aviator Careers 1–2000*.
author interview with Mrs. Charles C. Bassett, 28 January 1984.
author correspondence with Charles Bassett, III 3 September 1998.
author correspondence with Harvard University Archives 26 Dec 1998
• *BATCHELOR, HENRY*
Arthur, Reginald Wright. *Contact: Naval Aviator Careers 1–2000*.
author correspondence with Mr. Harold G. Rader, Worchester Academy 3/9/85.
author correspondence with Martha L. Mitchell, Archivist, Brown University, 26 February 1985.
author conversation with Harried Tyne 5 June 1991
• *BAUGHAM, JAMES*
Author interview with Mrs. Mildred Rumley 24 May 1992.
Data supplied by Mrs. Mildred Rumley.
• *BAYLIES, FRANK*
Franks, Norman, and Frank W. Bailey. *Over the Front*. London: Grub Street, 1992
Memorial Volume of the American Field Service In France.
New England Aviators. In Two Volumes. Boston and New York: Houghton Mifflin, 1919.
Over the Front by Franks and Bailey
author interviews with Frank and Haraold Baylies 17 February 1995
author correspondence with New Bedford City Library 15 May 1995
• *BAYNE, JAMES*
Grand Rapids *Herald*, various issues
• *BENNEY, PHILIP*
Memorial Volume of the American Field Service in France.
• *BENOIT, LEO E.* (BENWAY, ERNEST L.)
Bluebook of Aviation. Ronald Hoagland, Editor. Los Angeles: Hoagland Co., 1932
Boebert, Earl. "The Eagle Squadrons." *AAHS Journal*, 9, No. 1 (Spring 1964.)
Leo Benoit Passport issued 2 May 1917
Leo Benoit American Ambulance Service application via Bill Foley
author conversation with Eddie Guy 4 December 1997
author conversation with Harold Marconda 10 January 1998
author conversation with Doris Benway 10 January 1998
author conversation with David Benway 26 January 1998
various Benway documents via Doris and David Benway
E.L. Benway pilot log book.
• *BIDDLE, CHARLES*
Bluebook of Aviation. Roland Hoagland, ed. LA: Hoagland Co.,1932
"Cross & Cockade" Journal, Vol 13, #3
• *BIDDLE, JULIAN*
Aero Club Of America 1917
St. Mark's School In The War Against Germany, Albert Emerson Benson, Editor c. 1920 via Constance M. Biddle.
Yale in the World War. In two volumes. George Henry Nettleton, ed. New Haven: Yale University Press, 1925.
Author correspondence with Yale University, 25 June 1992.
Author interview with Constance M. Biddle 7/6/92
• *BIGELOW, STEPHEN*
An American for Lafayette. "The Dairies of E.C.C. Genet. Lafayette Escadrille."
Flammer, Philip M. *The Vivid Air*. "The Lafayette Escadrille." Athens: University of Georgia Press, 1981.
Harvard's Military Record in the World War,
Mason, Herbert Molloy. *The Lafayette Escadrille*. New York: Random House, 1964.
Parsons, Edwin C. *The Great Adventure*. Garden City, New York: Doubleday, Doran and Co., 1937
Wynne, H. Hugh. "Escadrille Lafayette." *Cross & Cockade Journal*, 2, No. 1 (1961)
Author correspondence and interviews with Stephanie Bigelow Neal.
• *BLAKE, CHARLES*
Blue Book Of Aviation. Ronald Hoagland, editor. Hogland Co, 1932
Author correspondence Dept of Vets Affairs, WA, DC 3 Aug 1990
• *BLUETHENTHAL, ARTHUR*
Memorial Volume Of The American Field Service In France.
Author interview with Arthur Bluethenthal, II 6 July 1992.
Data furnished by Mr. Arthur Bluethenthal 18 October 1992.
• *BOAL, PIERRE*
Hall & Nordhoff. *The Lafayette Flying Corps*.
• *BOGGS, ELLISON*
Florida death certificate for Ellison Boggs
• *BOOTH, WILLIAM VERNON, JR.*
Memoirs of the Harvard Dead in the War Against Germany. Vol III.
St. Mark's School In The War Against Germany.
The New York *Times*. July 16, 1918, 13:6
• *BOSWORTH, CLARENCE*
University of Vermont in the Great War. War Services Committee, 1924. (This source says Bosworth "served as pilot with the Lafayette Escadrille on the French front; awarded the Croix de Guerre.")
Author correspondence with U of Vermont 9 November 1992
Clarence Bosworth passport application via National Archives.
• *BOULIGNY, EDGAR*
A Dictionary of Louisiana Biography. Vol I. Glenn R. Conrad, editor. Louisiana Historical Association, 1981. via Michael Shaver.
Martin, Fontaine. *A History of the Bouligny Families & Allied Families*. The Center for Louisiana Studies, University of Southwestern Louisiana. via Michael Shaver.
Copy of Edgar Bouligny's Louisiana death certificate
author correspondence with New Orleans Public Library 21 Ap 1986
author correspondence with Susan Davenport, Chalmette National Cemetery, New Orleans, Louisiana
letter to author from Fontaine Martin 3 January 1997
• *BOYESEN, ALGERNON*
author correspondence with Columbia University 10 Jan 1995
New York *Times* 1 Jan 1904; 5 October 1913; 15 August 1915 via Marilyn Van Dyke
author correspondence with Columbia University 3 June 1997
Algernon Boyesen passport application via the National Archives.
• *BRADY, LESTER*
Military Records of Cornell in the World War. Ithaca: New York University, 1930.
author correspondence with Dr. Douglas Brady 28 April 1986
• *BRIDGMAN, RAY*
Flammer, Philip M. *Vivid Air*.
Mason, Herbert Molloy, Jr. *The Lafayette Escadrille*. New York: Random House, 1964
U.S.A.S. in WWI. Vol.III
Yale in the World War. in two vols. New Haven: Yale University Press, 1925
Connell, Dennis & Frank W. Bailey. "Escadrille Lafayette, N.124 Combat Log." *Cross & Cockade Journal*, 21, No. 4 (1980), 363–368.
New York *Post*. 23 September 1939.
New York *Times* 26 Nov. 1951
Carl Dolan interview conducted by Ken Boege, 1974

Author correspondence to New York University 3 February 1983
Author interviews & correspondence with Caroline Bridgman-Rees: 15 January 1984; 9 February 1988; 13 May 1988
Caroline Bridgman-Rees correspondence with author 22 February 1984; 1 June 1984; 8 March 1988
• *BROWN, JASPER*
author correspondence with Mrs. Jasper Brown 10 November 1984; 18 November 1984; 20 September 1989.
author conversation with J. Winthrop Brown, Jr. 13 November 1992
author conversation with Mrs. Ester Cornish 13 November 1992
• *BROWN, STAFFORD*
Gold Star Record of Massachusetts. Eben Putnam, Editor. Boston: The Commonwealth of Massachusetts, 1929.
Memorial Volume of the American Field Service In France.
New England Aviators.
University of Vermont in the Great War. War Service Committee of the University of Vermont, 1924.
author correspondence with Dartmouth College 11 September 1989
• *BUCKLEY, EVERETT*
State of California death certificate
Dennett, Carl *Prisoners Of The Great War.* Houghton Mifflin, 1919.
author conversation with Lola Clark 11/20/92
author conversation with Iva Riegel 11/20/92
author correspondence with University of Illinois 12/29/92
• *BUFFUM, THOMAS*
Harvard's Military Record in the World War.
Author correspondence with Harvard University
Author correspondence with the Gunnery School 9 March 1985
Author correspondence with Susanna Goodwin 16 August 1990
• *BULLARD, EUGENE*
Carisella & Ryan *The Black Swallow Of Death.* C. 1972, Marlborough House, Inc., Boston.
Dr. Gros letter to Capt. W.W. Hoffman, HQ, AEF 16 November 1917, Vol. 5 Austen Crehore Collection via Dr. Jim Parks.
• *BULLEN, RICHARD NIXON*
Johnson, Jean Lidnsay *When Midwest Millionaires Lived Like Kings, 1981*
The *Oconomowoc Enterprise* 21 July 1929
author interview with Tom Bullen 9 October 1991.
Richard Bullen passport application via the National Archives.
• *BULLEN, WILLIAM GRAHAM*
Johnson, Jean Lindsay *When Midwest Millionaires Lived Like Kings*, 1981
The *Oconomowoc Enterprise* 21 June 1929
author interview with Tom Bullen 9 October 1991
• *BUSH, PHILIP NELSON*
author correspondence with the Dept. of Veterans Affairs 3 August 1990.
The *Washington Post* 29 October 1942
author correspondence with University of Pennsylvania 12/22/92
• *BYERS, LOUIS LESLIE*
The Bluebook of Aviation. Roland W. Hoagland, editor. The Hoagland Company, 1932.
author correspondence with University of Pennsylvania 16 December 1992
Louis Byers Massachusetts death certificate
• *CAMPBELL, ANDREW COURTNEY*
Mason, Herbert Molloy. *The Lafayette Escadrille.* New York: Random House. 1964.
Parsons, Edwin C. *The Great Adventure.*
Thenault, *Capitaine* Georges. *l'Escadrille Lafayette.*
"Errata & Addendum." *Cross & Cockade Journal*, 5, No.2 (1964),191
Wynne, H. Hugh. "Escadrille Lafayette." Henry S. Jones interview. *Cross & Cockade Journal*, 2, No. 1 (1961).
White Lake Yacht Club *News* July 1927; October 1936
Author correspondence with Alumni Office, University of Virginia, Charlottesville.
Author conversation with Betty Miller, Mercersburg Academy, PA 4/15/90
The Mercersburg Academy Alumni Quarterly October 1917; July 1927
Various documents via Mr. Courtney Pitkin 4/14/90
Henry Jones interview conducted by Fred Andrews 17 September 1969
• *CAMPBELL, HUGH*
Arthur, Reginald Wright. *Contact: Naval Aviator Careers 1–200.*
The Navy Book of Distinguished Service. Harry Stringer, ed. Fasset Publishing Company, Washington, D.C., 1921.
New York *Times* 10 June 1920; 21 June 1920; 23 June 1920.
• *CARRERE, JOSEPH*
Carrere record-New York Adjutant General's Office via Marilyn Van Dyke.
• *CASSADY, THOMAS*
The Bluebook of Aviation. Ronald Hoagland, editor. Los Angeles: The Hoagland Company, 1932.
Franks, Norman & Frank Bailey *Over The Front.* London: Grub Street, 1992.
Who's Who In Aeronautics. Lester D. Gardner, editor. New York: Gardner, Moffatt Co., 1922.
• *CHADWICK, OLIVER*
Memoir Of The Harvard Dead In The War Against Germany, Vol 11.
New England Aviators.
author correspondence with Mrs. Anton V. Long 5 February 1994
• *CHAMBERLAIN, CYRUS*
Letters of Cyrus Foss Chamberlain. Privately printed. Minneapolis, 1918.
Princeton in the World War. Princeton University, 1932
author correspondence with Princeton University Alumni records 1 Feb 1993
• *CHAPMAN, CHARLES*
Decorations U.S. Army, 1862–1926 Washington: U.S. Govt. Printing Office, 1927.
Letters of Lieutenant Charles Wesley Chapman, Jr. Privately printed. Waterloo, Iowa, 1919.
author correspondence with Amherst College Archives 29 Jan 1992
author correspondence with Waterloo Public Library 21 Feb 1993
author conversation with Jean Chapman McKinstry 1 March 1993
• *CHAPMAN, VICTOR*
Birmingham, Stephen. *America's Secret Aristocracy.* Boston: Toronto: Little, Brown and Company, 1987
Dictionary Of American Biography. Vols. 4, 21. New York: Charles Scribner's sons, 1930.
Flammer, Philip M. *The Vivid Air.*
Hall, Bert, and John J. Niles. *One Man's War*: "The Story of the Lafayette Escadrille." New York: Henry Holt and Co., 1928.
Memoirs Of The Harvard Dead In The War Against Germany. Vol. 1.
Rockwell, Paul A. *American Fighters in the Foreign Legion.*
St. Paul's School in the Great War. Published by the Alumni Association, 1926.
Thenault, *Capitaine* Georges. *l'Escadrille Lafayette.*
Victor Chapman's Letters from France. New York: The MacMillan Co., 1917.
War Letters of Kiffin Yates Rockwell.
Weeks, Alice S. *Greater Love Hath No Man.* Boston: Bruce Humphries, 1939.
Owen Wister. "John Jay Chapman." *Atlantic Monthly*, May 1934.
The New York *Times* 5 November 1933.
Author correspondence with Conrad Chapman 27 October 1988.
Author interview with Mr. Robert Robertson Chapman, 2 October 1988.
• *CHARTON, LOUIS*
Rockwell, Paul A. *American Fighters of the Foreign Legion.*
• *CHATKOFF, HERMAN*
Rockwell, Paul A. *American Fighters in the Foreign Legion.*
Author correspondence with Dept of Vets Affairs 6 February 1991
• *CLAPP, ROGER*
Codman, Charles. *Contact.* Boston: Little, Brown, and Co., 1937.
Over there 1917–1918 Recollections of Wartime by Henry Forster. Privately published memoir.
• *COATSWORTH, CALEB*
Arthur, Reginald Wright. *Contact: Naval Aviator Careers 1–2000.*
author conversation & correspondence with James Coatsworth April 1985–April 1987; August 1998
• *COLLIER, EDWARD*
author correspondence with Phillips Exeter Academy 31 July 1996
• *COLLINS, PHELPS*
Detroit Michigan *News* 27 August 1931
author conversation with Thelma Ash 3 May 1995
Capt. Phelps Collins Burial file via National Archives.
• *CONNELLY, JAMES*
Decorations United States Army 1862–1926. Washington: United States Government Printing Office, 1927.
New York *Times* 2 February 1944
The Philadelphia *Inquirer* 2 February 1944
author interview with William Wallace Atterbury 8 March 1993
author interview with William Connelly 10 June 1995
• *COOK, ALAN*
Phillips Academy Andover in the Great War. New Haven: Yale University Press, 1919.

author interview with Edward Cook 21 March 1985
author correspondence with Alan S. Cook 26 October 1996, 13 November 1996.
• *COOKSON, LINN*
author correspondence with Iris Zimmer, Carlinville Public Library 2 February 1993
author interview with Lucy Powell 9 February 1993:
• *COREY, RUSSELL*
Arthur,Reginald Wright. *Contact: Naval Aviator Careers 1–2000.*
Yale in the World War. Yale University Press, 1925.
author interview with Angela Corey 20 October 1984
• *CORSI, EDWARD*
Karolevitz, Robert, and Ross Fenn. *Flight of Eagles.* Sioux Falls: Brevet Press, Inc., 1974.
various letters of Charles Corsi to Ross Fenn via Tom Walden
Franks, Norman, and Frank W. Bailey. *Over The Front.* London: Grub Street, 1992.
• *COTTON, JOHN*
The Amherst Memorial Volume. Claude M. Fuess, editor. Amherst, 1926.
author interview with Henricus Cotton 20 October 1985
• *COURT, ISADORE*
Isadore Court Record-New York Adjutant General's Office via Marilyn Van Dyke.
• *COWDIN, ELLIOT*
Flammer, Philip M. *Primus Inter Pares: A History Of The Lafayette Escadrille.* A dissertation, 1963.
Harvard's Military Record In The World War.
National Cyclopedia of American Biography. Clifton, N.J.:James T. White & Co.
St. Paul's School In The Great War. Published by The Alumni Association, 1926
New York *Times.* Jan. 7, 1933. 15:5; Jan 8, 1941. 19–1.
Paul Ayres Rockwell letter to James R. Nilo Dec. 8, 1959.
Letters of William Thaw to Herman Hoskier and Dr. Edmond Gros March 1924
State of Florida Death Certificate for Elliot Christopher Cowdin.
Author correspondence with Nancy Morgan Smith January 5, 1985.
Author interview with Mrs. Charles Morgan December 29, 1984.
• *CREHORE, AUSTEN*
Wellman, William A. *Go, Get' Em!* Boston: Page Company, 1918.
AAHS JOURNAL, VOL. 16, #4 "First International Air Race."
New York "Times" obituary 22 August 1962
author interview with June Crehore Gulick 7 February 1993
• *CUNNINGHAM, ARTHUR*
Harvard's Military Record in the World War.
New England Aviators 1914–1918.
author conversation with Laura (Cunningham) O'Brien 28 July 1991
author conversation with Mrs. John Cunningham 4 April 1992
• *CURTIS, FRAZIER*
New England Aviators 1914–1918. Boston: Houghton Mifflin, 1919
New York "Times" obituary 28 January 1940
San Diego "Union" obituary 29 January 1940
• *CUSHMAN, ALVAN*
Phillips Academy Andover in the Great War. Claude Fuess, editor. New Haven: Yale University Press, 1919.
the New York *Times* 21 February 1948
author correspondence with Lafayette College 12 March 1993
author correspondence with Phillips Academy, Andover, MA 15 Sept 1995
• *DAVIS, PHILIP*
Memoirs of the Harvard Dead in the War Against Germany.
New England Aviators 1914–1918. Boston: Houghton Mifflin, 1919.
• *DOCK, GEORGE*
author interview with Mrs. George Dock 24 November 1985; 13 May 1986
• *DOLAN, CHARLES*
Flammer, Philip M. *The Vivid Air.*
Aircraft Yearbook, 1920. New York: Doubleday, Page And Company, 1920.
New England Aviators.
Who's Who In Aviation. Chicago & New York: Ziff-Davis Publishing Company, 1942. Chicago Aero Commission.
Lockridge, Jim. "The Last of the First." *Latitude 20, The Hawaiian Air Magazine.* 7, No. 2 (1979).
Connell, Dennis & Frank W. Bailey. "Escadrille Lafayette, N.124, Combat Log." *Cross & Cockade Journal*, 21, No. 4 (1980),363–368.
"The Editor's Cockpit." *Cross & Cockade Journal*, 16, No. 4 (1975), 83.
Wynne, H. Hugh. "Escadrille Lafayette." *Cross & Cockade Journal*, 2, No. 1 (1961)
General Orders #27, Office First Army Air Service Commander, AEF 17 November 1918.
Carl Dolan interview conducted by Ken Boege, 1974.
Author correspondence with Charles H. Dolan 28 November 1979, 19 September 1980.
Author correspondence with Mrs. Romona Dolan 22 October 1982, 1 November 1982, 19 January 1984.
Author Correspondence with Charles H. Dolan III 17 August 1988.
Dr. S.H. Kleiser interview with Rear Admiral Edwin Parsons November 1962.
Dr. James Parks letter to author 18 April 1990
• *DONZE, ROBERT*
Blue Book of Aviation. Ronald Hoagland, editor. Los Angeles: The Hoagland Company, 1932.
author interview with Helen Vaughn 22 January 1988
• *DOOLITTLE, JAMES*
Flammer, Philip M. *The Vivid Air.*
Parsons, Edwin C. *The Great Adventure.*
Buffalo *Daily Courier* 27–29 July 1918
New York *Times.* 21 July 1917; 9 August 1917; 27 July 1918.
Author correspondence with John Sterling.
James Ralph Doolittle death certificate
Austen Crehore letter to Lafayette Flying Corps Membership via Dr. Jim Parks.
Henry Jones interview conducted by Fred Andrews 1969, 1970
• *DOWD, DENNIS*
Hervier, Paul-Louis. *The American Volunteers* with the Allies. Paris: Editions De La Nouvelle Tevue, 1918.
American Fighters In The Foreign Legion
Author correspondence with Columbia University 24 May 1993
author correspondence with Georgetown University 14 July 1993
The New York *Times.* 2 Nov 1916; 12 Aug 1933 via Marilyn VanDyke
Author correspondence with Margaret Dowd 7 July 1997
• *DOWD, MEREDITH*
Memorial Volume of the American Field Service In France.
author interview with Andrea Dowd Whitcraft 8 October 1993
• *DREW, SIDNEY*
Life and Letters of Sidney Rankin Drew. Mrs. Sidney Drew, editor. New York: The Cheltenham Press, 1921.
Aldrich, Nelson W, Jr. *Tommy Hitchcock, An American Hero.* C. 1984, Margaret Mellon Hitchcock.
• *DREXEL, JOHN*
Allen, Frederick Lewis *The Great Pierpont Morgan.* Harper & Row, 1948
Aero Club Of America 1917-New York: 237 Madison Ave.
Aircraft Yearbook 1919.
Berriman, A.E. *Aviation.* London:Methuen & Co.Ltd.,1913
Birmingham, Stephen. *America's Secret Aristocracy.* Boston & Toronto: Little, Brown, and Company, 1987
Dictionary Of American Biography. Vol.V. New York: Charles Scribners Sons, 1930.
Duffy, Sister Consuela Marie, SBS. *Katharine Drexel, a biography.* Mother Katharine Drexel Guild. Cornwells Heights, Pa, 1972
Flammer, Philip M. *The Vivid Air.*
Harris, Sherwood. *THE FIRST TO FLY.* "Aviation Pioneer Days." New York: Simon & Schuster, 1970
Mason, Herbert Molloy, Jr. *The Lafayette Escadrille.* New York: Random House, 1964
Myers, Gustavus. *History of the Great American Fortunes.* Three vols. Chicago: Charles H. Kerr & Company, 1910
The National Cyclopedia. Vol.II. New York: James T. White & Co, 1892.
See Them Flying "Houston Peterson's Air-Age Scrapbook 1909–1910." New York: Richard W. Baron 1969.
Thayer, Lt. Lucien H. *America's First Eagles.* "The Official History of the U.S. Air Service, A.E.F.(1917–1918)." Mesa, Arizona: Champlin Fighter Museum Press, 1983
War Records Of The Knickerbocker Club—1914–1918
Various issues of *New York Times* 1893–1955
Admiral Parsons interview by Dr. S.H. Kleiser November 1962.
Author correspondence with Neil Flanagan, Eton College, UK
Fred Andrews interview with Henry Jones 17 July 1971
• *DUGAN, WILLIAM*
An American for Lafayette. "The Diaries of E.C.C. Genet. Lafayette Escadrille."

Dugan, William E. His unpublished memoirs. Via Thomas Foulkes.
Flammer, Philip M. *The Vivid Air.*
Rockwell, Paul A. *American Fighters in the Foreign Legion.*
Technology's War Record. By Alumni Association Massachusetts Institute of Technology, 1920.
William Dugan letter from Zent, Costa Rica, 1 April, 1921. Via Thomas Foulkes.
Frank B. Dugan letter 28 January 1963 via Thomas Foulkes.
RAdm E.C. Parsons(Ret) letter 20 November 1962. Via Thomas Foulkes.
New York *Times*, various issues.
Author conversations and correspondence with G. Colburn Dugan December 1982, March 1983.
• *DULON, LOWELL*
Dr. Gros enlistment records from Crehore collection via Dr. Jim Parks
author correspondence with University of Wisconsin Alumni Association 8 October 1988
author correspondence with National Personnel Records Center, St. Louis, Missouri.
Dulon 1919 passport applicaton via National Archives
• *EATON, SHERBURNE*
Harvard's Military Record in the World War.
New England Aviators 1914–1918.
author correspondence with Harvard Alumni Records 12 February 85
author conversations & correspondence with Florence Eaton 18 March 1985
• *EDGAR, STUART*
author correspondence with Cornell University 12 November 1992.
author interview with Runion Colie, Jr. 3/28/93
Stuart Edgar burial records via the National Archives 1 July 1998
• *ELDREDGE, DONALD*
Culver in the World War. South Bend: Peerless Press, 1930.
author correspondence with South Bend Public Library 25 November 1985
author interview with Stewart Eldredge 1 December 1985
author correspondence with Elizabeth Eldrege 12 April 1993
• *ELLIOTT, CHESTER*
Harvard's Military Record in the World War.
author interview with Ethel (Elliott) Dill 17 February 1985
author correspondence with Gari Dill-Marlow 2 April 1993
• *ELY, DINSMORE*
Dinsmore Ely, One Who Served. Chicago: A.C. McClurg & Co., 1919.
New England Aviators.
Technology's War Record. By Alumni Association, Massachusetts Institute of Technology, 1920.
• *ENDICOTT, JOHN*
Harvard's Military Record in the World War.
author correspondence with Harvard University 2 June 1989
author correspondence with Harvard University 17 December 1996
• *EOFF, ROBERT*
author interview with Robert Eoff 17 November 1984
author interview with Fred Thon 21 April 1990
• *FAIRCHILD, EDWIN*
Military Records of Cornell in the World War.
Letters & correspondence of Edwin Bradley Fairchild
author interview with Alan Hansvold 17 November 1984
• *FAITH, CLARENCE*
Clarence Faith death certificate
author correspondence with Dept of Veterans Affairs 3 August 1990
• *FAUNTLEROY, CEDRIC*
author conversation with Jacquie Fauntleroy 7 June 1992
author conversation with John Fauntleroy Dalman 7 June 1992
• *FERGUSON, IAN*
author correspondence with Easton Library 11 November 1985
author correspondence with Juliana Painter 12 November 1985
author correspondence with Juliana Kimball 12 November 1985
author correspondence with Louise Ransom 10 November 1985
• *FLYNN, JOSEPH*
Sloan, Jr., James J. *Wings of Honor.* Atgen, Pa: Schiffer Publishing Ltd., 1994
Dr. Gros notes from Crehore collection via Dr. Parks
• *FORD, CHRISTOPHER*
Aero Club of America 1917.
Hamlen, Joseph. *Flight Fever.* Garden City, New York: Doubleday & Company, 1971
Who's Who In Aviation. Chicago: Ziff-Davis Publishing Company, 1942
Connell, Dennis & Frank W. Bailey. "Escadrille Lafayette N.124, Combat Log." *Cross & Cockade Journal*, 21, No. 4 (1980), 363–368
Author interview with Mrs. Theodore Vanneman 7 November 1982
Author interview with Marion Ford Perlik 12 February 1988
"Military History of Lt. Col. Christopher W. Ford, A.C., U.S. Army" via Marion Ford Perlik.
Various documents, including Christopher Ford's WWI logbook via Marion Ford Perlik
Official inventory of effects of Pilot Christopher Ford dated 29 October 1918 via Doughboy Historical Society
• *FORD, TOD*
William Henry Meeker, His Book. Privately Printed, 1917.
author correspondence with Dept of Vets Affairs 19 October 1991
author conversation with Aidan Gough 5 April 1991
• *FORSTER, HENRY*
Arthur, Reginald Wright. *Contact: Naval Aviator Careers 1–2000.*
Forster, Henry. *Over There 1917–1918 Recollections of Wartime.*
New England Aviators.
author interview with Hamilton Forster 28 March 1993
author interview with Bayard Forster 4/5/93
• *FOWLER, ERIC*
Memorial Volume of the American Field Service in France.
Princeton in the World War. Princeton University Press, 1932.
author interview with Eric A. Fowler 4 May 1993
information provided by Eric A. Fowler 4 May, 17 May 1993.
• *FREY, WILLIAM*
Dr. Gros records from Crehore collection via Dr. Jim Parks.
William Frey passport record via the National Archives.
• *GENET, EDMUND*
An American For Lafayette."The Diaries of E.C.C. Genet.Lafayette Escadrille.
Parsons, Edwin C. *The Great Adventure.*
Rockwell, Paul Ayres. *American Fighters In The Foreign Legion.*
War Letters of Edmond Genet.
Various documents via Roberta Arminio, Ossining Historical Society.
Author interview with De Witt Talmage Minich 29 January 1990
• *GIBSON, WILLIAM*
U.S. Pilots Stationed in the Zone of Advance. Compiled by Lt. Samual Vary, May 1919.
Who's Who in America 1924–25
author correspondence with Rensselaer Polytechnic Institute 23 January 1995
• *GILL, JOSEPH*
State of California death certificate
author correspondence with Dept of Veterans Affairs, Golden Gate National Cemetery, San Bruno, CA
author correspondence with Stanford University 29 March 1993
• *GILMORE, WILLIAM*
Arthur, Reginald Wright. *Contact: Naval Aviator Careers 1–2000.*
History of the American Field Service In France. Houghton Mifflin, 1920.
Technology's War Record. By Alumni Association, Massachusetts Institute of Technology, 1920.
Dr. Gros's records from Crehore collection via Dr. Jim Parks
Gilmore American Field Service enlistment sheet via Bill Foley
• *GLOVER, CLARENCE*
Paul Rockwell letter to Fred Andrews 9 September 1971
1928 letter from the "Vigilance Committee" via A. Crehore
22 October 1917 New York *Times* letter by Clarence Glover
Glover French Foreign Legion service record
• *GOURARD, REGINALD*
Reginald Gourard passport application via the National Archives
• *GREY, CHARLES*
American Aviators In The Great War 1914–1918. Millie Glasebrook, editor. Carson City, Nevada: The Glasebrook Foundation, 1984.
• *GRIEB, NORMAN*
Veil, Charles. *Adventure's a Wench.* New York: Grosset and Dunlap, 1934.
Yale in the World War. Yale University Press, 1925.
author correspondence with Yale University 25 May 1993.
author interview with Marsha Grieb Slothower 8 June 1993
author interview with H. Norman Grieb 10 June 1995
• *GRIER, JAMES*
Arthur, Reginald Wright. *Contact: Naval Aviator Careers 1–2000.*
Author correspondence with Historical Society of Pennsylvania 1 December 1995

• *GUEST, DAVID*
Yale In the World War. Yale University Press, 1925.
author correspondence with Yale University Alumni Ass. 9 May 1990
author conversation with Helen (Guest) Longley 3 March 1991; 2 February 1996
• *GUNDELACH, ANDRE*
Codman, Charles. *Contact* Boston: Little Brown, & Co., 1937.
author correspondence with Bob Gundelach 22 November 1995
• *GUY, DAVID*
Princeton in the World War. Princeton University Press, 1932.
author interview with Valerie Guy 2 October 1983
author correspondence with Valerie Guy 20 November 1984
• *HALL, JAMES*
Briand, Paul L., Jr. *In Search of Paradise*. "The Nordhoff-Hall Story." Honolulu: Mutual Publishing Paperback Series. C.1986 by Margaret Briand.
Flammer, Philip *The Vivid Air*.
Hall, James Norman, *My Island Home*. "An Autobiography." Boston: Little, Brown and Company, 1952.
Harvard's Military Record In The World War.
Parsons, Edwin C., *The Great Adventure*.
Rickenbacker, Edward V. *Fighting the Flying Circus*. New York: Frederick A. Stokes Company, 1919.
Bailey, Frank W. "The 103rd Aero." USAS (Formerly Lafayette Escadrille.) *Cross & Cockade Journal* 19, No.4,(1978) 289–343.
Connell, Dennis & Frank W. Bailey. "Escadrille Lafayette, N.124, Combat Log." *Cross & Cockade Journal*. 21, No. 4 (1980), 363–368.
Correspondence of Edwin Bradley Fairchild.
Granville Pollock letter to Austen Crehore 22 November 1938 via Dr. Jim Parks
Austen Crehore letter to Ted Parsons 25 April 1947 via Dr. Jim Parks
Author Correspondence with Nicholas Rutgers 19 November 1989
• *HALL, WESTON*
The Epic Of Flight. "Soldiers Of Fortune." Time/Life Books, Inc., 1981
Flammer, Philip M. *The Vivid Air*.
Hall, Bert. *En l'Air*. New York: The New Library, 1918
Hall, Bert and John J. Niles. *One Man's War*. New York: Henry Holt and Company, 1929.
King, David Wooster. *L.M. 8046*. New York: Duffield & Company, 1927.
Mason, Herbert Molloy, Jr. *The Lafayette Escadrille*. New York: Random House, 1964
Rockwell, Paul A. *American Fighters In The Foreign Legion*.
Thenault, *Capitaine* Georges. *l'Escadrille Lafayette*.
War Letters of Kiffin Yates Rockwell.
Weeks, Alice S. *Greater Love Hath No Man*. Boston: Bruce Humphries, Inc., 1939.
Whitehouse, Arch. *Legion of the Lafayette*. Garden City, New York: Doubleday and Co., 1962
Paul Ayres Rockwell interview conducted by Dr. Kleiser, 1961.
Letter by James Norman Hall to Willis B. Haviland 6 September 1943, via Barbara Haviland Lamm
Correspondence of William Thaw via Virginia Slade Murphy
Correspondence of Beth Chapline via Mrs. Robert Chapline.
Various documents courtesy the Harvey Higgins Historical Society, Higginsville, Missouri, via J.R. Santmyer, L/Col(Ret)
Sunday Portland *Oregonian* for April 5, 1936 via Bert Hall, Jr.
Bert Hall's scrapbook via Bert Hall, Jr.
Author interview with Mrs. Robert Chapline 16 February 1986
Author interview with Harley Hoffman 16 February 1986
Author interview with Roy Milne 8 March 1986
Author interview with Bert Hall, Jr. 9 March 1988.
Author interview with Milton E. Parker 21 March 1988
Fred Andrews interview with Henry S. Jones Spring 1969
• *HANFORD, ROBERT*
author correspondence with Georgetown University 14 July 1993
Robert Hanford passport application via the National Archives
• *HARRISON, JOHN*
Hall & Nordhoff. *The Lafayette Flying Corps*.
• *HAVILAND, WILLIS*
Arthur, Reginald Wright. *Contact: Naval Aviator Careers* 1–2000.
McConnell, James R. *Flying for France*. 1917.
Weeks, Alice. *Greater Love Hath No Man*. Boston: Bruce Humphries, 1939.
History of the American Field Service In France. Houghton Mifflin, 1920.
War Letters of Edmond Genet. Charles Scribners Sons 1918.
The Navy Book of Distinguished Service. Harry Stringer-Editor. Fasset Publishing Co., Wash.,D C, 1921.
Gordon, Dennis & Martin O'Conner. "Worried Landings." *Naval History*, 2, No. 1
Lamm, Barbara Haviland. "Dad And His Flying Machines." *The American Legion*, October 1979.
Prop Wash. Naval Air Station Whidbey Island, Wash. Sept. 12, 1944 via Barbara Haviland Lamm.
Shirley, Noel. "John Lansing Callan-Naval Aviation Pioneer." *Over The Front*, 2, No. 2, (1987), 180–185.
Atlanta *Constitution* December 2, 1944.
Corona *Daily Independent* December 1, 1944.
Correspondence of W.B. Haviland via Barbara Haviland Lamm.
Frederick Prince interview conducted by Dr. S.H. Kleiser 19 May 1961.
Author Interviews with Barbara Haviland Lamm and Willis Haviland Lamm February–March 1988.
• *HEILBUTH, JOHN*
History of the American Field Service In France. Boston: Houghton Mifflin Company, 1920.
Heilbuth American Ambulance enlistment sheet via Bill Foley
• *HEWITT, THOMAS*
An American for Lafayette: The Dairies of E.C.C. Genet, Lafayette Escadrille.
Flammer, Philip M. *The Vivid Air*.
Mason, Herbert Molloy. *The Lafayette Escadrille*. New York: Random House, 1964.
Dr. S.H. Kleiser interview conducted with Admiral Ted Parsons, Dec. 1962.
Air Travel Magazine, February 1918 "An Aerial Soldier Of Fortune."
Washington *Times*, various editions.
New York *Times*, various editions.
Various documents, USAF Academy Library, USAF Academy, Colorado, via Dr. James Parks
Correspondence by author with Law Alumni Office, The George Washington University.
Interment Record, Arlington National Cemetery: Hewitt, Thomas M.,Jr. #618041
Certificate of Death, District of Columbia: Thomas M. Hewitt
E.B. Fairchild letter to Austen Crehore 30 April 1928 via Dr. Jim Parks
• *HICKSON, LESLIE*
The Amhurst Memorial Volume. Claude M. Fuess, editor. Amherst College, 1926.
Hickson American Field Service enlistment sheet via Bill Foley
author correspondence with Amherst College Archives 26 Jan 95
author correspondence with Harvard University Archives 26 Jan 95
• *HILL, DUDLEY*
The U.S. Air Service In World War One. Vol. IV. Maurer Maurer, Editor. Office of Air Force History Headquarters USAF Washington, 1979
Bailey, Frank W. "The 103rd Aero-USAS (Formerly Lafayette Escadrille)." *Cross & Cockade Journal*., 19, No.4 (1978).
Author interviews with Mrs. Evaline Hill Wagner September/October 1982
26 January 1942 letter from Carl Dolan to Willis Haviland
Papers and effects of Dudley Hill.
• *HINKLE, EDWARD*
Parsons, Edwin. *I Flew With the Lafayette Escadrille*.
"In Memorium." *Cross & Cockade Journal*, 8, No.2,(1967),197.
Ronnie, Art. "Hinkle of N.124." *Cross & Cockade Journal*, 5, No.1 (1964),48–52.
Ronnie, Art. "WWI Pilot Receives French Medal." *Cross & Cockade Journal*, 4, No. 3 (1963),309.
Truth or Consequences local paper, 13 July 1963
El Paso *Times* 24 April 1966 article by Dale Walker; also 22 January 1967.
Cincinnati *Enquirer* 22 January 1967
taped interview conducted by Dr. S.H. Kleiser
taped interview conducted by Dale Walker
• *HITCHCOCK, THOMAS JR.*
Aldrich, Nelson W., Jr. *Tommy Hitchcock An American Hero*.C. 1984 Margaret Mellon Hitchcock
• *HOBBS, WARREN*
Memorial Volume of the American Field Service In France.
Author correspondence with Worcester Academy 23 September 1996
• *HOEBER, ROBERT*
author correspondence with Dept of Veterans Affairs 3 August 1990
author conversation with Charles Hoeber 20 December 1993; 26 January 1993
author correspondence with Robert Hoeber October 1995

• *HOLDEN, MILTON*
Holden American Field Service enlistment sheet via Bill Foley
The *New York Times*: Dec 3, 1926; July 18, 1927; August 31, 1927; Sept 1, 1927; February 17, 1932; February 24, 1932; April 6, 1932; April 7, 1932; May 31, 1974 via Marilyn Van Dyke
Milton Holden Florida certificate of death.
• *HORTON, DABNEY*
New England Aviators 1914–1918.
"Cross & Cockade Journal" Vol.9. No. 3
author correspondence with Dartmouth College
author conversation and correspondence with Elizabeth (Horton) DeLoach 3 March 1985; 2 February 1987
• *HOSKIER, RONALD*
An American for Lafayette. "The Diaries of E.C.C. Genet. Lafayette Escadrille."
Memoirs of the Harvard War Dead in the War Against Germany.
Mason, Herbert Molloy. *The Lafayette Escadrille*. New York: Random House. 1964.
Parsons, Edwin C. *The Great Adventure*.
Whitehouse, Arch. *Legion of the Lafayette*. Garden City, New York: Doubleday and Co., 1962.
Who's Who In America. 1914–1939.
Ronnie, Art. "Hinkle of N.124." *Cross & Cockade Journal*, 5, No. 1 (1964), 48–52.
Author interview with E.R. Hoskier 27 October 1988.
Author interview with Eric Wood 20 November 1988.
Author interview with William Martin 8 October 1989
Author correspondence with Neil Flanagan, Eton College, UK
• *HOUGH, EDWIN*
Hall & Nordhoff. *The Lafayette Flying Corps*.
• *HUFFER, JEAN*
"Over the Front" Journal, Vol. 2, No. 2
author interview with John Lyons 24 October 1993
author interview with Jacques Dardenne 4 March 1994
author interview and correspondence with William Huffer November 1995
• *HUGER, DANIEL*
Arthur, Reginald Wright. *Contact: Naval Aviator Careers 1–2000*.
author correspondence with Daniel E. Huger, Jr., 7 October 1998
• *HUGHES, EARL*
author interview with Kenmar Blass 29 January 1994
Earl Hughes death certificate via Michigan Dept of Health
• *HULL, MARK*
Hall & Nordhoff. *The Lafayette Flying Corps*.
Mark Hull passport application via the National Archives
• *JACOB, SERENO*
author conversations with Merrit Jacob 12 April 1987
State of Connecticut death certificate.
• *JOHNSON, CHARLES*
Cunningham, Mary B. *The Founding Family of St. Louis*. c. 1977 Cunningham & Blythe. via James M. Chouteau
Flammer, Philip M. *The Vivid Air*.
Mason, Herbert Molloy, Jr. *The Lafayette Escadrille*. New York: Random House, 1964
Rockwell, Paul. *American Fighters In The Foreign Legion 1914–18*.
Cross & Cockade Journal. Vol. 18, #2 "The Diary of H. Clyde Balsley."
"Chouteau Johnson and the Lafayette Escadrille" by Robert E. Parkin, Genealogical Research & Productions, via Auguste Chouteau.
"The Diary of H. Clyde Balsley." *Cross & Cockade Journal*.,18, No. 2 (1977),97–123.
Wynne, H. Hugh. "Escadrille Lafayette." *Cross & Cockade Journal*, 2, #1 (1961)
"Sixty-First Annual Report of the Association of Graduates of the United States Military Academy at West Point, New York June 11, 1930." Printed by the Moore Printing Company, Inc., Newburgh, N.Y. via USMA Archives, West Point, New York
Clyde Balsley letter to Austen Crehore 5 March 1928 via Dr. Jim Parks
C.H. Dolan letter to H. Clyde Balsley 30 October 1939
Paul Ayres Rockwell letter to James R. Nilo, 8 December 1959.
Author correspondence with Alumni Records, University of Virginia, Charlottesville, VA.
Author interview with Mrs. Dudley Hill 14 March 1982
• *JOHNSON, HARRY*
author correspondence with Lehigh University 17 June 1996
author conversation with Guy Roche Johnson III 8 November 1997
24 May 1918 combat report by Lt. Thomas Cassady via Arnold Johnson
• *JOHNSTON, ARCHIBALD*
Military Records of Cornell University in the World War. Ithaca, New York: Cornell University, 1930.
Who's Who in American Aeronautics. New York: The Gardner, Moffat Co., 1922.
Author interviews with Arch Johnston, Jr 18 April 1982, 30 April 1997.
• *JONES, CHARLES*
author interview with Charles Maury Jones, Jr. 13 January 1985
• *JONES, HENRY*
Flammer, Philip M. *The Vivid Air*.
History of the American Field Service. Vol.III. Boston: Houghton Mifflin Co., 1920.
Mason, Herbert Molloy. *The Lafayette Escadrille*. New York: Random House, 1964.
Stevenson, William Yorke *At The Front In A Flivver*. Boston: Houghton Mifflin, 1920.
Bailey, Frank W. "The 103rd Aero.USAS(Formerly Lafayette Escadrille." *Cross & Cockade Journal* 19, #4 (1978),289–342.
West Point "Assembly" bulletin October 1948.
Correspondence of Henry S. Jones
Correspondence by author with Miss Prudence Clark, Harford Historical Society.
Correspondence by author with Mrs. Louise Tiffany and Mrs. E.R. Potok.
Henry Jones interview conducted by Fred Andrews 1969, 1970.
• *JUDD, DAVID*
Arthur, Reginald Wright. *Contact: Naval Aviator Careers 1–2000*.
The Navy Book of Distinguished Service. Harry Stringer, ed. Fassett Publishing Company, Washington, D.C., 1921.
New England Aviators 1914–1918. Boston: Houghton Mifflin, 1919.
• *KENYON, HUGO*
History of the American Field Service In France. Houghton
"War Goes Aloft," an unpublished manuscript by C. Hoggson Faith
author correspondence with Hugo G.R. Kenyon 31 December 1991
"Providence Sunday Journal" 12 November 1953
• *KERWOOD, CHARLES*
Who's Who in Aviation 1942–1943. Chicago & New York: Ziff-Davis, 1942.
author interview with Mrs. Charles Kerwood 22 March 1982
• *KINSOLVING, CHARLES*
author interview with Charles Kinsolving 14 February 1978
Montery Peninsula *Herald* 25 December 1984
• *KIRKWOOD, WILLIAM*
History of the American Field Service In France.
author conversation with Wilhelmina (Kirkwood) Angley 8 March 92
author conversation with Miriam (Kirkwood) Moran 12 April 92
William F. Kirkwood Massachusetts death certificate
Various documents via Miriam (Kirkwood) Moran.
• *KOWALL, JOHN*
Dr. Gros's records from Crehore collection via Dr. Jim Parks
• *KRUIJFF, THEODORE DE*
deKruijff military burial record via the National Archives
• *KYLE, GEORGE*
Blue Book of Aviation 1932. Los Angeles: The Hoagland Co, 1932
author correspondence with David Kyle 31 July 1985
author correspondence with Donald Kyle 25 July 1985
• *LARNER, GORMAN*
Who's Who in American Aeronautics. Lester D, Gardner, ed. Gardner, Moffat Company, Inc. 1922.
Easton, Maryland *Star Democrat* 22 May 1984
author correspondence with Mrs. Jean Larner 112/16/98
• *LEE, HENRY*
Dr. Gros records from Crehore papers via Dr. Jim Parks
author correspondence with Pratt Institute 23 February 1995
Lee record-New York Adjutant General's Office via Marilyn Van Dyke.
author correspondence with the National Personnel Records Center, St. Louis, Mo. 10 June 1998
• *LEHR, MANDERSON*
author correspondence with Lewis Lehr 7 September 1988; 11 November 1993.
• *LEWIS, DAVID*
"The Wartime Experiences of David W. Lewis" compiled by Dr. James J. Parks. Cross & Cockade Journals Vol 22, #2; Vol 23, No 3; Vol 25, No 4; Over the Front, Vol 3, No 1.
author correspondence with Doris Lewis 25 November 1993.

• *LITTAUER, KENNETH*
Who's Who in American Aviation, 1922.
"Cross & Cockade Journal" Vol. 9, No 4; Vol 13, #1
• *LOOMIS, RALPH*
Arthur, Reginald Wright. *Contact: Naval Aviator Careers 1–2000.*
New England Aviators. Boston: Houghton Mifflin, 1919.
author interview with Ann Benson 12 May 1986
author interview with F. Payson Todd 21 August 1987
• *LOOMIS, WILLIAM*
New England Aviators. Boston: Houghton Mifflin, 1919.
Who's Who In Aviation.
author interview with Richard G. Loomis 11 August 1985
author correspondence with H.R. Loomis 4 November 1998
• *LOUGHRAN, EDWARD*
author interview with James Loughran 6 January 1994.
• *LOVELL, WALTER*
An American For Lafayette, The Dairies of E.C.C. Genet, Lafayette Escadrille.
Flammer, Philip M. *The Vivid Air.*
Friends of France, Houghton Mifflin, 1916
Harvard's Military Record in the World War.
Mason, Herbert Molloy, Jr. *The Lafayette Escadrille.*
New England Aviators 1914–1918.
Parsons, Edwin C. *I Flew With The Lafayette Escadrille.*
Bulletin of the American Field Service Ass'n.
Wynn, H. Hugh. "Escadrille Lafayette." *Cross & Cockade Journal*, 2, No. 1 (1961).
Taped interview of Harold B. Willis by Thomas G. Miller, Jr., 25 January 1961.
Author interview of Charles and Ann Lovell, Chestnut Hill, MA May 15, 1988.
• *LUDLAM, WILLIAM*
Dr. Gros records from Crehore papers via Dr. Jim Parks
William Ludlam passport application via the National Archives
• *LUFBERY, RAOUL*
An American for Lafayette. The Diaries of E.C.C. Genet, Lafayette Escadrille.
Dictionary Of American Biography, Vol.XI. New York: Charles Scribners Sons, 1930.
Driggs, Laurence L. *Heroes of Aviation.* Boston: Little, Brown and Co., 1918.
Flammer, Philip M. *The Vivid Air.*
Parsons, Edwin C. *The Great Adventure.*
Rockwell, Paul A. *American Fighters in the Foreign Legion.*
Thayer, Lt. Lucien H. *America's First Eagles.* The Official History of the U.S. Air Service A.E.F.(1917–1918) Champlin Fighter Museum Press, 1983.
Thenault, *Capitaine* Georges. *l'Escadrille Lafayette.* Paris: Librairie Hachette, 1939.
El Paso *Times* interview of Edward Foote Hinkle 24 April 1966.
Air Force Magazine. February 1974. "Carl Dolan, Last Survivor of the Lafayette Escadrille." by Capt. Paul E. Sjordal, USAF
Cavanaugh, R.L. "The 94th and its Nieuports. 15 April–11 June 1918." *Cross & Cockade Journal*. 21, No. 3. (1980), 193–222.
Connell, Dennis & Frank W. Bailey. "Escadrille Lafayette, N.124, Combat Log." *Cross & Cockade Journal*. 21, No. 4 (1980), 363–368.
Ronnie, Art. "Hinkle of N.124." *Cross & Cockade Journal.* 5, No. 1, (1964), 48–52.
Carl Dolan interview conducted by Kenneth Boege in 1974.
Author correspondence with Bertha Lufbery Haller 2 February 1987.
Author correspondence with Marcel Lufbery 6 February 1987.
Henry Jones interview conducted by Fred Andrews September 1969.
• *McCALL, GEORGE*
History of the American Field Service in France.
author correspondence with U of Pa archives 30 April 1987
author conversation with J. Bartow McCall 10 January 1997
Phi Kappa Sigma bulletin for 1930 via Bartow McCall
• *McCONNELL, JAMES*
An American for Lafayette."The Diaries of E.C.C. Genet, Lafayette Escadrille."
Flammer, Philip M. *The Vivid Air.*
Friends of France."The Field Service of the American Ambulance." Houghton Mifflin, 1916.
Mason, Herbert Molloy. *The Lafayette Escadrille.*
Memorial Volume of the American Field Service In France.
Parsons, Edwin C.*The Great Adventure.*
Rockwell, Paul A. *American Fighters in the Foreign Legion.*
Weeks, Alice. *Greater Love Hath No Man.* Boston: Bruce Humphries, Inc. 1939.
Who's Who In America. 1903–1937.
Bulletin of the American Field Service Association.
University of Virginia Alumni Association bulletin
University of Virginia *Alumni News.* March–April 1988.
Wynn, H. Hugh. "Escadrille Lafayette" *Cross & Cockade Journal*. 2, No.1 (Spring 1961)
National Service Journal. March 1917. "The Day's Work of an American Airman on the Somme." by James McConnell.
New York *Times* March 24, 25, 30, April 3, 1917.
Author interview with Dr. Kenneth Rockwell 12 February 1989
Author interview with Mrs. Evelyn Truitt 25 October 1988
Author interview with Mrs. Eleanor Vonerffa 3 March 1990
• *McCREARY, JAMES*
Hall & Nordhoff. *The Lafayette Flying Corps.*
• *McGINN, WILLIAM*
Dr. Gros records from Crehore collection via Dr. Jim Parks
William McGinn passport application via the National Archives
• *McKEE, HERSHEL*
author correspondence with Linda Jeanne McKee 11 August 1985
• *McKERNESS, WILLIAM*
Service Records Conn. Men & Women in the Armed Forces of the United States during World War 1917–1920. Vol. III
Soldiers who made the Supreme Sacrifice-World War I. Vol. IV Clara Newell, editor.
• *McMILLEN, JAMES*
Dade, George, and George Vecsey. *Getting off the Ground.* New York: E.P. Dutton, 1979.
author interview with James McMillen, Jr. June 12 1985
author interview with Jane McMillen Hayes 13 May 1985
author correspondence with Jane McMillen Hayes 24 Jan 1994
• *MACKE, GORDON*
Aero Club of America 1917. New York.
Dr. Gros records from Crehore collection via Dr. Jim Parks
author correspondence with Valparaiso University, 3 September 1996
Gordon Macke passport application via the National Archives
• *MacMONAGLE, DOUGLAS*
Friends of France. Houghton Mifflin Co., 1916
History of the American Field Service In France.
Parsons, Edwin C. *I Flew With The Lafayette Escadrille.*
Mason, Herbert Molloy *The Lafayette Escadrille.*
Memorial Volume of the American Field Service In France.
Whitehouse, Arch. *Legion of the Lafayette.* New York: Doubleday, 1962.
Flammer, Philip M. *The Vivid Air.* "The Lafayette Escadrille."
"Bulletin of the American Field Service Ass'n" No. 4, July 1921
Wynne, H. Hugh. "Escadrille Lafayette." *Cross & Cockade Journal.*" 2, No. 1 (1961).
Author correspondence with Carl Dolan 28 November 1979.
Author interview conducted with Douglas MacMonagle Moody 14 May, 1988.
26 September 1917 Memorial statement written by *Capitaine* Thenault via Douglas MacMonagle Moody.
Carl Dolan interview conducted by Ken Boege in 1974.
American Ambulance application No. 205
• *MAGLEY, GUY*
author conversations with Mary Magley 17 Dec 1996, 4 Jan 1997.
• *MALONE, CHARLES*
Thomas Malone letter to Lewis Crenshaw February 1928
State of California death certificate
author correspondence with Dept of Army, Arlington National Cemetery 8 August 1990.
• *MANNIERE, HAROLD*
Dr. Gros records from Crehore collection via Dr. Jim Parks.
author correspondence with Kingston New York public library 3 Oct 1996.
author conversation with Harold Huppert 9 October 1996
• *MARR, KENNETH*
Friends of France. The American Field Service Of The American Ambulance. Boston: Houghton Mifflin Co., 1916
Flammer, Philip M. *The Vivid Air.* "The Lafayette Escadrille."
History of the American Field Service in France.
Mason, Herbert Molloy, Jr. *The Lafayette Escadrille.*
Parsons, Edwin C. *The Great Adventure.*

Rickenbacker, Capt. E.V. *Fighting the Flying Circus*. New York: Frederick A. Stokes Company, 1919
"Air Service Combat Organization." *Cross & Cockade Journal*, 3, No. 1, (1962), 37.
Bailey, Frank W. "The U.S. 103rd Aero Squadron USAS (Formerly Lafayette Escadrille." *Cross & Cockade Journal*, 19, No.4 (1978), 289–342.
Cavanagh, R.L. "The 94th And its Nieuports 15 April–11 June 1918." *Cross & Cockade Journal*, 21, No. 3 (1980), 193–222.
Connell,. Dennis & Frank W. Bailey. "Escadrille Lafayette, N.124, Combat Log. *Cross & Cockade Journal*, 21, No. 4 (1980), 363–368.
Wynne, H. Hugh. "Escadrille Lafayette." *Cross & Cockade Journal*, 2, No. 1 (1961)
Author interview with Hugh Parshall 19 January 1985
Author interview with William Lanyon 2 March 1985
Author interview with Robert Lanyon 2 March 1985
Alice Marr letter to P.J. Carisella dated 7-25-62
California certificate of death for Kenneth Marr

• *MASSON, DIDIER*

Flammer, Philip M. *The Vivid Air*. "The Lafayette Escadrille."
Harris, Sherwood. *The First To Fly: Aviation's Pioneer Days*. New York: Simon & Schuster, 1970
Mason, Herbert Molloy, Jr. *The Lafayette Escadrille.*
Parsons, Edwin C. *I Flew With The Lafayette Escadrille.*
Soldier of Fortune. A volume in the Epic of Flight series. Time-Life Books, 1981
Wynne, H. Hugh. "Escadrille Lafayette." *Cross & Cockade Journal*, 2, No.1 (1961)
Correspondence between Ernest Jones and Didier Masson and Didier Masson, Jr. 1935–1950

• *MEEKER, WILLIAM*

Memoirs of Harvard Dead in the War Against Germany.
William Henry Meeker, His Book. Privately Printed, 1917

• *MILLER, ALVIN*

Hall & Nordhoff. *The Lafayette Flying Corps.*
Alvin Miller passport application via the National Archives

MILLER, GEORGE

Dr. Gros enlistment records from Crehore collection via Dr. Parks
George Miller passport application via the National Archives

• *MILLER, WALTER*

Memorial Volume of the American Field Service In France.
Walter Miller burial record via National Archives 10 June 1998

• *MILLS, GORDON*

Dr. Gros enlistment records from Crehore collection via Dr. Parks.

• *MOLTER, BENNET*

author correspondence with New Orleans public library 4 November 1986
author conversations with Bennett Molter, Jr. 9 August 1987.

• *MOORE, ROBERT*

"Southern Aviation Journal" 1 January 1931."'Bob' Moore."
Author correspondence with the Texas/Dallas Historical archives 25 October 1986.
Author correspondence with Genealogy Collection, Miami-Dade Public Library 25 October 1986.
Florida death certificate for Robert Louis Moore

• *MOSELEY, GEORGE*

Arthur, Reginald Wright. *Contact: Naval Aviator Careers 1–2000.*
Extracts from the War Letters of George Clark Moseley. C, 1923, George Clark Moseley.
Football Y Men 1872–1919. Men Of Yale Series, Vol. I. Albert Beecher Crawford, Editor. C.1962, Yale University.
Author interview with Thomas Moseley 7 November 1989
Author interview with Mrs. George Moseley 20 May 1984
Author interview with Mrs. Melissa Foulke 3 September 1993
Author interview with Bruce Moseley 31 October 1993

• *MOUVET, OSCAR*

Rockwell, Paul A. *American Fighters in the Foreign Legion.*
Oscar Mouvet Foreign Legion enlistment record

• *MUNSON, CURTIS*

Yale in the World War. Yale University Press, 1925.
"Cross & Cockade Journal." Vol. 23, No. 1
author correspondence with Yale University 10 November 1984
author interview with Mrs. Edward Roberts 16 June 1990
author interview with Anne Munson Calcagni 9 April 1993

• *NICHOLS, ALAN*

Letters Home From The Lafayette Flying Corps. Nancy Nichols, editor. San Francisco: J.D. Huff & Co, 1993
Memorial Volume of the American Field Service in France. American Field Service, 1921.

• *NORDHOFF, CHARLES*

Gordon, Dennis. *Lafayette Escadrille Pilot Biographies.*
State of California death certificate

• *OAKES, NATHAN*

Author interview with Marjorie King Oakes 20 October 1989
Nathan Oakes papers.

• *OVINGTON, CARTER*

author interview with Robert Boname 12 April 1996
author interview with William Oechler 12 April 1996
author correspondence with William Oechler 18 April 1996
author correspondence with Roland J. Marcel 16 May 1997

• *PADEN, DAVID*

author interview with Desmond Paden 3 March 1991
author correspondence with Evanston Historical Society 1 Dec 1987
author correspondence with the University of Michigan 5 May 1987
Paden's Florida death certificate

• *PALMER, HENRY*

Memoirs of the Harvard Dead in the War Against Germany.
Memorial Volume of the American Field Service in France.

• *PARKER, AUSTEN*

Arthur, Reginald Wright. *Contact: Naval Aviator Careers 1–2000.*
Military Records of Cornell University in the World War. Cornell University, 1930.
author interview with Jane Parker 22 October 1985

• *PARSONS, EDWIN*

An American For Lafayette "The Diaries of E.C.C. Genet, Lafayette Escadrille."
Flammer, Philip M. *The Vivid Air.*
Mason, Herbert Molloy, *The Lafayette Escadrille.*
New England Aviators.
Parsons, Edwin C. *The Great Adventure.*
Stevenson, William Yorke. *At The Front In A Flivver.* Boston & New York: Houghton Mifflin Co., 1917.
Walker, Dale. L. *Only The Clouds Remain.* New York: Alandale Press, 1980.
Whitehouse, Arch. *Legion of the Lafayette*. Garden City, New York: Doubleday and Co., 1962.
Who's Who in American Aviation.
Connell, Dennis & Frank W. Bailey. "Escadrille Lafayette, N.124, Combat Log." *Cross & Cockade Journal*, 21, No.4 (1980),363–368.
Wynne, H. Hugh. "Escadrille Lafayette." *Cross & Cockade Journal*, 2, No. 1 (1961), 74–81.
Admiral Parsons obituary. *Cross & Cockade Journal*, 9, No.3 (1968), 256.
Rear Admiral Parsons interview conducted by Dr. S.H. Kleiser 14 April 1961 and November of 1962.

• *PAVELKA, PAUL*

Bowe, John. *Soldiers of the Legion.* Chicago: Peterson Linotyping Co., 1918.
Flammer, Philip. *The Vivid Air.* "The Lafayette Escadrille."
Mason, Herbert Molloy. *The Lafayette Escadrille.*
Parsons, Edwin C. *The Great Adventure.*
Rockwell, Paul A. *American Fighters in the Foreign Legion.*
War Letters of Kiffin Yates Rockwell.
Weeks, Alice S. *Greater Love Hath No Man.* Boston: Bruce Humphries, 1939.
Brown, Philip C. "Pavelka of the Lafayette." *Cross & Cockade Journal*, 19, No.2 (1978), 97–108
Dounce, Harry Esty. "Paul Pavelka, A Soldier of Fortune." The New York *Sun*, December 30, 1917.
McConnell, James R. "The Day's Work of an American Airman on the Somme." *National Service Journal*, March 1917.
Author interview with Joseph Pavelka 19 October 1988.

• *PELTON, ALFRED*

author correspondence with Stewart K. Taylor
author conversation with Bishop College School 13 Sept 1996

• *PETERSON, DAVID*

Flammer, Philip, *The Vivid Air.* "The Lafayette Escadrille."
Mason, Herbert Molloy. *The Lafayette Escadrille.*
Parsons, Edwin C. *The Great Adventure.*
Rickenbacker, Capt. E.V. *Fighting The Flying Circus*. New York: Frederick A. Stokes Co., 1919.

Alumni Bulletin of Lehigh University: Vol.5, No.2,"Nov.1917"; Vol. 5, No.4,"May.1918"; Vol.6, No.1,"August 1918"; Vol.6,No.4,"May 1919."
"Insignia of the A.E.F. Aero Squadrons." *Daughters of the American Revolution Magazine*, LIV, No. 3, (March 1920).
Correspondence of Dr. Preston B. Peterson via David Peterson.
New York *Times*, various issues, 1917, 1919, 1928
Carl Dolan interview conducted by Ken Boege, 1974.
Letter to author from M.R. Zimmer & Associates, Inc. 1 January 1989
Author correspondence with Wayne County Historical Society September, 1988
• *POLLOCK, GRANVILLE*
author correspondence with the B of Vet Affairs 10 March 1990
Granville Pollock Texas certificate of death
• *PONDER, WILLIAM*
author correspondence with Janine Ponder Hall 10 Feb 1994
Dallas Texas "Journal" 26 May 1932
• *POTTER, THOMAS*
St. Paul's School in the Great War. Alumni Association, 1926
New York *Times* 24 April 1936 via Marilyn Van Dyke
Potter record New York Adjutant General's Office via Marilyn Van Dyke
Thomas Potter passport application via the National Archives
• *PRINCE, FREDERICK*
Aldrich, Nelson W., Jr. *Tommy Hitchcock, An American Hero*.
Beebe, Lucius. *The Big Spenders*. Doubleday & Company, Inc.,1966.
Flammer, Philip M. *The Vivid Air*.
Harvard's Military Record In The World War.
Mason, Herbert Molloy, Jr. *The Lafayette Escadrille*.
National Cyclopedia of American Biography. Clifton, New Jersey: James T. White & Co.
New England Aviators 1914–1918.
Norman Prince, A Volunteer Who Died For The Cause He Loved, Boston: Houghton Mifflin, 1917
War Records Of The Knickerbocker Club 1914–1918. 1922.
Wynne, H. Hugh. "Escadrille Lafayette." *Cross & Cockade Journal*, 2, No. 1 (1961).
Frederick Prince, Jr. interview conducted by Dr. S.H. Kleiser 19 May 1961.
Author interview with Mrs. Virginia Mitchell Prince 6 January 1985.
Carl Dolan interview conducted by Ken Boege.
New York *Times*, various issues from 1917, 1918, 1923, 1927, 1929, 1932, 1933, 1949, 1953, 1962.
Letters of William Thaw to Dr. Edmund L. Gros 1 March 1924; to Mr. H.C. Hoskier March 12, 14, 18, 24, 27, 1924. Letters from Mr. H.C. Hoskier to William Thaw March 14, 18, 24, 1924.
• *PRINCE, NORMAN*
Flammer, Philip M. *The Vivid Air*. "The Lafayette Escadrille."
Harvard's Military Record In The World War.
Mason, Herbert Molloy, Jr. *The Lafayette Escadrille*.
McConnell, James R. *Flying For France*. Doubleday, 1917
Memoirs of the Harvard Dead in the War Against Germany.
New England Aviators 1914–1918.
Norman Prince, A Volunteer Who Died For The Cause He Loved.
War Letters of Edmond Genet. Grace Ellery Channing, Editor. New York: Charles Scribner's Sons, 1918
War Letters of Kiffin Yates Rockwell.
War Records Of The Knickerbocker Club 1914–1918. Houghton Mifflin, 1922
Cowdin, Elliot C. "The Lafayette Escadrille." *Harvard Alumni Bulletin*.
Frederick Prince interview conducted by Dr. S.H. Kleiser 19 May 1961
Mrs. Virginia Mitchell Prince interview conducted by author 6 January 1985
• *PUTNAM, DAVID*
Franks, Norman & Frank Bailey. *Over The Front*. London: Grub Street, 1992.
Memoirs of the Harvard Dead in the War against Germany.
New England Aviators 1914–1918.
author correspondence with Barbara Putnam Van Winkelen 2 July 1995
• *RAND, RUFUS*
Williams College in the World War. New York: Shilling Press, 1926
author correspondence & conversations with Penny Rand Winton 25 January 1992; 11 April 1993; 20 May 1994
• *RANDALL, JOHN*
Author interview with Richard Lent 9 May 1991
• *READ, ROBERT*
Arthur, Reginald Wright. *Contact: Naval Aviator Careers 1–2000*.
author interview with Mrs. Robert E. Read 11 November 1984
Robert Read obituary from Glens Falls *Post Star* 7 September 1996 via Marilyn J. Van Dyke.
author conversation with Robert Read, Jr. 9 November 1998
• *RENO, LEONARD*
Arthur, Reginald Wright. *Contact: Naval Aviator Careers 1–2000*.
Author interview with Kenneth Peters 7 July 1985
author conversation and correspondence with E. Brook Peters 15 February 1996; 27 February 1997
• *RHENO, WALTER*
New England Aviators 1914–1918.
• *RIDLON, HUGH*
author interview with Henrietta Ridlon 14 March 1985
author interview with Donald Piccard 9 November 1996
author correspondence with Paul Piccard 19 November 1996
author conversation with Helen Ridlon 30 November 1996
author correspondence with Donald Piccard 7 january 1997
Florida death certificate for Hugh O.T. Ridlon
• *ROCKWELL, GEORGE*
Friends of France. The American Field Service of the American Ambulance. Boston & New York: Houghton Mifflin, 1916.
Men of Mark in Connecticut. Col. N.G. Osborn, ed. Hartford: William R. Goodspeed, 1910. via Robert Ford
Service Records Connecticut in World War One 1917–1920. Vol.III, p. 2847.
author correspondence with Dept of Veterans Affairs
author correspondence with Bill Foley 19 December 1994
author correspondence with Mattatuck Museum, Waterbury, Ct 9 December 1996
State of New Jersey death certificate.
author conversation with Carol Potter 10 July 1997
various Rockwell documents via Jean Burns 19 November 1997
• *ROCKWELL, KIFFIN*
Flammer, Philip *Vivid Air*.
Mason, Herbert Molloy.The Lafayette Escadrille.
McConnell, James R. *Flying for France*. Garden City, New York: Doubleday, Page & Co., 1917
Parsons, Edwin C. *The Great Adventure*.
Rockwell, Paul A. *American Fighters in the Foreign Legion*.
Victor Chapman's Letters from France. New York: The MacMillan Co., 1917.
Weeks, Alice S. *Greater Love Hath No Man*. Boston: Bruce Humphries, 1939.
War Letters of Kiffin Yates Rockwell. Garden City, New York: The Country Life Press, 1925.
Connell, Dennis & Frank W. Bailey. "Escadrille Lafayette, N.124, Combat Log." *Cross & Cockade Journal*, 21, No.4 (Winter 1980), 363–368.
House, R.B. "Kiffin Yates Rockwell." *The North Carolina Booklet*. April–July 1920. via Dr. Kenneth Rockwell
Wynne, H. Hugh. "Escadrille Lafayette." *Cross & Cockade Journal*,2, No.1 (Spring 1961).
Dr. Kenneth Rockwell interviews by author 11/13/88, 2/12/89
Dr. Kenneth Rockwell letter to author, 21 October 1989
• *ROCKWELL, ROBERT*
Conover, Charlotte Reeve. *Dayton and Montgomery County*. Vol.III. c. 1932. via William Rockwell
Flammer, Philip M. *The Vivid Air*. "The Lafayette Escadrille."
Mason, Herbert Molloy, Jr. *The Lafayette Escadrille*.
U.S. Air Service in WWI. Vol.III. U.S. Government Printing Office, Washington, D.C., 1979
Who's Who in Aviation 1942–1943. Chicago Aero Commission. Chicago: Ziff-Davis Publishing Co., 1942
Bailey, Frank W. "The 103rd Aero USAS (Formerly Lafayette Escadrille)."*Cross & Cockade Journal* , 19, No. 4, (1978)
Wynne, H. Hugh. "Escadrille Lafayette." *Cross & Cockade Journal*, 2, No. 1 (1961).
Cincinnati *Enquirer*. January 28, 1958 via Kenyon College
Dayton *Daily News*. January 26, 1958 via Kenyon College.
1942 News Notes of the Service via Kenyon College.
Frederick H. Prince interview conducted by Dr. S.H. Kleiser 19 May 1961
Author interview with Art Lehman, formerly U.S. 103rd Pursuit Squadron
Author interview with Frank Rockwell, 17 October 1982
Author interview with William Rockwell 17 October 1982
• *ROCLE, MARIUS*
author correspondence with Peggy Rocle Hugenberger August 1988

• *RODGERS, WILLIAM*
Culver in the World War. Culver, Indiana: Culver Legion, 1930.
author correspondence with Princeton University 19 April 1990
Florida death certificate
• *ROLF, JOHN*
Hall, James Norman, and Charles B. Nordhoff. *The Lafayette Flying Corps*. In two volumes. Boston: Houghton Mifflin, 1920.
John Rolf passport application via the National Archives
• *ROODE, CLIFFORD DE*
author interview with David deRoode 5 June 1988, 9 October 1996
• *ROSS, RAYMOND*
The Amherst Memorial Volume. Claude Fuess, editor. Amherst College, 1926.
author correspondence with Bill Foley 12 December 1994
author correspondence with Amherst College 10 February 1995.
• *ROTHARMEL, KENNETH*
author interview with Richard Rotharmel 21 January 1987
• *ROUNDS, LELAND*
author interview with Gail Remington Willetts 5 November 1990
author interview with Dr. Eldredge Lasell 5 January 1990.
• *RUMSEY, LAURENCE*
Aldrich, Nelson W.,Jr. *Tommy Hitchcock An American Hero*.
An American for Lafayette, The Diaries of E.C.C. Genet.
Flammer, Philip M. *The Vivid Air*. Athens: U of Ga Press, 1981
Hall, Bert *One Man's War*. New York: Henry Holt and Company, 1929
Harvard's Military Record in the World War.
Mason, Herbert Molloy, Jr. *The Lafayette Escadrille*.
Rumsey, Laurence. "Training an American Aviator in France." *National Service Journal*. May 1917. via Laurence Goodyear.
Buffalo New York newspaper articles via Laurence Goodyear
New York *Times* September 22, 1922.
Author interview with Laurence Goodyear 14 March 1982.
Author interview with Grace Smith 12 June 1984.
Admiral Edwin C. Parsons interview conducted by Dr. S.H. Kleiser, November 1962.
• *SAUL, JOSEPH*
Hall, James Norman, and Charles B. Nordhoff. *The Lafayette Flying Corps*. In two volumes. Boston: Houghton Mifflin, 1920.
• *SAXON, HAROLD*
author interview with Mary Ann Tuerffs 9 February 1991
author correspondence with Janet D. Saxon 1 November 1995
• *SCANLAN, LAWRENCE*
Rockwell, Paul A. *American Fighters In The Foreign Legion*.
• *SCHREIBER, EDWIN*
New York *Times* 8 October 1918
Schreiber service card, New York State Archives via Marilyn Van Dyke
• *SEAVER, HORACE*
author correspondence with Cooper Union College 8 July 1990
• *SHAFFER, WALTER*
Bluebook Of Aviation. Los Angeles: Hoagland Co, 1932
Shaffer, Major Walter J. *Exploding Flying Myths*. Atlanta: Williams Printing Co., 1968
author interview with Clarissa King 4 November 1984; 12 Aug 1994
• *SHIPLEY, WALTER*
Hall, James Norman, and Charles B. Nordhoff. *The Lafayette Flying Corps*. In two volumes. Boston: Houghton Mifflin, 1920.
• *SHONINGER, CLARENCE*
author correspondence with Yale University 3 December 1984
author conversation with Ruth Shoninger Olarte 15 July 1994; correspondence from Ruth Shoninger Olarte 21 July 1994
• *SINCLAIRE, REGINALD*
Information provided by Halsted Morris 3 November 1992
• *SITTERLY, GLENN*
author correspondence with Mrs. Glenn Sitterly 26 February 1984
author correspondence with Alumni Director, Beloit College 20 June 1994
• *SKINNER, SAMUEL*
Memoirs of Harvard Dead in the War Against Germany.
Granville Pollock letter to Austen Crehore 22 November 1938
• *SOUBIRAN, ROBERT*
Bowe, James. *Soldier of the Legion*. Chicago, 1918
Flammer, Philip M. *The Vivid Air*. "The Lafayette Escadrille."
Mason, Herbert Molloy, Jr. *The Lafayette Escadrille*.
Rockwell, Paul Ayers. *American Fighters In The Foreign Legion*.
Thenault, *Capitaine* Georges. *l'Escadrille Lafayette*. Paris: Librairie Hachette, 1939
Bailey, Frank W. "The 103rd Aero Squadron USAS (Formerly Lafayette Escadrille)." *Cross & Cockade Journal*, 19, No. 4 (1987), 289–342.
Cross & Cockade Journal, 9, No. 4 (1968), page 4.
26 January 1942 letter from Carl Dolan to Willis Haviland
Interview of Harold B. Willis conducted by Thomas G. Miller, Jr. 25 January 1961
Author interview with Rosemary Soubiran 24 March 1985
Author correspondence with Jacqueline Soubiran Rogers 30 October 1987
• *SPEERS, WALLACE*
Princeton in the World War.Princeton University Press, 1932.
author correspondence with Princeton University 3 April 1990
Wallace Speers passport application via the National Archives
• *SPENCER, DUMARESQ*
Phillips Academy Andover In The Great War. Claude Fuess, editor. New Haven: Yale University Press, 1924.
Yale In The World War. Yale University Press, 1925.
• *STANLEY, ALFRED*
author interview with DeWolf Stanley 8 Dec 1990, 8 Feb 1997
• *STARRET, FRANK*
New England Aviators 1914–1918, Houghton Mifflin, 1920.
• *STEARNS, RUSSELL*
New England Aviators 1914–1918. Houghton Mifflin, 1920.
author interview with L. Metcalf Walling 18 November 1890
author interview with Mary Barrett 16 December 1990
• *STEHLIN, JOSEPH*
author interview with Patricia Stehlin Boykin 11 May 1990
• *STICKNEY, HENRY*
author correspondence with Rutland Vermont Free Library 26 March 1986
Stickney records via the Veterans Administration.
• *STONE, DONALD*
Unpublished memoirs of George Dock
• *STONE, GERALD*
History of the American Field Service in France.
author correspondence with Dartmouth College 8 March 1985
author correspondence with Gerald Stone, Jr. 9 August 1990
• *SULLIVAN, UPTON*
Arthur, Reginald Wright. *Contact: Naval Aviator Careers 1–2000*. author correspondence with Hannah Sullivan 8 November 1983
• *TABER, LESLIE*
Arthur, Reginald Wright. *Contact: Naval Aviator careers 1–2000*.
History of the American Ambulance Field Service In France.
The Navy Book of Distinguished Service. Harry Stinger, editor.
author correspondence with Orrell Taber 26 February 1986
author correspondence with Bond Taber 16 February 1999
• *TAILER, WILLIAM*
Information via Myrna Sloan, Archivist, The Bryant Library, Roslyn, New York 4 September 1996
• *TAYLOR, ELMER*
Arthur, Reginald Wright. *Contact: Naval Aviator careers 1–2000*.
Navy Book of Distinguished Service. Harry Stringer, editor.
• *TERRES, HUGH*
Arthur, Reginald Wright. *Contact: Naval Aviator Careers 1–2000*. author correspondence with the University of Oxford 10 June 1996
• *THAW WILLIAM*
Allen, Richard Sanders, *Revolution In The Sky*."Those Fabulous Lockheeds: The Pilots Who Flew Them." Brattleboro, Vermont: The Stephen Greene Press, 1967.
Birmingham, Stephen, *Dutchess. The Story of Wallace Warfield Windsor*. Boston-Toronto: Little, Brown and Company, 1981
Bryan,J. III, and Charles J.V. Murphy, *The Windsor Story*. New York: William Morrow and Company, Inc. 1979.
Hart, Percival Gray, *History of the 135th Aero Squadron*, Chicago, 1939.
Kinert, Reed *American Racing Planes and Historic Air Races*. Chicago: Wilcox and Follett Company, 1952.
The National Cyclopedia Of American Biography
Norman Prince, A Volunteer Who Died For The Cause He Loved.
Parsons, Edwin C. *The Great Adventure*.
Vonderman, Don, *The Great Air Races*, New York: Doubleday, 1969.
Yale In The World War. Yale University Press, 1925.
Letters of William Thaw to Dr. Edmond L. Gros 1 March 1924; to Mr. H.C. Hoskier March 12, 14, 18, 24, 27, 1924. Letters from H.C. Hoskier to William Thaw March 14, 18, 24, 1924 all via Virginia Slade Murphy
Letter of William Thaw to Austen Crehore 19 October 1928 via Dr. Jim Parks

Various interviews by author with Alexander Blair Thaw, Virginia Slade Murphy, Mrs. Elma Slade Chess, Mrs. Cord Meyer
Various editions: *New York Times*; *New York Herald*; *Philadelphia Post Gazette*.
Newman, Irving J. "A Biography of Colonel Thaw II." *Popular Aviation Magazine*, November 1934.
Copies of death certificates for Marjorie Everts Thaw and William Thaw.
• *THOMPSON, CLIFTON*
author correspondence with Dartmouth University 9 February 1986
• *TRINKARD, CHARLES*
Rockwell, Paul Ayers. *American Fighters of the Foreign Legion.*
Charles Trinkard burial records via National Archives 1 July 1998
• *TUCKER, DUDLEY*
Forster, Henry. *Over There 1917–1918 Recollections of Wartime.* A privately published memoir.
Memoirs of the Harvard Dead in the War Against Germany.
• *TURNURE, GEORGE*
George Evans Turnure Memorial Volume. Privately printed. c. 1936
author conversation and correspondence with Lisa Turnure Oldenbburg 17 May 1991
author conversation and correspondence with Pamela Turnure Timmons 29 October 1991
• *TYSON, STEPHEN*
Memorial Volume of the American Field Service In France.
• *VAN FLEET, WILLIAM*
Arthur, Reginald Wright. *Contact: Naval Aviator Careers 1–2000.*
Bluebook of Aviation 1932. Ronald Hoagland, editor.
author interview with William Ireland 5 January 1985.
• *VEIL, CHARLES*
Bluebook of Aviation 1932. Ronald Hoagland, editor.
Charles Herbert Veil California death certificate.
author correspondence with East Palestine Memorial Library 22 November 1994.
author interview with Mrs. Doris Veil 23 December 1993
• *WAINWRIGHT, NEAL*
Harvard's Military Record in the World War.
author correspondence with Robert Crozer 9 April 1996
History of the American Field Service In France.
author correspondence with Harvard University 8 June 1991
author correspondence with National Archives 17 April 1996
the Boston *Evening Transcript* 19 March 1935 via Boston Pub Lib.
the Boston *Herald* 29 May 1935 via Concord, MA library.
author correspondence with Harvard University Archives 16 Aug 1996.
• *WALCOTT, BENJAMIN*
Above the French Lines. Letters of Stuart Walcott American Aviator. Princeton University Press, 1918.
Princeton in the World War. Princeton University Press, 1932.
author correspondence with Mrs. Robert Meech 30 October 1991
author correspondence with Erin Younger 30 July 1991
author conversation with Philip Walcott Meech 26 January 19972
• *WASS, WILLIAM*
William Van Wass California death certificate
author correspondence with Dept of Vets Affairs 2 Sept 1990
• *WELLMAN, WILLIAM*
New England Aviators 1914–1918. Houghton Mifflin, 1920.
Cross & Cockade Journal Vol 17, #3 "William A. Wellman"
Wellman, William A. *Go, Get' Em!.* Boston: The Page Co, 1918.
• *WELLS, FRANK*
Indiana death certificate for Frank Wells
Questionnaire from Crehore letters via Dr. Jim Parks
• *WHITMORE, HERMAN*
author interview with Jean Whitmore 13 May 1984, 18 October 1994
• *WHITMORE, JOHN*
Author correspondence with St. Paul's School 13 November 1990
• *WILCOX, CHARLES*
Blue Book of Aviation. Los Angeles: The Hoagland Company, 1932
author interview with Charles Ward 2 February 1991
• *WILD, MARCELLUS*
Arthur, Reginald Wright. *Contact: Naval Aviator Careers 1–2000.*
author conversation and correspondence with Mark Wild 6 April 1991
• *WILLARD, GEORGE*
Letters and documents of George Gale Willard.
• *WILLIS, HAROLD*
An American for Lafayette. The Diaries of E.C.C. Genet, Lafayette Escadrille.
Dennet, Carl P. *Prisoners of the Great War.* Boston: Houghton Mifflin, 1919
Flammer, Philip M. *The Vivid Air*, "The Lafayette Escadrille."
Friends of France. Boston: Houghton Mifflin, 1916
Harvard's Military Record in the World War.
Mason, Herbert Molloy, Jr. *The Lafayette Escadrille.*
New England Aviators 1914–1918. Boston: Houghton Mifflin, 1919
Parsons, Edwin C. *I Flew With The Lafayette Escadrille.*
Who's Who In Aviation 1942–1943. Ziff-Davis, 1942.
Wynne, H. Hugh. "Escadrille Lafayette." *Cross & Cockade Journal*, 2, No. 1 (1961).
Taped interview of Harold B. Willis by Thomas G. Miller, Jr. 25 January 1961.
Author interviews with Harold B. Willis, Jr. 11 March 1984 and 8 May 1986
• *WILLOUGHBY, WESTEL*
Princeton in the World War. Princeton: Princeton University Press, 1932.
author interview with Mrs. Ruth Willoughby 16 December 1990
author correspondence with Princeton University 12 Dec 1990
• *WILSON, JOSEPH*
author correspondence with Dora Johnson 11 May 1984
author correspondence with William Miller 20 September 1989
• *WILSON, PIERRE*
Dr. Gros records from Austen Crehore collection via Dr. Jim Parks
• *WINSLOW, ALAN*
Blue Book of Aviation. Los Angeles: The Hoagland Co, 1932.
author conversation with Marjorie Winslow Briggs 24 June 1984
• *WINTER, CHARLES*
Winslow, Carroll Dana. *With the French Flying Corps.* New York: Charles Scribners' Sons, 1917
Conversations & Correspondence with Penelope Winslow Baker May, September 1987.
• *WOODWARD, HENRY*
*A Year For France· War Letters of Houston Woodward.*New Haven: The Yale Publishing Association, 1919.
Memorial Volume of the American Field Service in France.
• *WORTHINGTON, WARWICK*
Hall, James Norman, and Charles B. Nordhoff. *The Lafayette Flying Corps.* In two volumes. Boston: Houghton Mifflin, 1920.
• *WRIGHT, HAROLD*
Hall, James Norman, and Charles B. Nordhoff. *The Lafayette Flying Corps.* In two volumes. Boston: Houghton Mifflin, 1920.
Paul Rockwell article from Chicago *Daily News* 27 August 1918.
• *YORK, WALTER*
author correspondence with Dartmouth College 29 January 1985
author correspondence with Dr. Richard York 3 January 1985
• *ZINN, FREDERICK*
Rockwell, Paul Ayers. *American Fighters in the Foreign Legion.*
author interview with Don Zinn 18 December 1983